Numerical Control over Complex Analytic Singularities

MEMOIRS
of the
American Mathematical Society

Number 778

Numerical Control over Complex Analytic Singularities

David B. Massey

May 2003 • Volume 163 • Number 778 (end of volume) • ISSN 0065-9266

American Mathematical Society
Providence, Rhode Island

2000 *Mathematics Subject Classification.*
Primary 32B15, 32C35, 32C18, 32B10.

Library of Congress Cataloging-in-Publication Data

Massey, David B., 1959–
Numerical control over complex analytic singularities / David B. Massey.
p. cm. — (Memoirs of the American Mathematical Society, ISSN 0065-9266 ; no. 778)
"Volume 163, number 778 (end of volume)."
Includes bibliographical references and index.
ISBN 0-8218-3280-8 (alk. paper)
1. Milnor fibration. 2. Hypersurfaces. 3. Analytic sheaves. 4. Analytic spaces. 5. Singularities (Mathematics) I. Title. II. Series.

QA3.A57 no. 778
[QA614.58]
510 s—dc21
[514′.74] 2003040369

Memoirs of the American Mathematical Society

This journal is devoted entirely to research in pure and applied mathematics.

Subscription information. The 2003 subscription begins with volume 161 and consists of six mailings, each containing one or more numbers. Subscription prices for 2003 are $555 list, $444 institutional member. A late charge of 10% of the subscription price will be imposed on orders received from nonmembers after January 1 of the subscription year. Subscribers outside the United States and India must pay a postage surcharge of $31; subscribers in India must pay a postage surcharge of $43. Expedited delivery to destinations in North America $35; elsewhere $130. Each number may be ordered separately; *please specify number* when ordering an individual number. For prices and titles of recently released numbers, see the New Publications sections of the *Notices of the American Mathematical Society.*

Back number information. For back issues see the *AMS Catalog of Publications.*

Subscriptions and orders should be addressed to the American Mathematical Society, P. O. Box 845904, Boston, MA 02284-5904, USA. *All orders must be accompanied by payment.* Other correspondence should be addressed to 201 Charles Street, Providence, RI 02904-2294, USA.

Memoirs of the American Mathematical Society is published bimonthly (each volume consisting usually of more than one number) by the American Mathematical Society at 201 Charles Street, Providence, RI 02904-2294, USA. Periodicals postage paid at Providence, RI. Postmaster: Send address changes to Memoirs, American Mathematical Society, 201 Charles Street, Providence, RI 02904-2294, USA.

This publication is indexed in *Science Citation Index*®, *SciSearch*®, *Research Alert*®, *CompuMath Citation Index*®, *Current Contents*®/*Physical, Chemical & Earth Sciences.*
Printed in the United States of America.

♾ The paper used in this book is acid-free and falls within the guidelines established to ensure permanence and durability.
Visit the AMS home page at `http://www.ams.org/`

10 9 8 7 6 5 4 3 2 1 08 07 06 05 04 03

For my mother, Mary Alice Massey,

and in memory of my grandparents:

Leslie Ellsworth Porter
Mary Frances Porter

William Walter Massey
Bessie Ann Massey

TABLE OF CONTENTS

ABSTRACT

The Milnor number is a powerful invariant of an isolated, complex, affine hypersurface singularity. It provides data about the local, ambient, topological-type of the hypersurface, and the constancy of the Milnor number throughout a family implies that Thom's a_f condition holds and that the local, ambient, topological-type is constant in the family. Much of the usefulness of the Milnor number is due to the fact that it can be effectively calculated in an algebraic manner.

The Lê cycles and numbers are a generalization of the Milnor number to the setting of complex, affine hypersurface singularities, where the singular set is allowed to be of arbitrary dimension. As with the Milnor number, the Lê numbers provide data about the local, ambient, topological-type of the hypersurface, and the constancy of the Lê numbers throughout a family implies that Thom's a_f condition holds and that the Milnor fibrations are constant throughout the family. Again, much of the usefulness of the Lê numbers is due to the fact that they can be effectively calculated in an algebraic manner.

In this work, we generalize the Lê cycles and numbers to the case of hypersurfaces inside arbitrary analytic spaces. We define the Lê-Vogel cycles and numbers, and prove that the Lê-Vogel numbers control Thom's a_f condition. We also prove a relationship between the Euler characteristic of the Milnor fibre and the Lê-Vogel numbers. Moreover, we give examples which show that the Lê-Vogel numbers are effectively calculable.

In order to define the Lê-Vogel cycles and numbers, we require, and include, a great deal of background material on Vogel cycles, analytic intersection theory, and the derived category. Also, to serve as a model case for the Lê-Vogel cycles, we recall our earlier work on the Lê cycles of an affine hypersurface singularity.

1991 *Mathematics Subject Classification.* 32B15, 32C35, 32C18, 32B10

Key words and phrases. Gap sheaf, Vogel cycle, Milnor fibre and number, Lê cycles and numbers, vanishing cycles, perverse sheaves, Thom's a_f condition, Lê-Vogel cycles and numbers

Received by the editor May 1, 2001

PREFACE

In 1983, I began work on my dissertation, "Families of Hypersurfaces with One-dimensional Singular Sets", at Duke University. In that paper, I attempted to describe two numbers which one could effectively calculate from the defining equation of a hypersurface with a one-dimensional singular set – two numbers which control the topology and geometry in a similar fashion to how the Milnor number controls the topology and geometry for isolated hypersurface singularities.

Since that time, my work has centered around finding numerical data which "control" various topological and geometric properties of complex analytic singularities.

In 1987, while at The University of Notre Dame, I defined the *Lê cycles and the Lê numbers* for non-isolated hypersurface singularities. The Lê numbers are a generalization of the Milnor number, and they control the singularities; the constancy of the Lê numbers in a family implies the constancy of the Milnor fibres in the family, and also implies Thom's a_f condition holds. My work on Lê numbers from 1987 to the present is contained in my recent monograph, **Lê Cycles and Hypersurface Singularities**.

In 1988, I came to Northeastern University, and was immediately asked by Terry Gaffney how to generalize the Lê numbers of a hypersurface to the case of complete intersections. My answer to this was that I thought the generalization would have two distinct pieces: the first piece should be a method for associating numbers to an arbitrary constructible complex of sheaves on a complex analytic space – one should recover the Lê numbers of a hypersurface by applying this new method to the complex of vanishing cycles of the defining equation of the hypersurface. The second piece should be to decide what complex of sheaves should play the role of the sheaf of vanishing cycles in the case of a complete intersection.

This first piece – finding a method for associating numbers to a constructible complex of sheaves – is described in my 1994 paper, "Numerical Invariants of Perverse Sheaves". As the title indicates, a number of results for Lê numbers only generalize nicely in the case where the underlying complex is actually a perverse sheaf.

After this first piece was completed, it became apparent that the second piece of the generalization was not to replace the sheaf of vanishing cycles by some other complex. Rather, one should continue to use the vanishing cycles a function, but the function could now have an arbitrarily singular domain. This changed the problem to one of finding a sufficiently algebraic characterization of the vanishing cycles – one that actually allows one to effectively produce the numbers that should control the singularities.

The first part of such an algebraic description of the vanishing cycles appears in my paper "Hypercohomology of Milnor Fibres", and the final piece appears in the paper "Critical Points of Functions on Singular Spaces".

Having completed all of the above pieces, I thought it would be a relatively simple matter to merge them into one coherent whole; thus, I began writing a second book during the 1995-1996 academic year. Little did I suspect that the details, refinements, and corrections would take so long. Even now, I have only just realized that the correct topological treatment should use the micro-local theory of Kashiwara and Schapira. Hence, I could be delayed longer by writing an appendix on micro-local theory, and by rewriting all of my Morse theory proofs in terms of the micro-local theory. As the micro-local theory would add another level of complexity to an already complicated work, I have decided to leave the topological part of this work for a later volume.

Thus, what appears in this book is the general algebraic machinery (Vogel cycles and gap cycles), a mildly rewritten version of **Lê Cycles and Hypersurface Singularities** in terms of this general set-up, the generalization of the Milnor number to functions on arbitrarily singular spaces, the generalization of the Lê numbers and cycles to functions on arbitrarily singular spaces, and new generalized Lê-Iomdine formulas and results on Thom's a_f condition. Moreover, the correct treatment of these topics requires the inclusion of appendices on intersection theory and the derived category.

One reason that I feel obligated to include the appendix on intersection theory is to correct the stupidest sentence that I have ever had published. In **Lê Cycles and Hypersurface Singularities**, I attempted to give a quick summary of the needed intersection theory; I wrote "If we have two irreducible subschemes V and W in an open subset $\mathcal{U}$ of some affine space, V and W are said to *intersect properly* in $\mathcal{U}$ provided that $\operatorname{codim} V \cap W = \operatorname{codim} V + \operatorname{codim} W$; when this is the case, the *intersection product* of $[V]$ and $[W]$ is defined by $[V] \cdot [W] = [V \cap W]$." This statement is, in general, quite false. Moreover, I knew that it was false, and certainly never used it anywhere in the book; I have no idea how I wrote such a ridiculous thing. Hopefully, the included appendix will eliminate any confusion that I may have caused.

Many people have contributed to the results which appear here. But, since I have thanked them in the individual works listed above, I will not give this extensive list here. However, it would be difficult to exaggerate the importance of Terry Gaffney's and Lê Dũng Tráng's contributions to this book. Not only have they helped me with or given me many results, but their continuing enthusiasm for my work is an incredible motivating force.

I should also thank two former graduate students at Northeastern University, Mike Green and Robert Gassler; conversations with them contributed greatly to my own understanding. Finally, I want to thank some friends who have helped me greatly in ways not directly related to mathematics: General and Mrs. Hannon, Jenn Hannon, John Hannon, and Jed Hannon (the entire Hannon family), Mike Roberts, Tim Roberts, and especially Chad Brazee.

David B. Massey
Boston, MA
April 26, 2000

OVERVIEW

The Milnor number of an affine complex analytic hypersurface with an isolated singularity has been a ridiculously successful invariant: it can be effectively calculated, it determines the homotopy-type of the Milnor fibre, and its constancy in a family controls much of the geometry and topology of the family.

It is no wonder that there have been myriad attempts to generalize the Milnor number to the cases where the singularity is non-isolated or where the underlying space is arbitrary. From the topological side, one might suspect that the Betti numbers or the Euler characteristic of the Milnor fibre might be reasonable substitutes for the Milnor number. From a differential geometry point-of-view, one can consider various notions of indices of vector fields. From the algebraic side, there are sheaf-theoretic generalizations of the Milnor number. A nice, but by no means complete, expository discussion of the Milnor number and its generalizations is contained in [**Te1**].

Our own work on generalizing Milnor numbers began with the Lê varieties, Lê cycles, and Lê numbers of a non-isolated affine hypersurface singularity; this work appeared in [**Mas6**], [**Mas8**], [**Mas9**], and [**Mas14**]. If $\mathcal{U}$ is an open subset of $\mathbb{C}^{n+1}$, $f : \mathcal{U} \to \mathbb{C}$ is an analytic function, and $\mathbf{z} := (z_0, \dots, z_n)$ is a linear choice of coordinates for $\mathbb{C}^{n+1}$, then the Lê numbers, $\lambda^0_{f,\mathbf{z}}, \lambda^1_{f,\mathbf{z}}, \dots, \lambda^n_{f,\mathbf{z}}$, have a number of very desirable properties.

Let s denote the dimension of the critical locus of f at a point $\mathbf{p} \in f^{-1}(0)$. Then, the Lê numbers, $\lambda^i_{f,\mathbf{z}}$, are zero for $i > s$, and if $s = 0$, then $\lambda^0_{f,\mathbf{z}}$ is precisely the Milnor number. More generally, all of the Lê numbers are effectively calculable, and the Milnor fibre has a handle decomposition in which the number of handles attached of a given index is given by the corresponding Lê number. The constancy of the Lê numbers in a family implies that Thom's a_f holds for the total space (the union of the members of the family) and that the Milnor fibrations are constant in the family. All of these properties, and more, are proved in Part II of this book.

There is only one question which is addressed by the results of this book: **how does one generalize the Lê numbers of an analytic function to the setting where the underlying space is no longer affine, but, rather, is an arbitrarily singular analytic space?**

Obviously, to answer this, we need to consider how the Lê numbers are defined. The Lê numbers are intersection numbers of the Lê cycles with affine linear subspaces defined by the coordinate choice $\mathbf{z}$. Hence, if we could generalize the Lê cycles, $\Lambda^i_{f,\mathbf{z}}$, then we would know how to generalize the Lê numbers.

The Lê cycles are defined by looking at the relative polar varieties ([**L-T2**], [**Te3**], and [**Te4**]) of f, with the correct cycle structure, and using them to give a "decomposition" of the Jacobian ideal of f into a collection of cycles. In a more general setting, this decomposition has been studied by Vogel [**Vo**] and van Gastel [**Gas1**], [**Gas2**]. The first problem, when the underlying space is arbitrary, is that there are three competing definitions for the polar cycles – all of which involve *gap sheaves*, a notion first studied in [**Si-Tr**].

Part I of this book discusses these three competing definitions in the general context of decomposing any ideal, not necessarily one related to the Jacobian of a function. In Part I, we prove that the Lê-Iomdine formulas, which are such an important result on Lê numbers, actually hold in a much more general setting.

In Part II, we give all of our previously known results on Lê numbers, but give the proofs in terms of our work in Part I.

In Part III, we deal with the second major problem when the ambient space, X, is singular: how does one define an analog of the Milnor number for an isolated critical point? For that matter, what does an "isolated critical point" even mean in this setting? As we shall see, the derived category and the vanishing cycles of the constant sheaf along f will be unavoidable tools for answering these questions.

It turns out that if the constant sheaf $\mathbb{C}^\bullet_X$ is perverse (up to a shift), then the whole theory becomes much easier; for instance, this would be the case if X was a local complete intersection. However, for arbitrary spaces, the fact that the complex links of strata can have non-trivial cohomology in more then one degree leads us to take perverse cohomology of the constant sheaf with all possible shifts. In a sense, perverse cohomology lets us decompose the relevant topological data about X into a collection of "positive" and "negative" pieces, and then when we work with these pieces, data from the various strata cannot cancel each other, because they all have the same sign.

A result that appears at the end of Part III is a generalization of the result of Lê and Saito that the constancy of the Milnor number in a family implies Thom's a_f condition [**Lê-Sa**]; of course, in our theorem, the underlying space is arbitrary.

With Parts I and III out of the way, and using Part II as a guide, it is relatively trivial to define our generalization of the Lê cycles and numbers in Part IV. We refer to these new gadgets as the *Lê-Vogel cycles*, or *LêVo cycles* for short. There are collections of LêVo cycles for various shifts – these shifts correspond to degrees in which the complex links of strata of X have non-trivial cohomology. The indexing on the shifts is set-up in such a way that local complete intersections have non-zero LêVo cycles only when the shift is zero.

In Part IV, we prove an incredibly general Lê-Saito type result and prove a result relating the LêVo numbers to the Euler characteristic of the Milnor fibre.

Appendices A and B contain background information (without proofs) on intersection theory and the derived category. The intersection theory that we need is very simple – we need only proper intersections of analytic cycles in affine space; this is what is described in Appendix A. Appendix B contains more information than we really need; it is sort of a working mathematicians guide to the derived category, perverse sheaves, and vanishing cycles.

Appendix C contains some extremely technical arguments which are needed in Part II, where we prove that constancy of the Lê numbers in a family implies the constancy of the Milnor fibrations. We believe that these arguments would only serve to obstruct the exposition in Part II; hence, we have relegated them to an appendix.

Finally, a word on what is not contained in this book. One will not find most of the extremely topological results on Lê numbers extended to the general setting of the LêVo numbers. This includes results on handle-decompositions, Morse inequalities, and the constancy of the LêVo numbers implying constancy of the Milnor fibrations. While we certainly believe that we can generalize most of these results, the correct treatment appears to require the micro-local of Kashiwara and Schapira [**K-S1**], [**K-S2**]. As this would add significant length and complexity to an already long and difficult work, we have elected to place such results in a future book.

Part I. ALGEBRAIC PRELIMINARIES: GAP SHEAVES AND VOGEL CYCLES

Chapter 0. INTRODUCTION

Throughout this book, our primary algebraic tool consists of a method for taking a coherent sheaf of ideals and decomposing it into pure-dimensional "pieces". Actually, we begin with an ordered set of generators for the ideal, and produce a collection of pure-dimensional analytic cycles, the **Vogel cycles**, which seem to contain a great deal of "geometric" data related to the original ideal. Part I of this book contains the construction of the Vogel cycles; it is, regrettably, very technical in nature. The Vogel cycles are defined using gap sheaves, together with the associated analytic cycles which they define, the gap cycles. A gap sheaf is a formal device which gives a scheme-theoretic meaning to the analytic closure of the difference of an initial scheme and an analytic set.

If the underlying space is not Cohen-Macaulay, the main technical problem is that there are, at least, three different reasonable definitions of the gap sheaves and cycles; we select as "the" definition the one that works most nicely in inductive proofs. We show, however, that if one re-chooses the functions defining the ideal in a suitably "generic" way, then all competing definitions for the gap cycles and Vogel cycles agree.

In Chapter 3 of this part, we prove some extremely general Lê-Iomdine-Vogel formulas; as we shall see in later chapters, these formulas are an amazingly effective tool for transforming problems about a given singularity into problems involving a singularity of smaller dimension.

The reader who wishes to bypass this technical portion of the book can jump to the Summary of Part I, which begins on page 31.

Chapter 1. GAP SHEAVES

Let W be analytic subset of an analytic space X and let α be a coherent sheaf of ideals in $\mathcal{O}_X$. At each point $\mathbf{x}$ of $V(\alpha)$, we wish to consider scheme-theoretically those components of $V(\alpha)$ which are not contained in $|W|$. This leads one to the notion of a *gap sheaf*. Our primary references for gap sheaves are [**Si-Tr**] and [**Fi**].

Let β be a second coherent sheaf of ideals in $\mathcal{O}_X$. We write $\alpha_{\mathbf{x}}$ for the stalk of α in $\mathcal{O}_{X,\mathbf{x}}$.

Definition 1.1. Let S be the multiplicatively closed set $\mathcal{O}_{X,\mathbf{x}} - \bigcup p$ where the union is over all $p \in \operatorname{Ass}(\mathcal{O}_{X,\mathbf{x}}/\alpha_{\mathbf{x}})$ with $|V(p)| \not\subseteq |W|$. Then, we define $\alpha_{\mathbf{x}}\neg W$ to equal $S^{-1}\alpha_{\mathbf{x}} \cap \mathcal{O}_{X,\mathbf{x}}$. Thus, $\alpha_{\mathbf{x}}\neg W$ is the ideal in $\mathcal{O}_{X,\mathbf{x}}$ consisting of the intersection of those (possibly embedded) primary ideals, q, associated to $\alpha_{\mathbf{x}}$ such that $|V(q)| \not\subseteq |W|$.

Now, we have defined $\alpha_{\mathbf{x}}\neg W$ in each stalk. By [**Si-Tr**], if we perform this operation simultaneously at all points of $V(\alpha)$, then we obtain a coherent sheaf of ideals called a *gap sheaf* ; we write this sheaf as $\alpha\neg W$. If β is any coherent sheaf of ideals such that $W = \operatorname{supp}(\mathcal{O}_X/\beta)$, then

$$\alpha\neg W = \bigcup_{k=0}^{\infty} (\alpha : \beta^k).$$

If $V = V(\alpha)$, we let $V\neg W$ denote the scheme $V(\alpha\neg W)$. It is important to note that the scheme $V\neg W$ does not depend on the structure of W as a scheme, but only as an analytic set. The scheme $V\neg W$ is sometimes referred to as the *analytic closure of* $V - W$ [**Fi**, p.41]; this is certainly the correct, intuitive way to think of $V\neg W$.

We find it convenient to extend this gap sheaf notation to the case of analytic sets (reduced schemes) and analytic cycles.

Hence, if Z and W are analytic sets, then we let $Z\neg W$ denote the union of the components of Z which are not contained in W; if $C = \sum m_i[V_i]$ is an analytic cycle in a complex manifold M and W is an analytic subset of M, then we define $C\neg W$ by

$$C\neg W = \sum_{V_i \not\subseteq W} m_i[V_i].$$

If α is a coherent sheaf of ideals in $\mathcal{O}_M$, C is a cycle in M, and W is an analytic subset of M, then clearly $[V(\alpha)\neg W] = [V(\alpha)]\neg W$ and $|C\neg W| = |C|\neg W$.

The following properties of gap sheaves are immediate from the definition.

Proposition 1.2.

i) $\alpha_{\mathbf{x}} = \beta_{\mathbf{x}}$ *for all* $\mathbf{x} \in X - W$ *if and only if* $\alpha\neg W = \beta\neg W$;

ii) *if* α_i *is a finite collection of coherent ideals in* $\mathcal{O}_X$, *then* $\cap(\alpha_i\neg W) = (\cap\alpha_i)\neg W$;

iii) if $V(\alpha)\cap(X-W)$ is reduced, then the sheaf of ideals of functions vanishing on the analytic
set $\overline{V(\alpha)\cap(X-W)}$ *is* $\alpha\neg W$.

Later, the reader may wonder why we do not define something analogous to a gap sheaf, but where we **keep** those components which are contained in a given analytic set, W, instead of throwing them away.

On the level of schemes, we can not make this approach work; the primary ideals in a primary decomposition (of a given ideal) which define varieties contained in W would not be independent of the decomposition. We could just take the isolated primary ideals which define varieties contained in W, but this disposes of too much algebraic structure. Similarly, we could consider $V\neg(V\neg W)$, which would not dispose of **all** embedded components, but would eliminate embedded components contained in both W and $V\neg W$.

However, even this device would not aid us much later; as we shall see – beginning with Definition 2.14 – we need to deal more with the intersection product on analytic cycles, and not so much with primary decompositions.

The following lemma is very useful for calculating $V\neg W$.

Lemma 1.3. *Let $(X,\mathcal{O}_X)$ be an analytic space, let α,β, and γ be coherent sheaves of ideals in $\mathcal{O}_X$, let $f,g\in\mathcal{O}_X$, and let W, Y, and Z be analytic subsets of X such that $Z\subseteq W$. Then,*

i) $\alpha\neg W=(\alpha\neg Z)\neg W$, *and thus, as schemes,* $V(\alpha)\neg W=(V(\alpha)\neg Z)\neg W$;

ii) $(\alpha+\beta)\neg W = (\alpha\neg Z+\beta)\neg W$, *and thus, as schemes,*

$$\big(V(\alpha)\cap V(\beta)\big)\neg W = \big(V(\alpha\neg Z)\cap V(\beta)\big)\neg W;$$

iii) if $V(\alpha+\gamma)\subseteq W$, *then* $\big((\alpha\cap\beta)+\gamma\big)\neg W = (\beta+\gamma)\neg W$, *and thus, as schemes,*

$$\Big(\big(V(\alpha)\cup V(\beta)\big)\cap V(\gamma)\Big)\neg W = \big(V(\beta)\cap V(\gamma)\big)\neg W;$$

iv) if $V(\alpha+<g>)\subseteq W$, *then* $(\alpha+<fg>)\neg W = (\alpha+<f>)\neg W$, *and thus, as schemes,*

$$\big(V(\alpha)\cap V(fg)\big)\neg W = \big(V(\alpha)\cap V(f)\big)\neg W.$$

v) $\alpha\neg(W\cup Y)=(\alpha\neg W)\neg Y$, *and thus, as schemes,*

$$V(\alpha)\neg(W\cup Y)=(V(\alpha)\neg W)\neg Y.$$

The analog of ii) for sets and cycles is also trivial to verify; that is,

$$\big(|V(\alpha)|\cap|V(\beta)|\big)\neg W = \Big(\big(|V(\alpha)|\neg Z\big)\cap|V(\beta)|\Big)\neg W,$$

and, if all intersections are proper,

$$([V(\alpha)]\cdot[V(\beta)])\neg W \;=\; \Big(([V(\alpha)]\neg Z)\cdot[V(\beta)]\Big)\neg W.$$

Proof. Statements i), ii), iii), and iv) are merely exercises in localization (see [**Mas3**]). Statement v) is trivial. □

Remark 1.4. While it is a trivial observation, it is frequently important and useful to note that, for any coherent sheaf of ideals, α, in $\mathcal{O}_X$ and for any $f \in \mathcal{O}_X$, $V(\alpha)\neg V(f)$ and $V(f)$ intersect properly and $V(f)$ contains no embedded subvarieties of $V(\alpha)\neg V(f)$; thus, the intersection product cycle $[V(\alpha)\neg V(f)]\cdot[V(f)]$ in X is well-defined (without having to mention an ambient manifold) and is equal to $[V(<\alpha \neg V(f)> + <f>)]$.

If $V(\alpha)$ and $V(f)$ intersect properly, then $[V(\alpha)] = [V(\alpha)\neg V(f)]$ and, hence,

$$[V(\alpha)]\cdot[V(f)] = [V(\alpha)\neg V(f)]\cdot[V(f)] = [V(<\alpha \neg V(f)> + <f>)].$$

Lemma 1.5. *Let X be purely d-dimensional and Cohen-Macaulay. Let $f_1, \dots, f_k \in \mathcal{O}_X$ and let W be an analytic subset of X. If $V(f_1, \dots, f_k) \neg W$ is purely $(d-k)$-dimensional, then it contains no embedded subvarieties.*

Proof. By definition, $V(f_1, \dots, f_k) \neg W$ can not have any embedded subvarieties contained in W. At points, $\mathbf{p}$, outside of W, $f_1, \dots, f_k$ determines a regular sequence in the Cohen-Macaulay ring $\mathcal{O}_{X,\mathbf{p}}$; hence, there are no embedded subvarieties outside of W. □

Example 1.6. For the remainder of this chapter, we wish to describe the blow-up of a space along an ideal; the description via gap sheaves is very nice.

Let $(X, \mathcal{O}_X)$ be an analytic space, and let $\mathbf{f} := (f_0, \dots, f_k)$ be an ordered $(k+1)$-tuple of elements of $\mathcal{O}_X$. Then, the *blow-up of X along* $\mathbf{f}$ consists of an analytic subspace $\mathrm{Bl}_{\mathbf{f}} X \subseteq X \times \mathbb{P}^k$, together with the projection morphism $\pi : \mathrm{Bl}_{\mathbf{f}} X \to X$, which is the restriction of the standard projection from $X \times \mathbb{P}^k$ to X. If we use $[w_0 : \cdots : w_k]$ for homogeneous coordinates on $\mathbb{P}^k$, then the blow-up is given as a scheme by

$$\mathrm{Bl}_{\mathbf{f}} X := V\big(\{w_i f_j - w_j f_i\}_{\ 0\leqslant i,j\leqslant k}\big) \neg \big(V(f_0, \dots, f_k)\times \mathbb{P}^k\big).$$

In order to describe the exceptional divisor as a cycle, we need to work on affine coordinate patches in $\mathbb{P}^k$. We shall describe both the blow-up and the exceptional divisor on each affine patch $\{w_j \neq 0\}$.

On the patch $\{w_j \neq 0\}$, we use coordinates $\widetilde{w}_i := w_i/w_j$ for all $i \neq j$. Then,

$$\{w_j \neq 0\} \cap \mathrm{Bl}_{\mathbf{f}} X \;=\; V(\{f_i - \widetilde{w}_i f_j\}_{i \neq j}) \urcorner \left(V(f_j) \times \mathbb{P}^k\right), \tag{1.7}$$

and the exceptional divisor, E, is the cycle defined on each affine patch in the following manner

$$\{w_j \neq 0\} \cap E \;:=\; \left[V(\{f_i - \widetilde{w}_i f_j\}_{i \neq j}) \urcorner \left(V(f_j) \times \mathbb{P}^k\right)\right] \cdot \left[V(f_j) \times \mathbb{P}^k\right]. \tag{1.8}$$

We have made these definitions with respect to a chosen $(k+1)$-tuple $\mathbf{f}$. In fact, the analytic isomorphism-type of the morphism $\pi : \mathrm{Bl}_{\mathbf{f}} X \to X$ only depends on the ideal, I, generated by the components $f_0, \ldots, f_k$; this isomorphism-type is referred to as the *blow-up of X along I*. Of course, the isomorphism-type of the exceptional divisor also depends only on the ideal I, and this isomorphism-type is simply called the *the exceptional divisor of the blow-up of X along I*.

Chapter 2. GAP CYCLES AND VOGEL CYCLES

Let X be a d-dimensional analytic space and let $\mathbf{f} := (f_0, \dots, f_k)$ be an ordered $(k+1)$-tuple of elements of $\mathcal{O}_X$. We will define a sequence of cycles, the *Vogel cycles* ([**Vo**], [**Gas1**], [**Gas2**]) of $\mathbf{f}$; these cycles provide effectively calculable data about the coherent sheaf of ideals $< f_0, \dots, f_k >$. Before we can define the Vogel cycles, we must first define the *gap varieties* and *gap cycles* of $\mathbf{f}$.

It will prove useful (in Part IV) to define gap and Vogel objects with respect to a given cycle. Hence, throughout Part I, we let M denote the cycle $\sum_l m_l[V_l]$ in X; we assume that this is a minimal presentation of M – that is, we assume that the V_l are distinct, irreducible analytic subsets of X and that none of the m_l equal zero. In addition, to avoid cancellation of contributions from various V_l, we assume that all of the m_l have the same sign, i.e., that $\pm M > 0$.

If X is a union of irreducible components $\{X_i\}$, we will define the gap and Vogel cycles in X as sums of the gap and Vogel cycles from each X_i; similarly, we will define gap and Vogel cycles with respect to M simply by taking weighted sums of the gap and Vogel cycles of the irreducible components. The case of an irreducible space X can be recovered from the cycle case by simply taking $M = [X]$. Thus, we find that we need to first define the gap varieties, gap cycles, and Vogel cycles in the case where X is irreducible.

However, even if we assume that the underlying space is irreducible, there is a further complication in the general setting: $\mathcal{O}_X$ may not be Cohen-Macaulay. This causes numerous problems, for we must worry about embedded subvarieties. To deal with this problem, we introduce three avatars of gap varieties and examine the relations between them.

We will define the (ordinary) gap varieties, $\{\Pi^i_{\mathbf{f}}\}_i$, the modified gap varieties, $\{\widetilde{\Pi}^i_{\mathbf{f}}\}_i$, and the inductive gap varieties, $\{\widehat{\Pi}^i_{\mathbf{f}}\}_i$. We shall use the inductive gap varieties to define the Vogel cycles, but need to make assumptions about the (ordinary) gap varieties in order for the definition to make sense; the modified gap varieties are merely a convenient tool for proving results about $\Pi^i_{\mathbf{f}}$ and $\widehat{\Pi}^i_{\mathbf{f}}$.

Definition 2.1. Assume that X is irreducible (though, not necessarily reduced). For all i, we define the *gap varieties*, the *modified gap varieties*, and the *inductive gap varieties* of $\mathbf{f}$, which we denote by $\Pi^i_{\mathbf{f}}$, $\widetilde{\Pi}^i_{\mathbf{f}}$, and $\widehat{\Pi}^i_{\mathbf{f}}$, respectively.

First, if $i < d - (k+1)$ or $i > d$, we set $\Pi^i_{\mathbf{f}} = \widetilde{\Pi}^i_{\mathbf{f}} = \widehat{\Pi}^i_{\mathbf{f}} = \emptyset$.

We define $\Pi^d_{\mathbf{f}} := X \neg V(\mathbf{f})$, and $\widetilde{\Pi}^d_{\mathbf{f}} = \widehat{\Pi}^d_{\mathbf{f}} := X \neg V(f_k)$.

For $d - (k+1) \leqslant i < d$, the *$i$-th gap variety of* $\mathbf{f}$, $\Pi^i_{\mathbf{f}}$, is defined as

$$\Pi^i_{\mathbf{f}} := V(f_{i+k+1-d}, \dots, f_k) \neg V(\mathbf{f}).$$

Note that $\Pi^{d-(k+1)}_{\mathbf{f}} = \emptyset$.

For $d-(k+1) < i < d$, the *i-th modified gap variety of* $\mathbf{f}$, $\widetilde{\Pi}^i_{\mathbf{f}}$, is defined as

$$\widetilde{\Pi}^i_{\mathbf{f}} := V(f_{i+k+1-d}, \dots, f_k) \neg V(f_{i+k-d});$$

we define $\widetilde{\Pi}^{d-(k+1)}_{\mathbf{f}} := \emptyset$.

For $d-(k+1) < i < d$, the *i-th inductive gap variety of* $\mathbf{f}$, $\widehat{\Pi}^i_{\mathbf{f}}$, is defined by downward induction (recall that $\widehat{\Pi}^d_{\mathbf{f}}$ is defined above)

$$\widehat{\Pi}^i_{\mathbf{f}} := \left(\widehat{\Pi}^{i+1}_{\mathbf{f}} \cap V(f_{i+k+1-d})\right) \neg V(f_{i+k-d});$$

we define $\widehat{\Pi}^{d-(k+1)}_{\mathbf{f}} := \emptyset$.

Naturally, we define the *i-th gap cycle*, *modified i-th gap cycle*, and *inductive i-th gap cycle* of $\mathbf{f}$ to be the cycles defined by these schemes, i.e., $\left[\Pi^i_{\mathbf{f}}\right]$, $\left[\widetilde{\Pi}^i_{\mathbf{f}}\right]$, and $\left[\widehat{\Pi}^i_{\mathbf{f}}\right]$, respectively.

If X is a union of irreducible components $\{X_j\}$ and $\mathbf{f} \in (\mathcal{O}_X)^{k+1}$, then we define the *i-th gap cycle of* $\mathbf{f}$ by $\left[\Pi^i_{\mathbf{f}}\right] := \sum_j \left[\Pi^i_{\mathbf{f}_{|X_j}}\right]$, the *i-th modified gap cycle of* $\mathbf{f}$ by $\left[\widetilde{\Pi}^i_{\mathbf{f}}\right] := \sum_j \left[\widetilde{\Pi}^i_{\mathbf{f}_{|X_j}}\right]$, and the *i-th inductive gap cycle of* $\mathbf{f}$ by $\left[\widehat{\Pi}^i_{\mathbf{f}}\right] := \sum_j \left[\widehat{\Pi}^i_{\mathbf{f}_{|X_j}}\right]$.

We define the *i-th gap set of* $\mathbf{f}$, the *i-th modified gap set of* $\mathbf{f}$, and the *i-th inductive gap set of* $\mathbf{f}$ to be $\left|\left[\Pi^i_{\mathbf{f}}\right]\right|$, $\left|\left[\widetilde{\Pi}^i_{\mathbf{f}}\right]\right|$, and $\left|\left[\widehat{\Pi}^i_{\mathbf{f}}\right]\right|$, respectively. We will write simply $\left|\Pi^i_{\mathbf{f}}\right|$, $\left|\widetilde{\Pi}^i_{\mathbf{f}}\right|$, and $\left|\widehat{\Pi}^i_{\mathbf{f}}\right|$, respectively.

Finally, we need to define gap cycles and sets with respect to the cycle M. We define the *i-th gap cycle of* $\mathbf{f}$ *with respect to* M by $\Pi^i_{\mathbf{f}}(M) := \sum_l m_l \left[\Pi^i_{\mathbf{f}_{|V_l}}\right]$, the *i-th modified gap cycle of* $\mathbf{f}$ *with respect to* M by $\widetilde{\Pi}^i_{\mathbf{f}}(M) := \sum_l m_l \left[\widetilde{\Pi}^i_{\mathbf{f}_{|V_l}}\right]$, and the *i-th inductive gap cycle of* $\mathbf{f}$ *with respect to* M by $\widehat{\Pi}^i_{\mathbf{f}}(M) := \sum_l m_l \left[\widehat{\Pi}^i_{\mathbf{f}_{|V_l}}\right]$.

Of course, we define the associated gap sets with respect to M to be the sets underlying the various gap cycles.

Note that we have not defined gap **varieties** for $\mathbf{f}$ unless X is irreducible.

The following proposition gives a number of basic results and interrelationships between the various gap varieties.

Proposition 2.2. Let X be irreducible. Then,

i) there is an inclusion of sets $\left|\widehat{\Pi}^i_{\mathbf{f}}\right| \subseteq \left|\widetilde{\Pi}^i_{\mathbf{f}}\right| \subseteq \left|\Pi^i_{\mathbf{f}}\right|$, $\left|\widehat{\Pi}^i_{\mathbf{f}}\right|$ is purely i-dimensional, and all components
of $\left|\widetilde{\Pi}^i_{\mathbf{f}}\right|$ and $\left|\Pi^i_{\mathbf{f}}\right|$ have dimension at least i;

ii) there is an equality of schemes $\Pi^i_{\mathbf{f}} = \left(\Pi^{i+1}_{\mathbf{f}} \cap V(f_{i+k+1-d})\right) \neg V(\mathbf{f})$;

iii) if $i \leqslant d$, then $\left|\Pi_{\mathbf{f}}^{i-1}\right| \subseteq \left|\Pi_{\mathbf{f}}^{i}\right|$ and $\left|\widehat{\Pi}_{\mathbf{f}}^{i-1}\right| \subseteq \left|\widehat{\Pi}_{\mathbf{f}}^{i}\right|$;

iv) the sets $V(f_{i+k+1-d})$ and $\left|\widehat{\Pi}_{\mathbf{f}}^{i+1}\right|$ intersect properly, and there is an equality of cycles

$$\left[\widehat{\Pi}_{\mathbf{f}}^{i}\right] = \left(\left[\widehat{\Pi}_{\mathbf{f}}^{i+1}\right] \cdot \left[V(f_{i+k+1-d})\right]\right) \neg V(f_{i+k-d});$$

v) if there is an equality of sets $\left|\widetilde{\Pi}_{\mathbf{f}}^{i}\right| = \left|\Pi_{\mathbf{f}}^{i}\right|$, then the schemes $\widetilde{\Pi}_{\mathbf{f}}^{i}$ and $\Pi_{\mathbf{f}}^{i}$ are equal up to embedded subvariety, and so there is an equality of cycles $\left[\widetilde{\Pi}_{\mathbf{f}}^{i}\right] = \left[\Pi_{\mathbf{f}}^{i}\right]$;

vi) if there is an equality of sets $\left|\widehat{\Pi}_{\mathbf{f}}^{j}\right| = \left|\Pi_{\mathbf{f}}^{j}\right|$ for all $j \geqslant i+1$, then $\left[\widehat{\Pi}_{\mathbf{f}}^{i}\right] \leqslant \left[\Pi_{\mathbf{f}}^{i}\right]$;

vii) if there is an equality of schemes $\Pi_{\mathbf{f}}^{i+1} = \widehat{\Pi}_{\mathbf{f}}^{i+1}$, then there is an equality of schemes $\widetilde{\Pi}_{\mathbf{f}}^{i} = \widehat{\Pi}_{\mathbf{f}}^{i}$.

Proof. i) is obvious from the definitions. ii) follows immediately from Lemma 1.3.ii (using $V(\mathbf{f})$ for both Z and W). v) is immediate.

Proof of iii): ii) implies $\left|\Pi_{\mathbf{f}}^{i-1}\right| \subseteq \left|\Pi_{\mathbf{f}}^{i}\right|$. That $\left|\widehat{\Pi}_{\mathbf{f}}^{i-1}\right| \subseteq \left|\widehat{\Pi}_{\mathbf{f}}^{i}\right|$ follows from the inductive definition.

Proof of iv): By definition, $\widehat{\Pi}_{\mathbf{f}}^{i+1}$ has no components or embedded subvarieties contained in $V(f_{i+k+1-d})$. Thus, $\left[\widehat{\Pi}_{\mathbf{f}}^{i+1} \cap V(f_{i+k+1-d})\right] = \left[\widehat{\Pi}_{\mathbf{f}}^{i+1}\right] \cdot \left[V(f_{i+k+1-d})\right]$. The desired conclusion follows.

Proof of vi): By downward induction on i. Note first that $\left[\widehat{\Pi}_{\mathbf{f}}^{d}\right] \leqslant \left[\Pi_{\mathbf{f}}^{d}\right]$, since they are, in fact, equal. Suppose now that $i < d$ and that there is an equality of sets $\left|\widehat{\Pi}_{\mathbf{f}}^{j}\right| = \left|\Pi_{\mathbf{f}}^{j}\right|$ for all $j \geqslant i+1$. From induction, we know that $\left[\widehat{\Pi}_{\mathbf{f}}^{i+1}\right] \leqslant \left[\Pi_{\mathbf{f}}^{i+1}\right]$. Thus,

$$\left[\Pi_{\mathbf{f}}^{i}\right] = \left[\Pi_{\mathbf{f}}^{i+1} \cap V(f_{i+k+1-d})\right] \neg V(\mathbf{f}) \geqslant \left[\Pi_{\mathbf{f}}^{i+1} \cap V(f_{i+k+1-d})\right] \neg V(f_{i+k-d}).$$

Since iv) tells us that $\left|\Pi_{\mathbf{f}}^{i+1}\right|$ intersects $V(f_{i+k+1-d})$ properly, we may apply [**Fu**, 8.2.a] to conclude that $\left[\Pi_{\mathbf{f}}^{i+1} \cap V(f_{i+k+1-d})\right] \geqslant \left[\Pi_{\mathbf{f}}^{i+1}\right] \cdot \left[V(f_{i+k+1-d})\right]$ (the presence of embedded varieties in $\Pi_{\mathbf{f}}^{i+1}$ can cause a strict inequality). Therefore,

$$\left[\Pi_{\mathbf{f}}^{i}\right] \geqslant \left(\left[\Pi_{\mathbf{f}}^{i+1}\right] \cdot \left[V(f_{i+k+1-d})\right]\right) \neg V(f_{i+k-d}).$$

Now, applying our inductive hypothesis and iv), we conclude that $\left[\widehat{\Pi}_{\mathbf{f}}^{i}\right] \leqslant \left[\Pi_{\mathbf{f}}^{i}\right]$.

Proof of vii): We have

$$\widehat{\Pi}_{\mathbf{f}}^{i} = \left(\widehat{\Pi}_{\mathbf{f}}^{i+1} \cap V(f_{i+k+1-d})\right) \neg V(f_{i+k-d}) = \left(\Pi_{\mathbf{f}}^{i+1} \cap V(f_{i+k+1-d})\right) \neg V(f_{i+k-d}).$$

By 1.3.i, this equals $\left(\left(\Pi_{\mathbf{f}}^{i+1} \cap V(f_{i+k+1-d})\right) \neg V(\mathbf{f})\right) \neg V(f_{i+k-d})$. By ii) of this proposition, this last expression equals $\Pi_{\mathbf{f}}^{i} \neg V(f_{i+k-d}) = \left(V(f_{i+k+1-d}, \dots, f_k) \neg V(\mathbf{f})\right) \neg V(f_{i+k-d})$. Applying 1.3.i again, we find that $\widehat{\Pi}_{\mathbf{f}}^{i} = V(f_{i+k+1-d}, \dots, f_k) \neg V(f_{i+k-d}) = \widetilde{\Pi}_{\mathbf{f}}^{i}$. □

Note that 2.2.i implies that $\left|\widetilde{\Pi}^i_{\mathbf{f}}\right|$ and $\left|\Pi^i_{\mathbf{f}}\right|$ are **purely** i-dimensional if and only if they are i-dimensional.

We wish to define the Vogel cycles now. However, before we can do this, we need to decide which of the different gap cycles to use to define the Vogel cycles. As a preliminary step, we first define the sets which will underlie the Vogel cycles.

Definition 2.3. Assume that X is irreducible. If $i \neq d$, then we define the *i-th Vogel set of* $\mathbf{f}$, $D^i_{\mathbf{f}}$, to be the union of the irreducible components of $\left|\Pi^{i+1}_{\mathbf{f}} \cap V(f_{i+k+1-d})\right|$ which are contained in $V(\mathbf{f})$; by 2.2.ii, this is equivalent to

$$D^i_{\mathbf{f}} = \overline{\left|\Pi^{i+1}_{\mathbf{f}} \cap V(f_{i+k+1-d})\right| \; - \; \left|\Pi^i_{\mathbf{f}}\right|}.$$

We set $D^d_{\mathbf{f}} = \begin{cases} \emptyset, & \text{if } \mathbf{f} \not\equiv 0 \\ X, & \text{if } \mathbf{f} \equiv 0. \end{cases}$

Note that, if $i < d-(k+1)$ or $i > d$, then $D^i_{\mathbf{f}} = \emptyset$.

If X is a union of irreducible components $\{X_j\}$, we define $D^i_{\mathbf{f}} := \bigcup_j D^i_{\mathbf{f}_{|X_j}}$.

We define the *i-th Vogel set of* $\mathbf{f}$ *with respect to* M to be $D^i_{\mathbf{f}}(M) := \bigcup_j D^i_{\mathbf{f}_{|V_l}}$.

Proposition 2.4. *Every component of $D^i_{\mathbf{f}}(M)$ has dimension at least i and $|M| \cap V(\mathbf{f}) = \bigcup_i D^i_{\mathbf{f}}(M)$. If $\Pi^i_{\mathbf{f}}(M)$ is i-dimensional and C is an i-dimensional irreducible component of $|M| \cap |V(\mathbf{f})|$, then $C \subseteq D^i_{\mathbf{f}}(M)$.*

If X is irreducible of dimension d, then, for all $i \leqslant d-1$,

$$\left|\Pi^{i+1}_{\mathbf{f}} \cap V(\mathbf{f})\right| = \bigcup_{k \leqslant i} D^k_{\mathbf{f}}.$$

Proof. We may work on each irreducible set, V_l, separately; therefore, we assume that we are in the case where X is irreducible and $M = [X]$.

That every component of $D^i_{\mathbf{f}}$ has dimension at least i follows immediately from the fact that each component of $\Pi^{i+1}_{\mathbf{f}}$ has dimension at least $i+1$ (by 2.2.i).

By definition $X = \left|\Pi^d_{\mathbf{f}}\right| \cup D^d_{\mathbf{f}}$. Hence, $V(\mathbf{f}) = \left|\Pi^d_{\mathbf{f}} \cap V(\mathbf{f})\right| \cup D^d_{\mathbf{f}}$ and so, the equation $V(\mathbf{f}) = \bigcup_i D^i_{\mathbf{f}}$ follows once we show the final claim of the proposition.

Suppose that $i \leqslant d-1$. Then,

$$\left|\Pi_{\mathbf{f}}^{i+1} \cap V(\mathbf{f})\right| = \left|\Pi_{\mathbf{f}}^{i+1} \cap V(f_{i+k+1-d}) \cap V(\mathbf{f})\right| = \left|\left(\Pi_{\mathbf{f}}^{i} \cup D_{\mathbf{f}}^{i}\right) \cap V(\mathbf{f})\right| = \left|\Pi_{\mathbf{f}}^{i} \cap V(\mathbf{f})\right| \cup D_{\mathbf{f}}^{i}.$$

As $\Pi_{\mathbf{f}}^{i}$ is eventually empty, the desired conclusion follows.

Finally, suppose that C is an i-dimensional irreducible component of $|V(\mathbf{f})|$ and $\Pi_{\mathbf{f}}^{i}$ is i-dimensional. Then, C is contained in a component C' of $|V(f_{i+k+2-d}, \dots, f_k)|$; such a C' necessarily has dimension at least $i+1$. Thus, C' cannot be contained in $V(\mathbf{f})$. It follows that C' is contained in $\Pi_{\mathbf{f}}^{i+1}$. Therefore,

$$C \subseteq C' \cap V(f_{i+k+1-d}) \subseteq \left|\Pi_{\mathbf{f}}^{i+1} \cap V(f_{i+k+1-d})\right| = \left|\Pi_{\mathbf{f}}^{i}\right| \cup D_{\mathbf{f}}^{i}.$$

If $\Pi_{\mathbf{f}}^{i}$ is i-dimensional, then – since $C \subseteq V(\mathbf{f})$ and is i-dimensional – it follows that $C \not\subseteq \left|\Pi_{\mathbf{f}}^{i}\right|$, and so $C \subseteq D_{\mathbf{f}}^{i}$. □

Below, we prove the *Dimensionality Lemma* in which we state as hypotheses/conclusions that "$\left|\Pi_{\mathbf{f}}^{i}(M)\right|$ is i-dimensional" and "$D_{\mathbf{f}}^{i}(M)$ is i-dimensional". Since sets cannot be negative-dimensional, for $i < 0$, we mean that the respective set is empty. Note that 2.4 implies that $D_{\mathbf{f}}^{i}(M)$ is **purely** i-dimensional if and only if it is i-dimensional.

Lemma 2.5 (**Dimensionality Lemma**). *The following are equivalent:*

i) *for all* i, $\left|\Pi_{\mathbf{f}}^{i}(M)\right|$ *is* i*-dimensional;*

ii) *for all* i, $\left|\Pi_{\mathbf{f}}^{i}(M)\right| = \left|\widetilde{\Pi}_{\mathbf{f}}^{i}(M)\right|$;

iii) *for all* i, $\left|\Pi_{\mathbf{f}}^{i}(M)\right| = \left|\widehat{\Pi}_{\mathbf{f}}^{i}(M)\right|$.

In addition, these equivalent conditions imply

iv) *for all* i, $D_{\mathbf{f}}^{i}(M)$ *is* i*-dimensional;*

and, for all $\mathbf{p} \in |M| \cap V(\mathbf{f})$, *there exists a neighborhood of* $\mathbf{p}$ *in which iv) implies i), ii), and iii).*

Proof. Again we may consider each component appearing M separately; hence, we may assume that X is irreducible and $M = [X]$.

As all the statements are set-theoretic, to cut down on notation, we shall omit the vertical lines around the various gap sheaves.

We will show that i) and iii) are each equivalent to ii), that i) implies iv), and that, near points of $V(\mathbf{f})$, iv) implies i).

i)⇒ ii): Assume i). From the definition of $\widetilde{\Pi}_{\mathbf{f}}^{i}$, what we need to show is: if C is a component of $V(f_{i+k+1-d}, \dots, f_k)$, then C is contained in $V(\mathbf{f})$ if and only if C is contained in $V(f_{i+k-d})$.

As $V(\mathbf{f}) \subseteq V(f_{i+k-d})$, one implication is trivial, and so what we must show is that if C is a component of $V(f_{i+k+1-d}, \ldots, f_k)$ and $C \subseteq V(f_{i+k-d})$, then $C \subseteq V(\mathbf{f})$.

Suppose not. As C is a component of $V(f_{i+k+1-d}, \ldots, f_k)$, the dimension of C is at least i. If $C \not\subseteq V(\mathbf{f})$, then – by definition – C is a component of $\Pi^i_{\mathbf{f}}$. But C is also contained in $V(f_{i+k-d})$, and so C is a component of $\Pi^i_{\mathbf{f}} \cap V(f_{i+k-d}) = \Pi^{i-1}_{\mathbf{f}} \cup D^{i-1}_{\mathbf{f}}$. As C is not contained in $V(\mathbf{f})$, we conclude that C is a component of $\Pi^{i-1}_{\mathbf{f}}$ of dimension at least i. This contradicts i).

ii)$\Rightarrow$ i): Assume ii). From Definition 2.1, $\Pi^i_{\mathbf{f}}$ is purely i-dimensional for $i \geqslant d$. Suppose that i_0 is the largest integer i (less than d) such that $\Pi^i_{\mathbf{f}}$ is not purely i-dimensional. Then, $\Pi^{i_0+1}_{\mathbf{f}}$ is purely (i_0+1)-dimensional and, by ii), the set $\Pi^{i_0+1}_{\mathbf{f}}$ is equal to

$$V(f_{i_0+k+2-d}, \ldots, f_k) \neg V(f_{i_0+k+1-d}).$$

Hence, the intersection $\Pi^{i_0+1}_{\mathbf{f}} \cap V(f_{i_0+k+1-d})$ is proper, and so $\Pi^{i_0+1}_{\mathbf{f}} \cap V(f_{i_0+k+1-d})$ is purely i_0-dimensional. As there is an equality of sets $\Pi^{i_0+1}_{\mathbf{f}} \cap V(f_{i_0+k+1-d}) = \Pi^{i_0}_{\mathbf{f}} \cup D^{i_0}_{\mathbf{f}}$, this contradicts the fact that $\Pi^{i_0}_{\mathbf{f}}$ is not purely i_0-dimensional.

iii)$\Rightarrow$ ii): Assume iii). Then ii) follows immediately from the fact that $\widehat{\Pi}^i_{\mathbf{f}} \subseteq \widetilde{\Pi}^i_{\mathbf{f}} \subseteq \Pi^i_{\mathbf{f}}$ (see 2.2.i).

ii)$\Rightarrow$ iii): Assume ii). The proof is by induction. iii) is certainly true by definition for $i \geqslant d$. Now, suppose that $\Pi^i_{\mathbf{f}} = \widehat{\Pi}^i_{\mathbf{f}}$ for $i \geqslant m$, where $m \leqslant d$. We need to show that $\Pi^{m-1}_{\mathbf{f}} = \widehat{\Pi}^{m-1}_{\mathbf{f}}$. We have

$$\widehat{\Pi}^{m-1}_{\mathbf{f}} = \big(\widehat{\Pi}^m_{\mathbf{f}} \cap V(f_{m+k-d})\big) \neg V(f_{m+k-1-d}) = \big(\Pi^m_{\mathbf{f}} \cap V(f_{m+k-d})\big) \neg V(f_{m+k-1-d}).$$

By combining the definition of $\Pi^m_{\mathbf{f}}$ as $V(f_{m+k+1-d}, \ldots, f_k) \neg V(\mathbf{f})$ with Lemma 1.3.ii, we conclude that

$$\big(\Pi^m_{\mathbf{f}} \cap V(f_{m+k-d})\big) \neg V(f_{m+k-1-d}) = V(f_{m+k-d}, \ldots, f_k) \neg V(f_{m+k-1-d})$$

and so, $\widehat{\Pi}^{m-1}_{\mathbf{f}} = \widetilde{\Pi}^{m-1}_{\mathbf{f}}$. By ii), this implies that $\widehat{\Pi}^{m-1}_{\mathbf{f}} = \Pi^{m-1}_{\mathbf{f}}$ and we are finished.

i)$\Rightarrow$ iv): Assume i), and suppose that i_0 is such that $D^{i_0}_{\mathbf{f}}$ is **not** purely i_0-dimensional. Then, $\Pi^{i_0+1}_{\mathbf{f}} \cap V(f_{i_0+k+1-d})$ is not purely i_0-dimensional. As $\Pi^{i_0+1}_{\mathbf{f}}$ is purely (i_0+1)-dimensional by assumption, it follows that $V(f_{i_0+k+1-d})$ contains a component, C, of $\Pi^{i_0+1}_{\mathbf{f}}$. As C is a component of $\Pi^{i_0+1}_{\mathbf{f}}$, C is not contained in $V(\mathbf{f})$.

Thus, C is a component of $\Pi^{i_0+1}_{\mathbf{f}} \cap V(f_{i_0+k+1-d}) = \Pi^{i_0}_{\mathbf{f}} \cup D^{i_0}_{\mathbf{f}}$ which is not contained in $V(\mathbf{f})$, and so C is an (i_0+1)-dimensional component of $\Pi^{i_0}_{\mathbf{f}}$ – this contradicts our assumption.

iv)$\Rightarrow$ i): Assume iv), and that we are interested in the germ of the situation at a point $\mathbf{p} \in V(\mathbf{f})$. Let i_0 be the smallest i such that $\Pi^i_{\mathbf{f}}$ is **not** purely i-dimensional. By Proposition 2.2.i, $\Pi^{i_0}_{\mathbf{f}}$ must have dimension at least i_0+1. Thus, since $\mathbf{p} \in V(\mathbf{f})$, $\Pi^{i_0}_{\mathbf{f}} \cap V(f_{i_0+k-d})$ has dimension at least i_0. But, as sets, $\Pi^{i_0}_{\mathbf{f}} \cap V(f_{i_0+k-d}) = \Pi^{i_0-1}_{\mathbf{f}} \cup D^{i_0-1}_{\mathbf{f}}$, and by assumption

$D_{\mathbf{f}}^{i_0-1}$ is purely $(i_0 - 1)$-dimensional. Therefore, we conclude that $\Pi_{\mathbf{f}}^{i_0-1}$ has dimension at least i_0 – a contradiction of the choice of i_0. □

Remark 2.6. Our phrasing of Lemma 2.5 is the most elegant, and is in the form that we will usually need. However, it is occasionally helpful to note that our proof does not require that one knows i), ii), or iii) for **all** i. Specifically, what our proof actually shows is that:

- if $\left|\Pi_{\mathbf{f}}^{i-1}(M)\right|$ is $(i-1)$-dimensional, then $\left|\Pi_{\mathbf{f}}^{i}(M)\right| = \left|\widetilde{\Pi}_{\mathbf{f}}^{i}(M)\right|$;

- if $\left|\Pi_{\mathbf{f}}^{i}(M)\right| = \left|\widetilde{\Pi}_{\mathbf{f}}^{i}(M)\right|$ for all $i > k$, then $\left|\Pi_{\mathbf{f}}^{k}(M)\right|$ is k-dimensional;

- if $\left|\Pi_{\mathbf{f}}^{i}(M)\right| = \left|\widehat{\Pi}_{\mathbf{f}}^{i}(M)\right|$, then $\left|\Pi_{\mathbf{f}}^{i}(M)\right| = \left|\widetilde{\Pi}_{\mathbf{f}}^{i}(M)\right|$;

- if $\left|\Pi_{\mathbf{f}}^{i}(M)\right| = \left|\widehat{\Pi}_{\mathbf{f}}^{i}(M)\right|$ for all $i \geqslant m$, and $\left|\Pi_{\mathbf{f}}^{m-1}(M)\right| = \left|\widetilde{\Pi}_{\mathbf{f}}^{m-1}(M)\right|$, then $\left|\Pi_{\mathbf{f}}^{m-1}(M)\right| = \left|\widehat{\Pi}_{\mathbf{f}}^{m-1}(M)\right|$; in particular, if $\left|\Pi_{\mathbf{f}}^{i}(M)\right| = \left|\widetilde{\Pi}_{\mathbf{f}}^{i}(M)\right|$ for all $i \geqslant m-1$, then $\left|\Pi_{\mathbf{f}}^{i}(M)\right| = \left|\widehat{\Pi}_{\mathbf{f}}^{i}(M)\right|$ for all $i \geqslant m-1$;

- if $\left|\Pi_{\mathbf{f}}^{i}(M)\right|$ is i-dimensional and $\left|\Pi_{\mathbf{f}}^{i+1}(M)\right|$ is $(i+1)$-dimensional, then $D_{\mathbf{f}}^{i}(M)$ is i-dimensional; and

- if $\mathbf{p} \in |M| \cap V(\mathbf{f})$, $\left|\Pi_{\mathbf{f}}^{i-1}(M)\right|$ is $(i-1)$-dimensional at $\mathbf{p}$, and $D_{\mathbf{f}}^{i-1}(M)$ is $(i-1)$-dimensional at $\mathbf{p}$, then $\left|\Pi_{\mathbf{f}}^{i}(M)\right|$ is i-dimensional at $\mathbf{p}$.

Definition 2.7. If the equivalent conditions i), ii), and iii) of Lemma 2.5 hold, we say that *the gap sets of* $\mathbf{f}$ *with respect to* M *have the correct dimension.*

If the equivalent conditions i), ii), iii), and iv) of Lemma 2.5 hold at a point $\mathbf{p} \in |M| \cap V(\mathbf{f})$, we say that *the Vogel sets of* $\mathbf{f}$ *with respect to* M *have the correct dimension at* $\mathbf{p}$. We say simply that *the Vogel sets of* $\mathbf{f}$ *with respect to* M *have the correct dimension* provided that they have correct dimension at all points of $|M| \cap V(\mathbf{f})$.

Remark 2.8. Note that, since every component of $D_{\mathbf{f}}^{i}(M)$ has dimension at least i (see 2.4), if the Vogel sets all have correct dimension at $\mathbf{p}$, then all the Vogel sets have correct dimension at points **near p**.

Note also that if the gap sets have correct dimension, then the Vogel sets have correct dimension. Moreover, since we are interested only in what happens near $V(\mathbf{f})$, the natural assumption for us to make seems like it should be that the Vogel cycles have correct dimension. However, our usual assumption will be that **gap sets** have correct dimension; for 2.5 tells us that, in a neighborhood of $V(\mathbf{f})$, these assumptions are equivalent, and requiring the gap sets to have the correct dimension saves us from having to state over and over again that we take a small neighborhood of a point of $V(\mathbf{f})$.

It is important to remember that one implication of the Vogel and gap sets having correct dimension is that $D_{\mathbf{f}}^{i}(M)$, $\Pi_{\mathbf{f}}^{i}(M)$, $\widetilde{\Pi}_{\mathbf{f}}^{i}(M)$, and $\widehat{\Pi}_{\mathbf{f}}^{i}(M)$ are all empty if $i < 0$, and $\Pi_{\mathbf{f}}^{0}(M) =$

$\widetilde{\Pi}^0_{\mathbf{f}}(M) = \widehat{\Pi}^0_{\mathbf{f}}(M) = \emptyset$ at points of $|M| \cap V(\mathbf{f})$.

Finally, consider the special case where $\mathbf{p}$ is an isolated point of $|M| \cap V(\mathbf{f})$. Then, 2.4 implies that, near $\mathbf{p}$, $D^i_{\mathbf{f}}(M) = \emptyset$ if $i \geqslant 1$, and $D^0_{\mathbf{f}}(M) = \{\mathbf{p}\}$. Thus, 2.5 implies that the gap sets and the Vogel sets have correct dimension at $\mathbf{p}$.

Proposition 2.9. *If X is irreducible and Cohen-Macaulay, and all of the gap sets of $\mathbf{f}$ have correct dimension, then, for all i, the schemes $\Pi^i_{\mathbf{f}}$, $\widetilde{\Pi}^i_{\mathbf{f}}$, and $\widehat{\Pi}^i_{\mathbf{f}}$ are equal.*

Proof. By 2.2.v, if the gap sets have correct dimension, then the schemes $\Pi^i_{\mathbf{f}}$ and $\widetilde{\Pi}^i_{\mathbf{f}}$ are equal up to embedded subvariety. By Lemma 1.5, $\Pi^i_{\mathbf{f}}$ and $\widetilde{\Pi}^i_{\mathbf{f}}$ have no embedded subvarieties; therefore, they are equal as schemes.

To prove that the scheme structure of $\widehat{\Pi}^i_{\mathbf{f}}$ agrees with the other two, we must, of course, use induction. Let d denote the dimension of X. For $i \geqslant d$, we know that $\Pi^i_{\mathbf{f}} = \widetilde{\Pi}^i_{\mathbf{f}} = \widehat{\Pi}^i_{\mathbf{f}}$.

Suppose, inductively, that $\Pi^{i+1}_{\mathbf{f}} = \widetilde{\Pi}^{i+1}_{\mathbf{f}} = \widehat{\Pi}^{i+1}_{\mathbf{f}}$. Then, 2.2.vii tells us that $\widetilde{\Pi}^i_{\mathbf{f}} = \widehat{\Pi}^i_{\mathbf{f}}$ and, by the first paragraph above, we know that this equals $\Pi^i_{\mathbf{f}}$. □

While we have been selecting $(k+1)$-tuples, $\mathbf{f}$, our primary object of interest is, in fact, the ideal $< \mathbf{f} >$ generated by the $f_0, \ldots, f_k$. As far as the ideal $< \mathbf{f} >$ is concerned, the functions comprising $\mathbf{f}$ may not be suitably generic. However, as we shall see, to obtain a well-behaved ordered collection of generators, one only needs to replace $(f_0, \ldots, f_k)$ by generic linear combinations of the f_i's themselves. However, the term "generic" here is used in a non-standard way; what we need is to replace f_0 by a generic linear combination, then – fixing this new f_0 – replace f_1 by a generic linear combination, and so on. Since "generic" should always mean open and dense in **some** topology, we will define a new, convenient one.

Definition 2.10. The *pseudo-Zariski topology* (pZ-topology) on a topological space $(X, \mathcal{T})$ is a new topological space $(X, \mathcal{T}_{pZ})$ given by $\mathcal{U} \in \mathcal{T}_{pZ}$ if and only if $\mathcal{U}$ is empty or is an open, dense subset in $(X, \mathcal{T})$. (One verifies easily that this, in fact, yields a topology on X.)

Given two topological spaces X and Y, let π_X and π_Y denote the projections from $X \times Y$ onto X and Y, respectively. The *inductive pseudo-Zariski topology* (IPZ-topology) on $X \times Y$ is given by: $\mathcal{W} \subseteq X \times Y$ is open in the IPZ-topology if and only if $\pi_X(\mathcal{W})$ is open in the pZ-topology on X and, for all $x \in \pi_X(\mathcal{W})$, $\pi_Y\big(\mathcal{W} \cap \pi_X^{-1}(x)\big)$ is open in the pZ-topology on Y. (It is trivial to verify that this is a topology on $X \times Y$, and that a non-empty open set in the IPZ-topology on $X \times Y$ is a dense set in the cross-product topology on $X \times Y$.)

Finally, given a finite number of topological spaces $X_1, X_2, \ldots, X_m$, the IPZ-topology on $X_1 \times X_2 \times \cdots \times X_m$ is given inductively by using the IPZ-topology on each product in the expression $\Big(\big((X_1 \times X_2) \times X_3\big) \times \cdots \times X_{m-1}\Big) \times X_m$.

A *generic linear reorganization* of a $(k+1)$-tuple $\mathbf{f}$ is a matrix product $\mathbf{f}A$, where the matrix A is invertible and is an element of some given generic subset in the IPZ-topology on the $(k+1)$-fold product $\mathbb{C}^{k+1} \times \cdots \times \mathbb{C}^{k+1}$ (where we consider each column of A to be contained in one copy of $\mathbb{C}^{k+1}$).

Note that, if $X_1 = X_2 = \cdots = X_m = \mathbb{C}^N$ (or $\mathbb{P}^N$), then the IPZ-topology on the product is more fine than the Zariski topology, but sets which are open in the IPZ-topology need **not** be open in the classical topology on the product.

Proposition/Definition 2.11. *If X is irreducible, then, for all $\mathbf{p} \in X$, for a generic linear reorganization, $\hat{\mathbf{f}}$, of $\mathbf{f}$, the gap sets of $\hat{\mathbf{f}}$ all have correct dimension at $\mathbf{p}$ and, for all i, there is an equality of schemes $\Pi^i_{\hat{\mathbf{f}}} = \widetilde{\Pi}^i_{\hat{\mathbf{f}}} = \widehat{\Pi}^i_{\hat{\mathbf{f}}}$ in a neighborhood of $\mathbf{p}$.*

Therefore, for $\mathbf{p} \in |M|$, for a generic linear reorganization, $\hat{\mathbf{f}}$, of $\mathbf{f}$, the gap sets of $\hat{\mathbf{f}}$ with respect to $|M|$ all have correct dimension at $\mathbf{p}$ and, for all i, there is an equality of cycles $\Pi^i_{\hat{\mathbf{f}}}(M) = \widetilde{\Pi}^i_{\hat{\mathbf{f}}}(M) = \widehat{\Pi}^i_{\hat{\mathbf{f}}}(M)$ at $\mathbf{p}$.

If we are working in the algebraic category, then we may produce such generic linear reorganizations globally.

We refer to a reorganization $\hat{\mathbf{f}}$ such that the above equality of cycles holds as an *agreeable reorganization of $\mathbf{f}$ (with respect to M at $\mathbf{p}$)* (for it makes the various cycle structures agree).

Proof. Assume that X is irreducible. We fix a point $\mathbf{p} \in X$. Our sole reason for stating the results "at $\mathbf{p}$" is that, at several places in the proof, we will need to know that certain analytic sets have a finite number of analytic components. This is, of course, guaranteed near a given point or in the algebraic category. Hence, throughout the proof, we will make no further reference to working in a neighborhood of $\mathbf{p}$, but will assume that all of the analytic sets that arise have a finite number of components.

We first show:

($\dagger$) for a generic linear reorganization, $\hat{\mathbf{f}}$, of $\mathbf{f}$, for all i, $V(\hat{f}_{i+k-d})$ contains no component or embedded subvariety of $\Pi^i_{\hat{\mathbf{f}}}$.

We produce the $(k+1)$-tuple $\hat{\mathbf{f}}$ one element at a time, by downward induction. If $\mathbf{f}$ is identically zero on X, then ($\dagger$) is trivial. So, suppose that one of the f_i does not vanish on X. Then, for a generic linear combination $\hat{f}_k := a_0 f_0 + \cdots + a_k f_k$, $\hat{f}_k$ does not vanish on X. Thus, $V(\hat{f}_k)$ contains no component or embedded subvariety of $\Pi^d_{\hat{\mathbf{f}}}$.

Now, suppose that we have made generic linear reorganizations of $\mathbf{f}$ to produce $\hat{\mathbf{f}}$, and that $V(\hat{f}_{i+k-d})$ contains no component or embedded subvariety of $\Pi^i_{\hat{\mathbf{f}}}$ for all $i \geqslant m$. Then, for every component or embedded subvariety, W, of $\Pi^{m-1}_{\hat{\mathbf{f}}}$, W is contained in $V(\hat{f}_{m+k-d}, \dots, \hat{f}_k)$, but there exists some $\hat{f}_j$ with $j < m+k-d$ such that $W \not\subseteq V(\hat{f}_j)$. Thus, a generic linear combination of the $\hat{f}$'s will not vanish on any component of embedded subvariety of $\Pi^{m-1}_{\hat{\mathbf{f}}}$. This proves ($\dagger$).

As $\Pi^{i-1}_{\hat{\mathbf{f}}} = \big(\Pi^i_{\hat{\mathbf{f}}} \cap V(\hat{f}_{i+k-d})\big) \neg V(\hat{\mathbf{f}})$ by 2.2.ii, ($\dagger$) implies that the Vogel sets of $\hat{\mathbf{f}}$ all have correct dimension. We show that $\Pi^i_{\hat{\mathbf{f}}} = \widetilde{\Pi}^i_{\hat{\mathbf{f}}} = \widehat{\Pi}^i_{\hat{\mathbf{f}}}$ by downward induction on i.

When $i = d$, the statement is clear. Assume now that $\Pi^{i+1}_{\hat{\mathbf{f}}} = \widetilde{\Pi}^{i+1}_{\hat{\mathbf{f}}} = \widehat{\Pi}^{i+1}_{\hat{\mathbf{f}}}$. Then,

$$\widehat{\Pi}^i_{\hat{\mathbf{f}}} = \big(\widehat{\Pi}^{i+1}_{\hat{\mathbf{f}}} \cap V(\hat{f}_{i+k+1-d})\big) \neg V(\hat{f}_{i+k-d}) = \big(\Pi^{i+1}_{\hat{\mathbf{f}}} \cap V(\hat{f}_{i+k+1-d})\big) \neg V(\hat{f}_{i+k-d}),$$

which, by 1.3.i, equals $\Big((\Pi^{i+1}_{\hat{\mathbf{f}}} \cap V(\hat{f}_{i+k+1-d}))\neg V(\hat{\mathbf{f}})\Big)\neg V(\hat{f}_{i+k-d})$.

Therefore, applying 2.2.ii, followed by (†), we conclude that $\widehat{\Pi}^i_{\hat{\mathbf{f}}} = \Pi^i_{\hat{\mathbf{f}}} \neg V(\hat{f}_{i+k-d}) = \Pi^i_{\hat{\mathbf{f}}}$.

As $\widehat{\Pi}^i_{\hat{\mathbf{f}}} = \Pi^i_{\hat{\mathbf{f}}}$ for all i, by applying 2.2.vii, we conclude that $\widetilde{\Pi}^i_{\hat{\mathbf{f}}} = \widehat{\Pi}^i_{\hat{\mathbf{f}}} = \Pi^i_{\hat{\mathbf{f}}}$. □

We now wish to endow the Vogel sets a cycle structure. First, we need the following easy proposition.

Proposition 2.12. *If X is irreducible, and $\left|\Pi^{j-1}_{\mathbf{f}}\right|$ is $(j-1)$-dimensional for all $j \geqslant i$, then $\left[\Pi^j_{\mathbf{f}}\right] = \left[\widehat{\Pi}^j_{\mathbf{f}}\right]$ for all $j \geqslant i$, and there is an equality of cycles given by*

$$\left[\widehat{\Pi}^i_{\mathbf{f}}\right] = \Big(\left[\widehat{\Pi}^{i+1}_{\mathbf{f}}\right] \cdot \left[V(f_{i+k+1-d})\right]\Big) \neg V(\mathbf{f}).$$

Therefore, on $X - V(\mathbf{f})$, all of $V(f_k), V(f_{k-1}), \ldots, V(f_{i+k+1-d})$ intersect properly and, on $X - V(\mathbf{f})$,

$$\left[\widehat{\Pi}^i_{\mathbf{f}}\right] = V(f_k) \cdot V(f_{k-1}) \cdot \ldots \cdot V(f_{i+k+1-d}).$$

Proof. Using Remark 2.6, we see that $\left|\Pi^{j-1}_{\mathbf{f}}\right|$ being $(j-1)$-dimensional for all $j \geqslant i$ implies that $\left[\Pi^j_{\mathbf{f}}\right] = \left[\widehat{\Pi}^j_{\mathbf{f}}\right]$ for all $j \geqslant i$.

Now, by 2.2.iv, the statement $\left[\widehat{\Pi}^i_{\mathbf{f}}\right] = \Big(\left[\widehat{\Pi}^{i+1}_{\mathbf{f}}\right] \cdot \left[V(f_{i+k+1-d})\right]\Big) \neg V(\mathbf{f})$ is equivalent to the set-theoretic statement $\left|\widehat{\Pi}^i_{\mathbf{f}}\right| = \Big(\left|\widehat{\Pi}^{i+1}_{\mathbf{f}}\right| \cap V(f_{i+k+1-d})\Big) \neg V(\mathbf{f})$. This set-theoretic statement follows easily from 2.5.iii; for it tells us that

$$\Big(\left|\widehat{\Pi}^{i+1}_{\mathbf{f}}\right| \cap V(f_{i+k+1-d})\Big) \neg V(\mathbf{f}) = \Big(\left|\Pi^{i+1}_{\mathbf{f}}\right| \cap V(f_{i+k+1-d})\Big) \neg V(\mathbf{f})$$

and 2.2.ii tells us that this equals $\left|\Pi^i_{\mathbf{f}}\right|$. Applying 2.5.iii again yields the desired equality of cycles. □

Remark 2.13. It is tempting to write that $\left[\widehat{\Pi}^i_{\mathbf{f}}\right] = \big(V(f_k)\cdot V(f_{k-1})\cdot \ldots \cdot V(f_{i+k+1-d})\big)\neg V(\mathbf{f})$. We could do this if we were willing to use intersection theory with non-proper intersections; this seems especially innocuous when the non-proper part of the intersection lies in a portion that we are going to throw away, as it does here. Nonetheless, we do not want wish to write any formulas involving intersection theory which are not discussed in Appendix A.

However, 2.12 does tell us that, on X, $\left[\widehat{\Pi}^i_{\mathbf{f}}\right]$ is the closure in X of this cycle on $X - V(\mathbf{f})$. This cycle structure turns out to be the correct one to use in order to endow the Vogel sets with a cycle structure.

However, in order to guarantee that the cycles we define actually have as their underlying sets the Vogel sets of $\mathbf{f}$, we only define the Vogel **cycles** when the gap sets (or Vogel sets) have the correct dimensions and, even then, we must restrict ourselves to what happens in a neighborhood of $V(\mathbf{f})$.

Definition 2.14. If X is irreducible, and $\left|\Pi^{j-1}_{\mathbf{f}}\right|$ is $(j-1)$-dimensional at each point in $V(\mathbf{f})$ for all $j \geqslant i$, then we define the *i-th Vogel cycle of* $\mathbf{f}$, $\Delta^i_{\mathbf{f}}$, to be the sum of the components of

$$\left(\left[\widehat{\Pi}^{i+1}_{\mathbf{f}}\right] \cdot \left[V(f_{i+k+1-d})\right]\right) - \left[\widehat{\Pi}^i_{\mathbf{f}}\right]$$

which intersect $V(\mathbf{f})$. In other words, if

$$\left(\left[\widehat{\Pi}^{i+1}_{\mathbf{f}}\right] \cdot \left[V(f_{i+k+1-d})\right]\right) - \left[\widehat{\Pi}^i_{\mathbf{f}}\right] = \sum_j p_j\left[W_j\right],$$

then $\Delta^i_{\mathbf{f}} = \sum\limits_{W_j \cap V(\mathbf{f}) \neq \emptyset} p_j\left[W_j\right]$.

If $\left|\Pi^{j-1}_{\mathbf{f}_{|V_l}}\right|$ is $(j-1)$-dimensional at each point in $V(\mathbf{f}_{|V_l})$ for all $j \geqslant i$ and for all components V_l of $|M|$, then we say that *the i-th Vogel cycle of* $\mathbf{f}$ *with respect to M is defined* and its definition is $\Delta^i_{\mathbf{f}}(M) := \sum\limits_l m_l \, \Delta^i_{\mathbf{f}_{|V_l}}$.

Note that the Dimensionality Lemma implies that there is no difference between saying that **all** the Vogel sets have correct dimension and that **all** the Vogel cycles are defined; we prefer to say that the Vogel cycles are defined, as the Vogel cycles are the objects in which we are most interested.

We have defined the Vogel cycles to consist of pieces which intersect $V(\mathbf{f})$; however, 2.12 yields immediately:

Proposition 2.15. *If $\Delta^i_{\mathbf{f}}(M)$ is defined, then each $\Delta^i_{\mathbf{f}_{|V_l}}$ is non-negative and purely i-dimensional. Moreover, $\left|\Delta^i_{\mathbf{f}}(M)\right| = D^i_{\mathbf{f}}(M) \subseteq |M| \cap V(\mathbf{f})$.*

Remark 2.16. If X is irreducible, Proposition 2.12 and Proposition 2.15, together with the Dimensionality Lemma, tell us how the Vogel cycles should be calculated; we will describe this now, omitting the square brackets for the cycles.

One begins with $\widehat{\Pi}^d_{\mathbf{f}} = X \neg V(f_k)$; thus, $\widehat{\Pi}^d_{\mathbf{f}}$ is either 0 or X. Next, one calculates the intersection $\widehat{\Pi}^d_{\mathbf{f}} \cdot V(f_k)$. This intersection cycle has components contained in $V(\mathbf{f})$ and components which are not contained in $V(\mathbf{f})$. By 2.12, the sum of the components which are not contained in $V(\mathbf{f})$ is precisely $\widehat{\Pi}_{\mathbf{f}}^{d-1}$ and the sum of the components which are contained in $V(\mathbf{f})$ is $\Delta_{\mathbf{f}}^{d-1}$. Having calculated $\widehat{\Pi}_{\mathbf{f}}^{d} \cdot V(f_k) = \widehat{\Pi}_{\mathbf{f}}^{d-1} + \Delta_{\mathbf{f}}^{d-1}$, we use our newly found $\widehat{\Pi}_{\mathbf{f}}^{d-1}$ in the next step: the calculation of $\widehat{\Pi}_{\mathbf{f}}^{d-1} \cdot V(f_{k-1})$. One proceeds downward inductively.

The subtle point in the above description is that, if one is working in a neighborhood of a point of $V(\mathbf{f})$, one may check **while** performing the calculation that the Vogel sets, $|\Delta_{\mathbf{f}}^i|$, have correct dimension. For, by splitting the intersections into pieces which are contained in $V(\mathbf{f})$, and pieces which are not, we are actually obtaining a cycle $\Delta_{\mathbf{f}}^i$ whose underlying set is precisely $D_{\mathbf{f}}^i$ (this follows from 2.2.ii). Thus, one proceeds with the inductive calculation described above, and then checks that the calculated $\Delta_{\mathbf{f}}^i$ have correct dimension, which then tells one that the calculation is actually correct.

Consider the special case where $\mathbf{p}$ is an isolated point of $|M| \cap V(\mathbf{f})$. As we saw in Remark 2.8, it is automatic that the Vogel cycles are defined at $\mathbf{p}$, and only $\Delta_{\mathbf{f}}^0$ can be non-zero.

Example 2.17. We continue to suppress the square brackets around cycles. Let $X = \mathbb{C}^5$ and let

$$\mathbf{f} = (f_0, f_1, f_2, f_3, f_4) = (-2ux^2,\ -2vx^2,\ -2wx^2,\ -3x^2 - 2x(u^2+v^2+w^2),\ 2y).$$

(The reason for the strange, seemingly pointless, coefficients is that we will use this example later in a different context. See Example II.2.4.) Then, $V(\mathbf{f}) = V(x, y)$ and $\widehat{\Pi}_{\mathbf{f}}^5 = \mathbb{C}^5$.

$$\widehat{\Pi}_{\mathbf{f}}^5 \cdot V(f_4) = \widehat{\Pi}_{\mathbf{f}}^5 \cdot V(-2y) = V(y).$$

As $V(y)$ is not contained in $V(\mathbf{f})$, $\widehat{\Pi}_{\mathbf{f}}^4 = V(y)$, and we continue.

$$\widehat{\Pi}_{\mathbf{f}}^4 \cdot V(f_3) = V(y) \cdot V(-3x^2 - 2x(u^2+v^2+w^2)) =$$
$$V(-3x - 2(u^2+v^2+w^2), y) + V(x, y) = \widehat{\Pi}_{\mathbf{f}}^3 + \Delta_{\mathbf{f}}^3.$$

$$\widehat{\Pi}_{\mathbf{f}}^3 \cdot V(f_2) = V(-3x - 2(u^2+v^2+w^2), y) \cdot V(-2wx^2) =$$
$$V(-3x - 2(u^2+v^2), w, y) + 2V(u^2+v^2+w^2, x, y) = \widehat{\Pi}_{\mathbf{f}}^2 + \Delta_{\mathbf{f}}^2.$$

$$\widehat{\Pi}_{\mathbf{f}}^2 \cdot V(f_1) = V(-3x - 2(u^2+v^2), w, y) \cdot V(-2vx^2) =$$
$$V(-3x - 2u^2, v, w, y) + 2V(u^2+v^2, w, x, y) = \widehat{\Pi}_{\mathbf{f}}^1 + \Delta_{\mathbf{f}}^1.$$

$$\widehat{\Pi}_{\mathbf{f}}^1 \cdot V(f_0) = V(-3x - 2u^2, v, w, y) \cdot V(-2ux^2) =$$
$$V(u, v, w, x, y) + 2V(u^2, v, w, x, y) = 5[\mathbf{0}] = \Delta_{\mathbf{f}}^0.$$

Hence, we find the Vogel sets all have correct dimension, and so the Vogel cycles are defined and $\Delta^3_{\mathbf{f}} = V(x,y)$, $\Delta^2_{\mathbf{f}} = 2V(u^2+v^2+w^2,x,y)$, $\Delta^1_{\mathbf{f}} = 2V(u^2+v^2,w,x,y)$, and $\Delta^0_{\mathbf{f}} = 5[\mathbf{0}]$.

Remark 2.18. Suppose that all the Vogel cycles of $\mathbf{f}$ are defined and $k+1>d$. Consider the *truncated d-tuple* $\mathbf{f}_{\mathrm{tr}} := (f_{k+1-d},\dots,f_k)$; we claim that, in a neighborhood of $V(\mathbf{f})$, $|V(\mathbf{f})| = |V(\mathbf{f}_{\mathrm{tr}})|$ and both $\mathbf{f}$ and $\mathbf{f}_{\mathrm{tr}}$ will produce the same D^i, Δ^i, Π^i, $\widetilde{\Pi}^i$, and $\widehat{\Pi}^i$ for all i (all of them will be empty for $i<0$).

It is immediate from the definitions that $\widetilde{\Pi}^i_{\mathbf{f}} = \widetilde{\Pi}^i_{\mathbf{f}_{\mathrm{tr}}}$ and $\widehat{\Pi}^i_{\mathbf{f}} = \widehat{\Pi}^i_{\mathbf{f}_{\mathrm{tr}}}$. We would know that, near $V(\mathbf{f})$, $\Pi^i_{\mathbf{f}} = \Pi^i_{\mathbf{f}_{\mathrm{tr}}}$ and, hence, that $D^i_{\mathbf{f}} = D^i_{\mathbf{f}_{\mathrm{tr}}}$ and $\Delta^i_{\mathbf{f}} = \Delta^i_{\mathbf{f}_{\mathrm{tr}}}$, if we could show that there is an equality of sets $|V(\mathbf{f})| = |V(\mathbf{f}_{\mathrm{tr}})|$.

This is easy; by definition of $\Pi^i_{\mathbf{f}}$, $|V(\mathbf{f}_{\mathrm{tr}})| = |V(\mathbf{f})| \cup |\Pi^0_{\mathbf{f}}|$. As we are assuming that $\Pi^0_{\mathbf{f}}$ is 0-dimensional (and, of course, has no components contained in $V(\mathbf{f})$), there is a neighborhood of $V(\mathbf{f})$ in which $|V(\mathbf{f})| = |V(\mathbf{f}_{\mathrm{tr}})|$.

Suppose that all the Vogel cycles of $\mathbf{f}$ are defined and $k+1<d$. Consider the *extended d-tuple* $\mathbf{f}_{\mathrm{ex}} := (f_0,\dots,f_0,\dots,f_k)$ (where there are $d-k$ occurrences of f_0); clearly, $|V(\mathbf{f})| = |V(\mathbf{f}_{\mathrm{ex}})|$, and $\mathbf{f}$ and $\mathbf{f}_{\mathrm{ex}}$ will produce the same D^i, Δ^i, Π^i, $\widetilde{\Pi}^i$, and $\widehat{\Pi}^i$ for all i (all of them will be empty for $i<d-(k+1)$).

Looking at the two cases above, we see that, if all the Vogel cycles are defined, the whole theory remains unchanged if we assume that $d=k+1$, i.e., if we assume that the dimension of the underlying space X is exactly equal to the number of functions in our tuple $\mathbf{f}$.

Proposition 2.19. *Suppose that X is irreducible of dimension $k+1$. Let $s := \dim_{\mathbf{p}} V(\mathbf{f})$ and suppose that $s \geqslant 0$.*

If $\left|\Pi^{j-1}_{\mathbf{f}}\right|$ is $(j-1)$-dimensional for all $j \geqslant s$, then, in a neighborhood of $\mathbf{p}$, all of $V(f_k)$, $V(f_{k-1})$, $\dots$, $V(f_{s+1})$ intersect properly,

$$\left[\widehat{\Pi}^{s+1}_{\mathbf{f}}\right] = V(f_k)\cdot V(f_{k-1})\cdot\dots\cdot V(f_{s+1}),$$

and $\Delta^s_{\mathbf{f}}$ equals the sum of those components of $V(f_k)\cdot V(f_{k-1})\cdot\dots\cdot V(f_{s+1})\cdot V(f_s)$ which are contained in $V(\mathbf{f})$.

In particular, if $\mathbf{p}$ is an isolated point in $V(\mathbf{f})$, then

$$\left(\Delta^0_{\mathbf{f}}\right)_{\mathbf{p}} = \Big(V(f_k)\cdot V(f_{k-1})\cdot\dots\cdot V(f_1)\cdot V(f_0)\Big)_{\mathbf{p}}.$$

Proof. If $\left|\Pi^{j-1}_{\mathbf{f}}\right|$ is $(j-1)$-dimensional for all $j \geqslant s$, then 2.12 tells us that, on $X - V(\mathbf{f})$, $V(f_k), V(f_{k-1}), \dots, V(f_{s+1})$ intersect properly and

$$\left[\widehat{\Pi}^{s+1}_{\mathbf{f}}\right] = V(f_k)\cdot V(f_{k-1})\cdot\dots\cdot V(f_{s+1}).$$

However, $\widehat{\Pi}_{\mathbf{f}}^{s+1}$ is purely $(s+1)$-dimensional, and every component of $V(f_k, \ldots, f_{s+1})$ has dimension at least $s+1$. As $s = \dim_{\mathbf{p}} V(\mathbf{f})$, it follows that there is a neighborhood of $\mathbf{p}$ in which the closure of $|V(f_k, \ldots, f_{s+1}) - V(\mathbf{f})| = |V(f_k, \ldots, f_{s+1})|$, and thus in which $V(f_k), V(f_{k-1}), \ldots, V(f_{s+1})$ intersect properly and

$$\left[\widehat{\Pi}_{\mathbf{f}}^{s+1}\right] = V(f_k) \cdot V(f_{k-1}) \cdot \ldots \cdot V(f_{s+1}).$$

Therefore,

$$\left[\widehat{\Pi}_{\mathbf{f}}^{s+1}\right] \cdot V(f_s) \;=\; V(f_k) \cdot V(f_{k-1}) \cdot \ldots \cdot V(f_{s+1}) \cdot V(f_s)$$

and 2.12 tells us that the components of this that are contained in $V(\mathbf{f})$ are precisely $\Delta_{\mathbf{f}}^{s}$.

Recalling Remark 2.8, if $\mathbf{p}$ is an isolated point in $V(\mathbf{f})$, then all of the gaps sets have correct dimension at $\mathbf{p}$, which implies that $\Pi_{\mathbf{f}}^{0}$ is empty. Thus,

$$\left(\Delta_{\mathbf{f}}^{0}\right)_{\mathbf{p}} = \Big(V(f_k) \cdot V(f_{k-1}) \cdot \ldots \cdot V(f_1) \cdot V(f_0)\Big)_{\mathbf{p}}$$

follows at once from the above. □

We now prove a theorem which gives the basic relation between Vogel cycles and the blow-up. In fact, we show that the Vogel cycles are representatives of the *Segre classes*, as defined in [**Fu**, §4.2]. In the generic case, this is Theorem 3.3 of [**Gas1**], and is also proved in Lemma 2.2 of [**G-G**]. However, we are interested in cases which may not be quite so generic.

Theorem 2.20. *Let X be an irreducible analytic subset of an analytic manifold $\mathcal{U}$, let $\pi : \mathrm{Bl}_{\mathbf{f}} X \to X$ denote the blow-up of X along $\mathbf{f}$ (see Example 1.6), and let $E_{\mathbf{f}}$ denote the corresponding exceptional divisor.*

If $E_{\mathbf{f}}$ properly intersects $\mathcal{U} \times \mathbb{P}^m \times \{\mathbf{0}\}$ in $\mathcal{U} \times \mathbb{P}^k$ for all m, then

i) the Vogel cycles of $\mathbf{f}$ are defined;

ii) there exists a neighborhood Ω of $V(\mathbf{f})$ such that, for all m, $\mathrm{Bl}_{\mathbf{f}} X$ intersects $\Omega \times \mathbb{P}^m \times \{\mathbf{0}\}$ properly in $\Omega \times \mathbb{P}^k$; and

iii) inside Ω, for all i,

$$\widehat{\Pi}_{\mathbf{f}}^{i+1} = \pi_*(\mathrm{Bl}_{\mathbf{f}} X \cdot (\mathcal{U} \times \mathbb{P}^{i+k+1-d} \times \{\mathbf{0}\}))$$

and

$$\Delta_{\mathbf{f}}^{i} = \pi_*(E_{\mathbf{f}} \cdot (\mathcal{U} \times \mathbb{P}^{i+k+1-d} \times \{\mathbf{0}\})),$$

where the intersection takes place in $\mathcal{U} \times \mathbb{P}^k$ and π_* denotes the proper push-forward.

Moreover, for all $\mathbf{p} \in X$, there exists an open neighborhood $\mathcal{W}$ of $\mathbf{p}$ in $\mathcal{U}$ such that, for a generic linear reorganization, $\tilde{\mathbf{f}}$, of $\mathbf{f}$, $E_{\tilde{\mathbf{f}}}$ properly intersects $\mathcal{W} \times \mathbb{P}^m \times \{\mathbf{0}\}$ inside $\mathcal{W} \times \mathbb{P}^k$

for all m. In the algebraic category, we may produce such generic linear reorganizations globally, i.e., such that $E_{\tilde{\mathbf{f}}}$ properly intersects $\mathcal{U} \times \mathbb{P}^m \times \{\mathbf{0}\}$ inside $\mathcal{U} \times \mathbb{P}^k$ for all m.

Proof. We show the last two statements first. As in 2.11, the reason that we can only make local statements in the analytic case is because we must worry about analytic sets having an infinite number of irreducible components. For all $\mathbf{p} \in X$, $\pi^{-1}(\mathbf{p})$ is compact, and so, any analytic set can have only a finite number of irreducible components which meet $\pi^{-1}(\mathbf{p})$. In the algebraic setting, we know that we have a finite number of irreducible components globally. For notational ease, we assume in the following paragraph, in the analytic case, that $\mathcal{U}$ is rechosen as small as necessary at each stage so that $\mathcal{U} \times \mathbb{P}^k$ contains a finite number of analytic components (of any specified analytic set) which intersect $\pi^{-1}(\mathbf{p})$; this will mean that we will write $\mathcal{U}$ in place of the open neighborhood $\mathcal{W}$ which appears in the statement of the theorem.

Now, as each point in each component of $E_{\mathbf{f}}$ cannot have all of its homogeneous coordinates equal to zero, for each component ν of $E_{\mathbf{f}}$, there exists a homogeneous coordinate $w_{k(\nu)}$ such that $V(w_{k(\nu)})$ properly intersects ν. Therefore, for generic $(a_{0,0}, \dots, a_{0,k}) \in \mathbb{C}^{k+1}$, the linear form $\widetilde{w}_k := a_{0,0}w_0 + \cdots + a_{0,k}w_k$ is such that $V(\widetilde{w}_k)$ contains no component of $E_{\mathbf{f}}$. We continue in this manner; for generic $(a_{1,0}, \dots, a_{1,k}) \in \mathbb{C}^{k+1}$, the linear form $\widetilde{w}_{k-1} := a_{1,0}w_0 + \cdots + a_{1,k}w_k$ is such that $V(\widetilde{w}_{k-1})$ contains no component of $E_{\mathbf{f}} \cap V(\widetilde{w}_k)$. Continuing, we produce a generic linear reorganization, $\widetilde{\mathbf{w}}$, of $\mathbf{w}$ such that, for all m, $E_{\mathbf{f}}$ properly intersects $V(\widetilde{w}_{m+1}, \dots, \widetilde{w}_k)$ inside $\mathcal{U} \times \mathbb{P}^k$. This proves the last two claims of the theorem.

We now prove i), ii), and iii) of the theorem.

We use $[w_0 : \cdots : w_k]$ as homogeneous coordinates on $\mathbb{P}^k$. Let $\eta : \mathrm{Bl}_{\mathbf{f}} X \to \mathbb{P}^k$ denote the restriction of the projection. Until the end of the proof, we shall simply write f_j in place of $f_j \circ \pi$; no confusion will arise, since it is clear that we must mean $f_j \circ \pi$ when the domain is contained in $\mathrm{Bl}_{\mathbf{f}} X$.

Certainly, π^{-1} induces an isomorphism from $\Pi^{i+1}_{\mathbf{f}} - V(\mathbf{f})$ to

$$\eta^{-1}(\mathbb{P}^{i+k+1-d} \times \{\mathbf{0}\}) - E_{\mathbf{f}} \;=\; \mathrm{Bl}_{\mathbf{f}} X \cap (\mathcal{U} \times \mathbb{P}^{i+k+1-d} \times \{\mathbf{0}\}) - E_{\mathbf{f}}.$$

Hence, $\Pi^{i+1}_{\mathbf{f}}$ is purely $(i+1)$-dimensional if and only if

$$\overline{\mathrm{Bl}_{\mathbf{f}} X \cap (\mathcal{U} \times \mathbb{P}^{i+k+1-d} \times \{\mathbf{0}\}) - E_{\mathbf{f}}}$$

is purely $(i+1)$-dimensional. But, every component of $\mathrm{Bl}_{\mathbf{f}} X \cap (\mathcal{U} \times \mathbb{P}^{i+k+1-d} \times \{\mathbf{0}\})$ has dimension at least $i+1$, while – by hypothesis – $E_{\mathbf{f}} \cap (\mathcal{U} \times \mathbb{P}^{i+k+1-d} \times \{\mathbf{0}\})$ is purely i-dimensional. Thus,

$$\overline{\mathrm{Bl}_{\mathbf{f}} X \cap (\mathcal{U} \times \mathbb{P}^{i+k+1-d} \times \{\mathbf{0}\}) - E_{\mathbf{f}}} = \mathrm{Bl}_{\mathbf{f}} X \cap (\mathcal{U} \times \mathbb{P}^{i+k+1-d} \times \{\mathbf{0}\}),$$

and every component has dimension at least $i+1$. As $E_{\mathbf{f}}$ is locally defined in $\mathrm{Bl}_{\mathbf{f}} X$ by a single equation and $E_{\mathbf{f}} \cap (\mathcal{U} \times \mathbb{P}^{i+k+1-d} \times \{\mathbf{0}\})$ is purely i-dimensional, it follows that $\mathrm{Bl}_{\mathbf{f}} X \cap (\mathcal{U} \times \mathbb{P}^{i+k+1-d} \times \{\mathbf{0}\})$ is purely $(i+1)$-dimensional, for all i, at all points which lie in $E_{\mathbf{f}}$. This proves ii) from the statement of the theorem, and proves that $\Pi_{\mathbf{f}}^{i+1}$ is purely $(i+1)$-dimensional, for all i, at all points of $V(\mathbf{f})$, and so the Vogel cycles are defined. This proves i).

Note that the Dimensionality Lemma and the above paragraphs imply that, in a neighborhood of any point $\mathbf{p} \in V(\mathbf{f})$,

$$(*) \qquad \mathrm{Bl}_{\mathbf{f}} X \cap V(w_{i+k+2-d}, \ldots, w_k) = \overline{\mathrm{Bl}_{\mathbf{f}} X \cap V(w_{i+k+2-d}, \ldots, w_k) - V(f_{i+k+1-d})}.$$

Let $\mathbf{p}$ be a point in $V(\mathbf{f})$. As the Vogel cycles are defined, there exists a neighborhood of $\mathbf{p}$ such that $X - V(\mathbf{f})$, $V(f_k) - V(\mathbf{f})$, ... , $V(f_{i+k+2-d}) - V(\mathbf{f})$ all intersect properly and π induces an isomorphism

$$\big[\mathrm{Bl}_{\mathbf{f}} X - E\big] \cdot \big[V(f_k) - E\big] \cdot \ldots \cdot \big[V(f_{i+k+2-d}) - E\big] \cong$$

$$\big[X - V(\mathbf{f})\big] \cdot \big[V(f_k) - V(\mathbf{f})\big] \cdot \ldots \cdot \big[V(f_{i+k+2-d}) - V(\mathbf{f})\big].$$

By the Dimensionality Lemma, no component of this intersection is contained in the set $V(f_{i+k+1-d})$, and so we conclude that $\widehat{\Pi}_{\mathbf{f}}^{i+1}$ is equal to

$$\pi_* \left(\overline{\big[\mathrm{Bl}_{\mathbf{f}} X - V(f_{i+k+1-d})\big] \cdot \big[V(f_k) - V(f_{i+k+1-d})\big] \cdot \ldots \cdot \big[V(f_{i+k+2-d}) - V(f_{i+k+1-d})\big]}\right).$$

We claim that this implies the first equality of the theorem:

$$(\dagger) \qquad \widehat{\Pi}_{\mathbf{f}}^{i+1} = \pi_*\big(\mathrm{Bl}_{\mathbf{f}} X \cdot V(w_{i+k+2-d}, \ldots, w_k)\big),$$

in a neighborhood of any point in $V(\mathbf{f})$.

To see this, note that $\mathrm{Bl}_{\mathbf{f}} X - V(f_{i+k+1-d}) \subseteq \{w_{i+k+1-d} \neq 0\}$. On the open set, $\mathcal{W} \subseteq \mathcal{U} \times \mathbb{P}^k$, where $f_{i+k+1-d} \neq 0$ and $w_{i+k+1-d} \neq 0$, there is an equality of schemes

$$\mathrm{Bl}_{\mathbf{f}} X = V\left(\frac{f_j}{f_{i+k+1-d}} - \frac{w_j}{w_{i+k+1-d}}\right)_{j \neq i+k+1-d}.$$

At points of $\mathcal{W}$, $\left\{\frac{f_j}{f_{i+k+1-d}} - \frac{w_j}{w_{i+k+1-d}}\right\}_{j \neq i+k+1-d}$ is easily seen to be a regular sequence. Therefore, on $\mathcal{W}$, the cycle $[\mathrm{Bl}_{\mathbf{f}} X]$ is equal to the intersection product of the cycles

$$\left[V\left(\frac{f_j}{f_{i+k+1-d}} - \frac{w_j}{w_{i+k+1-d}}\right)\right]_{j \neq i+k+1-d}.$$

Moreover, on $\mathcal{W}$, for $j \geqslant i+k+2-d$,

$$\left[V\left(\frac{f_j}{f_{i+k+1-d}} - \frac{w_j}{w_{i+k+1-d}}\right)\right] \cdot [V(f_j)] = \left[V\left(\frac{f_j}{f_{i+k+1-d}} - \frac{w_j}{w_{i+k+1-d}}, f_j\right)\right] =$$

$$[V(f_j, w_j)] \;=\; [V(f_j)]\cdot[V(w_j)] \;=$$

$$\left[V\left(\frac{f_j}{f_{i+k+1-d}} - \frac{w_j}{w_{i+k+1-d}},\; w_j\right)\right] \;=\; \left[V\left(\frac{f_j}{f_{i+k+1-d}} - \frac{w_j}{w_{i+k+1-d}}\right)\right]\cdot[V(w_j)].$$

Hence, on $\mathcal{W}$,

$$[\mathrm{Bl}_{\mathbf{f}}\, X]\cdot[V(f_k)]\cdot\ldots\cdot[V(f_{i+k+2-d})] \;= [\mathrm{Bl}_{\mathbf{f}}\, X]\cdot[V(w_k)]\cdot\ldots\cdot[V(w_{i+k+2-d})] \;=$$

$$[\mathrm{Bl}_{\mathbf{f}}\, X]\cdot[V(w_{i+k+2-d},\ldots,w_k)],$$

and so (†) follows from our previous paragraphs and $(*)$.

Now, by definition, $\Delta^i_{\mathbf{f}} + \widehat{\Pi}^i_{\mathbf{f}} = \widehat{\Pi}^{i+1}_{\mathbf{f}}\cdot V(f_{i+k+1-d})$. Applying (†) and the push-forward formula (see Appendix A.14) – which we may use since $V(f_{i+k+1-d}\circ\pi)$ properly intersects $\mathrm{Bl}_{\mathbf{f}}\, X \cap V(w_{i+k+2-d},\ldots,w_k)$ by (*) – we conclude that

$$\Delta^i_{\mathbf{f}} + \widehat{\Pi}^i_{\mathbf{f}} = \pi_*\big(V(f_{i+k+1-d}\circ\pi)\cdot \mathrm{Bl}_{\mathbf{f}}\, X\cdot V(w_{i+k+2-d},\ldots,w_k)\big).$$

By the Dimensionality Lemma, $\Delta^i_{\mathbf{f}}$ consists of those components of the proper push-forward which are contained in $V(f)$. Hence, we will have proved the second equality of the theorem if we can show that the components of $V(f_{i+k+1-d}\circ\pi)\cdot\mathrm{Bl}_{\mathbf{f}}\, X\cdot V(w_{i+k+2-d},\ldots,w_k)$ which are contained in $E_{\mathbf{f}}$ are equal to $E_{\mathbf{f}}\cdot V(w_{i+k+2-d},\ldots,w_k)$.

On the open set where $w_{i+k+1-d}\neq 0$, $E_{\mathbf{f}}$ is defined to be $V(f_{i+k+1-d}\circ\pi)\cdot\mathrm{Bl}_{\mathbf{f}}\, X$. Thus, it is enough to show that $V(f_{i+k+1-d}\circ\pi)\cdot\mathrm{Bl}_{\mathbf{f}}\, X\cdot V(w_{i+k+2-d},\ldots,w_k)$ has no components contained in $E_{\mathbf{f}}$ which are also contained in $V(w_{i+k+1-d})$. However, by hypothesis, $V(w_{i+k+1-d},\ldots,w_k)$ properly intersects $E_{\mathbf{f}}$, and so every component of

$$E_{\mathbf{f}}\cap V(w_{i+k+1-d}, w_{i+k+2-d},\ldots,w_k)$$

has dimension $i-1$. As every component of $V(f_{i+k+1-d}\circ\pi)\cdot\mathrm{Bl}_{\mathbf{f}}\, X\cdot V(w_{i+k+2-d},\ldots,w_k)$ has dimension at least i, we are finished. □

Remark 2.21. Note that the proof of 2.20 shows that, for each i, if $E_{\mathbf{f}}$ properly intersects $\mathcal{U}\times\mathbb{P}^{i+k+1-d}\times\{\mathbf{0}\}$ in $\mathcal{U}\times\mathbb{P}^k$, then $\Pi^{i+1}_{\mathbf{f}}$ is $(i+1)$-dimensional near $V(\mathbf{f})$ – the point being that we do **not** need to assume that we have proper intersections for **all** i.

The following corollary follows immediately from Theorem 2.20.

Corollary 2.22 (The Segre-Vogel Relation). *Let X be an analytic subset of an analytic manifold $\mathcal{U}$, and let $\pi:\mathcal{U}\times\mathbb{P}^k\to\mathcal{U}$ denote the projection. Assume that M is purely d-dimensional. For each V_l appearing in M, consider $\mathrm{Bl}_{\mathbf{f}}V_l\subseteq V_l\times\mathbb{P}^k\subseteq\mathcal{U}\times\mathbb{P}^k$, and let $E^l_{\mathbf{f}}$ denote the corresponding exceptional divisor. Let $\mathrm{Bl}_{\mathbf{f}}M := \sum_l m_l[\mathrm{Bl}_{\mathbf{f}}V_l]$ and $E_{\mathbf{f}}(M) := \sum_l m_l[E^l_{\mathbf{f}}]$.*

If $|E_{\mathbf{f}}(M)|$ properly intersects $\mathcal{U} \times \mathbb{P}^m \times \{\mathbf{0}\}$ in $\mathcal{U} \times \mathbb{P}^k$ for all m, then

i) *the Vogel cycles of* $\mathbf{f}$ *with respect to M are defined;*

ii) *there exists a neighborhood Ω of $|M| \cap V(\mathbf{f})$ such that, for all m, $|\mathrm{Bl}_{\mathbf{f}} M|$ intersects $\Omega \times \mathbb{P}^m \times \{\mathbf{0}\}$ properly in $\Omega \times \mathbb{P}^k$; and*

iii) *inside Ω, for all i,*

$$\widehat{\Pi}_{\mathbf{f}}^{i+1}(M) = \pi_*(\mathrm{Bl}_{\mathbf{f}} M \cdot (\mathcal{U} \times \mathbb{P}^{i+k+1-d} \times \{\mathbf{0}\}))$$

and

$$\Delta_{\mathbf{f}}^{i}(M) = \pi_*(E_{\mathbf{f}}(M) \cdot (\mathcal{U} \times \mathbb{P}^{i+k+1-d} \times \{\mathbf{0}\})),$$

where the intersection takes place in $\mathcal{U} \times \mathbb{P}^k$ and π_* denotes the proper push-forward.

Moreover, for all $\mathbf{p} \in |M| \cap X$, there exists an open neighborhood $\mathcal{W}$ of $\mathbf{p}$ in $\mathcal{U}$ such that, for a generic linear reorganization, $\tilde{\mathbf{f}}$, of $\mathbf{f}$, $|E_{\tilde{\mathbf{f}}}(M)|$ properly intersects $\mathcal{W} \times \mathbb{P}^m \times \{\mathbf{0}\}$ inside $\mathcal{W} \times \mathbb{P}^k$ for all m. In the algebraic category, we may produce such generic linear reorganizations globally, i.e., such that $|E_{\tilde{\mathbf{f}}}(M)|$ properly intersects $\mathcal{U} \times \mathbb{P}^m \times \{\mathbf{0}\}$ inside $\mathcal{U} \times \mathbb{P}^k$ for all m.

Corollary 2.23. *Let X be an analytic subset of an analytic manifold $\mathcal{U}$. Assume that M is purely $(k+1)$-dimensional.*

Then, continuing with the notation from the previous corollary, the multiplicity of $\{\mathbf{p}\} \times \mathbb{P}^k$ in $E_{\mathbf{f}}(M)$ is $\left(\Delta_{\tilde{\mathbf{f}}}^0(M)\right)_{\mathbf{p}}$, where $\tilde{\mathbf{f}}$ is a generic linear reorganization of $\mathbf{f}$.

Moreover, if $\mathbf{p}$ is an isolated point in $|M| \cap V(\mathbf{f})$, then $\{\mathbf{p}\} \times \mathbb{P}^k$ is the unique component of $E_{\mathbf{f}}(M)$ over $\mathbf{p}$ and the multiplicity of $\{\mathbf{p}\} \times \mathbb{P}^k$ in $E_{\mathbf{f}}(M)$ is $\left(\Delta_{\mathbf{f}}^0(M)\right)_{\mathbf{p}}$; if, in addition, there is a regular sequence $\hat{\mathbf{f}} := (\hat{f}_0, \ldots, \hat{f}_k)$ in $\mathcal{O}_{\mathcal{U}}$ such that $\mathbf{f} := \hat{\mathbf{f}}_{|X}$, then $\left(\Delta_{\mathbf{f}}^0(M)\right)_{\mathbf{p}} = \left(M \cdot V(\hat{\mathbf{f}})\right)_{\mathbf{p}}$.

Proof. We use the notation from 2.22.

That the multiplicity of $\{\mathbf{p}\} \times \mathbb{P}^k$ in $E_{\mathbf{f}}(M)$ is $\left(\Delta_{\tilde{\mathbf{f}}}^0(M)\right)_{\mathbf{p}}$ follows immediately from 2.22.iii, for we will be able to pick some copy of $\mathbb{P}^0 \subseteq \mathbb{P}^k$ so that, over $\mathbf{p}$, $|E_{\mathbf{f}}(M)| \cap (\mathcal{U} \times \mathbb{P}^0) = \{\mathbf{p}\} \times \mathbb{P}^0$.

If $\mathbf{p}$ is an isolated point in $|M| \cap V(\mathbf{f})$, then since $E_{\mathbf{f}}(M) \subseteq \mathcal{U} \times \mathbb{P}^k$ is purely k-dimensional, the only component of $E_{\mathbf{f}}(M)$ over $\mathbf{p}$ has to be $\{\mathbf{p}\} \times \mathbb{P}^k$, and so the proper intersection condition of 2.22 is automatically satisfied over a neighborhood of $\mathbf{p}$. Thus, the multiplicity of $\{\mathbf{p}\} \times \mathbb{P}^k$ in $E_{\mathbf{f}}(M)$ is $\left(\Delta_{\mathbf{f}}^0(M)\right)_{\mathbf{p}}$.

By 2.19, if we restrict to each V_l, then

$$\left(\Delta_{\mathbf{f}}^0\right)_{\mathbf{p}} = \Big(V(f_k) \cdot V(f_{k-1}) \cdot \ldots \cdot V(f_1) \cdot V(f_0)\Big)_{\mathbf{p}} =$$

$$\Big(V_l \cdot V(\hat{f}_k) \cdot V(\hat{f}_{k-1}) \cdot \ldots \cdot V(\hat{f}_1) \cdot V(\hat{f}_0)\Big)_{\mathbf{p}} = (V_l \cdot V(\hat{\mathbf{f}}))_{\mathbf{p}},$$

where we used that $\hat{\mathbf{f}}$ is a regular sequence for the last equality (see Appendix A, section 4 for this equality and the one before it). □

Definition 2.24. We call a generic linear reorganization of $\mathbf{f}$, such as appears in Corollary 2.22, a *Vogel reorganization of* $\mathbf{f}$ *with respect to* M.

A generic linear reorganization of $\mathbf{f}$ which is both agreeable and Vogel is called *unifying.*

Remark 2.25. Theorem 3.3 of [**Gas1**] actually shows that, by replacing $\mathbf{f}$ by a generic linear transformation applied to $\mathbf{f}$, one obtains a unifying $\tilde{\mathbf{f}}$; the point being that the linear transformation is actually *generic*, not just generic in the IPZ topology. However, as one can see in the proof of 2.20, proving that one can use an IPZ-generic transformation to obtain a suitable $\tilde{\mathbf{f}}$ is quite trivial, and is actually what one uses in examples.

Chapter 3. THE LE-IOMDINE-VOGEL FORMULAS

As in the previous chapter, X will denote an analytic space of dimension d contained in an analytic manifold $\mathcal{U}$, $\mathbf{f} := (f_0, \dots, f_k)$ will be an ordered $(k+1)$-tuple of elements of $\mathcal{O}_X$, and $M = \sum_l m_l[V_l]$ will be an analytic cycle in X such that $\pm M > 0$.

We wish to examine the effect on the Vogel cycles of adding scalar multiples of a large power of a new function $g : X \to \mathbb{C}$ to f_0. The formulas that we derive are a powerful tool for inductive proofs.

Throughout most of this chapter, we will be making the assumption that the Vogel cycles of $\mathbf{f}$ have correct dimension; as discussed in Remark 2.18, this means that we may as well assume that the number of elements of $\mathbf{f}$ is exactly d. Therefore, we will find it convenient to let $n := d - 1$, and then write that the dimension of X is $n+1$ and that $\mathbf{f} = (f_0, \dots, f_n)$. Moreover, as all of our results will concern gap and Vogel **cycles**, the contributions from various irreducible components of M will simply add, and so – for simplicity – we will make the assumption that X is irreducible (though not necessarily reduced) and prove most results in the case where $M = [X]$.

Since we will be assuming that the gap sets have correct dimension, $\Delta^0_{\mathbf{f}}$ will be purely 0-dimensional, and for any $\mathbf{p} \in X$, we write $\left(\Delta^0_{\mathbf{f}}\right)_{\mathbf{p}}$ for the coefficient (possibly zero) of $\mathbf{p}$ appearing in the cycle $\Delta^0_{\mathbf{f}}$.

The following lemma relates the Vogel cycles of the $(n+1)$-tuple $(f_0, \dots, f_n)$ to the Vogel cycles of the $(n+1)$-tuple $(f_1, \dots, f_n, g)$, where g is a new function. We think of this as relating the Vogel cycles of $\mathbf{f}$ to the Vogel cycles of $\mathbf{f}$ restricted to $V(g)$ – the elimination of f_0 corresponds to the drop in dimension of the ambient space. As we shall see later, this "restriction" lemma is an essential step in proving the Lê-Iomdine-Vogel (LIV) formulas.

Lemma 3.1 (The Restriction Lemma). *Let X be an irreducible analytic space of dimension $n+1$, and let $\mathbf{f} := (f_0, \dots, f_n) \in \left(\mathcal{O}_X\right)^{n+1}$. Let $g \in \mathcal{O}_X$, let $\mathbf{h} := (f_1, \dots, f_n, g)$, and let $\mathbf{p} \in V(\mathbf{f}, g)$.*

i) Suppose that $\Pi^1_{\mathbf{f}}$ is 1-dimensional at $\mathbf{p}$. Then, $\Pi^1_{\mathbf{f}}$ properly intersects $V(g)$ at $\mathbf{p}$ if and only if $V(\mathbf{h}) = V(\mathbf{f}, g)$ as germs of sets at $\mathbf{p}$.

ii) Suppose that the Vogel sets of $\mathbf{f}$ have correct dimension at $\mathbf{p}$, that $V(\mathbf{h}) = V(\mathbf{f}, g)$ as germs of sets at $\mathbf{p}$, and that $V(g)$ properly intersects $D^i_{\mathbf{f}}$ at $\mathbf{p}$ for all $i \geqslant 1$.

Then, $\dim_{\mathbf{p}} V(\mathbf{h}) = \left(\dim_{\mathbf{p}} V(\mathbf{f})\right) - 1$ provided that $\dim_{\mathbf{p}} V(\mathbf{f}) \geqslant 1$, $V(g)$ properly intersects $\widehat{\Pi}^i_{\mathbf{f}}$ at $\mathbf{p}$ for all i, the Vogel sets of $\mathbf{h}$ have correct dimension at $\mathbf{p}$, and, for all i such that $1 \leqslant i \leqslant n$, there are equalities of germs of cycles at $\mathbf{p}$ given by

$$\widehat{\Pi}^i_{\mathbf{h}} = \widehat{\Pi}^{i+1}_{\mathbf{f}} \cdot V(g) \quad \text{and} \quad \Delta^i_{\mathbf{h}} = \Delta^{i+1}_{\mathbf{f}} \cdot V(g).$$

In addition, when $i = 0$, we have the following equality of germs of cycles

$$\Delta^0_{\mathbf{h}} = \left(\widehat{\Pi}^1_{\mathbf{f}} \cdot V(g)\right) + \left(\Delta^1_{\mathbf{f}} \cdot V(g)\right).$$

Proof.

Proof of i): As germs of sets at $\mathbf{p}$,

$$V(f_1,\ldots,f_n,g)=\big(\Pi^1_{\mathbf{f}}\cup V(\mathbf{f})\big)\cap V(g)=\big(\Pi^1_{\mathbf{f}}\cap V(g)\big)\cup V(\mathbf{f},g).$$

Since $\Pi^1_{\mathbf{f}}$ is purely 1-dimensional at $\mathbf{p}$, that $\Pi^1_{\mathbf{f}}$ properly intersects $V(g)$ at $\mathbf{p}$ is equivalent to $\Pi^1_{\mathbf{f}}\cap V(g)$ being empty or equal to $\{\mathbf{p}\}$. As $\mathbf{p}\in V(\mathbf{f},g)$, we have proved i).

Proof of ii): By 2.4, $V(\mathbf{f})=\bigcup D^i_{\mathbf{f}}$. As the Vogel cycles have correct dimension and those of dimension at least one are properly intersected by $V(g)$ at $\mathbf{p}$, we conclude that $\dim_{\mathbf{p}}V(\mathbf{h})$ equals $\big(\dim_{\mathbf{p}}V(\mathbf{f})\big)-1$ provided that $\dim_{\mathbf{p}}V(\mathbf{f})\geqslant 1$.

To see that $V(g)$ properly intersects $\widehat{\Pi}^i_{\mathbf{f}}$ at $\mathbf{p}$, we work solely with germs of sets at $\mathbf{p}$. For $i\leqslant 0$, $\Pi^i_{\mathbf{f}}=\emptyset$, and so there is nothing to prove. We now proceed with a proof by contradiction. Let m be the smallest i such that $\Pi^i_{\mathbf{f}}$ does not properly intersect $V(g)$ at $\mathbf{p}$. Note that i) implies that $m\geqslant 2$, and we have that $\dim_{\mathbf{p}}\Pi^m_{\mathbf{f}}\cap V(g)=m$. Thus, $\dim_{\mathbf{p}}\Pi^m_{\mathbf{f}}\cap V(f_{m-1})\cap V(g)\geqslant m$. However, $\Pi^m_{\mathbf{f}}\cap V(f_{m-1})=\Pi^{m-1}_{\mathbf{f}}\cup D^{m-1}_{\mathbf{f}}$, and so we would have to have that either $\dim_{\mathbf{p}}\Pi^{m-1}_{\mathbf{f}}\cap V(g)\geqslant m-1$ or $\dim_{\mathbf{p}}D^{m-1}_{\mathbf{f}}\cap V(g)\geqslant m-1$; the first possibility is excluded by definition of m, and the second possibility is excluded by hypothesis. Thus, we have shown that $V(g)$ properly intersects $\widehat{\Pi}^i_{\mathbf{f}}$ at $\mathbf{p}$ for all i.

To show that the Vogel sets of $\mathbf{h}$ have correct dimension at $\mathbf{p}$, we once again work on the level of germs of sets. By definition, $\Pi^i_{\mathbf{h}}=V(f_{i+1},\ldots,f_n,g)\neg V(\mathbf{h})$. One of our assumptions is that $V(\mathbf{h})=V(\mathbf{f},g)$; hence, $\Pi^i_{\mathbf{h}}=V(f_{i+1},\ldots,f_n,g)\neg V(\mathbf{f},g)$. We apply 1.3.iii to obtain $\Pi^i_{\mathbf{h}}=\big(\Pi^{i+1}_{\mathbf{f}}\cap V(g)\big)\neg V(\mathbf{f},g)$. However, $V(\mathbf{f},g)$ contains no components of $\Pi^{i+1}_{\mathbf{f}}\cap V(g)$, for $\Pi^{i+1}_{\mathbf{f}}\cap V(g)$ is purely i-dimensional, while – as sets – $\Pi^{i+1}_{\mathbf{f}}\cap V(\mathbf{f})\cap V(g)=\big(\bigcup_{m\leqslant i}D^m_{\mathbf{f}}\big)\cap V(g)$, which has dimension less than i. Therefore, as germs of sets at $\mathbf{p}$, $\Pi^i_{\mathbf{h}}=\Pi^{i+1}_{\mathbf{f}}\cap V(g)$ and is purely i-dimensional, and so the Vogel sets of $\mathbf{h}$ have correct dimension at $\mathbf{p}$.

We wish to see that, for $1\leqslant i\leqslant n$, $\widehat{\Pi}^i_{\mathbf{h}}=\widehat{\Pi}^{i+1}_{\mathbf{f}}\cdot V(g)$ at $\mathbf{p}$. As we saw above, this equality holds for the underlying sets and neither set has a component contained in $V(\mathbf{f})$. Therefore, it is enough to show that the cycles $\widehat{\Pi}^i_{\mathbf{h}}$ and $\widehat{\Pi}^{i+1}_{\mathbf{f}}\cdot V(g)$ are equal on $X-V(\mathbf{f})$. Applying Remark 2.13, we find that both of these cycles on $X-V(\mathbf{f})$ are given by $V(f_{i+1})\cdot\ldots\cdot V(f_n)\cdot V(g)$.

Finally, for $0\leqslant i\leqslant n-1$, $\Delta^i_{\mathbf{h}}=\widehat{\Pi}^{i+1}_{\mathbf{h}}\cdot V(f_{i+1})-\widehat{\Pi}^i_{\mathbf{h}}$. Thus, for $1\leqslant i\leqslant n-1$,

$$\Delta^i_{\mathbf{h}}=\widehat{\Pi}^{i+2}_{\mathbf{f}}\cdot V(g)\cdot V(f_{i+1})-\widehat{\Pi}^{i+1}_{\mathbf{f}}\cdot V(g)=\big(\widehat{\Pi}^{i+1}_{\mathbf{f}}+\Delta^{i+1}_{\mathbf{f}}\big)\cdot V(g)-\widehat{\Pi}^{i+1}_{\mathbf{f}}\cdot V(g)=\Delta^{i+1}_{\mathbf{f}}\cdot V(g).$$

When $i=0$, we have

$$\Delta^0_{\mathbf{h}}=\widehat{\Pi}^1_{\mathbf{h}}\cdot V(f_1)=\widehat{\Pi}^2_{\mathbf{f}}\cdot V(g)\cdot V(f_1)=\big(\widehat{\Pi}^1_{\mathbf{f}}+\Delta^1_{\mathbf{f}}\big)\cdot V(g).$$

When $i=n$, $\Delta^n_{\mathbf{h}}=\widehat{\Pi}^{n+1}_{\mathbf{h}}\cdot V(g)-\widehat{\Pi}^n_{\mathbf{h}}=\widehat{\Pi}^{n+1}_{\mathbf{h}}\cdot V(g)-\widehat{\Pi}^{n+1}_{\mathbf{f}}\cdot V(g)$. We need to show that $\widehat{\Pi}^{n+1}_{\mathbf{h}}-\widehat{\Pi}^{n+1}_{\mathbf{f}}=\Delta^{n+1}_{\mathbf{f}}$.

If $\mathbf{f}\equiv 0$, then $\widehat{\Pi}^{n+1}_{\mathbf{f}}=0$, $\Delta^{n+1}_{\mathbf{f}}=[X]$, and – as $V(g)$ properly intersects $\Delta^{n+1}_{\mathbf{f}}$ – we conclude that $\mathbf{h}\not\equiv 0$ and so $\widehat{\Pi}^{n+1}_{\mathbf{h}}=[X]$. Thus, if $\mathbf{f}\equiv 0$, $\widehat{\Pi}^{n+1}_{\mathbf{h}}-\widehat{\Pi}^{n+1}_{\mathbf{f}}=\Delta^{n+1}_{\mathbf{f}}$. If $\mathbf{f}\not\equiv 0$, then $\widehat{\Pi}^{n+1}_{\mathbf{f}}=[X]$, $\Delta^{n+1}_{\mathbf{f}}=0$, and – as $V(g)$ properly intersects $\widehat{\Pi}^{n+1}_{\mathbf{f}}$ – we conclude that $\mathbf{h}\not\equiv 0$ and so $\widehat{\Pi}^{n+1}_{\mathbf{h}}=[X]$. Thus, if $\mathbf{f}\not\equiv 0$, $\widehat{\Pi}^{n+1}_{\mathbf{h}}-\widehat{\Pi}^{n+1}_{\mathbf{f}}=\Delta^{n+1}_{\mathbf{f}}$. □

Definition 3.2. Suppose that $\mathbf{p} \in \widehat{\Pi}^1_{\mathbf{f}} \cap V(g)$ and that $\dim_{\mathbf{p}} \widehat{\Pi}^1_{\mathbf{f}} = 1$. Let η be an irreducible component (with its reduced structure) of $\widehat{\Pi}^1_{\mathbf{f}}$ which passes through $\mathbf{p}$.

If $\eta \cap V(g)$ is zero-dimensional at $\mathbf{p}$, then we define *the gap ratio of η at $\mathbf{p}$* (for $\mathbf{f}$ with respect to g) to be the ratio of intersection numbers $\dfrac{\left(\eta \cdot V(f_{k+1-d})\right)_{\mathbf{p}}}{\left(\eta \cdot V(g)\right)_{\mathbf{p}}}$.

If $\eta \cap V(g)$ is not zero-dimensional at $\mathbf{p}$ (i.e., if $\eta \subseteq V(g)$), then we define *the gap ratio of η at $\mathbf{p}$* (for $\mathbf{f}$ with respect to g) to be 0.

A *gap ratio* (at $\mathbf{p}$ for $\mathbf{f}$ with respect to g) is any one of the gap ratios of any component of $\widehat{\Pi}^1_{\mathbf{f}} \cap V(g)$ through $\mathbf{p}$.

If $\mathbf{p} \in V(g)$, but $\mathbf{p} \notin \widehat{\Pi}^1_{\mathbf{f}}$, then we say that *all the gap ratios are zero.*

Finally, a *gap ratio at $\mathbf{p}$ for $\mathbf{f}$ with respect to g and the cycle M* is a gap ratio (at $\mathbf{p}$ for $\mathbf{f}$ with respect to g) of $\mathbf{f}_{|_{V_l}}$ for some V_l appearing in M.

Lemma 3.3. *Let X be an irreducible analytic space of dimension $n+1$, let $\mathbf{f} := (f_0, \dots, f_n) \in \left(\mathcal{O}_X\right)^{n+1}$, let $g \in \mathcal{O}_X$, and let $\mathbf{p} \in V(g)$. Let a be a non-zero complex number, and let $j \geqslant 1$ be an integer.*

If j is greater than or equal to the maximum gap ratio at $\mathbf{p}$ for $\mathbf{f}$ with respect to g, then, for all but (possibly) a finite number of complex a,

i) $\widehat{\Pi}^1_{\mathbf{f}}$ *properly intersects $V(f_0 + ag^j)$ at $\mathbf{p}$, and $\left(\Delta^0_{\mathbf{f}}\right)_{\mathbf{p}} = \left(\widehat{\Pi}^1_{\mathbf{f}} \cdot V(f_0 + ag^j)\right)_{\mathbf{p}}$.*

Moreover, if we have the strict inequality that j is greater than the maximum gap ratio at $\mathbf{p}$ for $\mathbf{f}$ with respect to g, then i) holds for all non-zero a; in particular, this is the case if $j \geqslant 1 + \left(\Delta^0_{\mathbf{f}}\right)_{\mathbf{p}}$.

ii) Suppose that $\Pi^1_{\mathbf{f}}$ is 1-dimensional at $\mathbf{p}$, and that $\mathbf{p} \in V(\mathbf{f}, g)$. Then, $\Pi^1_{\mathbf{f}}$ properly intersects $V(f_0 + ag^j)$ at $\mathbf{p}$ if and only if there is an equality of germs of sets at $\mathbf{p}$ given by $V(f_1, \dots, f_n, f_0 + ag^j) = V(\mathbf{f}, g)$.

iii) Suppose that $\mathbf{p} \in V(\mathbf{f}, g)$ and that, at $\mathbf{p}$, there is an equality of germs of sets given by $V(f_1, \dots, f_n, f_0 + ag^j) = V(\mathbf{f}, g)$, the Vogel sets of $\mathbf{f}$ all have correct dimension, and that, for all $i \geqslant 1$, $V(g)$ properly intersects each $D^i_{\mathbf{f}}$.

If $1 \leqslant i \leqslant n$, then, at $\mathbf{p}$, $\widehat{\Pi}^{i+1}_{\mathbf{f}}$ properly intersects $V(f_0 + ag^j)$, the Vogel sets of the $(n+1)$-tuple $(f_1, \dots, f_n, f_0 + ag^j)$ have correct dimension, and there is an equality of germs of cycles given by

$$\widehat{\Pi}^i_{(f_1, \dots, f_n, f_0 + ag^j)} = \widehat{\Pi}^{i+1}_{\mathbf{f}} \cdot V(f_0 + ag^j).$$

Proof.

i) Recall that $\widehat{\Pi}^1_{\mathbf{f}}$ is purely one-dimensional at $\mathbf{p}$. Thus, we may write the cycle $\left[\widehat{\Pi}^1_{\mathbf{f}}\right]$ as $\sum m_\nu[\nu]$, where each ν is a reduced, irreducible curve at $\mathbf{p}$. Let $\alpha_\nu(t)$ denote a local parameterization of ν such that $\alpha_\nu(0) = \mathbf{p}$. Then, to show that $\widehat{\Pi}^1_{\mathbf{f}}$ properly intersects $V(f_0 + ag^j)$ at $\mathbf{p}$, we need to show that, for all ν, $(f_0 + ag^j)_{|_{\alpha_\nu(t)}} \not\equiv 0$. To show that $\left(\Delta^0_{\mathbf{f}}\right)_{\mathbf{p}} = \left(\widehat{\Pi}^1_{\mathbf{f}} \cdot V(f_0 + ag^j)\right)_{\mathbf{p}}$, we need to show that $\left(\widehat{\Pi}^1_{\mathbf{f}} \cdot V(f_0)\right)_{\mathbf{p}} = \sum m_\nu\left([\nu] \cdot V(f_0 + ag^j)\right)_{\mathbf{p}}$; calculating intersection numbers as in A.9 of Appendix A, we find that what we need to show is that, for all ν, $\mathrm{mult}_t f_0(\alpha_\nu(t)) = \mathrm{mult}_t((f_0 + ag^j) \circ \alpha_\nu)(t)$. Thus, we may prove both the proper intersection statement and the intersection formula at the same time by proving this multiplicity statement.

Clearly, $\mathrm{mult}_t((f_0 + ag^j) \circ \alpha_\nu)(t) = \min\left\{\mathrm{mult}_t(f_0 \circ \alpha_\nu)(t),\ \mathrm{mult}_t(g^j \circ \alpha_\nu)(t)\right\}$, unless the lowest degree terms of $f_0(\alpha_\nu(t))$ and $-a(g^j \circ \alpha_\nu)(t)$ are precisely equal. As $\mathrm{mult}_t(f_0 \circ \alpha_\nu)(t) = \left([\nu] \cdot V(f_0)\right)_{\mathbf{p}}$ and $\mathrm{mult}_t(g^j \circ \alpha_\nu)(t) = j\left([\nu] \cdot V(g)\right)_{\mathbf{p}}$, we conclude that $\mathrm{mult}_t((f_0 + ag^j) \circ \alpha_\nu)(t) = \mathrm{mult}_t(f_0 \circ \alpha_\nu)(t)$ if j is greater than the maximum gap ratio, and that this equality holds when j equals the maximum gap ratio except for the finite number of values of a which would cause cancellation of the lowest degree terms. This proves i).

ii) This follows immediately by applying Lemma 3.1.i with the g of the lemma replaced by $f_0 + ag^j$.

iii) This follows immediately by applying Lemma 3.1.ii with the g of the lemma replaced by $f_0 + ag^j$. □

Theorem 3.4 (The Lê-Iomdine-Vogel formulas). *Suppose that each V_l appearing in M has dimension $n+1$. Let $\mathbf{f} := (f_0, \dots, f_n) \in \left(\mathcal{O}_X\right)^{n+1}$, let $g \in \mathcal{O}_X$, and let $\mathbf{p} \in |M| \cap V(\mathbf{f}, g)$. Let a be a non-zero complex number, let $j \geqslant 1$ be an integer, and let*

$$\mathbf{h} := (f_1, \dots, f_n, f_0 + ag^j).$$

Suppose that the Vogel cycles of $\mathbf{f}$ with respect to M are defined at $\mathbf{p}$, and that $V(g)$ properly intersects each of the Vogel cycles, $\Delta^i_{\mathbf{f}}(M)$, at $\mathbf{p}$ for all $i \geqslant 1$.

If j is greater than or equal to the maximum gap ratio at $\mathbf{p}$ for $\mathbf{f}$ with respect to g and M, then for all but (possibly) a finite number of complex a, in a neighborhood of $\mathbf{p}$:

i) there is an equality of sets given by $|M| \cap V(\mathbf{h}) = |M| \cap V(\mathbf{f}, g)$,

ii) $\dim_{\mathbf{p}}(|M| \cap V(\mathbf{h})) = \left(\dim_{\mathbf{p}}(|M| \cap V(\mathbf{f}))\right) - 1$ provided that $\dim_{\mathbf{p}}(|M| \cap V(\mathbf{f})) \geqslant 1$,

iii) the Vogel cycles of $\mathbf{h}$ with respect to M exist at $\mathbf{p}$, and

iv) $\Delta^0_{\mathbf{h}}(M) = \Delta^0_{\mathbf{f}}(M) + j\left(\Delta^1_{\mathbf{f}}(M) \cdot V(g)\right)$ and, for $1 \leqslant i \leqslant n-1$,

$$\Delta^i_{\mathbf{h}}(M) = j\left(\Delta^{i+1}_{\mathbf{f}}(M) \cdot V(g)\right).$$

Moreover, if we have the strict inequality that j is greater than the maximum gap ratio at $\mathbf{p}$ *for* $\mathbf{f}$ *with respect to g and M, then these equalities hold for all non-zero a; in particular, this is the case if* $j \geqslant 1 + \max_l\{(\Delta^0_{\mathbf{f}_{|V_l}})_{\mathbf{p}}\}$.

Proof. The assumption that all m_l have the same sign prevents cancellation of contributions from various V_l; thus, the assumption that $V(g)$ properly intersects each $\Delta^i_{\mathbf{f}}(M)$ implies that $V(g)$ properly intersects each $\Delta^i_{\mathbf{f}_{|V_l}}$ for all l. Therefore, we are reduced to considering the case of Lemma 3.3, where X is irreducible and M equals [X].

Now, the equality of sets in i) is precisely 3.3.ii; the statement concerning $\dim_{\mathbf{p}} V(\mathbf{h})$ follows from this equality of sets, combined with the facts that $V(\mathbf{f}) = \bigcup D^i_{\mathbf{f}}$ (see 2.4) and that $V(g)$ properly intersects the non-zero-dimensional Vogel cycles of $V(\mathbf{f})$.

Now, suppose that $0 \leqslant i \leqslant n-1$. By definition, $\widehat{\Pi}^i_{\mathbf{h}} + \Delta^i_{\mathbf{h}} = \widehat{\Pi}^{i+1}_{\mathbf{h}} \cdot V(f_{i+1})$. By 3.3.iii, this equals $\widehat{\Pi}^{i+2}_{\mathbf{f}} \cdot V(f_0 + ag^j) \cdot V(f_{i+1})$. By definition of the Vogel cycles, this equals $\big(\widehat{\Pi}^{i+1}_{\mathbf{f}} + \Delta^{i+1}_{\mathbf{f}}\big) \cdot V(f_0 + ag^j)$. As $|\Delta^{i+1}_{\mathbf{f}}| \subseteq V(f_0)$, $\Delta^{i+1}_{\mathbf{f}} \cdot V(f_0 + ag^j) = \Delta^{i+1}_{\mathbf{f}} \cdot V(ag^j) = j\big(\Delta^{i+1}_{\mathbf{f}} \cdot V(g)\big)$. Therefore, we have shown that

$$(\dagger) \qquad \widehat{\Pi}^i_{\mathbf{h}} + \Delta^i_{\mathbf{h}} = \big(\widehat{\Pi}^{i+1}_{\mathbf{f}} \cdot V(f_0 + ag^j)\big) + j\big(\Delta^{i+1}_{\mathbf{f}} \cdot V(g)\big).$$

If $i = 0$, then $\widehat{\Pi}^i_{\mathbf{h}} = 0$, and the first equality of iv) of the theorem follows from (†) and 3.3.i.

If $1 \leqslant i \leqslant n-1$, then 3.3.iii tells us that $\big(\widehat{\Pi}^{i+1}_{\mathbf{f}} \cdot V(f_0 + ag^j)\big) = \widehat{\Pi}^i_{\mathbf{h}}$; cancelling $\widehat{\Pi}^i_{\mathbf{h}}$ from each side of (†) yields the second equality of the theorem. □

Remark 3.5. A principal use of the LIV formulas is in families; one requires something about the constancy of the Vogel cycles of $\mathbf{f}$ in the family, and the LIV formulas imply the constancy of the Vogel cycles of a tuple of function with a smaller zero locus.

However, it is possible to use these formulas "in reverse" – to calculate the Vogel cycles of $\mathbf{h}_{(a,j)} := (f_1, \dots, f_n, f_0 + ag^j)$ and have them tell us about the Vogel cycles of $(f_0, \dots, f_n)$. The difficulty of applying the LIV formulas in this manner is that it is not so easy to know when j is greater than or equal to the maximum gap ratio. We discuss this problem below, using the notation from the theorem.

Suppose that the Vogel cycles of $\mathbf{f}$ are defined at $\mathbf{p}$, and that $V(g)$ properly intersects each of the Vogel cycles, $\Delta^i_{\mathbf{f}}$, at $\mathbf{p}$ for all $i \geqslant 1$. Assume that, in a neighborhood of $\mathbf{p}$, there is an equality of sets given by $V(\mathbf{h}_{(a,j)}) = V(\mathbf{f}, g)$ (we are still assuming that $a \neq 0$).

By assuming that $V(g)$ properly intersects $\Delta^i_{\mathbf{f}}$ for $i \geqslant 1$, we are assuming that we can calculate the Vogel sets of $\mathbf{f}$ in dimensions one and higher. While it would be nice to be able to proceed without this assumption, there seems to be no way to avoid it. Notice that, if we could calculate $(\Delta^0_{\mathbf{f}})_{\mathbf{p}}$, then we would know that the LIV formulas hold for $j > (\Delta^0_{\mathbf{f}})_{\mathbf{p}}$. However, $(\Delta^0_{\mathbf{f}})_{\mathbf{p}}$ is typically more difficult to calculate than $(\Delta^1_{\mathbf{f}} \cdot V(g))_{\mathbf{p}}$. So, we will assume that we can also calculate the intersection number $(\Delta^1_{\mathbf{f}} \cdot V(g))_{\mathbf{p}}$, and then consider the problem of how can one tell when j is large enough for the LIV formulas to hold using data gathered from $\Delta^0_{\mathbf{h}_{(a,j)}}$ and $(\Delta^1_{\mathbf{f}} \cdot V(g))_{\mathbf{p}}$.

Our best answer is that:

if $j > \big(\Delta^0_{\mathbf{h}_{(a,j)}}\big)_{\mathbf{p}} - j\big(\Delta^1_{\mathbf{f}}\big)_{\mathbf{p}}$, then the LIV formulas hold, and so

$$\big(\Delta^0_{\mathbf{f}}\big)_{\mathbf{p}} = \big(\Delta^0_{\mathbf{h}_{(a,j)}}\big)_{\mathbf{p}} - j\big(\Delta^1_{\mathbf{f}}\big)_{\mathbf{p}}.$$

To see this, note that the proof of 3.4 shows that $\big(\Delta^0_{\mathbf{h}_{(a,j)}}\big)_{\mathbf{p}} - j\big(\Delta^1_{\mathbf{f}}\big)_{\mathbf{p}} = \big(\widehat{\Pi}^1_{\mathbf{f}} \cdot V(f_0 + ag^j)\big)_{\mathbf{p}}$. We claim that, if $j > \big(\widehat{\Pi}^1_{\mathbf{f}} \cdot V(f_0 + ag^j)\big)_{\mathbf{p}}$, then $j > (\Delta^0_{\mathbf{f}})_{\mathbf{p}} = \big(\widehat{\Pi}^1_{\mathbf{f}} \cdot V(f_0)\big)_{\mathbf{p}}$ and so the LIV formulas hold. This is easy; calculating intersection numbers as in the proof of 3.3,

$$\big(\widehat{\Pi}^1_{\mathbf{f}} \cdot V(f_0 + ag^j)\big)_{\mathbf{p}} \geqslant \min\big\{\big(\widehat{\Pi}^1_{\mathbf{f}} \cdot V(f_0)\big)_{\mathbf{p}},\ j\big(\widehat{\Pi}^1_{\mathbf{f}} \cdot V(g)\big)_{\mathbf{p}}\big\}.$$

The desired conclusion follows.

One might hope that if $\big(\Delta^0_{\mathbf{h}_{(a,j+1)}}\big)_{\mathbf{p}} - \big(\Delta^0_{\mathbf{h}_{(a,j)}}\big)_{\mathbf{p}} = \big(\Delta^1_{\mathbf{f}}\big)_{\mathbf{p}}$ (which would be true if the LIV formulas held), then one could, in fact, conclude that the LIV formulas **do** hold. Unfortunately, the situation is slightly more complicated than this.

Let us call (a, j) an *exceptional pair* if there exists a component ν of $\widehat{\Pi}^1_{\mathbf{f}}$ at $\mathbf{p}$ such that

$$\big(\nu \cdot V(f_0 + ag^j)\big)_{\mathbf{p}} \neq \min\big\{\big(\nu \cdot V(f_0)\big)_{\mathbf{p}},\ j\big(\nu \cdot V(g)\big)_{\mathbf{p}}\big\}.$$

Looking at the proofs of 3.3 and 3.4, it is easy to see that, if (a, j) is **not** an exceptional pair, then $\big(\Delta^0_{\mathbf{h}_{(a,j+1)}}\big)_{\mathbf{p}} - \big(\Delta^0_{\mathbf{h}_{(a,j)}}\big)_{\mathbf{p}} \geqslant \big(\Delta^1_{\mathbf{f}}\big)_{\mathbf{p}}$ with equality if and only if j is greater than or equal to the maximum gap ratio. Hence, if it were not for the existence of exceptional pairs, one could simply make a table of values of $\big(\Delta^0_{\mathbf{h}_{(a,j)}}\big)_{\mathbf{p}}$ for fixed a and increasing j, and when a difference between successive entries is exactly $\big(\Delta^1_{\mathbf{f}}\big)_{\mathbf{p}}$, one would have identified the maximum gap ratio and would know that the LIV formulas hold beyond that value for j.

On the other hand, if (a, j) **is** an exceptional pair, it is quite possible that

$$\big(\Delta^0_{\mathbf{h}_{(a,j+1)}}\big)_{\mathbf{p}} - \big(\Delta^0_{\mathbf{h}_{(a,j)}}\big)_{\mathbf{p}} = \big(\Delta^1_{\mathbf{f}}\big)_{\mathbf{p}}$$

and still j is smaller than the maximum gap ratio. Of course, this can only happen once for each possible exceptional pair, and the number of exceptional pairs is certainly no more than the number of components of $\widehat{\Pi}^1_{\mathbf{f}}$ through $\mathbf{p}$. Thus, if we know the number of components of $\widehat{\Pi}^1_{\mathbf{f}}$ through $\mathbf{p}$, call this number c, and we make a table of values of $\big(\Delta^0_{\mathbf{h}_{(a,j)}}\big)_{\mathbf{p}}$, once we see a difference between successive values equalling $\big(\Delta^1_{\mathbf{f}}\big)_{\mathbf{p}}$ more than c times, we know that j is high enough for the LIV formulas to hold.

Alternatively, and only pseudo-rigorously, if one selects the constant a "randomly", then a will not be part of an exceptional pair and so, $\big(\Delta^0_{\mathbf{h}_{(a,j+1)}}\big)_{\mathbf{p}} - \big(\Delta^0_{\mathbf{h}_{(a,j)}}\big)_{\mathbf{p}} \geqslant \big(\Delta^1_{\mathbf{f}}\big)_{\mathbf{p}}$ with equality if and only if j is greater than or equal to the maximum gap ratio. This approach is particularly well-suited for computer calculation.

The following lemma is related to 3.3 and 3.4 and will be of use to us later.

Lemma 3.6 . *Suppose that each V_l appearing in M has dimension $n+1$, let $\mathbf{f} := (f_0, \dots, f_n) \in (\mathcal{O}_X)^{n+1}$, let $\mathbf{p} \in |M| \cap V(\mathbf{f})$, and suppose that the Vogel cycles of $\mathbf{f}$ with respect to M at $\mathbf{p}$ exist.*

Let a be a non-zero complex number, and let $j \geqslant 1$ be an integer. Let π denote the projection from $\mathbb{C} \times X$ to X, and let w denote the projection from $\mathbb{C} \times X$ to $\mathbb{C}$.

Then, the Vogel sets of $\mathbf{h} := (w^j, f_1 \circ \pi, \dots, f_n \circ \pi, f_0 \circ \pi + aw^j)$ with respect to $\mathbb{C} \times M$ have correct dimension at $(0, \mathbf{p})$, for all $i \leqslant n+1$, $\mathbb{C} \times \widehat{\Pi}^i_{\mathbf{f}}(M)$ properly intersects $V(f_0 \circ \pi + aw^j)$, and there is an equality of germs of cycles at $(0, \mathbf{p})$ given by

$$\widehat{\Pi}^i_{\mathbf{h}}(\mathbb{C} \times M) \;=\; \big(\mathbb{C} \times \widehat{\Pi}^i_{\mathbf{f}}(M)\big) \;\cdot\; V(f_0 \circ \pi + aw^j).$$

Proof. As usual, we instantly reduce ourselves to the case where X is irreducible and $M = [X]$.

That $\mathbb{C} \times \widehat{\Pi}^i_{\mathbf{f}}$ properly intersects $V(f_0 \circ \pi + aw^j)$ is obvious.

First, note that $V(\mathbf{h}) = \{0\} \times V(\mathbf{f})$. Suppose that $1 \leqslant i \leqslant n+1$. Then,

$$\Pi^i_{\mathbf{h}} = V(f_i \circ \pi, \dots, f_n \circ \pi, f_0 \circ \pi + aw^j) \,\neg\, \big(\{0\} \times V(\mathbf{f})\big).$$

We have $V(f_i \circ \pi, \dots, f_n \circ \pi) = \big(\mathbb{C} \times \Pi^i_{\mathbf{f}}\big) \cup V(\mathbf{f} \circ \pi)$, and $V(\mathbf{f} \circ \pi) \cap V(f_0 \circ \pi + aw^j) \subseteq \{0\} \times V(\mathbf{f})$. Applying 1.3.iii, we find that

$$\Pi^i_{\mathbf{h}} = \big(\mathbb{C} \times \Pi^i_{\mathbf{f}}\big) \cap V(f_0 \circ \pi + aw^j) \,\neg\, \big(\{0\} \times V(\mathbf{f})\big).$$

Now, near $\mathbf{p}$, $\big(\mathbb{C} \times \Pi^i_{\mathbf{f}}\big) \cap V(f_0 \circ \pi + aw^j)$ is purely i-dimensional, and – not only does it have no components contained in $\{0\} \times V(\mathbf{f})$ – in fact, it has no components contained in $\mathbb{C} \times V(\mathbf{f})$; for, by 2.4, the set $\Pi^i_{\mathbf{f}} \cap V(\mathbf{f})$ equals the union of all of the Vogel sets of dimension less than or equal to $i-1$. Therefore, the set $\Pi^i_{\mathbf{h}}$ equals the set $\big(\mathbb{C} \times \Pi^i_{\mathbf{f}}\big) \cap V(f_0 \circ \pi + aw^j)$ and, hence, the Vogel sets of $\mathbf{h}$ have correct dimension at $(0, \mathbf{p})$.

As we saw above, $\big|\widehat{\Pi}^i_{\mathbf{h}}\big| = \big|\Pi^i_{\mathbf{h}}\big|$ has no components contained in $\mathbb{C} \times V(\mathbf{f})$. Thus, to prove that the cycles $\widehat{\Pi}^i_{\mathbf{h}}$ and $\big(\mathbb{C} \times \widehat{\Pi}^i_{\mathbf{f}}\big) \cdot V(f_0 \circ \pi + aw^j)$ are equal, it is enough to prove the equality on $\big(\mathbb{C} \times X\big) - \big(\mathbb{C} \times V(\mathbf{f})\big)$. Once again, we apply Remark 2.13 and find that both cycles are equal to

$$V(f_i \circ \pi) \cdot \ldots \cdot V(f_n \circ \pi) \cdot V(f_0 \circ \pi + aw^j)$$

on $\big(\mathbb{C} \times X\big) - \big(\mathbb{C} \times V(\mathbf{f})\big)$. □

Chapter 4. SUMMARY OF PART I

Let W be analytic subset of an analytic space X and let α be a coherent sheaf of ideals in $\mathcal{O}_X$. Let V denote the scheme $V(\alpha)$. Then, the *gap sheaf* $V \neg W$ is the analytic closure of $V - W$; that is, $V \neg W$ is the scheme obtained from V by removing any components or embedded subvarieties contained in W.

Let X be a d-dimensional irreducible (though not necessarily reduced) analytic space and let $\mathbf{f} := (f_0, \dots, f_k) \in (\mathcal{O}_X)^{k+1}$. The *$i$-th gap variety of* $\mathbf{f}$, $\Pi^i_{\mathbf{f}}$, is defined as

$$\Pi^i_{\mathbf{f}} := V(f_{i+k+1-d}, \dots, f_k) \neg V(\mathbf{f}),$$

if $d-(k+1) < i < d$. Similarly, the *i-th modified gap variety of* $\mathbf{f}$, $\widetilde{\Pi}^i_{\mathbf{f}}$, is defined as

$$\widetilde{\Pi}^i_{\mathbf{f}} := V(f_{i+k+1-d}, \dots, f_k) \neg V(f_{i+k-d}),$$

if $d-(k+1) < i < d$. The *i-th inductive gap variety of* $\mathbf{f}$, $\widehat{\Pi}^i_{\mathbf{f}}$, is defined by downward induction

$$\widehat{\Pi}^d_{\mathbf{f}} = \begin{cases} X, & \text{if } \mathbf{f} \not\equiv 0 \\ \emptyset, & \text{if } \mathbf{f} \equiv 0 \end{cases}$$

and

$$\widehat{\Pi}^i_{\mathbf{f}} := \left(\widehat{\Pi}^{i+1}_{\mathbf{f}} \cap V(f_{i+k+1-d})\right) \neg V(f_{i+k-d}),$$

if $d-(k+1) < i < d$.

If X is irreducible and Cohen-Macaulay, and each $\Pi^i_{\mathbf{f}}$ is i-dimensional,then all three types of gap varieties are equal. If X is an arbitrary irreducible space, then, locally, we may replace each member of the tuple $\mathbf{f}$ by a "generic" linear combination of the elements of $\mathbf{f}$ to obtain a new tuple, a *generic linear reorganization of* $\mathbf{f}$, for which the gap sheaves, modified gap sheaves, and inductive gap sheaves are all equal.

If X is irreducible of dimension d and each $\Pi^i_{\mathbf{f}}$ is i-dimensional, then, on $X - V(\mathbf{f})$, $\left[\widehat{\Pi}^i_{\mathbf{f}}\right] = V(f_k) \cdot V(f_{k-1}) \cdot \dots \cdot V(f_{i+k+1-d})$; hence, on X, $\left[\widehat{\Pi}^i_{\mathbf{f}}\right]$ is the closure in X of this cycle on $X - V(\mathbf{f})$.

If X is a union of irreducible components, $X = \bigcup_j X_j$, then we do not define gap **sheaves**, but only gap **cycles**. Writing $\left[V\right]$ for the cycle defined by a scheme V, we define the *i-th gap cycle of* $\mathbf{f}$ by $\left[\Pi^i_{\mathbf{f}}\right] := \sum_j \left[\Pi^i_{\mathbf{f}_{|X_j}}\right]$, the *$i$-th modified gap cycle of* $\mathbf{f}$ by $\left[\widetilde{\Pi}^i_{\mathbf{f}}\right] := \sum_j \left[\widetilde{\Pi}^i_{\mathbf{f}_{|X_j}}\right]$, and the *$i$-th inductive gap cycle of* $\mathbf{f}$ by $\left[\widehat{\Pi}^i_{\mathbf{f}}\right] := \sum_j \left[\widehat{\Pi}^i_{\mathbf{f}_{|X_j}}\right]$.

More generally, if we have an analytic cycle $M := \sum_l m_l[V_l]$ in X, where all of the m_l have the same sign, then we define the various gap cycles relative to M by taking the sum of the appropriate gap cycles restricted to each of the V_l, weighted by the m_l. The requirement that all of the m_l have the same sign prevents the cancellation of contributions from the various V_l.

The modified gap varieties and cycles are merely an intermediate tool. The inductive gap varieties are what we actually use to define (below) our primary objects of study: the

Vogel cycles. However, the hypotheses that must be satisfied before we can define the Vogel cycles include, crucially, the hypothesis that each gap set $\left|\Pi^i_{\mathbf{f}}\right|$ has dimension i. Thus, while one can safely forget the definition of the modified gap varieties, both the gap varieties and inductive gap varieties are important for our future results.

If X is irreducible of dimension d and each $\Pi^i_{\mathbf{f}}$ is i-dimensional, then the *i-th Vogel cycle of* $\mathbf{f}$, $\Delta^i_{\mathbf{f}}$ is given by

$$\Delta^i_{\mathbf{f}} = \left(\left[\widehat{\Pi}^{i+1}_{\mathbf{f}}\right] \cdot \left[V(f_{i+k+1-d})\right]\right) - \left[\widehat{\Pi}^i_{\mathbf{f}}\right].$$

If X is a union of irreducible components, then the i-th Vogel cycle is obtained by summing the i-th Vogel cycles of all of the irreducible components (as in the definition of the gap cycles). Similarly, one obtains Vogel cycles with respect to a given cycle M by taking the weighted sum of the Vogel cycles of $\mathbf{f}$ restricted to each subvariety appearing in M.

If each $\Delta^i_{\mathbf{f}}$ is i-dimensional (which one can obtain locally by replacing $\mathbf{f}$ by a generic linear reorganization), then each Vogel cycle, $\Delta^i_{\mathbf{f}}$, is **purely** i-dimensional, non-negative, and is contained in $V(\mathbf{f})$. Moreover, $V(\mathbf{f}) = \bigcup_i \left|\Delta^i_{\mathbf{f}}\right|$. Thus, we think of the Vogel cycles as decomposing $V(\mathbf{f})$ on the level of cycles.

We proved the important *Segre-Vogel Relation*: Let X be an irreducible, d-dimensional, analytic subset of an analytic manifold $\mathcal{U}$, let $\mathbf{f} = (f_0, \dots, f_k) \in (\mathcal{O}_X)^{k+1}$, let $\pi : \mathrm{Bl}_{\mathbf{f}} X \to X$ denote the blow-up of X along $\mathbf{f}$, and let $E_{\mathbf{f}}$ denote the corresponding exceptional divisor.

If $E_{\mathbf{f}}$ properly intersects $\mathcal{U} \times \mathbb{P}^m \times \{\mathbf{0}\}$ in $\mathcal{U} \times \mathbb{P}^k$ for all m, then Vogel cycles of $\mathbf{f}$ are defined and, in a neighborhood of $V(\mathbf{f})$, for all i,

$$\widehat{\Pi}^{i+1}_{\mathbf{f}} = \pi_*(\mathrm{Bl}_{\mathbf{f}} X \cdot (\mathcal{U} \times \mathbb{P}^{i+k+1-d} \times \{\mathbf{0}\}))$$

and

$$\Delta^i_{\mathbf{f}} = \pi_*(E_{\mathbf{f}} \cdot (\mathcal{U} \times \mathbb{P}^{i+k+1-d} \times \{\mathbf{0}\})),$$

where the intersection takes place in $\mathcal{U} \times \mathbb{P}^k$ and π_* denotes the proper push-forward.

Moreover, for all $\mathbf{p} \in X$, there exists an open neighborhood $\mathcal{W}$ of $\mathbf{p}$ in $\mathcal{U}$ such that, for a generic linear reorganization, $\tilde{\mathbf{f}}$, of $\mathbf{f}$, $E_{\tilde{\mathbf{f}}}$ properly intersects $\mathcal{W} \times \mathbb{P}^m \times \{\mathbf{0}\}$ inside $\mathcal{W} \times \mathbb{P}^k$ for all m. In the algebraic category, we may produce such generic linear reorganizations globally, i.e., such that $E_{\tilde{\mathbf{f}}}$ properly intersects $\mathcal{U} \times \mathbb{P}^m \times \{\mathbf{0}\}$ inside $\mathcal{U} \times \mathbb{P}^k$ for all m.

What we have just stated is the Segre-Vogel Relation for an irreducible space X, as it appears in Theorem 2.20. We give a more general version with respect to a pure-dimensional cycle in Corollary 2.22.

Finally, we derived the Lê-Iomdine-Vogel (LIV) formulas: Let X be an irreducible analytic space of dimension $n+1$, let $\mathbf{f} := (f_0, \dots, f_n) \in \left(\mathcal{O}_X\right)^{n+1}$, let $g \in \mathcal{O}_X$, and let $\mathbf{p} \in V(\mathbf{f}, g)$. Let a be a non-zero complex number, let $j \geqslant 1$ be an integer, and let

$$\mathbf{h} := (f_1, \dots, f_n, f_0 + ag^j).$$

Suppose that the Vogel cycles of $\mathbf{f}$ are defined at $\mathbf{p}$, and that $V(g)$ properly intersects each of the Vogel cycles, $\Delta^i_{\mathbf{f}}$, at $\mathbf{p}$ for all $i \geqslant 1$.

If j is sufficiently large, then there is an equality of sets given by $V(\mathbf{h}) = V(\mathbf{f}, g)$, $\dim_{\mathbf{p}} V(\mathbf{h}) = \big(\dim_{\mathbf{p}} V(\mathbf{f})\big) - 1$ provided that $\dim_{\mathbf{p}} V(\mathbf{f}) \geqslant 1$, the Vogel cycles of $\mathbf{h}$ exist at $\mathbf{p}$, and

$$\Delta^0_{\mathbf{h}} = \Delta^0_{\mathbf{f}} + j\big(\Delta^1_{\mathbf{f}} \cdot V(g)\big)$$

and, for $1 \leqslant i \leqslant n-1$,

$$\Delta^i_{\mathbf{h}} = j\big(\Delta^{i+1}_{\mathbf{f}} \cdot V(g)\big).$$

In particular, if $j \geqslant 1 + \big(\Delta^0_{\mathbf{f}}\big)_{\mathbf{p}}$, then these conclusions hold. Once again, there is a more general version of this result with respect to the cycle M.

Fundamental Concepts from Part I:

Gap sheaf, $V \neg W$ 1.1

Gap variety (resp. modified, inductive), $\Pi^i_{\mathbf{f}}$ (resp. $\widetilde{\Pi}^i_{\mathbf{f}}$, $\widehat{\Pi}^i_{\mathbf{f}}$) 2.1

Vogel set, $D^i_{\mathbf{f}}$ 2.3

Correct dimension 2.7

Pseudo-Zariski topology, (resp. inductive) 2.10

Generic linear reorganization 2.10

Agreeable reorganization 2.11

Vogel cycle, $\Delta^i_{\mathbf{f}}$ 2.14

Segre-Vogel relation 2.22

Vogel reorganization 2.24

Unifying reorganization 2.24

Gap ratio 3.2

Lê-Iomdine-Vogel formulas 3.4

Part II. LÊ CYCLES AND HYPERSURFACE SINGULARITIES

Chapter 0. INTRODUCTION

The Lê numbers and Lê cycles generalize the data given by the Milnor number of an isolated hypersurface singularity. In this introduction, we wish to quickly review why the Milnor number of an isolated hypersurface singularity is important. We will then give some previously-known general results on non-isolated hypersurface singularities, and indicate the types of results that can be obtained by the machinery contained in the remainder of Part II. We shall also describe how the results from Part I enter into the development of this Lê cycle machinery.

Part II deals with the case of hypersurfaces in open subsets of **affine** space – this is the case described in [**Mas6**]. The case where the ambient space is itself allowed to be singular is much more difficult, and is the problem addressed in Parts III and IV. While we could, of course, conclude the affine results as a corollary of the more general case, we prefer to describe the affine situation first, and use it as a guide in developing the general case.

Let $\mathcal{U}$ be an open neighborhood of the origin in $\mathbb{C}^{n+1}$ and let $f : (\mathcal{U}, \mathbf{0}) \to (\mathbb{C}, 0)$ be an analytic function. Then, the Milnor fibration [**Mi3**], [**Lê7**], [**Ra**] of f at the origin is an object of primary importance in the study of the local, ambient topology of the hypersurface, $V(f) := f^{-1}(0)$, defined by f at the origin. Milnor defined his fibration on a sphere of radius ϵ; however, his Theorem 5.11 of [**Mi3**] leads one to consider a more convenient, equivalent, fibration which lives inside the open ball of radius ϵ. Hence, throughout Part II, we will use the Milnor fibration as defined below.

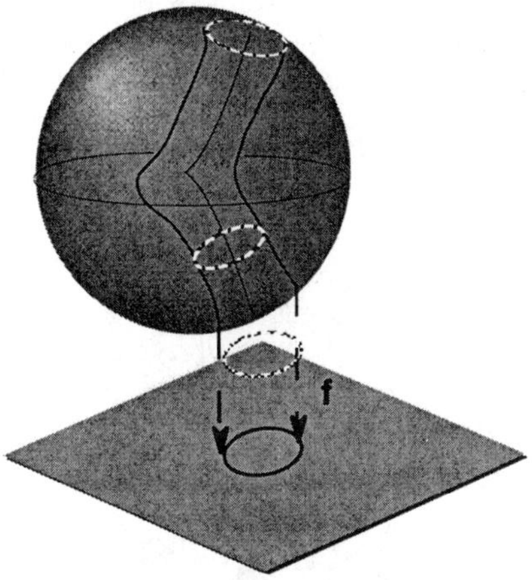

Figure 0.1. The Milnor Fibration inside a ball

For all $\epsilon > 0$, let $\overset{\circ}{B}_\epsilon$ denote the open ball of radius ϵ centered at the origin in $\mathbb{C}^{n+1}$. For all $\eta > 0$, let $\mathbb{D}_\eta$ denote the closed disc centered at the origin in $\mathbb{C}$, and let $\partial\mathbb{D}_\eta$ denote its boundary, which is a circle of radius η. Then, having fixed an analytic function, f, there exists $\epsilon_0 > 0$ such that, for all ϵ such that $0 < \epsilon \leqslant \epsilon_0$, there exists $\eta_\epsilon > 0$ such that, for all η such that $0 < \eta \leqslant \eta_\epsilon$, the restriction of f to a map $\overset{\circ}{B}_\epsilon \cap f^{-1}(\partial\mathbb{D}_\eta) \to \partial\mathbb{D}_\eta$ is a smooth, locally trivial fibration whose diffeomorphism-type is independent of the choice of ϵ and η.

This fibration is called the *Milnor fibration* of f at the origin and the fibre is the *Milnor fibre* of f at the origin, which we denote by $F_{f,\mathbf{0}}$. Hence, the Milnor fibre is a smooth complex n-manifold (of real dimension $2n$). The homotopy-type of the Milnor fibre is an invariant of the local, ambient topological-type of the hypersurface at the origin.

The Results of Milnor

We keep the notation from above; in particular, $\mathcal{U}$ is an open neighborhood of the origin in $\mathbb{C}^{n+1}$ and $f : (\mathcal{U}, \mathbf{0}) \to (\mathbb{C}, 0)$ is an analytic function (actually, for Milnor, f was required to be a polynomial). We will use Σf to denote the critical locus of the map f.

In [**Mi3**], Milnor proved the existence of the object that is now called the Milnor fibration. He also proved that the Milnor fibre, $F_{f,\mathbf{0}}$, has the homotopy-type of a finite n-dimensional CW-complex ([**Mi3**], Theorem 5.1). This implies that all of the homology groups are finitely-generated, are zero above degree n, and that $H_n(F_{f,\mathbf{0}})$ is free Abelian.

In addition, Milnor proved that if f has an isolated critical point at the origin, i.e., $\dim_{\mathbf{0}}\Sigma f = 0$, then $F_{f,\mathbf{0}}$ is $(n-1)$-connected ([**Mi3**], Lemma 6.4). Combining this with the previous result, it follows ([**Mi3**], Theorem 6.5) that, in the case of an isolated singularity, the Milnor fibre has the homotopy-type of a finite bouquet (one-point union) of n-spheres; the number of spheres in this bouquet is the *Milnor number* and is denoted by μ (or $\mu_{\mathbf{0}}(f)$, or some other such variant). In particular, the reduced homology is trivial except in degree n, and there the homology group is $\mathbb{Z}^{\mu}$. The Milnor number can be calculated algebraically by taking the dimension as a complex vector space of the algebra $\mathcal{O}_{\mathbf{0}}^{n+1}/J(f)$, where $\mathcal{O}_{\mathbf{0}}^{n+1}$ denotes the ring of analytic germs at the origin and $J(f)$ denotes the Jacobian ideal $\langle \frac{\partial f}{\partial z_0}, \dots, \frac{\partial f}{\partial z_n} \rangle$.

So that we can do an example, there is one final result of Milnor's that we wish to mention here. Suppose that f is a weighted homogeneous polynomial (i.e., there exist positive integers $r_0, \dots, r_n$ such that $f\left({z_0}^{r_0}, \dots, {z_n}^{r_n}\right)$ is a homogeneous polynomial). Then, ([**Mi3**], Lemma 9.4) the Milnor fibre, $F_{f,\mathbf{0}}$, is diffeomorphic to $f^{-1}(1)$.

Example 0.2. As an example, consider $f = xyz$, which defines a hypersurface in $\mathbb{C}^3$ consisting of the three coordinate planes. Thus, $V(f)$ is a hypersurface with a one-dimensional singular set consisting of the three coordinate axes.

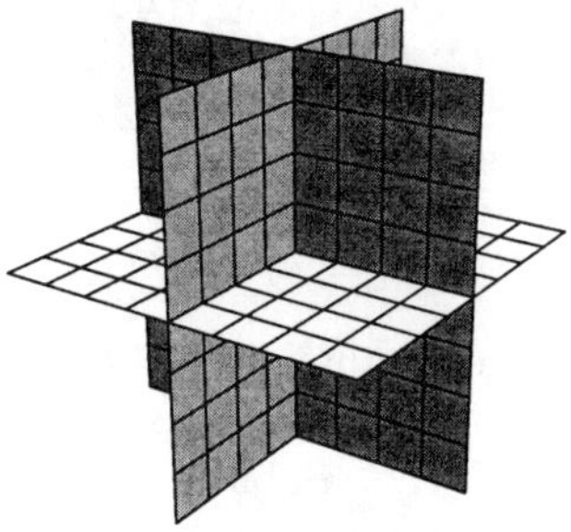

Figure 0.3. The coordinate hyperplanes

By the above result on weighted homogeneous polynomials, the Milnor fibre is diffeomorphic to the set of points where $xyz = 1$; but, this is where $x \neq 0$, $y \neq 0$, and $x = \frac{1}{yz}$.

Thus, $F_{f,\mathbf{0}}$ is homeomorphic to $\mathbb{C}^* \times \mathbb{C}^*$, where $\mathbb{C}^* = \mathbb{C} - 0$. In particular, $F_{f,\mathbf{0}}$ is homotopy-equivalent to the product of two circles, and so has non-zero homology in degrees 0, 1, and 2.

Further Results

We wish to consider another classic example: the Whitney umbrella.

Example 0.4. The Whitney umbrella is the hypersurface in $\mathbb{C}^3$ defined by the vanishing of $f = y^2 - zx^2$.

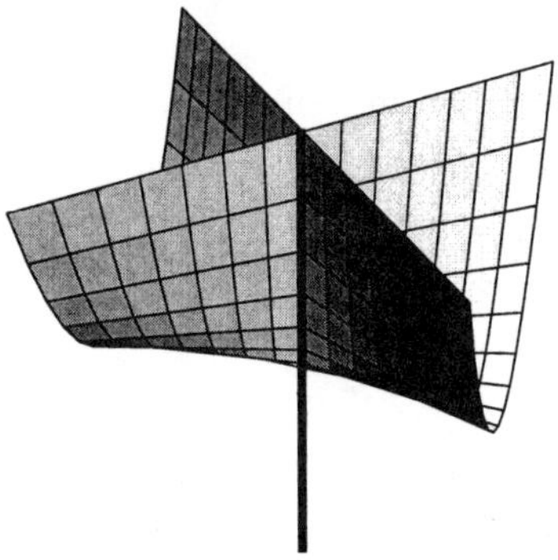

Figure 0.5. The Whitney umbrella

Here, we have drawn the picture over the real numbers – this is the rarely-seen picture that explains the word "umbrella" in the name of this example. The "handle" of this umbrella is not usually drawn when one is in the complex setting, for the inclusion of this line gives the impression that the local dimension of the hypersurface is not constant; something which is not possible over the complex numbers. A second reason why one rarely sees the above picture is that one frequently encounters the Whitney umbrella as a family of nodes degenerating to a cusp; this representation is achieved by making the analytic change of coordinates $z = x + t$ to obtain $f = y^2 - x^3 - tx^2$ (see Example 1.12).

To determine the homotopy-type of the Milnor fibre of the Whitney umbrella at the origin, we need a new result.

The result we need is that if we have an analytic function $g(z_0, \dots, z_n)$ and a variable y, disjoint from the z's, then the Milnor fibre of $y^2 + g(z_0, \dots, z_n)$ is homotopy-equivalent to the suspension of the Milnor fibre of g. By an abuse of language, one frequently says that the singularity of $y^2 + g(z_0, \dots, z_n)$ is the *suspension of the singularity* of g.

So, in our example, $F_{y^2-zx^2,\mathbf{0}}$ is homotopy-equivalent to the suspension of $F_{zx^2,\mathbf{0}}$. But, as zx^2 is homogeneous,

$$F_{zx^2,\mathbf{0}} \cong \{(z,x) \mid zx^2 = 1\} = \left\{ \left(\frac{1}{x^2}, x\right) \,\middle|\, x \neq 0 \right\} \cong \mathbb{C}^*.$$

Thus, $F_{y^2-zx^2,\mathbf{0}}$ is homotopy-equivalent to the suspension of a circle, i.e., the Milnor fibre of f at the origin is homotopy-equivalent to a 2-sphere.

The suspension result used above is a special case of a much more general result proved in various forms: first by Sebastiani and Thom in [**Se-Th**], then by Oka in [**Ok**] and Sakamoto in [**Sak**]. This result states that:

Theorem 0.6 (Sebastiani-Thom Result). If $f : (\mathcal{U}, \mathbf{0}) \to (\mathbb{C}, 0)$ and $g : (\mathcal{U}', \mathbf{0}) \to (\mathbb{C}, 0)$ are analytic functions, then the Milnor fibre of the function $h : (\mathcal{U} \times \mathcal{U}', \mathbf{0}) \to (\mathbb{C}, 0)$ defined by $h(\mathbf{w}, \mathbf{z}) := f(\mathbf{w}) + g(\mathbf{z})$ is homotopy-equivalent to the join, $F_{f,\mathbf{0}} * F_{g,\mathbf{0}}$, of the Milnor fibres of f and g.

This determines the homology of $F_{h,\mathbf{0}}$ in a simple way, since the reduced homology of the join of two spaces X and Y is given by

$$\widetilde{H}_{j+1}(X * Y) = \sum_{k+l=j} \widetilde{H}_k(X) \otimes \widetilde{H}_l(Y) \oplus \sum_{k+l=j-1} \operatorname{Tor}\left(\widetilde{H}_k(X), \widetilde{H}_l(Y)\right).$$

Returning now to Example 0.2 where $f = xyz$, we see that $F_{f,\mathbf{0}}$ need not have the homotopy-type of a bouquet of spheres when the singularity is non-isolated. However, there is the general result of Kato and Matsumoto [**K-M**]:

Theorem 0.7. *If $s := \dim_{\mathbf{0}} \Sigma f$, then $F_{f,\mathbf{0}}$ is $(n-s-1)$-connected; in particular, when $s = 0$, we recover the result of Milnor.*

Moreover, this is the best possible general bound on the connectivity of the Milnor fibre, as is shown by:

Example 0.8. Consider

$$g := z_0 z_1 \dots z_{s+1} + z_{s+2}^2 + \cdots + z_n^2;$$

we leave it as an exercise for the reader to verify, using our earlier methods, that this g has an s-dimensional critical locus at the origin and $F_{g,\mathbf{0}}$ has non-trivial homology in dimension $n - s$.

Lê's Attaching Result

The result of Kato and Matsumoto can be obtained from a more general result of Lê; a result which is one of few which allows calculations concerning the homology of the Milnor fibre for an arbitrary hypersurface singularity.

Let $\mathcal{U}$ be an open neighborhood of the origin in $\mathbb{C}^{n+1}$ and let $f : (\mathcal{U}, \mathbf{0}) \to (\mathbb{C}, 0)$ be an analytic function. Let $L : \mathbb{C}^{n+1} \to \mathbb{C}$ be a generic linear form. Then, it is easy to see that if $\dim_{\mathbf{0}}\Sigma f \geqslant 1$, then $\dim_{\mathbf{0}}\Sigma(f_{|_{V(L)}}) = (\dim_{\mathbf{0}}\Sigma f) - 1$.

Now, the main result of [**Lê1**] is:

Theorem 0.9. *The Milnor fibre $F_{f,\mathbf{0}}$ is obtained from the Milnor fibre $F_{f_{|_{V(L)}},\mathbf{0}}$ by attaching a certain number of n-handles (n-cells on the homotopy level); this number of attached n-handles is given by the intersection number $\left(\Gamma^1_{f,L} \cdot V(f)\right)_{\mathbf{0}}$, where $\Gamma^1_{f,L}$ denotes the relative polar curve of f with respect to L.*

We will define the polar curve and discuss how to calculate intersection numbers in Chapter 1, but we can already see that Kato and Matsumoto's result follows inductively from Theorem 0.9 since we already know Milnor's result for isolated singularities and because attaching handles of index k does not affect the connectivity in dimensions less than $k - 1$.

Not only does Lê's result imply Kato and Matsumoto's, but – assuming that $\left(\Gamma^1_{f,L} \cdot V(f)\right)_{\mathbf{0}}$ is effectively calculable – Lê's result enables the calculation of the Euler characteristic of the Milnor fibre, together with some Morse-type inequalities on the Betti numbers of the Milnor fibre; for instance, the n-th Betti number, $b_n\left(F_{f,\mathbf{0}}\right)$, must be less than or equal to $\left(\Gamma^1_{f,L} \cdot V(f)\right)_{\mathbf{0}}$.

Unfortunately, the Morse inequalities above are usually far from being equalities. Of course, the real value of Lê's result is that it allows one to calculate some important information even in the cases where the homotopy-type of the Milnor fibre cannot be determined by other means.

The Result of Lê and Ramanujam

As the homotopy-type of the Milnor fibre is an invariant of the local, ambient topological-type of the hypersurface at the origin, if one has a family of hypersurfaces with isolated singularities in which the local, ambient, topological-type is constant, then the Milnor number must remain constant in the family. In 1976, Lê and Ramanujam [**L-R**] proved the converse of this; we describe their result now.

Let $\overset{\circ}{\mathbb{D}}$ be an open disc about the origin in $\mathbb{C}$, let $\mathcal{U}$ be an open neighborhood of the origin in $\mathbb{C}^{n+1}$, and let $f : (\overset{\circ}{\mathbb{D}} \times \mathcal{U}, \overset{\circ}{\mathbb{D}} \times \mathbf{0}) \to (\mathbb{C}, 0)$ be an analytic function; we write f_t for the function defined by $f_t(\mathbf{z}) := f(t, \mathbf{z})$. Lê and Ramanujam proved:

Theorem 0.10. *Suppose that, for all small t, $\dim_{\mathbf{0}}\Sigma f_t = 0$ and that the Milnor number of f_t at the origin is independent of t. Then, for all small t,*

i) the fibre-homotopy type of the Milnor fibrations of f_t at the origin is independent of t;

and, if $n \neq 2$,

ii) the diffeomorphism-type of the Milnor fibrations of f_t and the local, ambient, topological-type of $V(f_t)$ at the origin are independent of t.

The Result of Lê and Saito

The result of Lê and Saito again deals with families of singularities, so we continue with $f : (\overset{\circ}{\mathbb{D}} \times \mathcal{U}, \overset{\circ}{\mathbb{D}} \times \mathbf{0}) \to (\mathbb{C}, 0)$ as above. The result of [**Lê-Sa**] tells one how limiting tangent spaces to nearby level hypersurfaces of f approach the singularity.

Theorem 0.11. *Suppose that, for all small t, $\dim_{\mathbf{0}}\Sigma f_t = 0$ and that the Milnor number of f_t at the origin is independent of t. Then, $\overset{\circ}{\mathbb{D}} \times \mathbf{0}$ satisfies Thom's a_f condition at the origin with respect to the ambient stratum, i.e., if $\mathbf{p}_i$ is a sequence of points in $\overset{\circ}{\mathbb{D}} \times \mathcal{U} - \Sigma f$ such that $\mathbf{p}_i \to \mathbf{0}$ and such that $T_{\mathbf{p}_i} V(f - f(\mathbf{p}_i))$ converges to some $\mathcal{T}$, then $\mathbb{C} \times \mathbf{0} = T_{\mathbf{0}}(\overset{\circ}{\mathbb{D}} \times \mathbf{0}) \subseteq \mathcal{T}$.*

Generalizing the Milnor Number

So, suppose we have a single analytic function, $f : (\mathcal{U}, \mathbf{0}) \to (\mathbb{C}, 0)$ with a critical locus of arbitrary dimension $s := \dim_{\mathbf{0}}\Sigma f$. What properties would we want generalized Milnor numbers of f at $\mathbf{0}$ to have?

First, associated to f, we want there to be $s+1$ numbers which are effectively calculable; call the numbers $\lambda^0_f, \dots, \lambda^s_f$. In the case of an isolated singularity, we want λ^0_f to be the Milnor number of f and all other λ^i_f to be zero.

For arbitrary s, we would like to generalize Milnor's result for isolated singularities and show that the Milnor fibre of f at the origin has a handle decomposition in which the number of attached handles of each index are given by the appropriate λ^i_f.

Finally, we would like to have generalizations of the results of Lê and Ramanujam and Lê and Saito to families of hypersurface singularities of arbitrary dimension.

The $\lambda^0_f, \dots, \lambda^s_f$ that we define to achieve these goals are called the *Lê numbers* of f. In order to define the Lê numbers, we shall apply the machinery of Part I to the Jacobian ideal of f; if we let $\mathbf{z} := (z_0, \dots, z_n)$ be coordinates on $\mathcal{U}$, we obtain an ordered $(n+1)$-tuple $J_{\mathbf{z}}(f) := \left(\frac{\partial f}{\partial z_0}, \dots, \frac{\partial f}{\partial z_n}\right)$, and the corresponding Vogel cycles, $\left\{\Delta^i_{J_{\mathbf{z}}(f)}\right\}_i$, from Part I (I.2.14) are the *Lê cycles*, $\left\{\Lambda^i_{f,\mathbf{z}}\right\}_i$, of f with respect to $\mathbf{z}$. We then obtain Lê numbers, $\lambda^0_{f,\mathbf{z}}, \dots, \lambda^s_{f,\mathbf{z}}$, by intersecting these Lê cycles with affine linear subspaces defined by the coordinate functions $(z_0, \dots, z_n)$.

Thus, our generalization of the Milnor number depends on a choice of coordinates. Nonetheless, as we shall see, these Lê numbers have all of the properties that we expect to have in a generalization of the Milnor number to functions with non-isolated critical loci on affine spaces.

Chapter 1. DEFINITIONS AND BASIC PROPERTIES

In this chapter, we define and prove some elementary results about the fundamental objects of study in Part II – the Lê cycles and Lê numbers. The Lê cycles are analytic cycles which, in a sense, decompose the critical locus of an analytic function. The Lê numbers are intersection numbers of the Lê cycles with certain affine linear subspaces.

To define the Lê cycles, we first need to define the relative polar cycles, which are the cycles associated to the relative polar varieties. The relative polar varieties were studied by Lê and Teissier in a number of places (see, for instance, [**Te4**], [**Te5**], and [**Te7**]). Lê and Teissier define the relative polar varieties of a function with respect to generic linear flags, and they usually assume that the flags have been chosen generically enough so that the relative polar varieties have many special properties.

However, the whole theory seems to behave more nicely if one does not require the flags to be quite so generic, and then works with possibly non-reduced schemes and cycles. This means that we will define the relative polar varieties and cycles in terms of gap varieties and cycles, and then the Lê cycles will be obtained from the corresponding Vogel cycles.

The key features of our definition of the relative polar varieties in terms of gap varieties are that the polar varieties are not necessarily reduced and that the dimension of the critical locus of the function is allowed to be arbitrary. The reader who is familiar with the works of Lê and Teissier ([**Te4**], [**Te5**], [**Te7**]) should note that we index by the generic dimension instead of the codimension.

There is one further difference between our presentation of the relative polar varieties and that of Lê and Teissier; instead of fixing a complete flag inside the ambient affine space, we fix a linear choice of coordinates $\mathbf{z} := (z_0, \dots, z_n)$ for $\mathbb{C}^{n+1}$. We do this because we frequently find it useful to have the linear functions $z_0, \dots, z_n$ at our disposal.

Let $\mathcal{U}$ be an open subset of $\mathbb{C}^{n+1}$, let $\mathbf{z} := (z_0, \dots, z_n)$ be a linear choice of coordinates for $\mathbb{C}^{n+1}$, and let $h : (\mathcal{U}, \mathbf{0}) \to (\mathbb{C}, \mathbf{0})$ be an analytic function. We write Σh for the critical locus of h, i.e., $\Sigma h := V\left(\frac{\partial h}{\partial z_0}, \dots, \frac{\partial h}{\partial z_n}\right)$

Definition 1.1. For $0 \leqslant k \leqslant n$, the *k-th (relative) polar variety*, $\Gamma^k_{h,\mathbf{z}}$, of h with respect to $\mathbf{z}$ is the scheme $V\left(\frac{\partial h}{\partial z_k}, \dots, \frac{\partial h}{\partial z_n}\right) \neg \Sigma h$ (see [**Mas7**], [**Mas8**], [**Mas11**]). If the choice of the coordinate system is clear, we will often simply write Γ^k_h.

If $\mathbf{f}$ equals the Jacobian $(n+1)$-tuple $J_{\mathbf{z}}(h) := \left(\frac{\partial h}{\partial z_0}, \dots, \frac{\partial h}{\partial z_n}\right)$, then $\Gamma^k_{h,\mathbf{z}}$ agrees with the gap variety, $\Pi^k_{\mathbf{f}}$, of $\mathbf{f}$ (see Part I, Definition 2.1).

Thus, on the level of ideals, $\Gamma^k_{h,\mathbf{z}}$ consists of those components of the scheme

$$V\left(\frac{\partial h}{\partial z_k}, \dots, \frac{\partial h}{\partial z_n}\right)$$

which are not contained in $|\Sigma h|$. Note, in particular, that $\Gamma^0_{h,\mathbf{z}}$ is empty.

Naturally, we define *the k-th polar cycle* of h with respect to $\mathbf{z}$ to be the analytic cycle $\left[\Gamma^k_{h,\mathbf{z}}\right]$. This agrees with our previous definition of the gap cycles (of the Jacobian tuple) from Part I, Definition 2.1.

Clearly, as sets, $\emptyset = \Gamma^0_{h,\mathbf{z}} \subseteq \Gamma^1_{h,\mathbf{z}} \subseteq \dots \subseteq \Gamma^{n+1}_{h,\mathbf{z}} = \mathcal{U}$. In fact, by I.2.2.ii, we have that :

Proposition 1.2. $\left(\Gamma^{k+1}_{h,\mathbf{z}} \cap V\left(\frac{\partial h}{\partial z_k}\right)\right) \neg \Sigma h = \Gamma^k_{h,\mathbf{z}}$ *as schemes, and thus all the components of the cycle* $\left[\Gamma^{k+1}_{h,\mathbf{z}} \cap V\left(\frac{\partial h}{\partial z_k}\right)\right] - \left[\Gamma^k_{h,\mathbf{z}}\right]$ *are contained in the critical set of the map* h.

As the ideal $\langle \frac{\partial h}{\partial z_k}, \dots, \frac{\partial h}{\partial z_n} \rangle$ is invariant under any linear change of coordinates which leaves $V(z_0, \dots, z_{k-1})$ invariant, we see that the scheme $\Gamma^k_{h,\mathbf{z}}$ depends only on h and the choice of the first k coordinates. At times, it will be convenient to subscript the k-th polar variety with only the first k coordinates instead of the whole coordinate system; for instance, we write Γ^1_{h,z_0} for the polar curve.

While it is immediate from the number of defining equations that every component of the analytic set $\left|\Gamma^k_{h,\mathbf{z}}\right|$ has dimension at least k, one usually requires that the coordinate system be suitably generic so that the dimension of $\Gamma^k_{h,\mathbf{z}}$ equals k. In this case, we have the following:

Proposition 1.3. *If* $\dim_{\mathbf{p}}\Gamma^k_{h,\mathbf{z}} = k$*, then* $\Gamma^k_{h,\mathbf{z}}$ *has no embedded subvarieties through the point* $\mathbf{p}$.

Proof. This is immediate from I.1.5. $\square$

Proposition 1.4. *If* $\dim_{\mathbf{p}}\Sigma h < k$*, then* $\Gamma^k_{h,\mathbf{z}}$ *and* $V\left(\frac{\partial h}{\partial z_k}, \dots, \frac{\partial h}{\partial z_n}\right)$ *are equal up to embedded subvariety and, hence, are equal as cycles at* $\mathbf{p}$. *If* $\dim_{\mathbf{p}}\Sigma h < k$ *and* $\dim_{\mathbf{p}}\Gamma^k_{h,\mathbf{z}} = k$*, then* $\Gamma^k_{h,\mathbf{z}}$ *and* $V\left(\frac{\partial h}{\partial z_k}, \dots, \frac{\partial h}{\partial z_n}\right)$ *are equal as schemes at* $\mathbf{p}$.

If $\mathbf{f}$ *equals the Jacobian* $(n+1)$*-tuple,* $J_{\mathbf{z}}(h)$*, and* $|\Gamma^k_{h,\mathbf{z}}|$ *is purely* k*-dimensional for all* k*, then, for all* k*,* $\Gamma^k_{h,\mathbf{z}}$ *and the various gap schemes,* $\Pi^k_{\mathbf{f}}$*,* $\tilde{\Pi}^k_{\mathbf{f}}$*, and* $\widehat{\Pi}^k_{\mathbf{f}}$ *are all equal as schemes.*

Proof. As schemes, $V := V\left(\frac{\partial h}{\partial z_k}, \dots, \frac{\partial h}{\partial z_n}\right)$ consists of the components not contained in Σh – these comprise $\Gamma^k_{h,\mathbf{z}}$ – together with those contained in Σh. By the number of defining equations, every isolated component of V must have dimension at least k. Thus, if $\dim_{\mathbf{p}}\Sigma h < k$, the only components of V which are contained in Σh must be embedded. Therefore,

$V\left(\frac{\partial h}{\partial z_k}, \dots, \frac{\partial h}{\partial z_n}\right)$ equals $\Gamma^k_{h,\mathbf{z}}$ up to embedded subvariety and, hence, they are equal as cycles.

But this certainly implies that $\Gamma^k_{h,\mathbf{z}}$ and V are equal as germs of sets at $\mathbf{p}$. Thus, if $\dim_{\mathbf{p}}\Gamma^k_{h,\mathbf{z}} = k$, then $\dim_{\mathbf{p}} V = k$, i.e., V is a local complete intersection at $\mathbf{p}$. Hence, V has no embedded subvarieties at $\mathbf{p}$. The second statement follows.

Since $\mathcal{O}_{\mathcal{U}}$ is Cohen-Macaulay, the final statement follows immediately from I.2.9 (since, by definition, $\Gamma^k_{h,\mathbf{z}} = \Pi^k_{\mathbf{f}}$). $\square$

Definition 1.5. If the intersection of $\Gamma^k_{h,\mathbf{z}}$ and $V(z_0 - p_0, \dots, z_{k-1} - p_{k-1})$ is purely zero-dimensional at a point $\mathbf{p} = (p_0, \dots, p_n)$ (i.e., either $\mathbf{p}$ is an isolated point of the intersection or $\mathbf{p}$ is not in the intersection), then we say that the *k-th polar number*, $\gamma^k_{h,\mathbf{z}}(\mathbf{p})$, *is defined* and we set $\gamma^k_{h,\mathbf{z}}(\mathbf{p})$ equal to the intersection number

$$\left(\Gamma^k_{h,\mathbf{z}} \cdot V(z_0 - p_0, \dots, z_{k-1} - p_{k-1})\right)_{\mathbf{p}}.$$

(We use the term polar **numbers**, instead of polar multiplicities, since we are not assuming that the coordinates are so generic that this intersection number gives the multiplicities.)

Note that, if $\gamma^k_{h,\mathbf{z}}$ is defined at $\mathbf{p}$, then it must be defined at all points near $\mathbf{p}$. Note also that, if $\gamma^k_{h,\mathbf{z}}(\mathbf{p})$ is defined, then $\Gamma^k_{h,\mathbf{z}}$ must be purely k-dimensional at $\mathbf{p}$ and so – by 1.3 – $\Gamma^k_{h,\mathbf{z}}$ has no embedded subvarieties at $\mathbf{p}$.

Remark 1.6. As sets,

$$\Sigma(h_{|_{V(z_0-p_0,\dots,z_{k-1}-p_{k-1})}}) = V\left(z_0 - p_0, \dots, z_{k-1} - p_{k-1}, \frac{\partial h}{\partial z_k}, \dots, \frac{\partial h}{\partial z_n}\right)$$

$$= V(z_0 - p_0, \dots, z_{k-1} - p_{k-1}) \cap \left(\Sigma h \cup \Gamma^k_{h,\mathbf{z}}\right).$$

Hence, if $\gamma^k_{h,\mathbf{z}}(\mathbf{p})$ is defined and $\mathbf{p} \in \Sigma h$, then

$$\Sigma(h_{|_{V(z_0-p_0,\dots,z_{k-1}-p_{k-1})}}) = V(z_0 - p_0, \dots, z_{k-1} - p_{k-1}) \cap \Sigma h$$

at $\mathbf{p}$.

We now wish to define the Lê cycles. Unlike the polar varieties and cycles, the Lê cycles are supported on the critical set of h itself. These cycles demonstrate a number of properties which generalize the data given by the Milnor number for an isolated singularity.

Definition 1.7. For $0 \leqslant k \leqslant n$, we define the *k-th* Lê *cycle of h with respect to* $\mathbf{z}$, $\left[\Lambda^k_{h,\mathbf{z}}\right]$, to be

$$\left[\Gamma^{k+1}_{h,\mathbf{z}} \cap V\left(\frac{\partial h}{\partial z_k}\right)\right] - \left[\Gamma^k_{h,\mathbf{z}}\right].$$

If the choice of coordinate system is clear, we will sometimes simply write $\left[\Lambda^k_h\right]$. Also, as we have given the Lê cycles no structure as schemes, we will usually omit the brackets and write $\Lambda^k_{h,\mathbf{z}}$ to denote the Lê cycle – unless we explicitly state that we are considering it as a set only.

Note that, as every component of $\Gamma^{k+1}_{h,\mathbf{z}}$ has dimension at least $k+1$, every component of $\Lambda^k_{h,\mathbf{z}}$ has dimension at least k. We say that the cycle $\left[\Lambda^k_{h,\mathbf{z}}\right]$ or the set $\left|\Lambda^k_{h,\mathbf{z}}\right|$ has *correct dimension at a point* $\mathbf{p}$ provided that $\left|\Lambda^k_{h,\mathbf{z}}\right|$ is purely k-dimensional at $\mathbf{p}$.

We define the *k-th* Lê *number of h at* $\mathbf{p}$ *with respect to* $\mathbf{z}$, $\lambda^k_{h,\mathbf{z}}(\mathbf{p})$, to equal the intersection number $\left(\Lambda^k_{h,\mathbf{z}} \cdot V(z_0 - p_0, \dots, z_{k-1} - p_{k-1})\right)_{\mathbf{p}}$, provided this intersection is purely zero-dimensional at $\mathbf{p}$. If this intersection is not purely zero-dimensional at $\mathbf{p}$, then we say that the k-th Lê number (of h at $\mathbf{p}$ with respect to $\mathbf{z}$) is *undefined.* Here, when $k=0$, we mean that

$$\lambda^0_{h,\mathbf{z}}(\mathbf{p}) = \left(\Lambda^0_{h,\mathbf{z}} \cdot \mathcal{U}\right)_{\mathbf{p}} =$$

$$\left[\Gamma^1_{h,\mathbf{z}} \cap V\left(\frac{\partial h}{\partial z_0}\right)\right]_{\mathbf{p}} = \left(\left[\Gamma^1_{h,\mathbf{z}}\right] \cdot \left[V\left(\frac{\partial h}{\partial z_0}\right)\right]\right)_{\mathbf{p}}.$$

(This last equality holds whenever $\Gamma^1_{h,\mathbf{z}}$ is one-dimensional at $\mathbf{p}$, for then $\Gamma^1_{h,\mathbf{z}}$ has no embedded subvarieties by 1.7 and $\Gamma^1_{h,\mathbf{z}} \cap V\left(\frac{\partial h}{\partial z_0}\right)$ is zero-dimensional. See Appendix A.4.)

Note that if $\lambda^k_{h,\mathbf{z}}(\mathbf{p})$ is defined, then $\lambda^k_{h,\mathbf{z}}$ is defined at all points near $\mathbf{p}$ and $\left|\Lambda^k_{h,\mathbf{z}}\right|$ must have correct dimension at $\mathbf{p}$. Also note that, since $\Gamma^{k+1}_{h,\mathbf{z}}$ and $\Gamma^k_{h,\mathbf{z}}$ depend only on the choice of the coordinates z_0 through z_k, the k-th Lê cycle, $\left[\Lambda^k_{h,\mathbf{z}}\right]$, depends only on the choice of $(z_0, \dots, z_k)$. Finally, note that if h is a polynomial, then since we are taking **linear** coordinates, we remain inside the algebraic category.

Remark 1.8. We have defined the Lê cycles as we did in all of our previous work (see, for instance, [**Mas6**]). We wish to see that this agrees with the Vogel cycles of $J_{\mathbf{z}}(f) := V\left(\frac{\partial h}{\partial z_0}, \dots, \frac{\partial h}{\partial z_n}\right)$ under reasonable hypotheses (see 1.9, below). Note, however, that it follows from the definitions that the sets underlying the Lê cycles, $|\Lambda^i_{f,\mathbf{z}}|$, are **always** equal to the Vogel sets $D^i_{J_{\mathbf{z}}(f)}$.

Proposition 1.9. *If* $\dim \Gamma^k_{f,\mathbf{z}}$ *is purely k-dimensional for all* $k \geqslant 0$, *then, for all* $k \geqslant 0$, *the k-th Lê cycle of f with respect to* $\mathbf{z}$ *is equal to the k-th Vogel cycle of the Jacobian* $(n+1)$*-tuple of f with respect to* $\mathbf{z}$, *i.e.,* $\Lambda^k_{f,\mathbf{z}} = \Delta^k_{J_{\mathbf{z}}(f)}$.

Proof. This follows immediately from the second paragraph of Proposition 1.4, Definition I.2.14, and Proposition I.2.12. □

While we shall defer most of our examples until later – when we will have more results to play with – it is instructive to include at least one at this early stage.

Example 1.10. Let $h = y^2 - x^3 - tx^2$; this is the Whitney umbrella of Example 0.4, but written as a family of nodes degenerating to a cusp. We fix the coordinate system $\mathbf{z} = (t, x, y)$ and will suppress any further reference to it.

We find

$$\Sigma h = V(-x^2, \ -3x^2 - 2tx, \ 2y) = V(x, y).$$

Thus, the critical locus of h is one-dimensional and consists of the t-axis.

Now the critical locus is one-dimensional, while the dimension of every component of $V\left(\frac{\partial h}{\partial y}\right)$ is at least two. Hence, $V\left(\frac{\partial h}{\partial y}\right)$ cannot possibly have any components contained in Σh and, therefore, we begin calculating polar varieties with Γ^2_h. We have simply

$$\Gamma^2_h = V\left(\frac{\partial h}{\partial y}\right) = V(2y) = V(y)$$

with no components to dispose of.

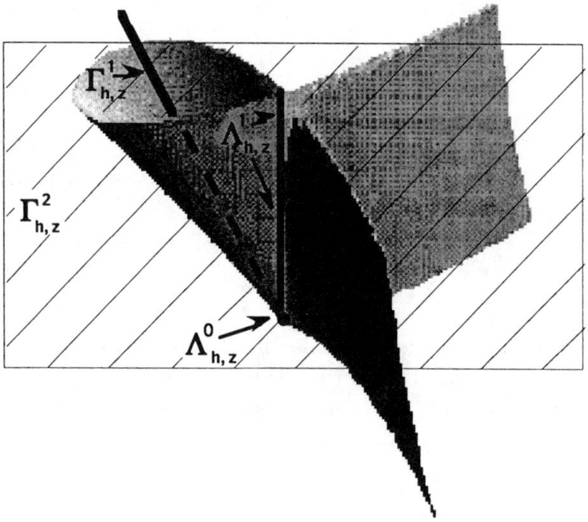

Figure 1.11. Polar and Lê cycles

Next, we have

$$\Gamma^2_h \cap V\left(\frac{\partial h}{\partial x}\right) = V(y) \cap V(-3x^2 - 2tx) = V(y, -3x^2 - 2tx).$$

Applying 1.2 and then I.1.3.iv, we find

$$\Gamma^1_h = \left(\Gamma^2_h \cap V\left(\frac{\partial h}{\partial x}\right)\right) \urcorner\, \Sigma h = V(y, -3x^2 - 2tx) \urcorner\, V(x,y) = V(y, -3x - 2t).$$

From the definition of the Lê cycles (1.11), we obtain

$$\Lambda^1_h = \left[V(y, -3x^2 - 2tx)\right] - \left[V(y, -3x - 2t)\right] =$$

$$\Big([V(y,x)] + [V(y, -3x - 2t)] \Big) - [V(y, -3x - 2t)] = [V(y,x)] \, .$$

Thus, Λ^1_h has as its underlying set the t-axis, and this axis occurs with multiplicity 1.

Now we find

$$\Lambda^0_h = \left[\Gamma^1_h \cap V\left(\frac{\partial h}{\partial t}\right)\right] = \left[V(y, -3x - 2t) \cap V(-x^2)\right] = 2\left[V(t,x,y)\right] = 2[\mathbf{0}].$$

Finally, we calculate the Lê numbers: $\lambda^1_h(\mathbf{0}) = \left(V(y,x) \cdot V(t)\right)_{\mathbf{0}} = 1$ and clearly $\lambda^0_h(\mathbf{0}) = 2$.

Proposition 1.12. *The* Lê *cycles are all non-negative and are contained in the critical set of h. Every component of $\left|\Lambda^k_{h,\mathbf{z}}\right|$ has dimension at least k. If $s = \dim_{\mathbf{p}}\Sigma h$ then, for all k with $s < k < n+1$, $\mathbf{p}$ is not contained in $\left|\Lambda^k_{h,\mathbf{z}}\right|$, i.e., $\left|\Lambda^k_{h,\mathbf{z}}\right|$ is empty at $\mathbf{p}$; thus, for $s < k < n+1$, $\lambda^k_{h,\mathbf{z}}(\mathbf{p})$ is defined and equal to* 0.

Proof. The first statement follows from 1.2. The second statement follows from the definition of the Lê cycles and the fact that every component of $\Gamma^{k+1}_{h,\mathbf{z}}$ has dimension at least $k+1$. The third statement follows from the first two. □

Due to the result of 1.12, we usually only consider $\lambda^0_{h,\mathbf{z}}(\mathbf{p}), \dots, \lambda^s_{h,\mathbf{z}}(\mathbf{p})$.

Proposition 1.13. *As sets, for all k, $\Gamma^{k+1}_{h,\mathbf{z}} \cap \Sigma h = \bigcup_{i \leqslant k} \Lambda^i_{h,\mathbf{z}}$. In particular, letting $k = s := \dim_{\mathbf{p}}\Sigma h$, as germs of sets at $\mathbf{p}$, $\Sigma h = \bigcup_{i \leqslant s} \Lambda^i_{h,\mathbf{z}}$.*

Proof. This follows immediately from I.2.4. □

Recall from Remark 1.6 that, if $\gamma^1_{h,\mathbf{z}}(\mathbf{p})$ is defined and $\mathbf{p} \in \Sigma h$, then

$$\Sigma(h_{|_{V(z_0 - p_0)}}) = V(z_0 - p_0) \cap \Sigma h$$

at $\mathbf{p}$. This is especially useful for inductive proofs when combined with the easy:

Proposition 1.14. *If* $s := \dim_{\mathbf{p}} \Sigma h \geqslant 1$, $\Lambda^i_{h,\mathbf{z}}$ *has correct dimension at* $\mathbf{p}$ *for all* $i \leqslant s-1$ *(by this, we mean to allow that* $\mathbf{p}$ *may not be contained in some of the* $\Lambda^j_{h,\mathbf{z}}$*'s), and* $\lambda^s_{h,\mathbf{z}}(\mathbf{p})$ *is defined, then* $\dim_{\mathbf{p}}(\Sigma h \cap V(z_0 - p_0)) = s-1$.

Proof. As we are assuming that $\Lambda^i_{h,\mathbf{z}}$ has correct dimension at $\mathbf{p}$ for all $i \leqslant s-1$, it follows from 1.13 that we have only to show that the hyperplane slice $V(z_0 - p_0)$ actually reduces the dimension of $\Lambda^s_{h,\mathbf{z}}$. But, this must be the case, since

$$\Lambda^s_{h,\mathbf{z}} \cap V(z_0 - p_0, \dots, z_{s-1} - p_{s-1})$$

is zero-dimensional at $\mathbf{p}$. $\square$

Proposition 1.15. *Fix an integer* $k \geqslant 0$. *Suppose that* $\mathbf{p} \in \Sigma h$. *If, for all* j *with* $0 \leqslant j \leqslant k$, $\Lambda^j_{h,\mathbf{z}}$ *is purely* j*-dimensional at* $\mathbf{p}$, *then, for all* j *such that* $1 \leqslant j \leqslant k+1$, $\left|\Gamma^j_{h,\mathbf{z}}\right|$ *is purely* j*-dimensional at* $\mathbf{p}$, *and the cycles*

$$\left[\Gamma^j_{h,\mathbf{z}} \cap V\left(\frac{\partial h}{\partial z_{j-1}}\right)\right] \quad \textit{and} \quad \left[\Gamma^j_{h,\mathbf{z}}\right] \cdot \left[V\left(\frac{\partial h}{\partial z_{j-1}}\right)\right]$$

are equal at $\mathbf{p}$.

In addition, if for all j *with* $0 \leqslant j \leqslant k$, $\Lambda^j_{h,\mathbf{z}}$ *is purely* j*-dimensional at* $\mathbf{p}$, *then, for* $0 \leqslant j \leqslant k+1$, *every* j*-dimensional (isolated) component of the critical locus of* h *through* $\mathbf{p}$ *is contained in* $\left|\Lambda^j_{h,\mathbf{z}}\right|$.

Proof. Since $\Gamma^0_{h,\mathbf{z}}$ is empty, we may inductively apply the final statement in Remark I.2.6 to conclude that $\left|\Gamma^j_{h,\mathbf{z}}\right|$ is purely j-dimensional for all $j \leqslant k+1$; it follows from 1.3 that $\Gamma^j_{h,\mathbf{z}}$ has no embedded subvarieties through $\mathbf{p}$ for $j \leqslant k+1$.

Now, for $1 \leqslant j \leqslant k+1$, $\left|\Gamma^j_{h,\mathbf{z}}\right|$ is purely j-dimensional at $\mathbf{p}$, and $\left|\Gamma^j_{h,\mathbf{z}} \cap V\left(\frac{\partial h}{\partial z_{j-1}}\right)\right| = \left|\Gamma^{j-1}_{h,\mathbf{z}}\right| \cup \left|\Lambda^{j-1}_{h,\mathbf{z}}\right|$ is purely $(j-1)$-dimensional at $\mathbf{p}$. Thus, $V\left(\frac{\partial h}{\partial z_{j-1}}\right)$ contains no components or embedded subvarieties of $\Gamma^j_{h,\mathbf{z}}$, and therefore we may apply Appendix A.4 to conclude that

$$\left[\Gamma^j_{h,\mathbf{z}} \cap V\left(\frac{\partial h}{\partial z_{j-1}}\right)\right] \quad \text{and} \quad \left[\Gamma^j_{h,\mathbf{z}}\right] \cdot \left[V\left(\frac{\partial h}{\partial z_{j-1}}\right)\right]$$

are equal at $\mathbf{p}$.

The last statement follows immediately from Proposition I.2.4 $\square$

In practice, we use the first part of 1.15 to calculate the Lê cycles as follows:

Assume for the moment that all the Lê cycles have the correct dimension, and let s denote the dimension of the the critical locus of h. Then, in a neighbor hood of the critical locus,

$$\left[\Gamma^{s+1}_{h,\mathbf{z}}\right] = \left[V\left(\frac{\partial h}{\partial z_{s+1}}, \dots, \frac{\partial h}{\partial z_n}\right)\right];$$

$$\left[\Gamma^{s+1}_{h,\mathbf{z}}\right] \cdot \left[V\left(\frac{\partial h}{\partial z_s}\right)\right] = \left[\Gamma^{s}_{h,\mathbf{z}}\right] + \left[\Lambda^{s}_{h,\mathbf{z}}\right];$$

$$\left[\Gamma^{s}_{h,\mathbf{z}}\right] \cdot \left[V\left(\frac{\partial h}{\partial z_{s-1}}\right)\right] = \left[\Gamma^{s-1}_{h,\mathbf{z}}\right] + \left[\Lambda^{s-1}_{h,\mathbf{z}}\right];$$

$$\vdots$$

$$\left[\Gamma^{2}_{h,\mathbf{z}}\right] \cdot \left[V\left(\frac{\partial h}{\partial z_1}\right)\right] = \left[\Gamma^{1}_{h,\mathbf{z}}\right] + \left[\Lambda^{1}_{h,\mathbf{z}}\right];$$

$$\left[\Gamma^{1}_{h,\mathbf{z}}\right] \cdot \left[V\left(\frac{\partial h}{\partial z_0}\right)\right] = \left[\Lambda^{0}_{h,\mathbf{z}}\right].$$

In each line above, one obtains $\left[\Gamma^{k}_{h,\mathbf{z}}\right]$ from the calculation in the previous line.

Now, to write the above equalities, we have used that the Lê cycles have correct dimension – but, in any case, the equalities are true for **sets** (using intersection and union of sets, of course). And so, after doing the above calculations, one verifies that the cycles that we have written above as the Lê cycles do, in fact, have the correct dimension and thus the equalities are correct. On the other hand, if the cycles that we have written above as the Lê cycles do **not** have the correct dimension, then the equalities above may be false.

Remark 1.16. As we will see in Example 2.1, in the case of an isolated singularity, $\lambda^0_{h,\mathbf{z}}$ is nothing other than the Milnor number.

In the general case, it is tempting to think of $\lambda^0_{h,\mathbf{z}}(\mathbf{p})$ as the local (generic) degree of the Jacobian map of h at $\mathbf{p}$, i.e., the number of points in

$$\overset{\circ}{B}_\epsilon \cap V\left(\frac{\partial h}{\partial z_0} - a_0, \ldots, \frac{\partial h}{\partial z_n} - a_n\right),$$

where $\overset{\circ}{B}_\epsilon$ is a small open ball centered at $\mathbf{p}$ and $\mathbf{a}$ is a generic point with length that is small compared to ϵ; unfortunately, there is no such local degree.

Consider the example $h = z_2^2 + (z_0 - z_1^2)^2$ and let $\mathbf{p}$ be the origin.

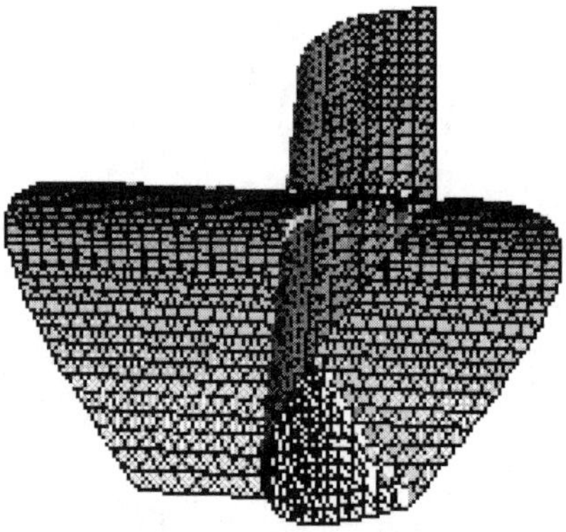

Figure 1.17. The hypersurface defined by h

Then,

$$\mathring{B}_\epsilon \cap V\left(\frac{\partial h}{\partial z_0} - a_0,\ \frac{\partial h}{\partial z_1} - a_1,\ \frac{\partial h}{\partial z_2} - a_2\right) =$$

$$\mathring{B}_\epsilon \cap V(2(z_0 - z_1^2) - a_0,\ 2(z_0 - z_1^2)(-2z_1) - a_1,\ 2z_2 - a_2).$$

The solutions to these equations are

$$z_0 = \frac{a_0}{2} + \frac{a_1^2}{4a_0^2}, \qquad z_1 = -\frac{a_1}{2a_0}, \qquad z_2 = \frac{a_2}{2}.$$

The number of solutions of these equations inside any small ball does not just depend on picking small, generic a_0, a_1, and a_2, but also depends on the relative sizes of a_0 and a_1. If a_1 is small relative to a_0, then there will be one solution inside the ball; if a_0 is small relative to a_1, then there will be no solutions inside the ball.

Do either of these numbers actually agree with $\lambda^0_{h,\mathbf{z}}(\mathbf{0})$? Yes; with these coordinates, $\lambda^0_{h,\mathbf{z}}(\mathbf{0}) = 1$. This can be seen from the above calculations together with the discussion below, which shows how "close" $\lambda^0_{h,\mathbf{z}}$ is to being the generic degree of the Jacobian map of h.

We claim that, if $\dim_{\mathbf{p}}\Gamma^1_{h,\mathbf{z}} = 1$, then $\lambda^0_{h,\mathbf{z}}(\mathbf{p})$ exists and equals the number of points in

$$\mathring{B}_\epsilon \cap V\left(\frac{\partial h}{\partial z_0} - a_0, \ldots, \frac{\partial h}{\partial z_n} - a_n\right),$$

where $\mathring{B}_\epsilon$ is a small open ball centered at $\mathbf{p}$, $a_0 \neq 0$ is small compared to ϵ, and $a_1, \ldots, a_n$ are generic, with length that is small compared to that of a_0.

To see this, note that this number of points equals the sum of the intersection numbers given by

$$\sum_{\mathbf{q}} \left(V\left(\frac{\partial h}{\partial z_1}, \dots, \frac{\partial h}{\partial z_n}\right) \cdot V\left(\frac{\partial h}{\partial z_0} - a_0\right)\right)_{\mathbf{q}}$$

where the sum is over all $\mathbf{q}$ in

$$\overset{\circ}{B}_\epsilon \cap V\left(\frac{\partial h}{\partial z_0} - a_0, \frac{\partial h}{\partial z_1}, \dots, \frac{\partial h}{\partial z_n}\right).$$

But, for $a_0 \neq 0$, these points, $\mathbf{q}$, do not occur on the critical locus of h, and so this sum equals

$$\sum_{\mathbf{q}} \left(\Gamma^1_{h,\mathbf{z}} \cdot V\left(\frac{\partial h}{\partial z_0} - a_0\right)\right)_{\mathbf{q}}.$$

This last sum is none other than

$$\lambda^0_{h,\mathbf{z}}(\mathbf{p}) = \left(\Gamma^1_{h,\mathbf{z}} \cdot V\left(\frac{\partial h}{\partial z_0}\right)\right)_{\mathbf{p}}.$$

It is also possible to give a more intuitive characterization of $\lambda^s_{h,\mathbf{z}}(\mathbf{p})$ where $s = \dim_{\mathbf{p}} \Sigma h$. Assuming that $\lambda^s_{h,\mathbf{z}}(\mathbf{p})$ exists, by moving to a generic point, it is trivial to show that

$$\lambda^s_{h,\mathbf{z}}(\mathbf{p}) = \sum_{\nu} n_\nu \overset{\circ}{\mu}_\nu,$$

where ν runs over all s-dimensional components of Σh at $\mathbf{p}$, n_ν is the local degree of the map $(z_0, \dots, z_{s-1})$ restricted to ν at $\mathbf{p}$, and $\overset{\circ}{\mu}_\nu$ denotes the generic transverse Milnor number of h along the component ν in a neighborhood of $\mathbf{p}$. In particular, if the coordinate system is generic enough so that n_ν is actually the multiplicity of ν at $\mathbf{p}$ for all ν, then $\lambda^s_{h,\mathbf{z}}(\mathbf{p})$ is merely the multiplicity of the Jacobian scheme of h (the scheme defined by the vanishing of the Jacobian ideal) at $\mathbf{p}$.

The next proposition tells us how the Lê numbers behave under the taking of hyperplane sections – a fundamental result; the statements concerning cycles could be derived from the Restriction Lemma (I.3.1), but we want to use polar and Lê numbers both in the hypotheses and the conclusions.

Proposition 1.18. *Suppose $\Sigma h \cap V(z_0 - p_0) = \Sigma(h_{|_{V(z_0-p_0)}})$, and use the coordinates $\tilde{\mathbf{z}} = (z_1, \dots, z_n)$ for $V(z_0 - p_0)$. Let $k \geqslant 1$ and suppose that $\gamma^k_{h,\mathbf{z}}(\mathbf{p})$ and $\lambda^k_{h,\mathbf{z}}(\mathbf{p})$ are defined.*

Then, $\gamma^{k-1}_{h_{|_{V(z_0-p_0)}},\tilde{\mathbf{z}}}(\mathbf{p})$ and $\lambda^{k-1}_{h_{|_{V(z_0-p_0)}},\tilde{\mathbf{z}}}(\mathbf{p})$ are defined,

$$\Gamma^k_{h_{|_{V(z_0-p_0)}},\tilde{\mathbf{z}}} = \Gamma^{k+1}_{h,\mathbf{z}} \cdot V(z_0 - p_0),$$

$$\Gamma^{k-1}_{h_{|_{V(z_0-p_0)}},\tilde{\mathbf{z}}} + \Lambda^{k-1}_{h_{|_{V(z_0-p_0)}},\tilde{\mathbf{z}}} = \left(\Gamma^k_{h,\mathbf{z}} + \Lambda^k_{h,\mathbf{z}}\right) \cdot V(z_0 - p_0),$$

and

$$\gamma^{k-1}_{h_{|_{V(z_0-p_0)}},\tilde{\mathbf{z}}}(\mathbf{p}) + \lambda^{k-1}_{h_{|_{V(z_0-p_0)}},\tilde{\mathbf{z}}}(\mathbf{p}) = \gamma^k_{h,\mathbf{z}}(\mathbf{p}) + \lambda^k_{h,\mathbf{z}}(\mathbf{p}).$$

In the special case when $k = 1$, it follows that if $\gamma^1_{h,\mathbf{z}}(\mathbf{p})$ and $\lambda^1_{h,\mathbf{z}}(\mathbf{p})$ are defined, then so is $\lambda^0_{h_{|_{V(z_0-p_0)}},\tilde{\mathbf{z}}}(\mathbf{p})$, and

$$\lambda^0_{h_{|_{V(z_0-p_0)}},\tilde{\mathbf{z}}}(\mathbf{p}) = \gamma^1_{h,\mathbf{z}}(\mathbf{p}) + \lambda^1_{h,\mathbf{z}}(\mathbf{p}).$$

Moreover, we conclude that if $k \geqslant 1$ and $\gamma^k_{h,\mathbf{z}}(\mathbf{p})$, $\lambda^k_{h,\mathbf{z}}(\mathbf{p})$, $\gamma^{k+1}_{h,\mathbf{z}}(\mathbf{p})$, and $\lambda^{k+1}_{h,\mathbf{z}}(\mathbf{p})$ are defined, then so are

$$\gamma^{k-1}_{h_{|_{V(z_0-p_0)}},\tilde{\mathbf{z}}}(\mathbf{p}),\ \lambda^{k-1}_{h_{|_{V(z_0-p_0)}},\tilde{\mathbf{z}}}(\mathbf{p}),\ \gamma^{k}_{h_{|_{V(z_0-p_0)}},\tilde{\mathbf{z}}}(\mathbf{p}),\quad \text{and}\ \lambda^{k}_{h_{|_{V(z_0-p_0)}},\tilde{\mathbf{z}}}(\mathbf{p}),$$

and

$$\Gamma^{k}_{h_{|_{V(z_0-p_0)}},\tilde{\mathbf{z}}} = \Gamma^{k+1}_{h,\mathbf{z}} \cdot V(z_0 - p_0),$$

$$\Lambda^{k}_{h_{|_{V(z_0-p_0)}},\tilde{\mathbf{z}}} = \Lambda^{k+1}_{h,\mathbf{z}} \cdot V(z_0 - p_0),$$

and so

$$\gamma^{k}_{h_{|_{V(z_0-p_0)}},\tilde{\mathbf{z}}}(\mathbf{p}) = \gamma^{k+1}_{h,\mathbf{z}}(\mathbf{p}),$$

and

$$\lambda^{k}_{h_{|_{V(z_0-p_0)}},\tilde{\mathbf{z}}}(\mathbf{p}) = \lambda^{k+1}_{h,\mathbf{z}}(\mathbf{p}).$$

Proof. Clearly, it suffices to prove the assertions for $\mathbf{p} = \mathbf{0}$. The assumption that $\gamma^k_{h,\mathbf{z}}(\mathbf{0})$ and $\lambda^k_{h,\mathbf{z}}(\mathbf{0})$ are defined is equivalent to

$$\dim_{\mathbf{0}} \Gamma^{k+1}_{h,\mathbf{z}} \cap V\left(\frac{\partial h}{\partial z_k}\right) \cap V(z_0, \ldots, z_{k-1}) \leqslant 0.$$

Hence, $\Gamma^{k+1}_{h,\mathbf{z}}$ is purely $(k+1)$-dimensional at the origin and thus has no embedded subvarieties (Proposition 1.3). Also, $\Gamma^{k+1}_{h,\mathbf{z}} \cap V\left(\frac{\partial h}{\partial z_k}\right)$ is purely k-dimensional at the origin and so, by Appendix A.4, we have an equality of cycles

$$\left[\Gamma^{k+1}_{h,\mathbf{z}} \cap V\left(\frac{\partial h}{\partial z_k}\right)\right] = \Gamma^{k+1}_{h,\mathbf{z}} \cdot V\left(\frac{\partial h}{\partial z_k}\right) = \Gamma^k_{h,\mathbf{z}} + \Lambda^k_{h,\mathbf{z}}.$$

In addition, we see that $\Gamma^{k+1}_{h,\mathbf{z}} \cap V\left(\frac{\partial h}{\partial z_k}\right) \cap V(z_0)$ is purely $(k-1)$-dimensional at the origin; we easily conclude that

$$(\dagger) \qquad \dim_{\mathbf{0}} \Gamma^{k+1}_{h,\mathbf{z}} \cap \Sigma h \cap V(z_0) \leqslant k - 1.$$

Now, let us consider the cycle $\Gamma^k_{h_{|_{V(z_0)}},\tilde{\mathbf{z}}}$. By definition,

$$\Gamma^k_{h_{|_{V(z_0)}},\tilde{\mathbf{z}}} = V\left(z_0, \frac{\partial h}{\partial z_{k+1}}, \dots, \frac{\partial h}{\partial z_n}\right) \neg \Sigma(h_{|_{V(z_0)}}).$$

Using Lemma I.1.3.ii and our hypothesis that $\Sigma h \cap V(z_0) = \Sigma(h_{|_{V(z_0)}})$, the equality above gives us

$$\Gamma^k_{h_{|_{V(z_0)}},\tilde{\mathbf{z}}} = \left(V(z_0) \cap \Gamma^{k+1}_{h,\mathbf{z}}\right) \neg (\Sigma h \cap V(z_0)) = \left(V(z_0) \cap \Gamma^{k+1}_{h,\mathbf{z}}\right) \neg \Sigma h.$$

But, $V(z_0)\cap\Gamma^{k+1}_{h,\mathbf{z}}$ is purely k-dimensional at the origin and, by (†), $\dim_{\mathbf{0}}\Gamma^{k+1}_{h,\mathbf{z}}\cap\Sigma h\cap V(z_0) \leqslant k-1$; therefore, Σh contains no isolated components of $V(z_0) \cap \Gamma^{k+1}_{h,\mathbf{z}}$ and so, as cycles,

$$\Gamma^k_{h_{|_{V(z_0)}},\tilde{\mathbf{z}}} = \Gamma^{k+1}_{h,\mathbf{z}} \cap V(z_0) = \Gamma^{k+1}_{h,\mathbf{z}} \cdot V(z_0).$$

We find

$$\Gamma^{k-1}_{h_{|_{V(z_0)}},\tilde{\mathbf{z}}} + \Lambda^{k-1}_{h_{|_{V(z_0)}},\tilde{\mathbf{z}}} = \Gamma^k_{h_{|_{V(z_0)}},\tilde{\mathbf{z}}} \cdot V\left(\frac{\partial h}{\partial z_k}\right) =$$

$$\Gamma^{k+1}_{h,\mathbf{z}} \cdot V(z_0) \cdot V\left(\frac{\partial h}{\partial z_k}\right) = \left(\Gamma^k_{h,\mathbf{z}} + \Lambda^k_{h,\mathbf{z}}\right) \cdot V(z_0).$$

That $\gamma^{k-1}_{h_{|_{V(z_0)}},\tilde{\mathbf{z}}}(\mathbf{0})$ and $\lambda^{k-1}_{h_{|_{V(z_0)}},\tilde{\mathbf{z}}}(\mathbf{0})$ are defined and that

$$\gamma^{k-1}_{h_{|_{V(z_0)}},\tilde{\mathbf{z}}}(\mathbf{0}) + \lambda^{k-1}_{h_{|_{V(z_0)}},\tilde{\mathbf{z}}}(\mathbf{0}) = \gamma^k_{h,\mathbf{z}}(\mathbf{0}) + \lambda^k_{h,\mathbf{z}}(\mathbf{0})$$

follows by intersecting the cycle $V(z_1, \dots, z_n)$ with each side of the above equality of cycles.

The remaining equalities follow easily – we leave them as an exercise. □

The following corollary is essentially a converse of the result stated in Remark 1.6.

Corollary 1.19. *Let $k \geqslant 0$. Suppose*

$$\Sigma h \cap V(z_0 - p_0, \dots, z_k - p_k) = \Sigma(h_{|_{V(z_0-p_0,\dots,z_k-p_k)}}),$$

and that $\gamma^i_{h,\mathbf{z}}(\mathbf{p})$ and $\lambda^i_{h,\mathbf{z}}(\mathbf{p})$ are defined for all $i \leqslant k$. Then, $\gamma^{k+1}_{h,\mathbf{z}}(\mathbf{p})$ is defined.

In particular, if $s := \dim_{\mathbf{p}} \Sigma h$, $\lambda^i_{h,\mathbf{z}}(\mathbf{p})$ is defined for $0 \leqslant i \leqslant s$, and, for all k such that $0 \leqslant k \leqslant n-1$,

$$\Sigma h \cap V(z_0 - p_0, \dots, z_k - p_k) = \Sigma(h_{|_{V(z_0-p_0,\dots,z_k-p_k)}}),$$

then $\gamma^i_{h,\mathbf{z}}(\mathbf{p})$ is defined for $0 \leqslant i \leqslant n$.

Proof. The last statement follows immediately from the first by induction, since $\gamma^0_{h,\mathbf{z}}(\mathbf{p})$ is always defined (and is zero).

It suffices to prove the first statement when $\mathbf{p} = \mathbf{0}$.

Case 1: If $\mathbf{0} \notin \Sigma h$, then near $\mathbf{0}$,

$$\Gamma^{k+1}_{h,\mathbf{z}} = V\left(\frac{\partial h}{\partial z_{k+1}}, \dots, \frac{\partial h}{\partial z_n}\right)$$

and so

$$\Gamma^{k+1}_{h,\mathbf{z}} \cap V(z_0, \dots z_k) = \Sigma(h_{|_{V(z_0,\dots,z_k)}}) = \Sigma h \cap V(z_0, \dots, z_k) = \emptyset.$$

Hence, $\gamma^{k+1}_{h,\mathbf{z}}(\mathbf{0})$ is defined and equal to zero.

Case 2: $\mathbf{0} \in \Sigma h$. The proof is by induction on k.

For $k = 0$, the claim is that if $\mathbf{0} \in \Sigma h$, $\lambda^0_{h,\mathbf{z}}(\mathbf{0})$ is defined, and $\Sigma h \cap V(z_0) = \Sigma(h_{|_{V(z_0)}})$, then $\dim_{\mathbf{0}} \Gamma^1_{h,\mathbf{z}} \cap V(z_0) \leqslant 0$. As $\mathbf{0} \in \Sigma h$ and $\lambda^0_{h,\mathbf{z}}(\mathbf{0})$ is defined, we must have that $\dim_{\mathbf{0}} \Gamma^1_{h,\mathbf{z}} \leqslant 1$. So, if $\dim_{\mathbf{0}} \Gamma^1_{h,\mathbf{z}} \cap V(z_0) \geqslant 1$, then $V(z_0)$ must contain a component of $\Gamma^1_{h,\mathbf{z}}$ through the origin. But, since $\Sigma h \cap V(z_0) = \Sigma(h_{|_{V(z_0)}})$,

$$\Gamma^1_{h,\mathbf{z}} \cap V(z_0) \subseteq V\left(z_0, \frac{\partial h}{\partial z_1}, \dots, \frac{\partial h}{\partial z_n}\right) = V\left(z_0, \frac{\partial h}{\partial z_0}, \dots, \frac{\partial h}{\partial z_n}\right).$$

Hence, any component of $\Gamma^1_{h,\mathbf{z}}$ contained in $V(z_0)$ must also be contained in Σh; this contradicts the definition of $\Gamma^1_{h,\mathbf{z}}$.

Suppose now that the corollary is true up to $k-1$, where $k \geqslant 1$. Suppose $\mathbf{0} \in \Sigma h$, $\Sigma h \cap V(z_0, \dots, z_k) = \Sigma(h_{|_{V(z_0,\dots,z_k)}})$, and that $\gamma^i_{h,\mathbf{z}}(\mathbf{0})$ and $\lambda^i_{h,\mathbf{z}}(\mathbf{0})$ are defined for all $i \leqslant k$. As $\mathbf{0} \in \Sigma h$ and $\gamma^i_{h,\mathbf{z}}(\mathbf{0})$ is defined for all $i \leqslant k$, Remark 1.6 implies that $\Sigma h \cap V(z_0, \dots, z_i) = \Sigma(h_{|_{V(z_0,\dots,z_i)}})$ for all $i \leqslant k-1$.

In particular, as $k \geqslant 1$, $\Sigma h \cap V(z_0) = \Sigma(h_{|_{V(z_0)}})$. Thus, we may apply Proposition 1.18 to conclude that $\gamma^i_{h_{|_{V(z_0)}},\tilde{\mathbf{z}}}(\mathbf{0})$ and $\lambda^i_{h_{|_{V(z_0)}},\tilde{\mathbf{z}}}(\mathbf{0})$ are defined for all $i \leqslant k-1$ and, as sets,

$$\Gamma^k_{h_{|_{V(z_0)}},\tilde{\mathbf{z}}} = \Gamma^{k+1}_{h,\mathbf{z}} \cap V(z_0).$$

Since $\Sigma(h_{|_{V(z_0)}}) \cap V(z_1, \dots, z_k) = \Sigma(h_{|_{V(z_0,\dots,z_k)}})$, we are in a position to apply our inductive hypothesis to $h_{|_{V(z_0)}}$.

We conclude that $\gamma^k_{h_{|_{V(z_0)}},\tilde{\mathbf{z}}}(\mathbf{0})$ is defined, i.e.,

$$\dim_{\mathbf{0}} \Gamma^k_{h_{|_{V(z_0)}},\tilde{\mathbf{z}}} \cap V(z_1, \dots, z_k) \leqslant 0.$$

As $\Gamma^k_{h_{|_{V(z_0)}},\tilde{\mathbf{z}}} = \Gamma^{k+1}_{h,\mathbf{z}} \cap V(z_0)$, the proof is finished. □

We shall need the following relation between three intersection numbers. For isolated singularities, this formula appears in the proof of Proposition II.1.2 of [**Te2**] – our argument is essentially the same.

Proposition 1.20. *Let* $\mathbf{p} \in \Sigma h$. *Then,* $\lambda^0_{h,\mathbf{z}}(\mathbf{p})$ *is defined if and only if* $\dim_{\mathbf{p}}\Gamma^1_{h,\mathbf{z}} \leqslant 1$.

Moreover, if $\gamma^1_{h,\mathbf{z}}(\mathbf{p})$ *is defined, then* $\lambda^0_{h,\mathbf{z}}(\mathbf{p})$ *is defined, the dimension of* $\Gamma^1_{h,\mathbf{z}} \cap V(h-h(\mathbf{p}))$ *at* $\mathbf{p}$ *is at most zero, and*

$$\left(\Gamma^1_{h,\mathbf{z}} \cdot V(h - h(\mathbf{p}))\right)_{\mathbf{p}} = \lambda^0_{h,\mathbf{z}}(\mathbf{p}) + \gamma^1_{h,\mathbf{z}}(\mathbf{p}).$$

Proof. $\Gamma^1_{h,\mathbf{z}}$ consists of those components of $V\left(\frac{\partial h}{\partial z_1}, \dots, \frac{\partial h}{\partial z_n}\right)$ which are not contained in $|\Sigma h|$. Thus, $V\left(\frac{\partial h}{\partial z_0}\right)$ contains no components of $\Gamma^1_{h,\mathbf{z}}$. Therefore, $\Gamma^1_{h,\mathbf{z}}$ is purely one-dimensional at $\mathbf{p}$ if and only if $\Gamma^1_{h,\mathbf{z}} \cap V\left(\frac{\partial h}{\partial z_0}\right)$ is purely zero-dimensional at $\mathbf{p}$, i.e., if and only if $\lambda^0_{h,\mathbf{z}}(\mathbf{p})$ is defined.

If $\gamma^1_{h,\mathbf{z}}(\mathbf{p})$ is defined, then $\Gamma^1_{h,\mathbf{z}}$ must be purely one-dimensional at $\mathbf{p}$ and so $\lambda^0_{h,\mathbf{z}}(\mathbf{p})$ is defined, by the above.

The remainder of the proof is an argument which first appeared in Proposition II.1.2 of [**Te2**], and then appeared again in Proposition 1.3 of [**Lê1**]; this argument shows that an easy application of the chain rule yields $\dim_{\mathbf{p}}\Gamma^1_{h,\mathbf{z}} \cap V(h - h(\mathbf{p})) \leqslant 0$, and

$$\left(\Gamma^1_{h,\mathbf{z}} \cdot V(h - h(\mathbf{p}))\right)_{\mathbf{p}} = \lambda^0_{h,\mathbf{z}}(\mathbf{p}) + \gamma^1_{h,\mathbf{z}}(\mathbf{p}).$$

For convenience, we assume that $\mathbf{p} = \mathbf{0}$ and that $h(\mathbf{0}) = 0$.

Suppose $\Gamma^1_{h,\mathbf{z}} = \sum m_W[W]$ as cycles. We know that we can calculate the intersection number of a curve and a hypersurface by parameterizing the curve and looking at the multiplicity of the composition of the defining function of the hypersurface with the parameterization. So, for each component W, pick a local analytic parameterization $\alpha(t)$ of W such that $\alpha(0) = \mathbf{0}$. We must show two things: that $h(\alpha(t))$ is not identically zero, and that

$$\text{mult}_t h(\alpha(t)) = \text{mult}_t \left(\frac{\partial h}{\partial z_0}\right)_{|\alpha(t)} + \text{mult}_t z_0(\alpha(t)).$$

As we already know that the righthand side of the above equality is finite, we have only to prove that the equality holds in order to conclude that $h(\alpha(t))$ is not identically zero. But this is easy:

$$\text{mult}_t h(\alpha(t)) = 1 + \text{mult}_t \frac{d}{dt}\{h(\alpha(t))\} = 1 + \text{mult}_t \left\{\left(\frac{\partial h}{\partial z_0}\right)_{|\alpha(t)} \cdot \alpha_0'(t)\right\},$$

where the remaining terms that come from the chain rule are zero since $\alpha(t)$ parameterizes a component of the polar curve. Thus,

$$\text{mult}_t h(\alpha(t)) = 1 + \text{mult}_t \left(\frac{\partial h}{\partial z_0}\right)_{|\alpha(t)} + \text{mult}_t \alpha_0'(t)$$

$$= \text{mult}_t \left(\frac{\partial h}{\partial z_0}\right)_{|\alpha(t)} + \text{mult}_t \alpha_0(t) = \text{mult}_t \left(\frac{\partial h}{\partial z_0}\right)_{|\alpha(t)} + \text{mult}_t z_0(\alpha(t))$$

and we are finished. □

Of course, what we want to know is that, for a generic choice of coordinates, $\mathbf{z}$, the polar numbers and the Lê numbers are actually defined. Our results in Part I – specifically, I.2.11 and I.2.22 – tell us that, by replacing $\mathbf{z}$ by a generic linear reorganization, we can guarantee that the polar and Lê cycles exist and have the correct dimension. However, we must still do some work to know that the polar and Lê **numbers** are defined generically. To show this, we pick $\mathbf{z}$ generically with respect to a certain type of stratification of the hypersurface defined by h. Below, we define the type of stratification that we need, together with a proposition guaranteeing its existence.

Definition 1.21. A *good stratification* for h at a point $\mathbf{p} \in V(h)$ is an analytic stratification, $\mathfrak{G}$, of the hypersurface $V(h)$ in a neighborhood, $\mathcal{U}$, of $\mathbf{p}$ such that the smooth part of $V(h)$ is a stratum and so that the stratification satisfies Thom's a_h condition with respect to $\mathcal{U} - V(h)$. That is, if $\mathbf{q}_i$ is a sequence of points in $\mathcal{U} - V(h)$ such that $\mathbf{q}_i \rightarrow \mathbf{q} \in S \in \mathfrak{G}$ and $T_{\mathbf{q}_i} V(h - h(\mathbf{q}_i))$ converges to some hyperplane $\mathcal{T}$, then $T_q S \subseteq \mathcal{T}$.

Proposition 1.22 (Hamm and Lê [**H-L**]). *For all* $h : (\mathcal{U}, \mathbf{0}) \rightarrow (\mathbb{C}, 0)$, *for all* $\mathbf{p} \in V(h)$, *there exists a good stratification for* h *at* $\mathbf{p}$.

The notion defined below, that of *prepolar coordinates*, is crucial throughout the remainder of Part II. It provides a generic condition on linear choices of coordinates which implies that all the Lê numbers and polar numbers are defined. Moreover, prepolarity seems to be the right condition to obtain many topological results. The importance of this definition cannot be overstated.

Definition 1.23. Suppose that $\{S_\alpha\}$ is a good stratification for h in a neighborhood, $\mathcal{U}$, of the origin. Let $\mathbf{p} \in V(h)$. Then, a hyperplane, H, in $\mathbb{C}^{n+1}$ through $\mathbf{p}$ is a *prepolar slice* for h at $\mathbf{p}$ with respect to $\{S_\alpha\}$ provided that H transversely intersects all the strata of $\{S_\alpha\}$– except perhaps the stratum $\{\mathbf{p}\}$ itself – in a neighborhood of $\mathbf{p}$.

If H is a prepolar slice for h at $\mathbf{p}$ with respect to $\{S_\alpha\}$, then, as germs of sets at $\mathbf{p}$, $\Sigma(h_{|_H}) = (\Sigma h) \cap H$ and $\dim_{\mathbf{p}} \Sigma(h_{|_H}) = (\dim_{\mathbf{p}} \Sigma h) - 1$ provided $\dim_{\mathbf{p}} \Sigma h \geqslant 1$; moreover, $\{H \cap S_\alpha\}$ is a good stratification for $h_{|_H}$ at $\mathbf{p}$ (see [**H-L**]).

By 2.1.3 of [**H-L**], for a fixed good stratification for h, prepolar slices are generic.

We say simply that H is a *prepolar slice* for h at $\mathbf{p}$ provided that there exists a good stratification with respect to which H is a prepolar slice.

Let $(z_0, \dots, z_n)$ be a linear choice of coordinates for $\mathbb{C}^{n+1}$, let $\mathbf{p} \in V(h)$, and let $\{S_\alpha\}$ be a good stratification for h at $\mathbf{p}$.

For $0 \leqslant i \leqslant n$, $(z_0, \dots, z_i)$ is a *prepolar-tuple* for h at $\mathbf{p}$ with respect to $\{S_\alpha\}$ if and only if $V(z_0 - p_0)$ is a prepolar slice for h at $\mathbf{p}$ with respect to $\{S_\alpha\}$ and for all j such that $1 \leqslant j \leqslant i$, $V(z_j - p_j)$ is a prepolar slice for $h_{|V(z_0-p_0,\dots,z_{j-1}-p_{j-1})}$ at $\mathbf{p}$ with respect to $\{S_\alpha \cap V(z_0 - p_0, \dots, z_{j-1} - p_{j-1})\}$.

As prepolar slices are generic, a generic linear reorganization of $\mathbf{z}$ will produce a prepolar-tuple.

Naturally, we say that $(z_0, \dots, z_i)$ is a *prepolar-tuple* for h at $\mathbf{p}$ provided that there exists a good stratification for h at $\mathbf{p}$ with respect to which $(z_0, \dots, z_i)$ is a prepolar-tuple.

Finally, we say that the coordinates $(z_0, \dots, z_n)$ are *prepolar* for h if and only if for all $\mathbf{p} \in V(h)$, if s denotes $\dim_{\mathbf{p}} \Sigma h$, then $(z_0, \dots, z_{s-1})$ is a prepolar-tuple for h at $\mathbf{p}$ (if $s = 0$ or $\mathbf{p} \notin \Sigma h$, we mean that there is no condition on the coordinates.)

Note that, as prepolar for h is a condition at **all** points in Σh, it is **not** immediate that such coordinates exist (we shall, however, prove this in 10.2.)

Remark 1.24. It will be helpful to interpret good stratifications and prepolar-tuples in terms of conormal geometry and blowing-up the Jacobian tuple $J_{\mathbf{z}}(h)$. Let $\mathrm{Bl}_{J_{\mathbf{z}}(h)} \mathcal{U} \xrightarrow{\pi} \mathcal{U}$ denote the blow-up; $\mathrm{Bl}_{J_{\mathbf{z}}(h)} \mathcal{U} \subseteq \mathcal{U} \times \mathbb{P}^n$. Let E denote the corresponding exceptional divisor. We identify the projectivized cotangent space $\mathbb{P}(T^*\mathcal{U})$ with $\mathcal{U} \times \mathbb{P}^n$.

Then, a good stratification for h at the origin is a stratification $\{S_\alpha\}$ of $V(h)$ in a neighborhood of $\mathbf{0}$ such that the smooth part of $V(h)$ is a stratum and such that, for all S_α, $\pi^{-1}(S_\alpha) \subseteq \mathbb{P}(T^*_{S_\alpha}\mathcal{U})$.

The tuple $(z_0, \dots, z_k)$ is prepolar at the origin with respect to $\{S_\alpha\}$ if and only if, for all i with $0 \leqslant i \leqslant k$, for all S_α with $\dim S_\alpha \geqslant i+1$,

$$\mathbb{P}(T^*_{S_\alpha}\mathcal{U}) \cap \big(V(z_0, \dots, z_i) \times \mathbb{P}^i \times \{\mathbf{0}\}\big) = \emptyset.$$

We will show that by choosing coordinates which are prepolar, one guarantees the existence of the Lê and polar numbers. First, we need a lemma. We use the notation from Remark 1.24.

Lemma 1.25. *Suppose that $(z_0, \dots, z_k)$ is a prepolar tuple for h at the origin. Then, over a neighborhood of the origin, for all i with $0 \leqslant i \leqslant k$,*

$$E \cap \big(V(z_0, \dots, z_{i-1}) \times \mathbb{P}^i \times \{\mathbf{0}\}\big) \subseteq \{\mathbf{0}\} \times \mathbb{P}^i \times \{\mathbf{0}\}.$$

(When $i = 0$, we mean that $E \cap \big(\mathcal{U} \times \mathbb{P}^0 \times \{\mathbf{0}\}\big) \subseteq \{\mathbf{0}\} \times \mathbb{P}^0 \times \{\mathbf{0}\}$.)

Proof. If $(z_0, \dots, z_k)$ is a prepolar tuple, then $(z_0, \dots, z_i)$ is a prepolar tuple for all i such that $0 \leqslant i \leqslant k$; thus, we only need to prove the claim when $i = k$. We use our characterizations from Remark 1.24. Let $\{S_\alpha\}$ be a good stratification for h at the origin.

We first show that it suffices to prove the claim with z_{i-1} replaced by z_i; more precisely we show that:

(†) for all S_α, in a neighborhood of the origin, for all i such that $0 \leqslant i \leqslant n$,

$$\mathbb{P}(T^*_{S_\alpha}\mathcal{U}) \cap \big(V(z_0, \ldots, z_{i-1}) \times \mathbb{P}^i \times \{\mathbf{0}\}\big) \subseteq$$

$$\big(V(z_0, \ldots, z_i) \times \mathbb{P}^i \times \{\mathbf{0}\}\big) \cup \big(V(z_0, \ldots, z_{i-1}) \times \mathbb{P}^{i-1} \times \{\mathbf{0}\}\big).$$

(When $i = 0$, we mean that $\mathbb{P}(T^*_{S_\alpha}\mathcal{U}) \cap \big(\mathcal{U} \times \mathbb{P}^i \times \{\mathbf{0}\}\big) \subseteq \big(V(z_0) \times \mathbb{P}^0 \times \{\mathbf{0}\}\big)$.)

Suppose we have an analytic curve

$$\beta(t) := (\mathbf{r}(t), [a_0(t)dz_0 + \cdots + a_i(t)dz_i]) \in \mathbb{P}(\overline{T^*_{S_\alpha}\mathcal{U}}) \cap \big(V(z_0, \ldots, z_{i-1}) \times \mathbb{P}^i \times \{\mathbf{0}\}\big)$$

such that $\mathbf{r}(0) = \mathbf{0}$ and such that, for all $t \neq 0$, $\mathbf{r}(t) \in S_\alpha$.

Then, for $t \neq 0$, $\mathbf{r}'(t) = (0, \ldots, 0, r_i'(t), \ldots, r_n'(t)) \in T_{\mathbf{r}(t)}S_\alpha$ and

$$0 \equiv (a_0(t)dz_0 + \cdots + a_i(t)dz_i)(\mathbf{r}'(t)) = a_i(t)r_i'(t).$$

Thus, either $a_i(t) \equiv 0$ or $r_i'(t) \equiv 0$. Since $\mathbf{r}(0) = \mathbf{0}$, if $r_i'(t) \equiv 0$, then $r_i(t) \equiv 0$. This proves (†).

Now, over a neighborhood of the origin,

$$(*)\ E \cap \big(V(z_0, \ldots, z_{i-1}) \times \mathbb{P}^i \times \{\mathbf{0}\}\big) \subseteq \bigcup_{S_\alpha \subseteq \Sigma h} \big(\pi^{-1}(S_\alpha) \cap \big(V(z_0, \ldots, z_{i-1}) \times \mathbb{P}^i \times \{\mathbf{0}\}\big)\big) \subseteq$$

$$\bigcup_{S_\alpha \subseteq \Sigma h} \big(\mathbb{P}(T^*_{S_\alpha}\mathcal{U}) \cap \big(V(z_0, \ldots, z_{i-1}) \times \mathbb{P}^i \times \{\mathbf{0}\}\big)\big).$$

We proceed by induction on k.

When $k = i = 0$, $(*)$ and (†) combined yield that

$$E \cap \big(\mathcal{U} \times \mathbb{P}^0 \times \{\mathbf{0}\}\big) \subseteq \bigcup_{S_\alpha \subseteq \Sigma h} \big(\mathbb{P}(T^*_{S_\alpha}\mathcal{U}) \cap \big(V(z_0) \times \mathbb{P}^0 \times \{\mathbf{0}\}\big)\big)$$

over a neighborhood of the origin. However, the characterization of z_0 being prepolar that we gave in Remark 1.24 tells us that this last quantity equals

$$\mathbb{P}(T^*_{\mathbf{0}}\mathcal{U}) \cap \big(V(z_0) \times \mathbb{P}^0 \times \{\mathbf{0}\}\big) = \{\mathbf{0}\} \times \mathbb{P}^0 \times \{\mathbf{0}\}.$$

This proves the claim when $k = 0$.

Now, suppose that the lemma is true for k; we wish to see that it is also true for $k+1 = i+1$. Combining $(*)$ and (†) again, over a neighborhood of the origin, we find

$$E \cap \big(V(z_0, \ldots, z_k) \times \mathbb{P}^{k+1} \times \{\mathbf{0}\}\big) \subseteq \bigcup_{S_\alpha \subseteq \Sigma h} \big(\mathbb{P}(T^*_{S_\alpha}\mathcal{U}) \cap \big(V(z_0, \ldots, z_k) \times \mathbb{P}^{k+1} \times \{\mathbf{0}\}\big)\big) \subseteq$$

$$\Big\{ \bigcup_{S_\alpha \subseteq \Sigma h} \big(\mathbb{P}(T^*_{S_\alpha}\mathcal{U}) \cap (V(z_0, \dots, z_{k+1}) \times \mathbb{P}^{k+1} \times \{\mathbf{0}\})\big)\Big\} \cup$$

$$\Big\{ \bigcup_{S_\alpha \subseteq \Sigma h} \big(\mathbb{P}(T^*_{S_\alpha}\mathcal{U}) \cap (V(z_0, \dots, z_{k}) \times \mathbb{P}^{k} \times \{\mathbf{0}\})\big)\Big\}.$$

Prepolarity, as described in Remark 1.24, implies that the image under π of this last quantity is contained in

$$\Big(V(z_0, \dots, z_{k+1}) \cap \bigcup_{\substack{S_\alpha \subseteq \Sigma h \\ \dim S_\alpha \leqslant k+1}} S_\alpha\Big) \cup \Big(V(z_0, \dots, z_{k}) \cap \bigcup_{\substack{S_\alpha \subseteq \Sigma h \\ \dim S_\alpha \leqslant k}} S_\alpha\Big).$$

As $(z_0, \dots, z_{k+1})$ is prepolar, near the origin, both of the above intersections are contained in $\{\mathbf{0}\}$. Thus, we conclude that, over a neighborhood of the origin,

$$E \cap \big(V(z_0, \dots, z_k) \times \mathbb{P}^{k+1} \times \{\mathbf{0}\}\big) \subseteq \{\mathbf{0}\} \times \mathbb{P}^{k+1} \times \{\mathbf{0}\}. \quad \square$$

In the following theorem, we continue to use the notation from Remark 1.24. The characterization here of the Lê cycles in terms of blowing-up was first shown to us by T. Gaffney (without the description of how generic the coordinates must be). We generalize this result in IV.1.10.

Theorem 1.26. *Let* $\mathbf{p} \in V(h)$ *and let* $s = \dim_{\mathbf{p}} \Sigma h$.

If $(z_0, \dots, z_k)$ *is a prepolar tuple for* h *at* $\mathbf{p}$*, then there exists a neighborhood* Ω *of* $\mathbf{p}$ *such that, for all* i *such that* $0 \leqslant i \leqslant k$*, the exceptional divisor* E *properly intersects* $\Omega \times \mathbb{P}^i \times \{\mathbf{0}\}$ *in* $\Omega \times \mathbb{P}^n$.

Hence, if z_0 *is prepolar for* h *at* $\mathbf{p}$*, then* $\gamma^1_{h,z_0}(p)$ *is defined.*

Moreover, if $(z_0, \dots, z_n)$ *is a prepolar tuple for* h *at* $\mathbf{p}$*, then, for all* i*, the Lê numbers and polar numbers* $\lambda^i_{h,\mathbf{z}}(\mathbf{p})$ *and* $\gamma^i_{h,\mathbf{z}}(\mathbf{p})$ *exist and, in a neighborhood of* Σh,

$$\Gamma^{i+1}_{h,\mathbf{z}} = \pi_*(\mathrm{Bl}_{J_{\mathbf{z}}(h)}\mathcal{U} \cdot (\Omega \times \mathbb{P}^i \times \{\mathbf{0}\}))$$

and

$$\Lambda^i_{h,\mathbf{z}} = \pi_*(E \cdot (\Omega \times \mathbb{P}^i \times \{\mathbf{0}\})),$$

where the intersection takes place in $\Omega \times \mathbb{P}^n$ *and* π_* *denotes the proper push-forward.*

Proof. For convenience, we will assume that $\mathbf{p} = \mathbf{0}$. Fix a good stratification $\{S_\alpha\}$ for h at $\mathbf{0}$.

We show the first statement in the theorem by induction on k. We need to show that if $(z_0, \dots, z_k)$ is a prepolar tuple, then there is a neighborhood of the origin over which $E \cap (\mathcal{U} \times \mathbb{P}^i \times \{\mathbf{0}\})$ is purely i-dimensional for all $i \leqslant k$. As $(z_0, \dots, z_k)$ being a prepolar tuple implies that $(z_0, \dots, z_i)$ is prepolar for all $i \leqslant k$, it suffices to show that if $(z_0, \dots, z_k)$ is a prepolar tuple, then there is a neighborhood of the origin over which $E \cap (\mathcal{U} \times \mathbb{P}^k \times \{\mathbf{0}\})$ is purely k-dimensional.

The lemma implies that, near $\mathbf{0}$,

$$E \cap (\mathcal{U} \times \mathbb{P}^0 \times \{\mathbf{0}\}) \subseteq \mathbb{P}(T^*_{\{\mathbf{0}\}}\mathcal{U}) \cap (V(z_0) \times \mathbb{P}^0 \times \{\mathbf{0}\}) = \{\mathbf{0}\} \times \mathbb{P}^0 \times \{\mathbf{0}\}.$$

This proves the desired result when $k = 0$.

Suppose now that E properly intersects $\mathcal{U} \times \mathbb{P}^k \times \{\mathbf{0}\}$ over the origin, but does not properly intersect $\mathcal{U} \times \mathbb{P}^{k+1} \times \{\mathbf{0}\}$ over the origin. Let C be a component of $E \cap \big(\mathcal{U} \times \mathbb{P}^{k+1} \times \{\mathbf{0}\}\big)$ which has dimension at least $k+2$ and such that $\mathbf{0} \in \pi(C)$. We will use $w_0, \dots, w_n$ as homogeneous coordinates on $\mathbb{P}^n$.

Our inductive hypothesis implies that $\mathbf{0} \notin \pi(C \cap V(w_{k+1}))$. We shall derive a contradiction by using our other hypotheses to prove that $\mathbf{0} \in \pi(C \cap V(w_{k+1}))$.

Certainly, $\mathbf{0} \in \pi\big(C \cap (V(z_0, \dots, z_k) \times \mathbb{P}^{k+1} \times \{\mathbf{0}\})\big)$, and the lemma implies that

$$C \cap (V(z_0, \dots, z_k) \times \mathbb{P}^{k+1} \times \{\mathbf{0}\}) \subseteq \{\mathbf{0}\} \times \mathbb{P}^{k+1} \times \{\mathbf{0}\}.$$

Therefore, as each component of $C \cap (V(z_0, \dots, z_k) \times \mathbb{P}^{k+1} \times \{\mathbf{0}\})$ has dimension at least 1, it follows that each component must intersect $V(w_{k+1})$. This is a contradiction, and establishes the first statement of the theorem.

If z_0 is prepolar for h at $\mathbf{0}$, then $\Sigma h \cap V(z_0) = \Sigma(h_{|_{V(z_0)}})$, and we have just shown that E properly intersects $\mathcal{U} \times \mathbb{P}^0 \times \{\mathbf{0}\}$ over a neighborhood of the origin. By Remark I.2.21, this implies that $\Gamma^1_{h,\mathbf{z}}$ is purely 1-dimensional at $\mathbf{0}$, and now Corollary 1.19 allows us to conclude that $\gamma^1_{h,z_0}(p)$ is defined.

If $(z_0, \dots, z_n)$ is prepolar for h at the origin, then the Segre-Vogel Relation (Corollary I.2.22) allows us to conclude that there is a neighborhood, Ω, of the origin such that

$$\Gamma^{i+1}_{h,\mathbf{z}} = \pi_*(\mathrm{Bl}_{J_{\mathbf{z}}(h)}\mathcal{U} \cdot (\Omega \times \mathbb{P}^i \times \{\mathbf{0}\}))$$

and

$$\Lambda^i_{h,\mathbf{z}} = \pi_*(E \cdot (\Omega \times \mathbb{P}^i \times \{\mathbf{0}\})).$$

Lemma 1.25 tells us that, as germs of sets at the origin,

$$\Lambda^i_{h,\mathbf{z}} \cap V(z_0, \dots, z_{i-1}) = \pi(E \cap (V(z_0, \dots, z_{i-1}) \times \mathbb{P}^i \times \{\mathbf{0}\})) \subseteq \{\mathbf{0}\}.$$

Therefore, all of the Lê numbers are defined, and Corollary 1.19 implies that all of the polar numbers are also defined. □

Remark 1.27. While it is true that prepolar coordinates occur generically and guarantee the existence of the Lê numbers, it is **not** true that all sets of prepolar coordinates yield the same Lê numbers.

If $\dim_{\mathbf{p}} \Sigma h = 1$, then $\{V(h) - \Sigma h, \Sigma h - \mathbf{p}, \mathbf{p}\}$ is a good stratification for h in a neighborhood of $\mathbf{p}$; hence, $V(z_0 - p_0)$ is a prepolar slice if and only if $\dim_{\mathbf{p}} \Sigma(h_{|_{V(z_0 - p_0)}}) = 0$. This is the case if and only if $\gamma^1_{h,\mathbf{z}}(\mathbf{p})$ and $\lambda^1_{h,\mathbf{z}}(\mathbf{p})$ are defined.

Now, consider the example from Remark 1.16. The coordinates $\mathbf{z} = (z_0, z_1, z_2)$ are prepolar for $h = z_2^2 + (z_0 - z_1^2)^2$ at the origin, and $\lambda^0_{h,\mathbf{z}}(\mathbf{0}) = 1$ and $\lambda^1_{h,\mathbf{z}}(\mathbf{0}) = 2$. However, the coordinates $\mathbf{z}$ are really not very generic, as Σh is smooth at the origin, but $V(z_0)$ intersects Σh with multiplicity 2 at $\mathbf{0}$. The generic values of λ^0_h and λ^1_h (that is, the values with respect to generic coordinates) are 0 and 1, respectively.

Note that the alternating sum of the Lê numbers is the same for the non-generic and generic coordinates. As we shall see in Chapter 3, this is a general fact: as long as the coordinates are prepolar, the alternating sum of the Lê numbers is independent of the coordinates and is, in fact, equal to the reduced Euler characteristic of the Milnor fibre. We know of no algebraic way to prove this independence.

It is reasonable to ask why we do not strengthen our notion of prepolar in order to disallow examples such as the one above, where the Lê numbers do not have their generic values. The answer is that later (in Proposition 10.2) we shall show that, given h and a point $\mathbf{p} \in V(h)$, one may pick generic coordinates which are prepolar for h **at every point in a neighborhood of p**. This result allows us to give another characterization of the Lê cycles in Chapter 10. Example 2.4 shows that this result would be false if we were to strengthen the notion of prepolar to require the Lê numbers to obtain their generic values at each point in this open neighborhood of $\mathbf{p}$.

Finally, note that there **are** generic values for the Lê numbers; this follows easily from the characterization of the Lê cycles given in Theorem 1.26, combined with Kleiman's Transversality Lemma [**Kl**].

We conclude this chapter with four results which do not seem to be of fundamental importance, but which are fairly surprising.

Proposition 1.28. *Suppose that* $\dim_{\mathbf{p}} \Sigma h = 1$ *and* $V(z_0 - p_0)$ *is a prepolar slice for* h *at* $\mathbf{p}$. *If* $V(z_0 - p_0)$ *does not transversely intersect the set* $|\Sigma h|$ *at* $\mathbf{p}$ *(in particular, if* $|\Sigma h|$ *is not smooth at* $\mathbf{p}$*), then* $\lambda^0_{h,\mathbf{z}}(\mathbf{p}) \neq 0$.

Proof. Despite the different appearance of the statement, this is precisely what Lê proves in [**Lê12**]. □

Proposition 1.29. *Let* $k \geqslant 1$. *Suppose that* $\Lambda^0_{h,\mathbf{z}}, \dots, \Lambda^{k-1}_{h,\mathbf{z}}$ *have correct dimension at* $\mathbf{p}$. *Suppose, for all pairs of distinct irreducible germs,* V *and* W, *of* Σh *through* $\mathbf{p}$, *that* $\dim_{\mathbf{p}}(V \cap W) \leqslant k-1$. *Finally, suppose that* $\lambda^k_{h,\mathbf{z}}(\mathbf{p}) = 0$. *Then,* $\lambda^j_{h,\mathbf{z}}(\mathbf{p}) = 0$ *for all* $j \leqslant k$.

Proof. One applies 2.3 of [**La**] to the case where the irreducible normal variety is $\mathbb{C}^{n+1}$ and the subvariety locally defined by $n-k$ equations is $V\left(\frac{\partial h}{\partial z_{k+1}}, \dots, \frac{\partial h}{\partial z_n}\right)$, which equals $\Gamma^{k+1}_{h,\mathbf{z}} \cup \Sigma h$ as a set.

Let V be an irreducible component of Σh at $\mathbf{p}$ and let $\{W_i\}_i$ be the remaining irreducible components of Σh at $\mathbf{p}$.

Then, the lemma of Lazarsfeld says that if

$$\mathbf{p} \in \left(\Gamma^{k+1}_{h,\mathbf{z}} \cap V\right) \cup \left(\left(\bigcup_i W_i\right) \cap V\right),$$

then

$$\dim_{\mathbf{p}} \left(\Gamma^{k+1}_{h,\mathbf{z}} \cap V\right) \cup \left(\left(\bigcup_i W_i\right) \cap V\right) \geqslant k.$$

Now the proposition follows easily from 1.13. □

Proposition 1.30. *Let $s = \dim_{\mathbf{p}} \Sigma h$ and suppose that $\lambda^j_{h,\mathbf{z}}(\mathbf{p})$ and $\gamma^j_{h,\mathbf{z}}(\mathbf{p})$ exist for all $j \leqslant s$. Suppose that the critical locus of h at $\mathbf{p}$ is itself singular and denote the singular set of the critical locus by $\Sigma\Sigma h$. Then, every $(s-1)$-dimensional component of $\Sigma\Sigma h$ through $\mathbf{p}$ is contained in the set $|\Lambda^{s-1}_{h,\mathbf{z}}|$.*

Proof. Let C be an $(s-1)$-dimensional component of $\Sigma\Sigma h$ at $\mathbf{p}$. As all the Lê and polar numbers exist, we may inductively apply Proposition 1.18, together with Remark 1.6 and Proposition 1.14, to conclude that $\mathbf{p}$ is a singular point of the one-dimensional critical locus of $h_{|_{V(z_0-p_0,\dots,z_{s-2}-p_{s-2})}}$.

Using $(z_{s-1}, \dots, z_n)$ as coordinates for $h_{|_{V(z_0-p_0,\dots,z_{s-2}-p_{s-2})}}$, we also conclude from Proposition 1.18 that, at $\mathbf{p}$,

$$\lambda^1_{h_{|_{V(z_0-p_0,\dots,z_{s-2}-p_{s-2})}}} \quad \text{and} \quad \gamma^1_{h_{|_{V(z_0-p_0,\dots,z_{s-2}-p_{s-2})}}}$$

exist – which, for a one-dimensional critical locus, is equivalent to $V(z_{s-1} - p_{s-1})$ being prepolar for $h_{|_{V(z_0-p_0,\dots,z_{s-2}-p_{s-2})}}$ at $\mathbf{p}$.

Therefore, by Proposition 1.28, $\lambda^0_{h_{|_{V(z_0-p_0,\dots,z_{s-2}-p_{s-2})}}}(\mathbf{p}) \neq 0$. This is equivalent to saying that $\mathbf{p} \in \Gamma^1_{h_{|_{V(z_0-p_0,\dots,z_{s-2}-p_{s-2})}}}$ and now, by applying Proposition 1.18 once more, we find that $\mathbf{p} \in \Gamma^s_{h,\mathbf{z}}$.

As we may apply this same argument at each point of C near $\mathbf{p}$, we find that $C \subseteq \Gamma^s_{h,\mathbf{z}} \cap \Sigma h$. Finally, as the Lê numbers are defined, each of the Lê cycles has correct dimension at $\mathbf{p}$ and so the result follows from Proposition 1.13. □

Proposition 1.30. *Let $s = \dim_{\mathbf{p}} \Sigma h$, suppose that $\lambda^j_{h,\mathbf{z}}(\mathbf{p})$ and $\gamma^j_{h,\mathbf{z}}(\mathbf{p})$ exist for all $j \leqslant s$, and suppose that $\lambda^{s-1}_{h,\mathbf{z}}(\mathbf{p}) = 0$. Then, $\lambda^j_{h,\mathbf{z}}(\mathbf{p}) = 0$ for all $j \leqslant s-1$.*

Proof. The result follows from Proposition 1.29, using $i = s-1$, since the preceding proposition proves: if there exist two irreducible components, V and W, of Σh at $\mathbf{p}$ such that $\dim_{\mathbf{p}}(V \cap W) = s-1$, then $\mathbf{p} \in \Lambda^{s-1}_{h,\mathbf{z}}$ and so $\lambda^{s-1}_{h,\mathbf{z}}(\mathbf{p}) \neq 0$. □

Chapter 2. ELEMENTARY EXAMPLES

Example 2.1. If $\mathbf{0}$ is an isolated singularity of h, then regardless of the coordinate system $\mathbf{z}$, it follows from Proposition 1.11 that the only possibly non-zero Lê number is $\lambda^0_{h,\mathbf{z}}(\mathbf{0})$. Moreover, as $V\left(\frac{\partial h}{\partial z_0}, \frac{\partial h}{\partial z_1}, \dots, \frac{\partial h}{\partial z_n}\right)$ is zero-dimensional, $V\left(\frac{\partial h}{\partial z_1}, \dots, \frac{\partial h}{\partial z_n}\right)$ is one-dimensional with no components contained in Σh and with no embedded subvarieties. Therefore,

$$\Gamma^1_{h,\mathbf{z}} = V\left(\frac{\partial h}{\partial z_1}, \dots, \frac{\partial h}{\partial z_n}\right)$$

and so

$$\lambda^0_{h,\mathbf{z}}(\mathbf{0}) = \left(\Gamma^1_{h,\mathbf{z}} \cdot V\left(\frac{\partial h}{\partial z_0}\right)\right)_{\mathbf{0}} = \left(V\left(\frac{\partial h}{\partial z_1}, \dots, \frac{\partial h}{\partial z_n}\right) \cdot V\left(\frac{\partial h}{\partial z_0}\right)\right)_{\mathbf{0}} =$$

$$\left[V\left(\frac{\partial h}{\partial z_0}, \dots, \frac{\partial h}{\partial z_n}\right)\right]_{\mathbf{0}} = \text{the Milnor number of } h \text{ at } \mathbf{0}.$$

Example 2.2. Here, we generalize Example 1.12. Let $h = y^2 - x^a - tx^b$, where $a > b > 1$. We fix the coordinate system $\mathbf{z} = (t, x, y)$ and will suppress any further reference to it.

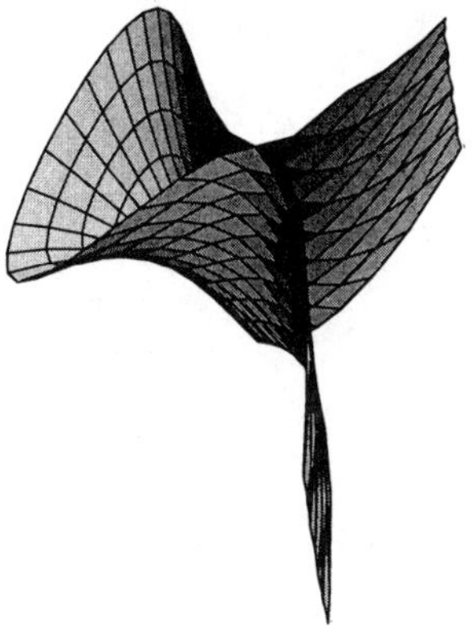

Figure 2.3. Generalization of nodes degenerating to a cusp

We find:

$$\Sigma h = V(-x^b, \ -ax^{a-1} - btx^{b-1}, \ 2y) = V(x, y).$$

$$\Gamma^2_h = V\left(\frac{\partial h}{\partial y}\right) = V(2y) = V(y).$$

$$\Gamma^2_h \cdot V\left(\frac{\partial h}{\partial x}\right) = V(y)\cdot V(-ax^{a-1} - btx^{b-1}) =$$

$$V(y)\cdot (V(-ax^{a-b} - bt) + V(x^{b-1})) = V(-ax^{a-b} - bt, y) + (b-1)V(x,y)$$
$$= \Gamma^1_h + \Lambda^1_h.$$

$$\Gamma^1_h \cdot V\left(\frac{\partial h}{\partial t}\right) = V(-ax^{a-b} - bt, y)\cdot V(-x^b) = bV(t,x,y) = b[\mathbf{0}] = \Lambda^0_h.$$

Thus, $\lambda^0_h(\mathbf{0}) = b$ and $\lambda^1_h(\mathbf{0}) = b-1$.

Notice that the exponent a does not appear; this is because

$$h = y^2 - x^a - tx^b = y^2 - x^b(x^{a-b} - t)$$

which, after an analytic coordinate change at the origin, equals $y^2 - x^b u$.

Example 2.4 (The FM Cone). Let $h = y^2 - x^3 - (u^2+v^2+w^2)x^2$ and fix the coordinates (u,v,w,x,y).

$$\Sigma h = V(-2ux^2,\ -2vx^2,\ -2wx^2,\ -3x^2 - 2x(u^2+v^2+w^2),\ 2y) = V(x,y).$$

As Σh is three-dimensional, we begin our calculation with Γ^4_h.

$$\Gamma^4_h = V(-2y) = V(y).$$

$$\Gamma^4_h \cdot V\left(\frac{\partial h}{\partial x}\right) = V(y)\cdot V(-3x^2 - 2x(u^2+v^2+w^2)) =$$

$$V(-3x - 2(u^2+v^2+w^2), y) + V(x,y) = \Gamma^3_h + \Lambda^3_h.$$

$$\Gamma^3_h \cdot V\left(\frac{\partial h}{\partial w}\right) = V(-3x - 2(u^2+v^2+w^2), y)\cdot V(-2wx^2) =$$

$$V(-3x - 2(u^2+v^2), w, y) + 2V(u^2+v^2+w^2, x, y) = \Gamma^2_h + \Lambda^2_h.$$

$$\Gamma^2_h \cdot V\left(\frac{\partial h}{\partial v}\right) = V(-3x - 2(u^2+v^2), w, y)\cdot V(-2vx^2) =$$

$$V(-3x - 2u^2, v, w, y) + 2V(u^2+v^2, w, x, y) = \Gamma^1_h + \Lambda^1_h.$$

$$\Gamma^1_h \cdot V\left(\frac{\partial h}{\partial u}\right) = V(-3x - 2u^2, v, w, y)\cdot V(-2ux^2) =$$

$$V(u,v,w,x,y) + 2V(u^2, v, w, x, y) = 5[\mathbf{0}] = \Lambda^0_h.$$

Hence, $\Lambda_h^3 = V(x,y)$, $\Lambda_h^2 = 2V(u^2+v^2+w^2,x,y)$ = a cone (as a set), $\Lambda_h^1 = 2V(u^2+v^2,w,x,y)$, and $\Lambda_h^0 = 5[\mathbf{0}]$. Thus, at the origin, $\lambda_h^3 = 1$, $\lambda_h^2 = 4$, $\lambda_h^1 = 4$, and $\lambda_h^0 = 5$.

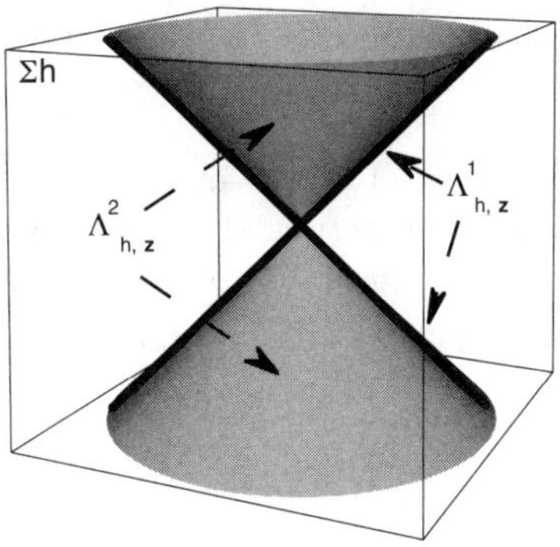

Figure 2.5. The critical locus of h

Though λ_h^1 is independent of a generic choice of coordinates, Λ_h^1 depends on the coordinates – for, by symmetry, if we re-ordered u, v, and w, then Λ_h^1 would change correspondingly. Moreover, one can check that this is a generic problem.

Such "non-fixed" Lê cycles arise from the absolute polar varieties ([**L-T2**], [**Te4**], [**Te5**]) of the higher dimensional Lê cycles (this follows from two results in Part IV: Theorems 1.10 and 3.2). For instance, in the present case, Λ_h^2 is a cone, and its one-dimensional polar variety varies with the choice of coordinates, but generically always consists of two lines; this is the case for Λ_h^1 as well.

Example 2.6. Let $h = xyz$, so that $V(h)$ consists of the coordinate planes in $\mathbb{C}^3$. (See Example 0.2.) Then, $\Sigma h = V(x,y) \cup V(x,z) \cup V(y,z)$ = union of the three coordinate axes.

The coordinates (x,y,z) are extremely non-generic, so choose some other generic coordinates $\tilde{\mathbf{z}} = (\tilde{z}_0, \tilde{z}_1, \tilde{z}_2)$. Then, the set $|\Lambda_{h,\tilde{\mathbf{z}}}^1| = \Sigma h$. Hence,

$$\lambda_{h,\tilde{\mathbf{z}}}^1(\mathbf{0}) = \left(\Lambda_{h,\tilde{\mathbf{z}}}^1 \cdot V(\tilde{z}_0)\right)_{\mathbf{0}} = \sum_{\mathbf{p}} \left(\Lambda_{h,\tilde{\mathbf{z}}}^1 \cdot V(\tilde{z}_0 - \xi)\right)_{\mathbf{p}} = \sum_{\mathbf{p}} \lambda_{h,\tilde{\mathbf{z}}}^1(\mathbf{p}),$$

where the sum is over all $\mathbf{p} \in \overset{\circ}{B}_\epsilon \cap \Lambda_{h,\tilde{\mathbf{z}}}^1 \cap V(\tilde{z}_0 - \xi)$ for small ϵ and $0 < \xi \ll \epsilon$; this set consists of three points and, by symmetry, λ^1 must be the same at each of these three points. We wish to use Proposition 1.27 to calculate $\lambda_{h,\tilde{\mathbf{z}}}^1(\mathbf{p})$.

As each $\mathbf{p} \in \Sigma h$, it follows from 1.10 that $\gamma_{h,\tilde{\mathbf{z}}}^1$ is supported only at $\Lambda_{h,\tilde{\mathbf{z}}}^0$, which is generically zero-dimensional. Thus, our points $\mathbf{p}$ are such that $\gamma_{h,\tilde{\mathbf{z}}}^1(\mathbf{p}) = 0$, and it follows from 1.27 that $\lambda_{h,\tilde{\mathbf{z}}}^1(p) = \lambda^0_{h_{|_{V(\tilde{z}_0 - p_0)}},\hat{\mathbf{z}}}(\mathbf{p})$, where $\hat{\mathbf{z}}$ denotes the restriction of the coordinates $\tilde{\mathbf{z}}$ to $V(\tilde{z}_0 - p_0)$.

Now, $h_{|_{V(\tilde{z}_0-p_0)}}$ has an isolated singularity at each of our three points $\mathbf{p}$; thus, at each of these three $\mathbf{p}$'s, $\lambda^0_{h_{|_{V(\tilde{z}_0-p_0)}},\tilde{\mathbf{z}}}(\mathbf{p})$ equals the Milnor number of $h_{|_{V(\tilde{z}_0-p_0)}}$ at $\mathbf{p}$, and this is easily seen to equal 1. It follows, finally, that $\lambda^1_{h,\tilde{\mathbf{z}}}(\mathbf{0}) = 3$.

The generic value of $\lambda^0_{h,\tilde{\mathbf{z}}}(\mathbf{0})$ is somewhat messier to calculate, and is just as easy to treat in the more general case given in Example 2.8. (However, the answer is that $\lambda^0_{h,\tilde{\mathbf{z}}}(\mathbf{0}) = 2$.)

Example 2.7. Let $\mathcal{U}$ be an open subset of $\mathbb{C}^{n+1}$, let $\mathbf{z} = (z_0, \dots, z_n)$ be the coordinates for $\mathbb{C}^{n+1}$, and $h : \mathcal{U} \to \mathbb{C}$ be any analytic function. The coordinates $\mathbf{z}$ may be non-generic for h. We wish to see how to calculate $\lambda^0_{h,\tilde{\mathbf{z}}}$ for a generic linear choice of $\tilde{\mathbf{z}}$.

So, let $\tilde{\mathbf{z}}$ be a generic linear choice of coordinates for $\mathbb{C}^{n+1}$, and let a_{ij} denote $\frac{\partial z_i}{\partial \tilde{z}_j}$.

Now,

$$\Gamma^1_{h,\tilde{\mathbf{z}}} = V\left(\frac{\partial h}{\partial \tilde{z}_1}, \dots, \frac{\partial h}{\partial \tilde{z}_n}\right) \neg\ \Sigma h =$$

$$V\left(a_{01}\frac{\partial h}{\partial z_0} + \cdots + a_{n1}\frac{\partial h}{\partial z_n},\ \dots\ ,\ a_{0n}\frac{\partial h}{\partial z_0} + \cdots + a_{nn}\frac{\partial h}{\partial z_n}\right) \neg\ \Sigma h.$$

By performing elementary row operations, we find that the ideal

$$\left\langle a_{01}\frac{\partial h}{\partial z_0} + \cdots + a_{n1}\frac{\partial h}{\partial z_n},\ \dots\ ,\ a_{0n}\frac{\partial h}{\partial z_0} + \cdots + a_{nn}\frac{\partial h}{\partial z_n}\right\rangle$$

is generated by

$$\frac{\partial h}{\partial z_0} + b_0\frac{\partial h}{\partial z_n},\ \frac{\partial h}{\partial z_1} + b_1\frac{\partial h}{\partial z_n},\ \dots,\ \frac{\partial h}{\partial z_{n-1}} + b_{n-1}\frac{\partial h}{\partial z_n},$$

where $b_0, \dots, b_{n-1}$ are generic $\neq 0$.

Thus,

$$\Gamma^1_{h,\tilde{\mathbf{z}}} = V\left(\frac{\partial h}{\partial z_0} + b_0\frac{\partial h}{\partial z_n}, \frac{\partial h}{\partial z_1} + b_1\frac{\partial h}{\partial z_n}, \dots, \frac{\partial h}{\partial z_{n-1}} + b_{n-1}\frac{\partial h}{\partial z_n}\right) \neg\ \Sigma h,$$

and $\Lambda^0_{h,\tilde{\mathbf{z}}}$ is given by intersecting this with $a_{00}\frac{\partial h}{\partial z_0} + \cdots + a_{n0}\frac{\partial h}{\partial z_n}$.

It is important to note that we are **not** claiming that the cycle $\Gamma^1_{h,\tilde{\mathbf{z}}}$ can be calculated by considering the cycle

$$V\left(\frac{\partial h}{\partial z_0} + b_0\frac{\partial h}{\partial z_n}\right) \cdot V\left(\frac{\partial h}{\partial z_1} + b_1\frac{\partial h}{\partial z_n}\right) \cdot \ldots \cdot V\left(\frac{\partial h}{\partial z_{n-1}} + b_{n-1}\frac{\partial h}{\partial z_n}\right)$$

and then disposing of any portions of the cycle which are contained in Σh. There could easily be a problem with embedded subvarieties.

Example 2.8. We can use the above example to calculate the generic value of λ^0 in Example 2.6. Actually, we can just as easily do a more general calculation.

Let $h = z_0z_1\dots z_n$, so that $V(h)$ is the union of the coordinate planes in $\mathbb{C}^{n+1}$ and $\Sigma h = \bigcup_{i\neq j} V(z_i, z_j) =$ the union of intersections of pairs of the different coordinate planes. We wish to show, for a generic choice of coordinates, $\tilde{\mathbf{z}}$, that $\lambda^0_{h,\tilde{\mathbf{z}}}(\mathbf{0}) = n$.

By the above, we find that $\Gamma^1_{h,\mathbf{z}}$ equals

$$V\big(z_1z_2\dots z_{n-1}(z_n + b_0z_0),\ z_0z_2\dots z_{n-1}(z_n+b_1z_1),\ \dots,\ z_0z_1\dots z_{n-2}(z_n + b_{n-1}z_{n-1})\big) \neg \Sigma h.$$

Applying 1.2.iii repeatedly, we conclude that

$$\Gamma^1_{h,\mathbf{z}} = V(z_n + b_0z_0,\ z_n + b_1z_1,\ \dots,\ z_n + b_{n-1}z_{n-1}).$$

Finally, by intersecting this with

$$V\left(\frac{\partial h}{\partial \tilde{z}_0}\right) = V(a_{00}z_1z_2\dots z_n + \dots + a_{n0}z_0z_1\dots z_{n-1})$$

we obtain the desired result that $\lambda^0_{h,\tilde{\mathbf{z}}}(\mathbf{0}) = n$.

We shall obtain this same result, but by inductive methods, in Example 5.2.

Example 2.9. Let h be an analytic map in the variables x and y, and suppose that $h = P\prod Q_i^{\alpha_i}$, where P and $\prod Q_i^{\alpha_i}$ are relatively prime and $\alpha_i \geqslant 2$, i.e., h gives a non-reduced curve singularity. We wish to calculate the Lê numbers of h at the origin.

Let $z_0 = ax + by$, where $a \neq 0$, and let $z_1 = y$. Then,

$$\left[V\left(\frac{\partial h}{\partial z_1}\right)\right] = \left[V\left(\frac{\partial h}{\partial x}\left(\frac{-b}{a}\right) + \frac{\partial h}{\partial y}\right)\right] =$$

$$\sum\left[V\left(Q_i^{\alpha_i-1}\right)\right] + \left[V\left(\frac{\frac{\partial h}{\partial x}\left(\frac{-b}{a}\right) + \frac{\partial h}{\partial y}}{\prod Q_i^{\alpha_i-1}}\right)\right] = \Lambda^1_{h,\mathbf{z}} + \Gamma^1_{h,\mathbf{z}}.$$

Thus, whenever

$$\left[V\left(\frac{\frac{\partial h}{\partial x}\left(\frac{-b}{a}\right) + \frac{\partial h}{\partial y}}{\prod Q_i^{\alpha_i-1}}\right)\right]$$

has no components contained in the critical locus of h (an easy argument shows that this is the case for a generic choice of (a,b)), we have that

$$\lambda^1_{h,\mathbf{z}} = \sum(\alpha_i - 1)\left(V(Q_i)\cdot V(ax+by)\right)_{\mathbf{0}}$$

and

$$\lambda^0_{h,\mathbf{z}} = \left(V\left(\frac{\frac{\partial h}{\partial x}\left(\frac{-b}{a}\right) + \frac{\partial h}{\partial y}}{\prod Q_i^{\alpha_i-1}}\right)\cdot V\left(\frac{\partial h}{\partial x}\right)\right)_{\mathbf{0}},$$

where we have used that $V\left(\frac{\partial h}{\partial z_0}\right) = V\left(\frac{\partial h}{\partial x}\right)$.

Note that the formula

$$\lambda^1_{h,\mathbf{z}} = \sum(\alpha_i - 1)\left(V(Q_i)\cdot V(ax+by)\right)_{\mathbf{0}}$$

agrees with our earlier formula from the end of Remark 1.19,

$$\lambda^1_{h,\mathbf{z}} = \sum_\nu n_\nu \overset{\circ}{\mu}_\nu,$$

since we clearly have $n_\nu = (V(Q_i) \cdot V(ax+by))_{\mathbf{0}}$ and $\overset{\circ}{\mu}_\nu = \alpha_i - 1$.

Example 2.10. In this example, we show that – unlike the Milnor number – the Lê numbers in a family need **not** be upper-semicontinuous. While this may seem to be mildly disturbing at first, the example makes it clear what can happen; if a high-dimensional Lê number jumps up, then the lower-dimensional Lê numbers are free to jump up or down.

Let $f_t(x,y,z) = z^2 - y^3 - txy^2$. The coordinates (x,y,z) are prepolar at the origin for f_t for all t; we fix this set of coordinates and will suppress further reference to them.

For $t_0 \neq 0$, we are back in the situation of Example 2.2, with $a = 3$ and $b = 2$; therefore, $\lambda^0_{f_{t_0}}(\mathbf{0}) = 2$ and $\lambda^1_{f_{t_0}}(\mathbf{0}) = 1$.

On the other hand, the hypersurface defined by f_0 is a cross-product singularity; hence, $\lambda^0_{f_0}(\mathbf{0}) = 0$, and one trivially finds that $\lambda^1_{f_0}(\mathbf{0}) = 2$.

Thus, at $t = 0$, λ^1 jumps up to 2 from its generic value of 1; this allows the behavior of $\lambda^0_{f_t}(\mathbf{0})$ to be about as "bad" as possible; the generic value of λ^0 is 2, while the special value is 0.

The situation is not completely uncontrolled – as we shall see in Corollary 4.16, if we have a family f_t, then the tuple of Lê numbers

$$\left(\lambda^s_{f_t,\mathbf{z}}(\mathbf{0}), \lambda^{s-1}_{f_t,\mathbf{z}}(\mathbf{0}), \ldots, \lambda^0_{f_t,\mathbf{z}}(\mathbf{0})\right)$$

is lexigraphically upper-semicontinuous in the t variable.

Chapter 3. A HANDLE DECOMPOSITION OF THE MILNOR FIBRE

In this chapter, we give a handle decomposition of the Milnor fibre of an analytic function with a critical locus of arbitrary dimension. This decomposition is more refined than that obtained by iteratively applying Lê's attaching result (Theorem 0.9).

Throughout this chapter, $h : \mathcal{U} \to \mathbb{C}$ will be an analytic function on an open subset of $\mathbb{C}^{n+1}$. If $\mathbf{p} \in V(h)$, then we let $F_{h,\mathbf{p}}$ denote the Milnor fibre of h at $\mathbf{p}$ (more generally, for all $\mathbf{p} \in \mathcal{U}$, $F_{h,\mathbf{p}}$ denotes the Milnor fibre of $f - f(\mathbf{p})$ at $\mathbf{p}$).

Our main tool is a proposition based on the argument of Lê and Perron [**L-P**], which is the same argument that is used in [**Ti**], [**Va1**], and [**Va2**].

In what follows, if we have a pair of topological spaces $X \subseteq Y$, then we say that Y *is obtained from* X *by canceling* k m*-handles* provided that X has a handle decomposition in which the handles of highest index are of index m and Y is obtained from X by attaching k $(m+1)$-handles each of which cancels with an m-handle of X (in terms of Morse functions, this says that Y is obtained from X by passing through k critical points – all of index $m+1$ – and these critical points cancel with k critical points in X each of index m. See [**Mi1**] and [**Sm**].) In particular, this implies that the cohomology groups of X and Y are identical except in degree m, where we have $H^m(X) \cong \mathbb{Z}^k \oplus H^m(Y)$.

Proposition 3.1. *Let* $\mathbf{p} \in V(h)$*. If* $V(z_0 - p_0)$ *is a prepolar slice for* h *at* $\mathbf{p}$*, and* $n \neq 2$*, then the Milnor fibre of* h *at* $\mathbf{p}$ *is obtained – up to diffeomorphism – from the product of a disk with the Milnor fibre of* $h_{|_{V(z_0-p_0)}}$ *at* $\mathbf{p}$ *by first attaching* $\gamma^1_{h,\mathbf{z}}(\mathbf{p})$ n*-handles, which cancel against* $\gamma^1_{h,\mathbf{z}}(\mathbf{p})$ $(n-1)$*-handles of* $\overset{\circ}{\mathbb{D}} \times F_{h_{|_{V(z_0-p_0)}},\mathbf{p}}$*, and then attaching* $\lambda^0_{h,\mathbf{z}}(\mathbf{p})$ *more* n*-handles.*

If $n = 2$*, we have the same conclusion except that the canceling is only up to homotopy.*

Proof. Essentially, this is Proposition 4.2 of [**Mas7**], except that here we have weakened the hypothesis on the genericity of the hyperplane slice. We use the coodinates $(z_0, \ldots, z_n)$ for our ambient space. Clearly, it suffices to prove the claim for $\mathbf{p} = \mathbf{0}$. We will follow the argument of [**Va2**].

By Proposition C.6.iii, we may use neighborhoods of the form $\mathbb{D}_\delta \times B_\epsilon$, $0 < \delta \ll \epsilon$, to define the Milnor fibre of h at the origin up to homotopy. Choose ϵ and δ such that B_ϵ is a Milnor ball for $h_{|_{V(z_0)}}$ and $\Gamma^1_{h,z_0} \cap (\mathbb{D}_\delta \times \partial B_\epsilon) = \emptyset$ – we may accomplish this last equality since Theorem 1.26 implies that $\dim_{\mathbf{0}}(\Gamma^1_{h,z_0} \cap V(z_0)) \leqslant 0$. Choose η such that $(B_\epsilon, \mathbb{D}_\eta)$ is a Milnor pair for $h_{|_{V(z_0)}}$ at the origin (see Appendix C.5). Let $\Psi := (h, z_0)$ and let Δ denote $\Psi(\Gamma^1_{h,z_0})$ in $\mathbb{C}^2$ (Δ is the *Cerf diagram* of h with respect to z_0). Δ is given its fitting ideal structure, which is possibly non-reduced (see [**Loo**]).

Choose $(\alpha, \beta) \in \mathbb{C}^2 - \Delta$ sufficiently small and let $h_\beta := h_{|_{V(z_0-\beta)}}$. Let $\mathbb{D}$ be a small disc in $\mathbb{D}_\eta \times \{\beta\}$ centered at (α, β), and let A be the region in $\mathbb{D}_\eta \times \{\beta\}$ formed by joining to $\mathbb{D}$ small discs centered at each of the points of $\Delta \cap V(z_0 - \beta)$, where the joining is via thickened paths which avoid $(0, \beta)$ (see Figure 3.2). Note that, counted with multiplicity, there are $(\Gamma^1_{h,z_0} \cdot V(z_0))_{\mathbf{0}}$ points in $\Delta \cap V(z_0 - \beta)$.

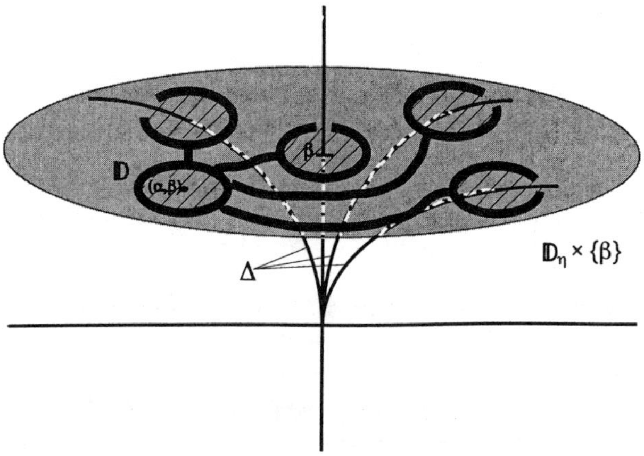

Figure 3.2. The Cerf diagram, and the sets A and C

Then, the argument of Lê and Perron [**L-P**] and Vannier [**Va1**] shows that the Milnor fibre of h is obtained from $W := h_\beta^{-1}(A) \cap (\{\beta\} \times B_\epsilon)$ by attaching $\lambda^0_{h,z_0}(\mathbf{0}) = \left(\Gamma^1_{h,z_0} \cdot V\left(\frac{\partial h}{\partial z_0}\right)\right)_{\mathbf{0}}$ n-handles – this part of their argument does not depend on the dimension of Σh. (Though, as we are not assuming that the polar curve is reduced, the details of the generalization of the isotopy result used by Lê and Perron need to be checked – this is done in [**Ti**].)

The problem is thus to show that W is obtained from the product of a disc with the Milnor fibre of $h_{|_{V(z_0)}}$ by canceling $\gamma^1_{h,z_0}(\mathbf{0}) = (\Gamma^1_{h,z_0} \cdot V(z_0))_{\mathbf{0}}$ $(n-1)$-handles, even in the case where the critical locus of h does not have dimension one.

Let $C \subseteq \mathbb{D}_\eta \times \{\beta\}$ be formed by taking a small disc around $(0, \beta)$ and joining it to $\mathbb{D}$ with a thickened path (see Figure 3.2). Let $U := h_\beta^{-1}(C) \cap (\{\beta\} \times B_\epsilon)$. Then, as $A \cup C$ is a strong deformation retract of $\mathbb{D}_\eta$, $U \cup V$ is homotopy-equivalent to $h_\beta^{-1}(\mathbb{D}_\eta) \cap (\{\beta\} \times B_\epsilon)$, which is in turn diffeomorphic to a real $2n$-ball, B^{2n}. Moreover, $U \cap W$ is diffeomorphic to $h_\beta^{-1}(\mathbb{D})$ which is diffeomorphic to the product of a disc and the Milnor fibre of $h_{|_{V(z_0)}}$.

Now, $U \cup W$ is obtained from U by attaching $\gamma^1_{h,z_0}(\mathbf{0})$ n-handles (that is, one handle of index n for each point of $\Delta \cap V(z_0 - \beta)$ in A, counted with multiplicity). As $U \cup W$ is contractible, this implies that U has the homotopy-type of a bouquet of $\gamma^1_{h,z_0}(\mathbf{0})$ $(n-1)$-spheres.

Consider the Mayer-Vietoris sequence of U and W. As $U \cup W$ is contractible, we have that $H_k(U \cap W) \cong H_k(U) \oplus H_k(W)$ for all $k \geqslant 1$, where we know that U has the homotopy-type of a bouquet of $(n-1)$-spheres, and $U \cap W$ is diffeomorphic to the product of a disc and the Milnor fibre of $h_{|_{V(z_0)}}$, which has the homotopy-type of an $(n-1)$-dimensional CW complex. Thus, we see that W has the homotopy-type of an $(n-1)$-dimensional CW complex.

But now, since $h_\beta^{-1}(\mathbb{D})$ is diffeomorphic to the product of a disc with the Milnor fibre of $h_{|_{V(z_0)}}$, and since W itself is obtained by attaching $\gamma^1_{h,z_0}(\mathbf{0})$ n-handles to $h_\beta^{-1}(\mathbb{D})$, we see that up to homotopy these $\gamma^1_{h,z_0}(\mathbf{0})$ n-handles must be canceling $\gamma^1_{h,z_0}(\mathbf{0})$ $(n-1)$-handles. But, if $n \geqslant 3$, then the real dimension of W is greater than or equal to 6, and so this handle cancellation is up to diffeomorphism [**Mi1**], [**Sm**].

Finally, if $n = 1$, then – by the classification of surfaces – we have that the handle cancellation is up to diffeomorphism. □

By an inductive application of Proposition 3.1 to each hyperplane slice, we arrive at Theorem 4.3 of [**Mas7**]. This theorem describes a handle decomposition of the Milnor fibre.

Theorem 3.3. *Let $\mathcal{U}$ be an open subset of $\mathbb{C}^{n+1}$, let $h : \mathcal{U} \to \mathbb{C}$ be an analytic map, let $\mathbf{p} \in V(h)$, let s denote $\dim_{\mathbf{p}}\Sigma h$, and let $\mathbf{z} = (z_0, \dots, z_{s-1})$ be prepolar for h at $\mathbf{p}$.*

If $s \leqslant n-2$, then $F_{h,\mathbf{p}}$ is obtained up to diffeomorphism from a real $2n$-ball by successively attaching $\lambda^{n-k}_{h,\mathbf{z}}(\mathbf{p})$ k-handles, where $n - s \leqslant k \leqslant n$;

if $s = n-1$, then $F_{h,\mathbf{p}}$ is obtained up to diffeomorphism from a real $2n$-manifold with the homotopy-type of a bouquet of $\lambda^{n-1}_{h,\mathbf{z}}(\mathbf{p})$ circles by successively attaching $\lambda^{n-k}_{h,\mathbf{z}}(\mathbf{p})$ k-handles, where $2 \leqslant k \leqslant n$

Hence, the reduced Euler characteristic of the Milnor fibre of h at $\mathbf{p}$ is given by

$$\widetilde{\chi}(F_{h,\mathbf{p}}) = \sum_{i=0}^{s} (-1)^{n-i} \lambda^i_{h,\mathbf{z}}(\mathbf{p})$$

and the reduced Betti numbers, $\tilde{b}_i(F_{h,\mathbf{p}})$, satisfy Morse inequalities with respect to the Lê numbers, i.e., for all k with $n - s \leqslant k \leqslant n$,

$$(-1)^k \sum_{i=n-s}^{k} (-1)^i \tilde{b}_i(F_{h,\mathbf{p}}) \leqslant (-1)^k \sum_{i=n-s}^{k} (-1)^i \lambda^{n-i}_{h,\mathbf{z}}(\mathbf{p})$$

and

$$(-1)^k \sum_{i=k}^{n} (-1)^i \tilde{b}_i(F_{h,\mathbf{p}}) \leqslant (-1)^k \sum_{i=k}^{n} (-1)^i \lambda^{n-i}_{h,\mathbf{z}}(\mathbf{p})$$

Proof. By induction on s. When $s = 0$, the result follows from Milnor's work. Now, assume the result for $s - 1$. As before, we consider only the case where $\mathbf{p} = \mathbf{0}$.

As $V(z_0)$ is prepolar, we may apply Proposition 3.1 to conclude that the Milnor fibre of h at $\mathbf{0}$ is obtained from the product of a disk with the Milnor fibre of $h_{|_{V(z_0)}}$ at $\mathbf{0}$ by first attaching $\gamma^1_{h,\mathbf{z}}(\mathbf{0})$ n-handles, which cancel against $\gamma^1_{h,\mathbf{z}}(\mathbf{0})$ $(n-1)$-handles of $\overset{\circ}{\mathbb{D}} \times F_{h_{|_{V(z_0)}},\mathbf{0}}$, and then attaching $\lambda^0_{h,\mathbf{z}}(\mathbf{0})$ more n-handles – up to diffeomorphism if $n \neq 2$ and up to homotopy otherwise.

But, $\tilde{\mathbf{z}} = (z_1, \dots, z_{s-1})$ is prepolar for $h_{|_{V(z_0)}}$ at the origin and, by our inductive hypothesis, the Milnor fibre of $h_{|_{V(z_0)}}$ at the origin is obtained by successively attaching $\lambda^{n-1-k}_{h_{|_{V(z_0)}},\tilde{\mathbf{z}}}(\mathbf{0})$ k-handles for $(n-1) - (s-1) \leqslant k \leqslant n-1$. By Proposition 1.18, if $n - 1 - k \neq 0$, then

$\lambda^{n-1-k}_{h_{|V(z_0)}, \tilde{\mathbf{z}}}(\mathbf{0}) = \lambda^{n-k}_{h,\mathbf{z}}(\mathbf{0})$, and $\lambda^{0}_{h_{|V(z_0)}, \tilde{\mathbf{z}}}(\mathbf{0}) = \gamma^1_{h,\mathbf{z}}(\mathbf{0}) + \lambda^1_{h,\mathbf{z}}(\mathbf{0})$. The conclusions concerning handles follow.

The Morse inequalities follow formally. □

Siersma's main result in [**Si2**] allows us to improve this result in a special case.

Corollary 3.4 *Let $\mathcal{U}$ be an open subset of $\mathbb{C}^{n+1}$, let $h : \mathcal{U} \to \mathbb{C}$ be an analytic map, let $\mathbf{p} \in V(h)$, and let s denote $\dim_{\mathbf{p}}\Sigma h$. Suppose that $(z_0, \dots, z_{s-1})$ is prepolar for h at $\mathbf{p}$, and suppose that $\lambda^s_{h,\mathbf{z}}(\mathbf{p}) = 1$.*

Then, either

$$\lambda^0_{h,\mathbf{z}}(\mathbf{p}) = \lambda^1_{h,\mathbf{z}}(\mathbf{p}) = \cdots = \lambda^{s-1}_{h,\mathbf{z}}(\mathbf{p}) = 0,$$

or the single $(n-s)$-handle in the handle decomposition of the previous theorem gets canceled – up to homotopy – by the attaching of one of the $\lambda^{s-1}_{h,\mathbf{z}}(\mathbf{p})$ $(n-s+1)$-handles.

Proof. The proof is exactly the inductive proof of 3.3, except in the first step one applies the result of [**Si**].

The function $h_{|V(z_0-p_0,\dots,z_{s-2}-p_{s-2})}$ has a one-dimensional critical locus at $\mathbf{p}$. Using $(z_{s-1}, \dots, z_n)$ as coordinates for $V(z_0 - p_0, \dots, z_{s-2} - p_{s-2})$, we conclude from 1.18 that

$$\lambda^1_{h_{|V(z_0-p_0,\dots,z_{s-2}-p_{s-2})}}(\mathbf{p}) = \lambda^s_{h,\mathbf{z}}(\mathbf{p}) = 1$$

and

$$\lambda^0_{h_{|V(z_0-p_0,\dots,z_{s-2}-p_{s-2})}}(\mathbf{p}) = \gamma^{s-1}_{h,\mathbf{z}}(\mathbf{p}) + \lambda^{s-1}_{h,\mathbf{z}}(\mathbf{p}).$$

Hence, $h_{|V(z_0-p_0,\dots,z_{s-2}-p_{s-2})}$ is an isolated line singularity in the sense of Siersma [**Si2**], and his result is that either $\lambda^0_{h_{|V(z_0-p_0,\dots,z_{s-2}-p_{s-2})}}(\mathbf{p}) = 0$ or that one only has homology in middle dimension, i.e., the one possible $(n-s)$-handle must get canceled up to homotopy.

The equality

$$\lambda^0_{h_{|V(z_0-p_0,\dots,z_{s-2}-p_{s-2})}}(\mathbf{p}) = \gamma^{s-1}_{h,\mathbf{z}}(\mathbf{p}) + \lambda^{s-1}_{h,\mathbf{z}}(\mathbf{p}) = 0$$

corresponds to the case where

$$\lambda^0_{h,\mathbf{z}}(\mathbf{p}) = \lambda^1_{h,\mathbf{z}}(\mathbf{p}) = \cdots = \lambda^{s-1}_{h,\mathbf{z}}(\mathbf{p}) = 0,$$

since $\lambda^{s-1}_{h,\mathbf{z}}(\mathbf{p}) = 0$ implies that $\mathbf{p} \notin \Gamma^{s-1}_{h,\mathbf{z}}$ and, by 1.13, $\Gamma^{s-1}_{h,\mathbf{z}} \cap \Sigma h = \bigcup_{i \leqslant s-2} \Lambda^i_{h,\mathbf{z}}$. Hence, if $\mathbf{p} \notin \Gamma^{s-1}_{h,\mathbf{z}}$, then it follows that all the lower Lê numbers are also zero. □

One might question whether the above result can possibly be correct. What about the case where $\lambda^s_{h,\mathbf{z}}(\mathbf{p}) = 1$, $\lambda^{s-1}_{h,\mathbf{z}}(\mathbf{p}) = 0$, and one of the lower λ's is not zero? In such a case, there would be no way to cancel the $(n-s)$-handle. Note, however, that 1.30 rules out the possibility of the existence of this case.

Chapter 4. GENERALIZED LÊ-IOMDINE FORMULAS

In this chapter, we generalize the formula of Lê and Iomdine (see [**Lê4**], [**Io**], [**Mas8**], [**Mas11**], [**M-S**], and [**Si3**]) to functions with an arbitrary-dimensional critical locus; on the level of cycles, this will be a special case of the Lê-Iomdine-Vogel formulas from I.3.4.

The Lê-Iomdine formulas that we present here tell us how the Lê numbers of a hypersurface singularity are related to the Lê numbers of a certain "sequence of hypersurface singularities" – a sequence which "approaches" the original singularity, but such that the critical loci of the terms in the sequence are of one dimension smaller than the original. These formulas have a large number of applications.

The statement that we give here has an improvement in a certain bound over what we proved in [**Mas14**]; in the case of a one-dimensional critical locus, this is the form of the statement as it appears in [**M-S**] and [**Si3**]. To give this improved bound, we need a definition. Throughout this chapter, we concentrate our attention at the origin.

We are about to introduce the *polar ratios.* These quantities first appeared in Proposition 3.5.2 of [**Te4**], and the fact that they are invariants of the "equi-singularity type" appears in Theorem 6 of [**Te 5**]. In [**Te 5**], Teissier investigates a number of questions of "Iomdine/Sebastiani-Thom type" in the case of isolated hypersurface singularities.

Definition 4.1. Suppose that Γ^1_{h,z_0} is purely one-dimensional at the origin. Let η be an irreducible component of Γ^1_{h,z_0} (with its reduced structure) such that $\eta \cap V(z_0)$ is zero-dimensional at the origin.

Then, the *polar ratio* of η (for h at $\mathbf{0}$ with respect to z_0) is

$$\frac{(\eta \cdot V(h))_{\mathbf{0}}}{(\eta \cdot V(z_0))_{\mathbf{0}}} = \frac{\left(\eta \cdot V\left(\frac{\partial h}{\partial z_0}\right)\right)_{\mathbf{0}} + (\eta \cdot V(z_0))_{\mathbf{0}}}{(\eta \cdot V(z_0))_{\mathbf{0}}} = \frac{\left(\eta \cdot V\left(\frac{\partial h}{\partial z_0}\right)\right)_{\mathbf{0}}}{(\eta \cdot V(z_0))_{\mathbf{0}}} + 1.$$

(The equalities follow from our proof of Proposition 1.20.)

If $\eta \cap V(z_0)$ is not zero-dimensional at the origin (i.e., if $\eta \subseteq V(z_0)$), then we say that the polar ratio of η equals 1.

A *polar ratio* (of h at $\mathbf{0}$ with respect to z_0) is any one of the polar ratios of any component of the polar curve (if the polar curve is empty at $\mathbf{0}$, we say that *the maximum polar ratio equals* 1).

If $\mathbf{0} \in \Sigma$ and all of the Lê cycles have the correct dimension at $\mathbf{0}$ (in particular, if the Lê numbers are defined at $\mathbf{0}$), this definition is (up to adding 1) a particular case of I.3.2.

Let $\mathbf{f} := \left(\frac{\partial h}{\partial z_0}, \dots, \frac{\partial h}{\partial z_n}\right)$ and $g := z_0$. Then, $\Gamma^1_{h,z_0} = \Pi^1_{\mathbf{f}}$, and there is an equality of sets $|\Pi^1_{\mathbf{f}}| = |\widehat{\Pi}^1_{\mathbf{f}}|$ by the Dimensionality Lemma (I.2.5). Let η be an irreducible component of Γ^1_{h,z_0}. Looking at I.3.2 (and using $d := n+1$ and $k := n$), we see that the polar ratio of η (for h) is precisely the gap ratio of η (for $\mathbf{f}$) plus one.

Remark 4.2. The case where h is a homogeneous polynomial of degree d is particularly easy to analyze. Provided that Γ^1_{h,z_0} is one-dimensional at the origin, each component of the polar curve is a line, and so the polar ratios are all 1 or d.

We are going to consider functions of the form $h + az_0^j$, where a is a non-zero complex number and j is suitably large. Clearly, however, the coordinate z_0 is extremely non-generic for $h+az_0^j$. Hence, if we are using the coordinates $(z_0, z_1, \ldots, z_n)$ for h, we use the coordinates $(z_1, z_2, \ldots, z_n, z_0)$ for $h + az_0^j$. The purpose of this "rotation" of the coordinate system is merely to get the z_0 coordinate out of the way. Normally, if h has an s-dimensional critical locus at the origin, then $h + az_0^j$ will have an $(s-1)$-dimensional critical locus at the origin; thus, it is only the choice of the coordinates $z_0, \ldots, z_{s-1}$ that we care about for h, and the coordinates $z_1, \ldots, z_{s-1}$ for $h + az_0^j$.

Lemma 4.3. *Let $j \geqslant 2$. Let $h : (\mathcal{U}, \mathbf{0}) \to (\mathbb{C}, 0)$ be an analytic function, let s denote $\dim_{\mathbf{0}} \Sigma h$, and assume that $s \geqslant 1$. Let $\mathbf{z} = (z_0, \ldots, z_n)$ be a linear choice of coordinates such that $\lambda^i_{h,\mathbf{z}}(\mathbf{0})$ is defined for all $i \leqslant s$. Let a be a non-zero complex number, and use the coordinates $\tilde{\mathbf{z}} = (z_1, \ldots, z_n, z_0)$ for $h + az_0^j$.*

If j is greater than or equal to the maximum polar ratio for h then, for all but a finite number of complex a,

i) $\dim_{\mathbf{0}} \Gamma^1_{h,\mathbf{z}} \cap V\left(\frac{\partial h}{\partial z_0} + jaz_0^{j-1}\right) = 0$;

ii) $\lambda^0_{h,\mathbf{z}}(\mathbf{0}) = \left(\Gamma^1_{h,\mathbf{z}} \cdot V\left(\frac{\partial h}{\partial z_0} + jaz_0^{j-1}\right)\right)_{\mathbf{0}}$;

iii) $\Sigma(h + az_0^j)$ *is $(s-1)$-dimensional at the origin and equal to $\Sigma h \cap V(z_0)$ as germs of sets at $\mathbf{0}$;*

iv) if $i \geqslant 1$, then we have an equality of cycles

$$\Gamma^i_{h+az_0^j,\tilde{\mathbf{z}}} = \Gamma^{i+1}_{h,\mathbf{z}} \cdot V\left(\frac{\partial h}{\partial z_0} + jaz_0^{j-1}\right)$$

near the origin;

v) if w is a variable disjoint from those of h, then

$$\left(\Gamma^1_{h+aw^j, w-z_0} \cdot V(h + aw^j)\right)_{\mathbf{0}} = j\lambda^0_{h,\mathbf{z}}(\mathbf{0}).$$

Moreover, if we have the strict inequality that j is greater than the maximum polar ratio for h, then the above equalities hold for all non-zero a; in particular, this is the case if $j \geqslant 2 + \lambda^0_{h,\mathbf{z}}(\mathbf{0})$.

Proof. Parts i), ii), iii), and iv) follow immediately by applying Lemma I.3.3, where the X, $\mathbf{f}$, g, a, j, and $\mathbf{p}$ of Lemma I.3.3 are replaced by $\mathcal{U}$, $\left(\frac{\partial h}{\partial z_0}, \ldots, \frac{\partial h}{\partial z_n}\right)$, z_0, $j \cdot a$, $j-1$, and $\mathbf{0}$, respectively. We will prove v) by applying Lemma I.3.6.

Let $\mathbf{f} := \left(\frac{\partial h}{\partial z_0}, \dots, \frac{\partial h}{\partial z_n}\right)$, and let $\mathbf{g} := \left(w^{j-1}, \frac{\partial h}{\partial z_1}, \dots, \frac{\partial h}{\partial z_n}, \frac{\partial h}{\partial z_0} + jaw^{j-1}\right)$. Then, I.3.6 tells us that the Vogel sets of $\mathbf{g}$ have correct dimension at $\mathbf{0}$, for all $i \leqslant n+1$, $\mathbb{C} \times \widehat{\Pi}^i_{\mathbf{f}}$ properly intersects $V\left(\frac{\partial h}{\partial z_0} + jaw^{j-1}\right)$, and there is an equality of germs of cycles at $\mathbf{0}$ given by

$$(\dagger) \qquad \widehat{\Pi}^i_{\mathbf{g}} = \left(\mathbb{C} \times \widehat{\Pi}^i_{\mathbf{f}}\right) \cdot V\left(\frac{\partial h}{\partial z_0} + jaw^{j-1}\right).$$

As the Vogel sets of $\mathbf{f}$ and $\mathbf{g}$ have correct dimension at $\mathbf{0}$, we may use the Dimensionality Lemma and Proposition I.2.9 to conclude that we may replace each of the inductive gap varieties in (†) by the ordinary gap varieties.

Now, let $L = w - z_0$ and use $(L, z_0, \dots, z_n)$ as coordinates for $\mathbb{C} \times \mathcal{U}$. Then,

$$\Gamma^1_{h+aw^j, w-z_0} = \Gamma^1_{h+a(L+z_0)^j, L} =$$

$$V\left(\frac{\partial h}{\partial z_0} + ja(L+z_0)^{j-1}, \frac{\partial h}{\partial z_1}, \dots, \frac{\partial h}{\partial z_n}\right) \neg\, \Sigma(h + a(L+z_0)^j)$$

which, back in $(w, z_0, \dots, z_n)$ coordinates, is equal to

$$V\left(\frac{\partial h}{\partial z_1}, \dots, \frac{\partial h}{\partial z_n}, \frac{\partial h}{\partial z_0} + jaw^{j-1}\right) \neg\, \Sigma(h + aw^j).$$

Thus, we see that $\Gamma^1_{h+aw^j, w-z_0} = \Pi^1_{\mathbf{g}}$, and so (†) tells us that

$$(*) \qquad \Gamma^1_{h+aw^j, w-z_0} = \left(\mathbb{C} \times \Gamma^1_{h,\mathbf{z}}\right) \cdot V\left(\frac{\partial h}{\partial z_0} + jaw^{j-1}\right).$$

Now, let us assume for the moment that $\gamma^1_{h+aw^j, w-z_0}(\mathbf{0})$ exists. Then, we may apply Proposition I.1.20 to conclude that

$$\left(\Gamma^1_{h+aw^j, w-z_0} \cdot V(h + aw^j)\right)_{\mathbf{0}} =$$

$$\left(\Gamma^1_{h+aw^j, w-z_0} \cdot V\left(\frac{\partial(h + aw^j)}{\partial(w - z_0)}\right)\right)_{\mathbf{0}} + \left(\Gamma^1_{h+aw^j, w-z_0} \cdot V(w - z_0)\right)_{\mathbf{0}} =$$

$$\left(\Gamma^1_{h+aw^j, w-z_0} \cdot V(jaw^{j-1})\right)_{\mathbf{0}} + \left(\Gamma^1_{h+aw^j, w-z_0} \cdot V(w - z_0)\right)_{\mathbf{0}}.$$

By (∗), this is equal to

$$(j-1)\lambda^0_{h,\mathbf{z}}(\mathbf{0}) + \left(V(w - z_0) \cdot \left(\mathbb{C} \times \left(\Gamma^1_{h,\mathbf{z}} \cdot V\left(\frac{\partial h}{\partial z_0} + jaz_0^{j-1}\right)\right)\right)\right)_{\mathbf{0}},$$

which, by i) and ii), is equal to $(j-1)\lambda^0_{h,\mathbf{z}}(\mathbf{0}) + \lambda^0_{h,\mathbf{z}}(\mathbf{0}) = j\lambda^0_{h,\mathbf{z}}(\mathbf{0})$.

Thus, we would be finished if we could show that $\gamma^1_{h+aw^j, w-z_0}(\mathbf{0})$ exists, but, as we saw above, $\Gamma^1_{h+aw^j, w-z_0} \cap V(w - z_0)$ is zero-dimensional at the origin. □

Our next result will be to obtain the *generalized Lê-Iomdine formulas*; these formulas are a stunningly useful tool for reducing questions on general hypersurface singularities to the much easier case of isolated hypersurface singularities. The formulas tell how the Lê numbers of h change when a large power of one of the variables is added. By 4.3.iii, this modification of the function h will have a critical locus of dimension one smaller than that of h itself. Proceeding inductively, one arrives at the case of an isolated singularity.

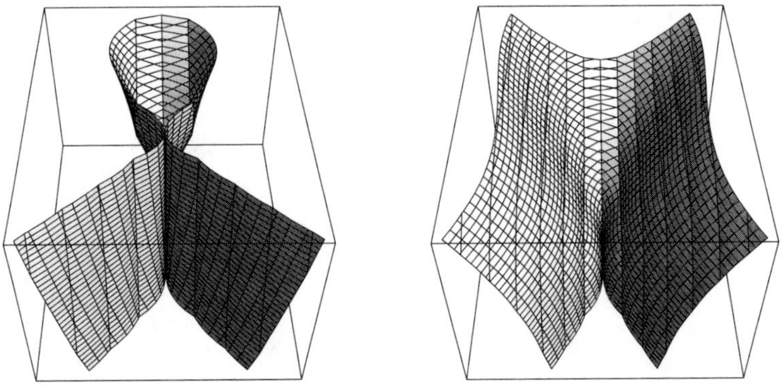

Figure 4.4. The effect of adding a large power of a variable

Theorem 4.5 (Lê-Iomdine formulas). *Let $j \geqslant 2$, let $h : (\mathcal{U}, \mathbf{0}) \to (\mathbb{C}, 0)$ be an analytic function, let s denote $\dim_{\mathbf{0}} \Sigma h$, and assume that $s \geqslant 1$. Let $\mathbf{z} = (z_0, \ldots, z_n)$ be a linear choice of coordinates such that $\lambda^i_{h,\mathbf{z}}(\mathbf{0})$ is defined for all $i \leqslant s$. Let a be a non-zero complex number, and use the coordinates $\tilde{\mathbf{z}} = (z_1, \ldots, z_n, z_0)$ for $h + az_0^j$.*

If j is greater than or equal to the maximum polar ratio for h then, for all but a finite number of complex a, $\Sigma(h + az_0^j) = \Sigma h \cap V(z_0)$ as germs of sets at $\mathbf{0}$, $\dim_{\mathbf{0}} \Sigma(h + az_0^j) = s - 1$, $\lambda^i_{h+az_0^j, \tilde{\mathbf{z}}}(\mathbf{0})$ exists for all $i \leqslant s - 1$, and

$$\lambda^0_{h+az_0^j, \tilde{\mathbf{z}}}(\mathbf{0}) = \lambda^0_{h,\mathbf{z}}(\mathbf{0}) + (j-1)\lambda^1_{h,\mathbf{z}}(\mathbf{0}),$$

and, for $1 \leqslant i \leqslant s - 1$,

$$\lambda^i_{h+az_0^j, \tilde{\mathbf{z}}}(\mathbf{0}) = (j-1)\lambda^{i+1}_{h,\mathbf{z}}(\mathbf{0}).$$

Moreover, if we have the strict inequality that j is greater than the maximum polar ratio for h, then the above equalities hold for all non-zero a; in particular, this is the case if $j \geqslant 2 + \lambda^0_{h,\mathbf{z}}(\mathbf{0})$.

Proof. This follows immediately from the Lê-Iomdine-Vogel formulas (I.3.4) by letting X, $\mathbf{f}$, g, a, j, and $\mathbf{p}$ of I.3.4 be replaced by $\mathcal{U}$, $\left(\frac{\partial h}{\partial z_0}, \ldots, \frac{\partial h}{\partial z_n}\right)$, z_0, $j \cdot a$, $j - 1$, and $\mathbf{0}$, respectively. $\square$

By applying the Lê-Iomdine formulas inductively, we immediately conclude

Corollary 4.6. *Let $h : (\mathcal{U}, \mathbf{0}) \to (\mathbb{C}, 0)$ be an analytic function, let s denote $\dim_{\mathbf{0}} \Sigma h$, and let $\mathbf{z} = (z_0, \dots, z_n)$ be a linear choice of coordinates such that $\lambda^i_{h,\mathbf{z}}(\mathbf{0})$ is defined for all $i \leqslant s$. Then, for $0 \ll j_0 \ll j_1 \ll \cdots \ll j_{s-1}$,*

$$h + z_0^{j_0} + z_1^{j_1} + \cdots + z_{s-1}^{j_{s-1}}$$

has an isolated singularity at the origin, and its Milnor number is given by

$$\mu(h + z_0^{j_0} + z_1^{j_1} + \cdots + z_{s-1}^{j_{s-1}}) = \sum_{i=0}^{s} \left(\lambda^i_{h,\mathbf{z}}(\mathbf{0}) \prod_{k=0}^{i-1} (j_k - 1) \right) =$$

$$\lambda^0_{h,\mathbf{z}}(\mathbf{0}) + (j_0 - 1)\lambda^1_{h,\mathbf{z}}(\mathbf{0}) + (j_1 - 1)(j_0 - 1)\lambda^2_{h,\mathbf{z}}(\mathbf{0}) + \dots$$
$$+ (j_{s-1} - 1) \dots (j_1 - 1)(j_0 - 1)\lambda^s_{h,\mathbf{z}}(\mathbf{0}).$$

As another quick application of the Lê-Iomdine formulas, we have the following Plücker formula.

Corollary 4.7. *Let h be a homogeneous polynomial of degree d in $n+1$ variables, let $s = \dim_{\mathbf{0}} \Sigma h$, and suppose that $\lambda^i_{h,\mathbf{z}}(\mathbf{0})$ exists for all $i \leqslant s$. Then,*

$$\sum_{i=0}^{s} (d-1)^i \lambda^i_{h,\mathbf{z}}(\mathbf{0}) = (d-1)^{n+1}.$$

Proof. By Remark 4.2, the maximum polar ratio is d. By an inductive application of the Lê-Iomdine formulas, we arrive at a function,

$$f := h + a_0 z_0^d + a_1 z_1^d + \cdots + a_{s-1} z_{s-1}^d,$$

with an isolated singularity at the origin and such that the Milnor number of $f = \lambda^0_f(\mathbf{0}) = \sum_{i=0}^{s} (d-1)^i \lambda^i_{h,\mathbf{z}}(\mathbf{0})$. Now, by [**M-O**], this Milnor number is precisely $(d-1)^{n+1}$. □

In Chapter 9, we will see that the Lê cycles are actually Segre cycles. Knowing this, the above Plücker formula is a special case of a much more general result of Van Gastel [**Gas1**, 1.2.c].

Remark 4.8. In general, Corollary 4.7 makes it slightly easier to calculate the Euler characteristic of the Milnor fibre of a homogeneous polynomial. In the case of a one-dimensional critical locus, 4.7 tells us that, if we know the degree and λ^1, then we know λ^0 and, hence, the Euler characteristic (see also [**M-S**] and [**Si3**]).

Being able to calculate the Euler characteristic of the Milnor fibre of a homogeneous singularity implies in many cases that we can also calculate the Euler characteristic of the Milnor fibre of a weighted-homogeneous polynomial.

To see this, let $f : \mathbb{C}^{n+1} \to \mathbb{C}$ be a weighted homogeneous polynomial. Then, there exist positive integers $r_0, \dots, r_n$ such that, if $\pi : \mathbb{C}^{n+1} \to \mathbb{C}^{n+1}$ is given by

$$\pi(z_0, \dots, z_n) = (z_0^{r_0}, \dots, z_n^{r_n}),$$

then $h := f \circ \pi$ is homogeneous.

We may define, up to diffeomorphism, the Milnor fibre of h at the origin by using the "weighted" ball

$$\overset{\circ}{B}{}'_\epsilon := \{(z_0, \dots, z_n) \mid |z_0|^{2r_0} + \cdots + |z_n|^{2r_n} < \epsilon^2\};$$

that is, for $0 \ll |\xi| \ll \epsilon \ll 1$, we define $F_{h,\mathbf{0}}$ to be $\overset{\circ}{B}{}'_\epsilon \cap h^{-1}(\xi)$. That this yields the same diffeomorphism-type as the standard ball is well-known; see, for instance, [**G-M2**, II.2]. Clearly, the restriction of π induces a map from $F_{h,\mathbf{0}}$ to the Milnor fibre (using standard balls), $F_{f,\mathbf{0}}$, of f at the origin; denote this map by $\tilde{\pi} : F_{h,\mathbf{0}} \to F_{f,\mathbf{0}}$.

Now, consider the stratification of $\mathbb{C}^{n+1}$ derived from the hyperplane arrangement given by all the coordinate hyperplanes. That is, let I denote the indexing set $\{0, \dots, n\}$, and for each $J \subseteq I$, let w_J denote the intersection of hyperplanes (a.k.a. the *flat*) given by

$$w_J := V(z_j \mid j \in J),$$

and let S_J denote the Whitney stratum

$$S_J := w_J - \bigcup_{J \subsetneq K} w_K.$$

Near a given point, a representative of the Milnor fibre of a given function will transversely intersect all strata of any fixed Whitney stratification; this follows from the fact that, locally, the stratified critical values of an analytic function are isolated. Thus, the stratification $\{S_J\}$ determines Whitney stratifications $\{S_J \cap F_{h,\mathbf{0}}\}$ and $\{S_J \cap F_{f,\mathbf{0}}\}$ of $F_{h,\mathbf{0}}$ and $F_{f,\mathbf{0}}$, respectively, and with these stratifications, $\tilde{\pi}$ becomes a stratified map. Moreover, the restriction of $\tilde{\pi}$ to a map from $S_J \cap F_{h,\mathbf{0}}$ to $S_J \cap F_{f,\mathbf{0}}$ is a topological covering map with fibre equal to $\prod_{i \notin J} r_i$ points.

Hence,

$$\chi(F_{h,\mathbf{0}}) = \sum_J \chi(S_J \cap F_{h,\mathbf{0}}) = \sum_J \left(\prod_{i \notin J} r_i\right) \chi(S_J \cap F_{f,\mathbf{0}}).$$

Some elementary combinatorics shows that this last quantity is equal to

$$\sum_J c_J \left(\sum_{J \subseteq K} \chi(S_K \cap F_{f,\mathbf{0}})\right),$$

where

$$c_J := (-1)^{|J|} \sum_{L \subseteq J} \left((-1)^{|L|} \prod_{i \notin L} r_i \right).$$

The advantage of this last form is that

$$\sum_{J \subseteq K} \chi(S_K \cap F_{f,\mathbf{0}}) = \chi(F_{f_{|w_J}, \mathbf{0}}).$$

Therefore, we have that

$$\chi(F_{h,\mathbf{0}}) = \sum_J c_J \chi(F_{f_{|w_J}, \mathbf{0}}),$$

where c_J is as above.

It follows that

$$\chi(F_{h,\mathbf{0}}) = (r_0 \dots r_n)\chi(F_{f,\mathbf{0}}) + \sum_{J \neq \emptyset} c_J \chi(F_{f_{|w_J}, \mathbf{0}}),$$

and so, finally, we arrive at the formula

$$\chi(F_{f,\mathbf{0}}) = \frac{\chi(F_{h,\mathbf{0}}) - \sum_{J \neq \emptyset} c_J \chi(F_{f_{|w_J}, \mathbf{0}})}{r_0 \dots r_n}.$$

This formula is inductively useful since, if $J \neq \emptyset$, then $f_{|w_J}$ is a weighted-homogeneous polynomial in fewer variables (compare with [**Di**]). Note that, in this formula, we need not consider the term where $J = \{0, \dots, n\}$, for then $w_J = \mathbf{0}$ and hence $\chi(F_{f_{|w_J}, \mathbf{0}}) = 0$.

This is particularly useful in the case where f is a weighted-homogeneous polynomial with a one-dimensional critical locus and each restriction to a flat, $f_{|w_J}$, also has a one-dimensional (or zero-dimensional) critical locus.

Example 4.9. For instance, consider the case of a possibly non-reduced, weighted-homogeneous plane curve singularity. Suppose that the irreducible factorization of $f(z_0, z_1)$ is $z_0^a z_1^b \prod f_i^{m_i}$, where we allow for the case where a or b equals 0. Let $\pi(z_0, z_1) = (z_0^{r_0}, z_1^{r_1})$, and let h denote the homogeneous polynomial $f \circ \pi$. Let h_i denote the homogeneous polynomial $f_i \circ \pi$. Let d be the degree of h and let d_i be the degree of h_i.

Then, the formula of 4.7 becomes

$$\chi(F_{f,\mathbf{0}}) = \frac{\chi(F_{h,\mathbf{0}}) + (r_0 - 1)r_1\chi(F_{f_{|V(z_0)}, \mathbf{0}}) + (r_1 - 1)r_0\chi(F_{f_{|V(z_1)}, \mathbf{0}})}{r_0 r_1}.$$

Now, $\chi(F_{f_{|V(z_k)}, \mathbf{0}}) = 0$ if $f_{|V(z_k)} \equiv 0$ and simply equals the multiplicity of $f_{|V(z_k)}$ otherwise. In addition, as h is homogeneous, we may calculate $\chi(F_{h,\mathbf{0}})$ from 4.6 by knowing only $\lambda_h^1(\mathbf{0})$, which we may calculate as in 2.9.

We find easily that

$$r_0 r_1 \chi(F_{f,\mathbf{0}}) = \begin{cases} -d \sum d_i, & \text{if } a \neq 0, b \neq 0 \\ d(r_0 - \sum d_i), & \text{if } a = 0, b \neq 0 \\ d(r_1 - \sum d_i), & \text{if } a \neq 0, b = 0 \\ d(r_0 + r_1 - \sum d_i), & \text{if } a = b = 0. \end{cases}$$

Example 4.10. We can also apply the formula of 4.8 in harder cases. Consider the swallowtail singularity; this is given as the zero locus of

$$f = 256z_0^3 - 27z_1^4 - 128z_0^2z_2^2 + 144z_0z_1^2z_2 + 16z_0z_2^4 - 4z_1^2z_2^3$$

(see, for instance, [**Te3**]).

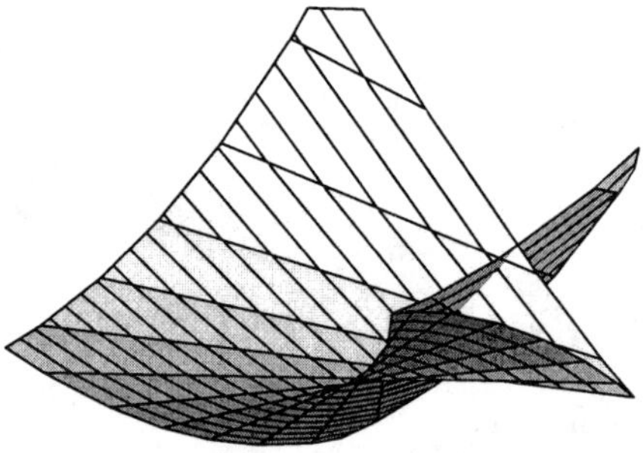

Figure 4.11. The swallowtail singularity

If $\pi(z_0, z_1, z_2) = (z_0^4, z_1^3, z_2^2)$, then $h = f \circ \pi$ is homogeneous of degree 12. Using the notation of 4.8, we find

$$c_{\{0\}} = (-1)(4\cdot 3\cdot 2 - 3\cdot 2) = -18.$$

$$f_{|w_{\{0\}}} = -27z_1^4 - 4z_1^2z_2^3 = z_1^2(-27z_1^2 - 4z_2^3).$$

Hence, from 4.9, we find that

$$\chi(F_{f_{|w_{\{0\}}},\mathbf{0}}) = -8.$$

Similarly,

$$c_{\{1\}} = (-1)(4\cdot 3\cdot 2 - 4\cdot 2) = -16.$$

$$f_{|w_{\{1\}}} = 256z_0^3 - 128z_0^2z_2^2 + 16z_0z_2^4 = 16z_0(16z_0^2 - 8z_0z_2^2 + z_2^4) =$$

$$16z_0(4z_0 - z_2^2)^2.$$

$$\chi(F_{f_{|w_{\{1\}}},\mathbf{0}}) = -3.$$

and

$$c_{\{2\}} = (-1)(4\cdot 3\cdot 2 - 4\cdot 3) = -12.$$

$$f_{|w_{\{2\}}} = 256z_0^3 - 27z_1^4.$$

$$\chi(F_{f_{|w_{\{2\}}},\mathbf{0}}) = -5.$$

We also find

$$f_{|w_{\{0,1\}}} \equiv 0,$$

$$\begin{aligned} c_{\{0,2\}} &= 4\cdot 3\cdot 2 - 3\cdot 2 - 4\cdot 3 + 3 = 9,\\ f_{|w_{\{0,2\}}} &= -27z_1^4,\\ \chi(F_{f_{|w_{\{0,2\}}},\mathbf{0}}) &= 4, \end{aligned}$$

and

$$\begin{aligned} c_{\{1,2\}} &= 4\cdot 3\cdot 2 - 4\cdot 2 - 4\cdot 3 + 4 = 8,\\ f_{|w_{\{1,2\}}} &= 256z_0^3,\\ \chi(F_{f_{|w_{\{0,2\}}},\mathbf{0}}) &= 3. \end{aligned}$$

Having made these calculations, it still remains for us to calculate $\chi(F_{h,\mathbf{0}})$. We can do this using 4.7, provided that we know that h has a one-dimensional critical locus and provided that we can calculate $\lambda_h^1(\mathbf{0})$. While this calculation can be made by hand, it is rather tedious; a computer algebra program – such as *Macaulay*, a public domain program written by Michael Stillman and Dave Bayer – can tell us that not only does h have a one-dimensional critical locus, but that the multiplicity of the Jacobian scheme at the origin is 83, i.e., $\lambda_h^1(\mathbf{0}) = 83$. Therefore,

$$\chi(F_{h,\mathbf{0}}) = \lambda_h^0(\mathbf{0}) - \lambda_h^1(\mathbf{0}) + 1 = d\lambda_h^1(\mathbf{0}) - (d-1)^{n+1} + 1 =$$

$$12\cdot 83 - 11^3 + 1 = 336.$$

Finally,

$$\chi(F_{f,\mathbf{0}}) = \frac{336 - (-18)(-8) - (-16)(-3) - (-12)(-5) - 9\cdot 4 - 8\cdot 3}{4\cdot 3\cdot 2} = 1.$$

Remark 4.12. In the above example, we resorted to a computer calculation at one point. If we are willing to use a computer algebra program at each step, then there is a much easier way to calculate the Euler characteristic of the Milnor fibre in the case of a one-dimensional critical locus – whether the function, h, is a weighted-homogeneous polynomial or not. This is the method that we describe in [**M-S**].

Any computer program which can calculate the multiplicities of ideals in a polynomial ring, given a set of generators, can calculate the Lê numbers of a polynomial. (A number of programs have this capability, but by far the most efficient that we know of is *Macaulay*.)

Given such a program and a polynomial, h, with a one-dimensional singular set, one proceeds as follows to calculate the Lê numbers, λ^0 and λ^1, at the origin with respect to a generic set of coordinates.

As we saw in 1.16, λ^1 is nothing other than the multiplicity of the Jacobian scheme of h at the origin. So, one can have the program calculate it.

Now, we need a hyperplane that is generic enough so that its intersection number (at the origin) with the (reduced) singular set is, in fact, equal to the multiplicity of the singular set. Usually, one knows the singular set (as a set) well enough to know such a hyperplane. (Alternatively, there are programs which can find the singular set for you – though how they present the answer is not always helpful.) We shall assume now, in addition to having λ^1, that we also have such a hyperplane, $V(L)$, for some linear form, L .

By the work of Iomdine [**Io**] and Lê [**Lê4**] (or our generalization in 4.5 or [**M-S**]), we have that: for all k sufficiently large, $h+L^k$ has an isolated singularity at the origin and the Milnor number $\mu(h+L^k)$ equals $\lambda^0+(k-1)\lambda^1$. But, the Milnor number is again nothing other than the multiplicity of the Jacobian scheme at the origin, and so we may use our program to calculate it. Thus, we can find λ^0 – provided that we have an effective method for knowing when we have chosen k large enough so that the formula of Iomdine and Lê holds.

However, we have such a method. If $h+L^k$ has an isolated singularity, let μ_k denote its Milnor number. (Given a particular k, one must either check by hand whether $h+L^k$ has an isolated singularity or have a program do it. *Macaulay* will tell you the dimension of the singular set in the course of calculating the multiplicity of the Jacobian scheme at the origin.) A quick look at the proof of the Iomdine-Lê formula in 4.5 shows that the formula holds provided that

$$\mu_k-(k-1)\lambda^1 \leqslant k-2.$$

Therefore, to find λ^0, one starts with a relatively small k and checks whether

$$\mu_k \leqslant k-2+(k-1)\lambda^1.$$

If the inequality is false, pick a larger k. Eventually, the inequality will hold and then

$$\lambda^0=\mu_k-(k-1)\lambda^1.$$

This is the same argument that we used in Remark I.3.5 in the more general setting of the Lê-Iomdine-Vogel formulas.

As we also mentioned in I.3.5, there is an alternative method for calculating not only the Lê numbers but also the maximum polar ratio of h. Here, one needs to make sure that the linear form L has been chosen generically enough so that there are no exceptional pairs (see I.3.5). As we are assuming the use of a computer, all that one needs to do is select the linear form "randomly".

Then, to find the maximum polar ratio, one calculates μ_k for successive values of k – looking for a difference of λ^1. Once this occurs, k is at least the maximum polar ratio and, as before, we conclude that

$$\lambda^0=\mu_k-(k-1)\lambda^1.$$

Note, moreover, that this method works whether λ^1 equals the multiplicity of the Jacobian scheme at the origin or not. As λ^1 at least the multiplicity of the Jacobian scheme at the origin, with equality being the generic case, it follows that if $\mu_{k+1}-\mu_k$ equals the multiplicity of the Jacobian scheme at the origin, then λ^1 equals the multiplicity of the Jacobian scheme at the origin and $\lambda^0 = \mu_k - (k-1)\lambda^1$.

While this method requires one to calculate at least two Milnor numbers, μ_k, it will still be a more efficient way of calculating λ^0 – provided that the maximum polar ratio is significantly smaller than λ^0 itself. This would be the case, for instance, if the polar curve had a large number of components.

Consider again the swallowtail of Example 4.10 defined by

$$h = 256z_0^3 - 27z_1^4 - 128z_0^2z_2^2 + 144z_0z_1^2z_2 + 16z_0z_2^4 - 4z_1^2z_2^3.$$

We use *Macaulay* to find that the multiplicity of the Jacobian scheme at the origin equals 5. (Alternatively, we know that the singularities of the swallowtail consist of a smooth curve of ordinary double points plus a multiplicity two curve of cusps; hence, the multiplicity of the Jacobian scheme at the origin $= 1 + 2 \cdot 2 = 5$.) Now, using the notation above and letting $L = z_2$ (this is "random" enough this time, but the reader is invited to check this by picking a "more random" linear form), we find

$k =$	2	3	4	5	6	7
$\mu_k =$	6	12	18	24	30	35

From the table, we see that the maximum polar ratio is at most 6, λ^1 is, in fact, equal to 5, and $\lambda^0 = 30 - (6-1)5 = 5$. Hence, the Euler characteristic of the Milnor fibre of h equals $\lambda^0 - \lambda^1 + 1 = 1$, which agrees with our previous calculation.

As the swallowtail is such an important singularity, one might wonder how close the Morse inequalities of Theorem 3.3 are to being equalities in this example. The answer is: not very. A. Suciu informs us that the degree 1 and 2 homology groups of the Milnor fibre of the swallowtail at the origin are both free Abelian of rank 2.

Why are the Lê numbers off by so much from the Betti numbers? It is because the Lê numbers record information that the Betti numbers do not.

In the case of the swallowtail, λ^1 records the information that there is an entire cusp of cusp singularities coming into the origin plus a line of quadratic singularities. Thus, λ^1 equals (the multiplicity of the cusp)(the Milnor number of the cusp) plus (the Milnor number of the quadratic singularity) $= (2)(2) + 1 = 5$. Now, as λ^1 is forced to be 5, λ^0 also has to be 5 – in order to make $\lambda^0 - \lambda^1 + 1$ come out to equal the Euler characteristic.

Remark 4.13. For a projective hypersurface, X, defined by a homogeneous polynomial, h, in the projective coordinates $(z_0 : \cdots : z_n)$, one may ask if there is some reasonable notion of **global** Lê numbers.

It is easy to see that if one takes affine patches on X, calculates λ^0 at points of each patch with respect to generic coordinates, and adds together the finite number of non-zero results that one gets, then the answer is precisely $\lambda_h^1(\mathbf{0})$ (in the ordinary, affine sense) with respect to generic coordinates. It seems reasonable, then, to define the global λ_X^i to be $\lambda_h^{i+1}(\mathbf{0})$.

If one makes this definition, it might initially look as though $\lambda^0_h(\mathbf{0})$ should provide a new interesting invariant of X. However, Corollary 4.7 tells us that $\lambda^0_h(\mathbf{0})$ can be calculated from the higher Lê numbers together with the degree of X.

We shall now prove a uniform version of the generalized Lê-Iomdine formulas for one-parameter families of germs of hypersurface singularities at the origin, i.e., $\overset{\circ}{\mathbb{D}}$ will be an open disc about the origin in $\mathbb{C}$, $\mathcal{U}$ will be an open neighborhood of the origin in $\mathbb{C}^{n+1}$ and $f : (\overset{\circ}{\mathbb{D}} \times \mathcal{U}, \overset{\circ}{\mathbb{D}} \times \mathbf{0}) \to (\mathbb{C}, 0)$ will be an analytic function; naturally, we write f_t for the function defined by $f_t(\mathbf{z}) := f(t, \mathbf{z})$.

First, we need a lemma

Lemma 4.14. *For all i and for all $\mathbf{p} = (t_0, z_0, \dots, z_n)$ near the origin such that $t_0 \neq 0$,*

$$\Gamma^i_{f_{t_0}, \mathbf{z}} = \Gamma^{i+1}_{f,(t,\mathbf{z})} \cap V(t - t_0) = \Gamma^{i+1}_{f,(t,\mathbf{z})} \cdot V(t - t_0)$$

as cycles at $\mathbf{p}$, regardless of how generic $(t, z_0, \dots, z_n)$ may be.

Proof. Fix any good stratification, $\mathfrak{G}$, for f at the origin. The stratified critical values of the function t are isolated; hence, in a neighborhood of the origin, the map t restricted to each of the strata of $\mathfrak{G}$ can have only 0 as a critical value (see [**Mas11**, Prop. 1.3]). Therefore, for all small $t_0 \neq 0$, $V(t - t_0)$ is a prepolar slice of f at $\mathbf{p}$. In particular, $\Sigma f \cap V(t-t_0) = \Sigma(f_{|_{V(t-t_0)}})$.

Thus,

$$\Gamma^i_{f_{t_0}, \mathbf{z}} = V\left(t - t_0, \frac{\partial f}{\partial z_i}, \dots, \frac{\partial f}{\partial z_n}\right) \neg\, \Sigma(f_{t_0}) =$$

$$V\left(t - t_0, \frac{\partial f}{\partial z_i}, \dots, \frac{\partial f}{\partial z_n}\right) \neg\, \Sigma f \cap V(t - t_0) =$$

$$V\left(t - t_0, \frac{\partial f}{\partial z_i}, \dots, \frac{\partial f}{\partial z_n}\right) \neg\, \Sigma f = \left(\Gamma^{i+1}_{f,(t,\mathbf{z})} \cap V(t - t_0)\right) \neg\, \Sigma f,$$

where the last equality uses 1.2.i.

But, we claim that this equals $\Gamma^{i+1}_{f,(t,\mathbf{z})} \cap V(t - t_0)$ up to embedded subvariety for small $t_0 \neq 0$. For otherwise, $\Gamma^{i+1}_{f,(t,\mathbf{z})} \cap V(t - t_0)$ would have a component contained in Σf for an infinite number of small $t_0 \neq 0$, which would imply that $\Gamma^{i+1}_{f,(t,\mathbf{z})}$ has a component contained in Σf – a contradiction of the definition of $\Gamma^{i+1}_{f,(t,\mathbf{z})}$. Therefore, $\Gamma^i_{f_{t_0}, \mathbf{z}} = \Gamma^{i+1}_{f,(t,\mathbf{z})} \cap V(t - t_0)$ up to embedded subvariety, and the conclusion follows easily. □

As in Theorem 4.5, when we use the coordinates $\mathbf{z} = (z_0, \dots, z_n)$ for f_t, we use the rotated coordinates $\tilde{\mathbf{z}} = (z_1, z_2, \dots, z_n, z_0)$ for $f_t + z_0^j$.

Theorem 4.15 (Uniform Lê-Iomdine formulas). *Let $s := \dim_{\mathbf{0}} \Sigma f_0$, and suppose that $s \geqslant 1$. Suppose that $\lambda^i_{f_t, \mathbf{z}}(\mathbf{0})$ is defined for all $i \leqslant s$ and for all small t. Then, there exist*

$\tau > 0$ and j_0 such that, for all $j \geqslant j_0$ and for all $t \in \overset{\circ}{\mathbb{D}}_\tau$, $\dim_{\mathbf{0}}\Sigma(f_0 + z_0^j) = s-1$, $\lambda^i_{f_t+z_0^j,\tilde{\mathbf{z}}}(\mathbf{0})$ *is defined for all $i \leqslant s-1$, and*

i) $\lambda^0_{f_t+z_0^j,\tilde{\mathbf{z}}}(\mathbf{0}) = \lambda^0_{f_t,\mathbf{z}}(\mathbf{0}) + (j-1)\lambda^1_{f_t,\mathbf{z}}(\mathbf{0})$;

ii) $\lambda^i_{f_t+z_0^j,\tilde{\mathbf{z}}}(\mathbf{0}) = (j-1)\lambda^{i+1}_{f_t,\mathbf{z}}(\mathbf{0})$, *for $1 \leqslant i \leqslant s-1$;*

iii) $\Sigma(f_t + z_0^j) = \Sigma f_t \cap V(z_0)$ *near* $\mathbf{0}$.

Proof. Given 4.5, all that we must show is that $\{\lambda^0_{f_{t_0},\mathbf{z}}(\mathbf{0})\}_{t_0}$ is bounded for small t_0. Clearly, it suffices to show that $\{\lambda^0_{f_{t_0},\mathbf{z}}(\mathbf{0})\}_{t_0}$ is bounded for small $t_0 \neq 0$. Of course, what we actually show is that, for small $t_0 \neq 0$, $\lambda^0_{f_{t_0},\mathbf{z}}(\mathbf{0})$ is independent of t_0.

For small $t_0 \neq 0$, we may apply the lemma to conclude

$$\Lambda^0_{f_{t_0},\mathbf{z}} = \Gamma^1_{f_{t_0},\mathbf{z}} \cdot V\left(\frac{\partial f}{\partial z_0}\right) =$$

$$\Gamma^2_{f,(t,\mathbf{z})} \cdot V(t-t_0) \cdot V\left(\frac{\partial f}{\partial z_0}\right) = \left(\Gamma^1_{f,(t,\mathbf{z})} + \Lambda^1_{f,(t,\mathbf{z})}\right) \cdot V(t-t_0).$$

Thus, $\Gamma^1_{f,(t,\mathbf{z})} + \Lambda^1_{f,(t,\mathbf{z})}$ has a one-dimensional component, $n_\nu[\nu]$, which coincides with $\mathbb{C} \times \mathbf{0}$ near $\mathbf{0}$ (and so, must actually be a component of $\Lambda^1_{f,(t,\mathbf{z})}$) and such that

$$\lambda^0_{f_{t_0},\mathbf{z}}(\mathbf{0}) = \left(n_\nu[\nu] \cdot V(t-t_0)\right)_{(t_0,\mathbf{0})} = n_\nu$$

for all small non-zero t_0. The conclusion follows. □

As we saw in Example 2.10, the Lê numbers in a family are not individually upper-semicontinuous. However, we do have the following.

Corollary 4.16. *Using the notation of the theorem, the tuple of Lê numbers*

$$\left(\lambda^s_{f_t,\mathbf{z}}(\mathbf{0}), \lambda^{s-1}_{f_t,\mathbf{z}}(\mathbf{0}), \dots, \lambda^0_{f_t,\mathbf{z}}(\mathbf{0})\right)$$

is lexigraphically upper-semicontinuous in the t variable, i.e., for all t small, either

$$\lambda^s_{f_0,\mathbf{z}}(\mathbf{0}) > \lambda^s_{f_t,\mathbf{z}}(\mathbf{0})$$

or

$$\lambda^s_{f_0,\mathbf{z}}(\mathbf{0}) = \lambda^s_{f_t,\mathbf{z}}(\mathbf{0}) \quad \text{and} \quad \lambda^{s-1}_{f_0,\mathbf{z}}(\mathbf{0}) > \lambda^{s-1}_{f_t,\mathbf{z}}(\mathbf{0})$$

or

$$\vdots$$

or

$$\lambda^s_{f_0,\mathbf{z}}(\mathbf{0}) = \lambda^s_{f_t,\mathbf{z}}(\mathbf{0}),\ \lambda^{s-1}_{f_0,\mathbf{z}}(\mathbf{0}) = \lambda^{s-1}_{f_t,\mathbf{z}}(\mathbf{0}), \dots,\ \lambda^1_{f_0,\mathbf{z}}(\mathbf{0}) = \lambda^1_{f_t,\mathbf{z}}(\mathbf{0}),$$

$$\textit{and } \lambda^0_{f_0,\mathbf{z}}(\mathbf{0}) \geqslant \lambda^0_{f_t,\mathbf{z}}(\mathbf{0}).$$

Proof. By applying 4.15 inductively, as in 4.6, we find that, if $0 \ll j_0 \ll j_1 \ll \dots \ll j_{s-1}$, then, for all small t, $f_t + z_0^{j_0} + z_1^{j_1} + \dots + z_{s-1}^{j_{s-1}}$ has an isolated singularity at the origin, and its Milnor number is given by

$$\mu(f_t + z_0^{j_0} + z_1^{j_1} + \dots + z_{s-1}^{j_{s-1}}) =$$

$$\lambda^0_{f_t,\mathbf{z}}(\mathbf{0}) + (j_0 - 1)\lambda^1_{f_t,\mathbf{z}}(\mathbf{0}) + (j_1 - 1)(j_0 - 1)\lambda^2_{f_t,\mathbf{z}}(\mathbf{0}) + \dots$$
$$+ (j_{s-1} - 1)\dots(j_1 - 1)(j_0 - 1)\lambda^s_{f_t,\mathbf{z}}(\mathbf{0}).$$

Now, as the Milnor number is upper-semicontinuous, the conclusion is immediate. □

Before we leave this chapter, we want to see how adding a large power of z_0 affects the prepolarity condition.

Proposition 4.17. *Let $\mathfrak{G}$ be a good stratification for h at $\mathbf{0}$ and let $V(z_0)$ be a prepolar slice with respect to $\mathfrak{G}$ at the origin. Suppose $a \neq 0$ and that j is such that $\Sigma(h + az_0^j) = \Sigma h \cap V(z_0)$ as sets. Then,*

$$\mathfrak{G}' = \{V(h + az_0^j) - \Sigma h \cap V(z_0)\} \cup \{G \cap V(z_0) \mid G \textit{ is a singular stratum of } \mathfrak{G}\}$$

is a good stratification for $h + az_0^j$ at $\mathbf{0}$.

Proof. Suppose we have $\mathbf{p}_i \notin \Sigma h \cap V(z_0)$ such that $\mathbf{p}_i \to \mathbf{p} \in G \cap V(z_0)$, where G is a singular stratum, and such that $T_{\mathbf{p}_i} V(h + az_0^j - (h + az_0^j)_{|\mathbf{p}_i}) \to \mathcal{T}$. We wish to show that $T_{\mathbf{p}}(G \cap V(z_0)) = T_{\mathbf{p}} G \cap T_{\mathbf{p}} V(z_0) \subseteq \mathcal{T}$.

If $\mathcal{T} = T_{\mathbf{p}} V(z_0)$ of $G = \{\mathbf{0}\}$, then we are finished. So suppose that $\mathcal{T} \neq T_{\mathbf{p}} V(z_0)$and $G \neq \{\mathbf{0}\}$. Then, for all but a finite number of i, $T_{\mathbf{p}_i} V(h + az_0^j - (h + az_0^j)_{|\mathbf{p}_i}) \neq T_{\mathbf{p}_i} V(z_0 - z_0(\mathbf{p}_i))$ and $\mathbf{p}_i \notin \Sigma h$. Hence,

$$T_{\mathbf{p}_i} V(h - h(\mathbf{p}_i), z_0 - z_0(\mathbf{p}_i)) = T_{\mathbf{p}_i} V(h + az_0^j - (h + az_0^j)_{|\mathbf{p}_i}, z_0 - z_0(\mathbf{p}_i)) =$$

$$T_{\mathbf{p}_i} V(h + az_0^j - (h + az_0^j)_{|\mathbf{p}_i}) \cap T_{\mathbf{p}_i} V(z_0 - z_0(\mathbf{p}_i)) \to \mathcal{T} \cap T_{\mathbf{p}} V(z_0)$$

where, by taking a subsequence, we may assume that $T_{\mathbf{p}_i} V(h - h(\mathbf{p}_i))$ approaches some hyperplane, T. As $\mathfrak{G}$ is a good stratification for h, $T_{\mathbf{p}} G \subseteq T$. Moreover, as $V(z_0)$ transversely intersects G,

$$T_{\mathbf{p}_i} V(h - h(\mathbf{p}_i), z_0 - z_0(\mathbf{p}_i)) \to T \cap T_{\mathbf{p}} V(z_0)$$

and thus $T \cap T_{\mathbf{p}} V(z_0) = \mathcal{T} \cap T_{\mathbf{p}} V(z_0)$. Therefore,

$$T_{\mathbf{p}} G \cap T_{\mathbf{p}} V(z_0) \subseteq T \cap T_{\mathbf{p}} V(z_0) = \mathcal{T} \cap T_{\mathbf{p}} V(z_0) \subseteq \mathcal{T}. \quad \square$$

Corollary 4.18. *Let $k \geqslant 0$ and suppose $(z_0, \ldots, z_k)$ is prepolar for h at the origin. If j is such that*

$$(*) \qquad \dim_{\mathbf{0}} \Gamma^{i+1}_{h,\mathbf{z}} \cap V\left(\frac{\partial h}{\partial z_0} + jaz_0^{j-1}\right) \cap V(z_1, \ldots, z_i) \leqslant 0$$

for all i with $0 \leqslant i \leqslant k$, then $(z_1, \ldots, z_k)$ is prepolar for $h + az_0^j$ at $\mathbf{0}$.

Proof. When $i = 0$, $(*)$ yields $\dim_{\mathbf{0}} \Gamma^{i+1}_{h,\mathbf{z}} \cap V\left(\frac{\partial h}{\partial z_0} + jaz_0^{j-1}\right) = 0$ and so, as sets,

$$\Sigma(h + az_0^j) = V\left(\frac{\partial h}{\partial z_0} + jaz_0^{j-1}\right) \cap V\left(\frac{\partial h}{\partial z_1}, \ldots, \frac{\partial h}{\partial z_n}\right) =$$

$$V\left(\frac{\partial h}{\partial z_0} + jaz_0^{j-1}\right) \cap (\Sigma h \cup \Gamma^1_{h,\mathbf{z}}) =$$

$$(\Sigma h \cap V(z_0)) \cup \left(\Gamma^1_{h,\mathbf{z}} \cap V\left(\frac{\partial h}{\partial z_0} + jaz_0^{j-1}\right)\right) = \Sigma h \cap V(z_0).$$

Thus, the hypothesis of 4.17 is satisfied and we apply it; this leaves us with only the problem of showing that each successive hyperplane slice transversely intersects the smooth part, i.e., as germs of sets at the origin, for all i with $0 \leqslant i \leqslant k$,

$$\Sigma(h + az_0^j{}_{|V(z_1,\ldots,z_i)}) = V\left(z_1, \ldots, z_i, \frac{\partial h}{\partial z_0} + jaz_0^{j-1}\right) \cap V\left(\frac{\partial h}{\partial z_{i+1}}, \ldots, \frac{\partial h}{\partial z_n}\right) =$$

$$V\left(z_1, \ldots, z_i, \frac{\partial h}{\partial z_0} + jaz_0^{j-1}\right) \cap (\Sigma h \cup \Gamma^{i+1}_{h,\mathbf{z}}) =$$

$$(\Sigma h \cap V(z_0, \ldots, z_i)) \cup \left(\Gamma^{i+1}_{h,\mathbf{z}} \cap V\left(\frac{\partial h}{\partial z_0} + jaz_0^{j-1}\right) \cap V(z_1, \ldots, z_i)\right)$$

which, by $(*)$, equals $\Sigma h \cap V(z_0, \ldots, z_i)$. $\square$

Proposition 4.19. *Let $k \geqslant 0$ and suppose $(z_0, \ldots, z_k)$ is prepolar for h at the origin. Then, for all large j,*

$$\dim_{\mathbf{0}} \Gamma^{i+1}_{h,\mathbf{z}} \cap V\left(\frac{\partial h}{\partial z_0} + jaz_0^{j-1}\right) \cap V(z_1, \ldots, z_i) \leqslant 0$$

for all i with $0 \leqslant i \leqslant k$ and so $(z_1, \ldots, z_k)$ is prepolar for $h + az_0^j$ at $\mathbf{0}$.

Proof. As $(z_0, \ldots, z_k)$ is prepolar for h, we may apply 1.26 to conclude that $\gamma^{i+1}_{h,\mathbf{z}}(\mathbf{0})$ exists for all i with $0 \leqslant i \leqslant k$, i.e.,

$$\dim_{\mathbf{0}} \Gamma^{i+1}_{h,\mathbf{z}} \cap V(z_0, z_1, \ldots, z_i) \leqslant 0.$$

It follows immediately that

$$\dim_{\mathbf{0}} \Gamma^{i+1}_{h,\mathbf{z}} \cap V(z_1, \dots, z_i) \leqslant 1.$$

Therefore,

$$\dim_{\mathbf{0}} \Gamma^{i+1}_{h,\mathbf{z}} \cap V\left(\frac{\partial h}{\partial z_0} + jaz_0^{j-1}\right) \cap V(z_1, \dots, z_i) \not\leqslant 0$$

if and only if $V\left(\frac{\partial h}{\partial z_0} + jaz_0^{j-1}\right)$ contains a component of $\Gamma^{i+1}_{h,\mathbf{z}} \cap V(z_1, \dots, z_i)$ through the origin. But, if a component W of $\Gamma^{i+1}_{h,\mathbf{z}} \cap V(z_1, \dots, z_i)$ through the origin were contained in both $V\left(\frac{\partial h}{\partial z_0} + j_1 a z_0^{j_1-1}\right)$ and $V\left(\frac{\partial h}{\partial z_0} + j_2 a z_0^{j_2-1}\right)$ for $j_1 \neq j_2$, then z_0 would have to equal 0 along that component – a contradiction, as $\dim_{\mathbf{0}} W \cap V(z_0) \leqslant 0$. The conclusion follows. □

Chapter 5. LÊ NUMBERS AND HYPERPLANE ARRANGEMENTS

The Plücker formula of Corollary 4.7 states: Let h be a homogeneous polynomial of degree d in $n+1$ variables, let $s = \dim_{\mathbf{0}}\Sigma h$, and suppose that $\lambda^i_{h,\mathbf{z}}(\mathbf{0})$ exists for all $i \leqslant s$. Then, $\sum_{i=0}^{s}(d-1)^i\lambda^i_{h,\mathbf{z}}(\mathbf{0}) = (d-1)^{n+1}$.

This formula allows us to calculate the Lê numbers for a central hyperplane arrangement in a purely combinatorial manner from the lattice of flats of the arrangement (see [**O-T**] and below). It was experimentally observed by D. Welsh and G. Ziegler that there was a fairly trivial relationship between the Lê numbers of the arrangement and the Möbius function (again, see [**O-T**] and below). This relationship generalizes to matroid-based polynomial identities (see [**MSSVWZ**]).

In this chapter, we give the combinatorial characterization of the Lê numbers for central hyperplane arrangements and prove the relation between the Lê numbers and the Möbius function.

A *central hyperplane arrangement in* $\mathbb{C}^{n+1}$ is simply the zero-locus of an analytic function $h : \mathbb{C}^{n+1} \to \mathbb{C}$ where h is a product of d linear forms on $\mathbb{C}^{n+1}$ (here, we are not necessarily assuming that the forms are distinct). Though this may appear to be fairly trivial as a hypersurface singularity, this apparent simplicity is deceiving – the study of hyperplane arrangements is quite complex and touches on many areas of mathematics (see, for instance, [**O-R**], [**O-S**],[**O-T**]).

Example 5.1. Suppose we have such an h. In this case, $V(h)$ equals the union of hyperplanes, $\{H_i\}_{i\in I}$, where I is the indexing set $\{1,\dots,d'\}$, each H_i occurs with some multiplicity $m_i := \operatorname{mult} H_i$, and $\sum m_i = d$ (in particular, if h is reduced, then each $m_i = 1$ and $d' = d$).

There is an obvious good, Whitney stratification of $V(h)$ obtained from the "flats" of the hyperplane arrangement; the collection of flats is given by $\{w_J\}_{J\subseteq I}$, where

$$w_J := \bigcap_{i\in J} H_i.$$

If we now take the stratification $\{S_J\}_{J\subseteq I}$, where

$$S_J = w_J - \bigcup_{J \subsetneq K} w_K,$$

then clearly h is analytically trivial along the strata, and therefore one has trivially a Whitney stratification. In words, the strata are intersections of the hyperplanes minus smaller intersections of hyperplanes.

We wish to calculate the Lê numbers of h at the origin with respect to generic coordinates $\mathbf{z}$. As h is analytically trivial along the strata, it is easy to see that, as sets, the Lê cycles are given by the unions of the flats of correct dimension. Hence, as cycles, for all k,

$$\Lambda^k_{h,\mathbf{z}} = \sum_{\dim S_J = k} a_J[w_J]$$

for some a_J. By 1.18, a_J may be calculated by taking any $\mathbf{p} \in S_J$ and a normal slice N to S_J in $\mathbb{C}^{n+1}$ at $\mathbf{p}$, and then $a_J = \lambda^0_{h_{|N}}(\mathbf{p})$, where we use generic coordinates. After a translation to make the point $\mathbf{p}$ the origin, we see that $h_{|N}$ at $\mathbf{p}$ is again (up to multiplication by units) a product of linear forms of degree $e_J := \sum_{i \in J} m_i$.

Therefore, we may use 4.7 to calculate the Lê numbers of h at the origin by a downward induction on the dimension of the flats. (In the following, it looks nicer if we suppress the subscripts.) We denote a hyperplane in the arrangement by H, a flat by w or v, and define

$$e(w) := \sum_{w \subseteq H} \operatorname{mult} H.$$

Next, we define the *vanishing Möbius function*, η, by downward induction on the dimension of the flats. For a hyperplane, H, in the arrangement, define

$$\eta(H) := \operatorname{mult} H - 1;$$

for a smaller dimensional flat, w, 4.7 tells us that we need

$$\eta(w) := (e(w) - 1)^{n+1-\dim w} - \sum_{v \supsetneq w} \eta(v) \cdot (e(w) - 1)^{\operatorname{codim}_v w}.$$

This equality is equivalent to

$$\sum_{v \supseteq w} (e(w) - 1)^{\dim v} \eta(v) = (e(w) - 1)^{n+1}.$$

Finally, having calculated the vanishing Möbius function, one has that, for all i,

$$\lambda^i_{h,\mathbf{z}}(\mathbf{0}) = \sum_{\dim w = i} \eta(w).$$

By 3.3, knowing the Lê numbers of the hyperplane arrangement gives us the Euler characteristic of the Milnor fibre together with Morse inequalities on the Betti numbers. (Another method for computing the Euler characteristic of the Milnor fibre from the data provided by the containment relations among the flats, i.e., by knowing the *intersection lattice*, is given in [**O-T**].)

Example 5.2. We wish to see what the above method gives us in the case of a generic central arrangement of d hyperplanes in $\mathbb{C}^{n+1}$ (see [**O-R**]). Here, "generic" means as generic as possible considering that all the hyperplanes pass through the origin – that is, each hyperplane occurs with multiplicity 1, and if w is a flat of dimension k, and $k \neq 0$, then w is the intersection of precisely $n+1-k$ hyperplanes of the arrangement; in terms of the above discussion, this says that if $w \neq \mathbf{0}$, then $e(w) = n+1-k$. We assume that $d > n+1$ for, otherwise, after a change of coordinates, $h = z_0 z_1 \dots z_{d-1}$ and the Milnor fibre is diffeomorphic to the $(d-1)$-fold product of $\mathbb{C}^*$'s.

For a generic arrangement, it is easy to see that for all j-dimensional flats $w \neq \mathbf{0}$, the number of k-dimensional flats containing w is given by $\binom{n+1-j}{k-j}$, provided that $k \geqslant j$. One also knows that, if $k \geqslant 1$, then the number of k-dimensional flats containing the origin is given by $\binom{d}{n+1-k}$. This is all the information that one needs to calculate the vanishing Möbius function, η, from the formula

$$\eta(w) := (e(w)-1)^{n+1-\dim w} - \sum_{v \supsetneq w} \eta(v) \cdot (e(w)-1)^{\operatorname{codim}_v w}$$

together with the fact that for all hyperplanes, H, in the arrangement we have $\eta(H) = 0$.

It is an amusing exercise to prove that this implies that, if $\dim w = j \neq 0$, then $\eta(w) = n - j$. Alternatively, this also follows from Example 2.8. (The above is the inductive proof of the formula of 2.8 that is referred to in that example.)

Therefore, for a generic central arrangement of d hyperplanes in $\mathbb{C}^{n+1}$, we have with respect to generic coordinates

$$\lambda_h^n(\mathbf{0}) = 0,$$

$$\lambda_h^{n-1}(\mathbf{0}) = \sum_{\dim w = n-1} \eta(w) = \binom{d}{2}(1),$$

$$\vdots$$

$$\lambda_{h,\mathbf{z}}^i(\mathbf{0}) = \sum_{\dim w = i} \eta(w) = \binom{d}{n+1-i}(n-i),$$

$$\vdots$$

$$\lambda_h^1(\mathbf{0}) = \sum_{\dim w = 1} \eta(w) = \binom{d}{n}(n-1).$$

So, finally,

$$\lambda_h^0(\mathbf{0}) = (d-1)^{n+1} - \sum_{i=1}^{n} (d-1)^i \lambda_{h,\mathbf{z}}^i(\mathbf{0}) =$$

$$(d-1)^{n+1} - \sum_{i=1}^{n} (d-1)^i \binom{d}{n+1-i}(n-i) =$$

$$(d-1)\binom{d-1}{n},$$

where the last equality is an exercise in combinatorics.

Now, by our earlier work, since we know the Lê numbers, we know the Euler characteristic of the Milnor fibre, $F_{h,\mathbf{0}}$, together with Morse inequalities on the Betti numbers, $b_i(F_{h,\mathbf{0}})$. But, in this special case, it is not difficult to obtain the Betti numbers precisely.

By an observation of D. Cohen [**Co1**], if $d > n+1$, a generic central arrangement of d hyperplanes in $\mathbb{C}^{n+1}$ is obtained by taking repeated hyperplane sections of a generic hyperplane arrangement of d hyperplanes in $\mathbb{C}^d$. It follows that for $i \leqslant n-1$, $b_i(F_{h,\mathbf{0}}) = \binom{d-1}{i}$. Therefore, we have only to calculate $b_n(F_{h,\mathbf{0}})$; but, since we know the Euler characteristic, this is easy, and we find – after some more combinatorics- that

$$b_n(F_{h,\mathbf{0}}) = (d-n)\binom{d-1}{n},$$

which agrees with the results of [**Co1**] and [**O-R**].

Note that the Morse inequalities of 3.3 can be far from equalities; for instance, the two easiest inequalities are

$$(d-n)\binom{d-1}{n} = b_n(F_{h,\mathbf{0}}) \leqslant \lambda^0_{h,\mathbf{z}}(\mathbf{0}) = (d-1)\binom{d-1}{n}$$

and

$$d-1 = b_1(F_{h,\mathbf{0}}) \leqslant \lambda^{n-1}_{h,\mathbf{z}}(\mathbf{0}) = \binom{d}{2} = \frac{d(d-1)}{2}.$$

Now, we wish to describe the relation between the Lê numbers of a central arrangement and the Möbius function – this is the result which is generalized in [**MSSVWZ**].

Let h be the product of d distinct linear forms on $\mathbb{C}^{n+1}$, so that each hyperplane in the arrangement $V(h)$ occurs with multiplicity 1. Let $\mathcal{A}$ denote the collection of hyperplanes which are components of $V(h)$. We use the variable H to denote hyperplanes in $\mathcal{A}$. We use the letters v and w to denote flats of arbitrary dimension. Finally, in agreement with our notation in 5.1, let $e_{\mathcal{A}}(v)$ = the number of hyperplanes of $\mathcal{A}$ which contain the flat v.

As we saw in 5.1 and 5.2, the Lê numbers of a central hyperplane arrangement can be described in terms of a function $\eta_{\mathcal{A}}$ defined inductively on the flats by: for all $H \in \mathcal{A}$, $\eta_{\mathcal{A}}(H) = 0$, and for all flats w,

$$\sum_{w \subseteq v} (e_{\mathcal{A}}(w) - 1)^{\dim v} \eta_{\mathcal{A}}(v) = (e_{\mathcal{A}}(w) - 1)^{n+1}.$$

The *Möbius function*, $\mu_{\mathcal{A}}$, on $\mathcal{A}$ (see [**O-T**]) is defined inductively on the flats by: $\mu_{\mathcal{A}}(\mathbb{C}^{n+1}) = 1$ and for all flats $v \subsetneq w$,

$$\sum_{\substack{\text{flats } u \\ v \subseteq u \subseteq w}} \mu_{\mathcal{A}}(u) = 0.$$

Here, we subscript by η, e, and μ by $\mathcal{A}$ because our proof is by induction on the ambient dimension, and the inductive step requires slicing $\mathcal{A}$ by hyperplanes, N, **not** contained in $\mathcal{A}$. This will produce new arrangements inside the ambient space N. So it is important that we indicate which arrangement is under consideration.

More notation now, related to the slicing. We will be taking two kinds of hyperplane slices. N will denote a prepolar hyperplane slice through the origin in $\mathbb{C}^{n+1}$, i.e., a hyperplane slice which contains no flats of $\mathcal{A}$ other than the origin. We will also use normal slices to the one-flats; if v is a one-dimensional flat and $p_v \in v - 0$, N_v will denote a normal slice to v at p_v – that is, N_v is a hyperplane in $\mathbb{C}^{n+1}$ which transversely intersects v at p_v. We use $\mathcal{A} \cap N$ to denote the obvious induced arrangement in N (which is identified with $\mathbb{C}^n$). The arrangement $\mathcal{A} \cap N_v$ is considered as a central arrangement where p_v becomes the origin and all hyperplanes not containing p_v are ignored. Note that the number of hyperplanes in the arrangement $\mathcal{A} \cap N_v$ is $e_{\mathcal{A}}(v)$.

An arrangement is *essential* provided that the origin is a flat of the arrangement (hence, the arrangement is not trivially a product).

What we want to show is that, if $\mathcal{A}$ is a an essential, central hyperplane arrangement, then

$$\eta_{\mathcal{A}}(\mathbf{0}) = (d-1)(-1)^{n+1}\mu_{\mathcal{A}}(\mathbf{0}) = (d-1)|\mu_{\mathcal{A}}(\mathbf{0})|.$$

To induct, we will first need the following three easy lemmas on η, μ, which describe the effects of slicing. We leave the first two as exercises using the inductive definitions of $\eta_{\mathcal{A}}$ and$\mu_{\mathcal{A}}$ given above. However, we prove the third.

Lemma 5.3.

$$\eta_{\mathcal{A}\cap N}(\mathbf{0}) = \frac{\eta_{\mathcal{A}}(\mathbf{0})}{d-1} + \sum_{\dim v=1} \eta_{\mathcal{A}}(v)$$

and, if v is a one-dimensional flat,

$$\eta_{\mathcal{A}\cap N_v}(p_v) = \eta_{\mathcal{A}}(v).$$

Lemma 5.4.

$$\mu_{\mathcal{A}\cap N}(\mathbf{0}) = -\sum_{\dim v\geq 2} \mu_{\mathcal{A}}(v).$$

and, if v is a one-dimensional flat,

$$\mu_{\mathcal{A}\cap N_v}(p_v) = \mu_{\mathcal{A}}(v).$$

Lemma 5.5.

$$d\mu_{\mathcal{A}}(\mathbf{0}) + \sum_{\dim v=1} (d - e_{\mathcal{A}}(v))\mu_{\mathcal{A}}(v) = 0.$$

Proof. By one of Weisner's formulas (see Lemma 2.40 of [**O-T**]), for all $H \in \mathcal{A}$,

$$\sum_{v\cap H=\mathbf{0}} \mu_{\mathcal{A}}(v) = 0.$$

Hence,

$$0 = \sum_H \left[\mu_{\mathcal{A}}(\mathbf{0}) + \sum_{\substack{\dim v=1 \\ v\not\subseteq H}} \mu_{\mathcal{A}}(v)\right] = d\mu_{\mathcal{A}}(\mathbf{0}) + \sum_{\dim v=1} (d - e(v))\mu_{\mathcal{A}}(v). \quad \square$$

Now, we can prove

Theorem 5.6. *If $\mathcal{A}$ is a an essential, central hyperplane arrangement consisting of d hyperplanes in $\mathbb{C}^{n+1}$, then*

$$\eta_{\mathcal{A}}(\mathbf{0}) = (d-1)(-1)^{n+1}\mu_{\mathcal{A}}(\mathbf{0}) = (d-1)|\mu_{\mathcal{A}}(\mathbf{0})|.$$

Proof. The proof is by induction on the ambient dimension. The formula is stupidly true when $n = 1$.

Now, suppose the formula is true for ambient dimension n. Then,

$$\eta_{\mathcal{A}\cap N}(\mathbf{0}) = (d-1)(-1)^{n}\mu_{\mathcal{A}\cap N}(\mathbf{0}).$$

Hence, by Lemmas 5.3 and 5.4,

$$\frac{\eta_{\mathcal{A}}(\mathbf{0})}{d-1} + \sum_{\dim v=1} \eta_{\mathcal{A}}(v) = (d-1)(-1)^{n}\left(-\sum_{\dim v\geqslant 2}\mu_{\mathcal{A}}(v)\right).$$

This gives us

$$\frac{\eta_{\mathcal{A}}(\mathbf{0})}{d-1} + \sum_{\dim v=1} \eta_{\mathcal{A}\cap N_v}(p_v) = (d-1)(-1)^{n+1}\sum_{\dim v\geqslant 2}\mu_{\mathcal{A}}(v).$$

Using our inductive hypothesis again, we have

$$\frac{\eta_{\mathcal{A}}(\mathbf{0})}{d-1} + \sum_{\dim v=1} (e_{\mathcal{A}}(v)-1)(-1)^{n}\mu_{\mathcal{A}\cap N_v}(p_v) = (d-1)(-1)^{n+1}\sum_{\dim v\geqslant 2}\mu_{\mathcal{A}}(v).$$

Therefore,

$$\frac{\eta_{\mathcal{A}}(\mathbf{0})}{d-1} + \sum_{\dim v=1} (e_{\mathcal{A}}(v)-1)(-1)^{n}\mu_{\mathcal{A}}(v) = (d-1)(-1)^{n+1}\sum_{\dim v\geq 2}\mu_{\mathcal{A}}(v)$$

and so

$$\eta_{\mathcal{A}}(\mathbf{0}) = (d-1)(-1)^{n+1}\left[(d-1)\sum_{\dim v\geq 2}\mu_{\mathcal{A}}(v) + \sum_{\dim v=1}(e_{\mathcal{A}}(v)-1)\mu_{\mathcal{A}}(v)\right] =$$

$$(d-1)(-1)^{n+1}\left[(d-1)\big(-\mu_{\mathcal{A}}(\mathbf{0}) - \sum_{\dim v=1}\mu_{\mathcal{A}}(v)\big) + \sum_{\dim v=1}(e_{\mathcal{A}}(v)-1)\mu_{\mathcal{A}}(v)\right]$$

$$= (d-1)(-1)^{n+1}\left[\mu_{\mathcal{A}}(\mathbf{0}) - d\mu_{\mathcal{A}}(\mathbf{0}) - \sum_{\dim v=1}(d-e_{\mathcal{A}}(v))\mu_{\mathcal{A}}(v)\right].$$

Now apply Lemma 5.5. □

Our inductive proof given above is somewhat unsatisfactory, for it gives us no geometric insight as to why the theorem is true. Should there be a geometric explanation for the identity in 5.6? Probably so. The result of Orlik and Solomon [**O-S**] is that $|\mu_{\mathcal{A}}(\mathbf{0})|$ is the $(n+1)$-st Betti number of the complement of the arrangement $\mathcal{A}$ in $\mathbb{C}^{n+1}$. How this could be used to prove 5.6 still escapes us.

Chapter 6. THOM'S a_f CONDITION

In this chapter, we will use the Lê numbers to provide conditions under which a submanifold of affine space satisfies Thom's a_f condition with respect to the ambient stratum. Let us recall the definition of the a_f condition (see [**Mat**]).

Definition 6.1. Let $\mathcal{U}$ be an open subset of some affine space, let $f : \mathcal{U} \to \mathbb{C}$ be an analytic function, and let $M \subseteq V(f)$ be a submanifold of $\mathcal{U}$. *Thom's a_f condition is satisfied* between $\mathcal{U} - \Sigma f$ and M (or *along* M, or *by the pair* $(\mathcal{U} - \Sigma f, M)$) if, whenever $\mathbf{p}_i \in \mathcal{U} - \Sigma f$, $\mathbf{p}_i \to \mathbf{p} \in M$, and $T_{\mathbf{p}_i} V(f - f(\mathbf{p}_i)) \to \mathcal{T}$, then $T_{\mathbf{p}} M \subseteq \mathcal{T}$.

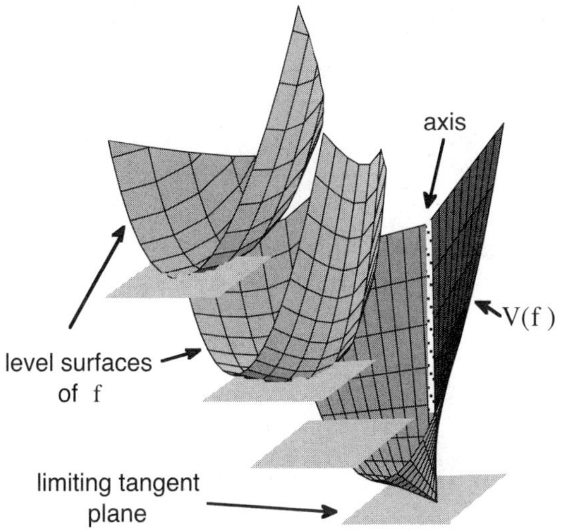

Figure 6.2. Failure of a_f along an axis

We are about to prove five different results – all of the form: if the Lê numbers are constant, then Thom's a_f condition holds. The order in which we must prove these results is interesting.

First, we give a proof of Lê and Saito's result that a constant Milnor number at the origin in a one-parameter family implies the Thom condition along the parameter axis. We then use Lê and Saito's result, combined with the generalized Lê-Iomdine formulas, to prove that the constancy of the Lê numbers at the origin in a one-parameter family implies the Thom condition along the parameter axis [**Mas14**]. We use this parameterized version to prove a non-parameterized version: if the Lê numbers of a single function are constant along a submanifold, then Thom's condition is satisfied along the submanifold. This non-parameterized version allows us to prove a multi-parameter version of Lê and Saito's result: if we have a family of isolated hypersurface singularities with constant Milnor number parameterized along a submanifold, then that submanifold satisfies Thom's a_f condition with respect to the ambient stratum. Finally, we use this last result to prove our best result – the multi-parameter version of the Lê number result above: if we have a family of hypersurface singularities with constant Lê numbers parameterized along a submanifold, then that submanifold satisfies Thom's a_f condition with respect to the ambient stratum.

In all of the results described above, it is extremely important that **our assumptions on the genericity of the coordinate system will be solely that the Lê numbers exist.** This is a dimensional requirement which is very easy to check. This should be contrasted with the results of [**H-M**] and [**HMS**].

First, we need a well-known lemma.

Lemma 6.3. *Let* $\overset{\circ}{\mathbb{D}}$ *be an open disc about the origin in* $\mathbb{C}$, *let* $\mathcal{U}$ *be an open neighborhood of the origin in* $\mathbb{C}^{n+1}$, *and let* $f : (\overset{\circ}{\mathbb{D}} \times \mathcal{U}, \overset{\circ}{\mathbb{D}} \times \mathbf{0}) \to (\mathbb{C}, 0)$ *be an analytic function; we write* f_t *for the function defined by* $f_t(\mathbf{z}) := f(t, \mathbf{z})$. *Suppose that* $\dim_{\mathbf{0}} \Sigma f_0 = 0$.

Then, for all small t, *the Milnor number of* f_t *at the origin is independent of* t *if and only if there exists an open neighborhood,* $\mathcal{W}$, *of the origin in* $\overset{\circ}{\mathbb{D}} \times \mathcal{U}$ *such that*

$$\mathcal{W} \cap V\left(\frac{\partial f}{\partial z_0}, \dots, \frac{\partial f}{\partial z_n}\right) = \mathcal{W} \cap (\mathbb{C} \times \{\mathbf{0}\}).$$

Proof. There are many proofs of this fact. We shall use intersection numbers.

The Milnor number of f_0 at the origin, $\mu_{\mathbf{0}}(f_0)$, equals the multiplicity of the origin in the cycle $\left[V\left(\frac{\partial f_0}{\partial z_0}, \dots, \frac{\partial f_0}{\partial z_n}\right)\right]$. Because f_0 has an isolated critical point at the origin, $V\left(t, \frac{\partial f}{\partial z_0}, \dots, \frac{\partial f}{\partial z_n}\right)$ is a local complete intersection, and so we have an equality of cycles

$$\left[V\left(\frac{\partial f_0}{\partial z_0}, \dots, \frac{\partial f_0}{\partial z_n}\right)\right] = \left[V\left(t, \frac{\partial f}{\partial z_0}, \dots, \frac{\partial f}{\partial z_n}\right)\right] = [V(t)] \cdot \left[V\left(\frac{\partial f}{\partial z_0}, \dots, \frac{\partial f}{\partial z_n}\right)\right].$$

Therefore,

$$\mu_{\mathbf{0}}(f_0) =$$

$$\left(V(t) \cdot V\left(\frac{\partial f}{\partial z_0}, \dots, \frac{\partial f}{\partial z_n}\right)\right)_{\mathbf{0}} = \sum_{\mathbf{p} \in \overset{\circ}{B}_\epsilon} \left(V(t-\eta) \cdot V\left(\frac{\partial f}{\partial z_0}, \dots, \frac{\partial f}{\partial z_n}\right)\right)_{\mathbf{p}}$$

$$= \mu_{\mathbf{0}}(f_\eta) + R,$$

where $\overset{\circ}{B}_\epsilon$ is a sufficiently small open ball around the origin in $\mathbb{C}^{n+2}$, η is chosen small with respect to ϵ – that is, there exists $\delta_\epsilon > 0$ such that we may use any η satisfying $0 < |\eta| < \delta_\epsilon$ – and R denotes the sum of the remaining terms, i.e., the terms coming from points $\mathbf{p}$ which are not in $\mathbb{C} \times \{\mathbf{0}\}$. Note that the sum is actually finite since we are really summing over

$\mathbf{p} \in \overset{\circ}{B}_\epsilon \cap V(t-\eta) \cap V\left(\frac{\partial f}{\partial z_0}, \dots, \frac{\partial f}{\partial z_n}\right)$. As all the intersection numbers are non-negative, R being zero is equivalent to there being no remaining terms, i.e., equivalent to $(\eta, \mathbf{0})$ being the only point in $\overset{\circ}{B}_\epsilon \cap V(t-\eta) \cap V\left(\frac{\partial f}{\partial z_0}, \dots, \frac{\partial f}{\partial z_n}\right)$.

The desired conclusion follows immediately, where the set $\mathcal{W}$ in the statement can be taken to be $\overset{\circ}{B}_\epsilon \cap \left(\overset{\circ}{\mathbb{D}}_{\delta_\epsilon} \times \mathcal{U}\right)$. □

Recall now the result of Lê and Saito [**Lê-Sa**] as stated in the introduction.

Theorem 6.4 (Lê-Saito [**Lê-Sa**]). *Let $\overset{\circ}{\mathbb{D}}$ be an open disc about the origin in $\mathbb{C}$, let $\mathcal{U}$ be an open neighborhood of the origin in $\mathbb{C}^{n+1}$, and let $f : (\overset{\circ}{\mathbb{D}} \times \mathcal{U}, \overset{\circ}{\mathbb{D}} \times \mathbf{0}) \to (\mathbb{C}, 0)$ be an analytic function; we write f_t for the function defined by $f_t(\mathbf{z}) := f(t, \mathbf{z})$.*

Suppose that $\dim_{\mathbf{0}} \Sigma f_0 = 0$ and that, for all small t, the Milnor number of f_t at the origin is independent of t. Then, $\overset{\circ}{\mathbb{D}} \times \{\mathbf{0}\}$ satisfies Thom's a_f condition at the origin with respect to the ambient stratum, i.e., if $\mathbf{p}_i$ is a sequence of points in $\overset{\circ}{\mathbb{D}} \times \mathcal{U} - \Sigma f$ such that $\mathbf{p}_i \to \mathbf{0}$ and such that $T_{\mathbf{p}_i} V(f - f(\mathbf{p}_i))$ converges to some $\mathcal{T}$, then $\mathbb{C} \times \mathbf{0} = T_{\mathbf{0}}(\overset{\circ}{\mathbb{D}} \times \{\mathbf{0}\}) \subseteq \mathcal{T}$.

Proof. We begin by noting that the existence of good stratifications as given in Proposition 1.22 implies that Thom's a_f is satisfied, near the origin, by $\overset{\circ}{\mathbb{D}} \times \{\mathbf{0}\} - \mathbf{0}$ with respect to the ambient stratum.

Now, consider the blow-up of $\overset{\circ}{\mathbb{D}} \times \mathcal{U}$ by the Jacobian ideal of f:

$$\begin{array}{ccc} & Bl_{J(f)}\left(\overset{\circ}{\mathbb{D}} \times \mathcal{U}\right) \subseteq \left(\overset{\circ}{\mathbb{D}} \times \mathcal{U}\right) \times \mathbb{P}^{n+1} & \\ \pi_1 \swarrow & & \searrow \pi_2 \\ \overset{\circ}{\mathbb{D}} \times \mathcal{U} & & \mathbb{P}^{n+1} \end{array}$$

We first wish to show that the fibre $\pi_1^{-1}(\mathbf{0})$ has dimension at most n.

The point $\mathbf{q} := [1 : 0 : \dots : 0] \in \mathbb{P}^{n+1}$ corresponds to the hyperplane $V(t)$. As $\mu_{\mathbf{0}}(f_t)$ is independent of t, the lemma implies that, in a neighborhood of the origin,

$$\pi_1(\pi_2^{-1}(\mathbf{q})) \subseteq \overset{\circ}{\mathbb{D}} \times \{\mathbf{0}\}.$$

However, as we noted above, the a_f condition holds generically on $\overset{\circ}{\mathbb{D}} \times \{\mathbf{0}\}$. Therefore, near $\mathbf{0}$, either $\pi_1(\pi_2^{-1}(\mathbf{q}))$ is empty or consists only of the origin. But, the dimension of every component of $\pi_2^{-1}(\mathbf{q})$ is at least $\dim Bl_{J(f)}\left(\overset{\circ}{\mathbb{D}} \times \mathcal{U}\right) - \dim \mathbb{P}^{n+1} = n + 2 - (n+1) = 1$. Thus, $\mathbf{0} \notin \pi_1(\pi_2^{-1}(\mathbf{q}))$, i.e., $\mathbf{q} \notin \pi_2(\pi_1^{-1}(\mathbf{0}))$. It follows that $\pi_1^{-1}(\mathbf{0})$ is a proper subset of $\mathbb{P}^{n+1}$ and, hence, has dimension at most n.

But, every component of the exceptional divisor $E := \pi_1^{-1}(\Sigma f)$ has dimension $n+1$. Therefore, above an open neighborhood of the origin, E equals the topological closure of

$E - \pi_1^{-1}(\mathbf{0})$, which is contained in $(\overset{\circ}{\mathbb{D}} \times \{\mathbf{0}\}) \times (\{0\} \times \mathbb{P}^n)$ since the a_f condition holds generically on the t-axis. It follows that $\pi_2(\pi_1^{-1}(\mathbf{0})) \subseteq \{0\} \times \mathbb{P}^n$, i.e., that the a_f condition holds along $\overset{\circ}{\mathbb{D}} \times \{\mathbf{0}\}$ at the origin. □

Our first generalization of the result of Lê and Saito is:

Theorem 6.5. *Let $\overset{\circ}{\mathbb{D}}$ be an open disc about the origin in $\mathbb{C}$, let $\mathcal{U}$ be an open neighborhood of the origin in $\mathbb{C}^{n+1}$, and let $f : (\overset{\circ}{\mathbb{D}} \times \mathcal{U}, \overset{\circ}{\mathbb{D}} \times \mathbf{0}) \to (\mathbb{C}, 0)$ be an analytic function; we write f_t for the function defined by $f_t(\mathbf{z}) := f(t, \mathbf{z})$.*

Let $s = \dim_{\mathbf{0}} \Sigma f_0$. Suppose that, for all small t, for all i with $0 \leqslant i \leqslant s$, $\lambda^i_{f_t,\mathbf{z}}(\mathbf{0})$ is defined and is independent of t. Then, $\overset{\circ}{\mathbb{D}} \times \mathbf{0}$ satisfies Thom's a_f condition at the origin with respect to the ambient stratum, i.e., if $\mathbf{p}_i$ is a sequence of points in $\overset{\circ}{\mathbb{D}} \times \mathcal{U} - \Sigma f$ such that $\mathbf{p}_i \to \mathbf{0}$ and such that $T_{\mathbf{p}_i} V(f - f(\mathbf{p}_i))$ converges to some $\mathcal{T}$, then $\mathbb{C} \times \mathbf{0} = T_{\mathbf{0}}(\overset{\circ}{\mathbb{D}} \times \mathbf{0}) \subseteq \mathcal{T}$.

Proof. The proof is by induction on s. For $s = 0$, the theorem is exactly that of Lê and Saito in 6.4.

Now, suppose that $s \geqslant 1$ and that, for all small t, for all i with $0 \leqslant i \leqslant s$, $\lambda^i_{f_t,\mathbf{z}}(\mathbf{0})$ is defined and is independent of t, but that there exists a sequence $\mathbf{p}_i$ of points in $\overset{\circ}{\mathbb{D}} \times \mathcal{U} - \Sigma f$ such that $\mathbf{p}_i \to \mathbf{0}$, such that $T_{\mathbf{p}_i} V(f - f(\mathbf{p}_i))$ converges to some $\mathcal{T}$, and $T_{\mathbf{0}}(\overset{\circ}{\mathbb{D}} \times \mathbf{0}) \not\subseteq \mathcal{T}$.

As the collection of such limiting $\mathcal{T}$ is analytic, we may apply the curve selection lemma (see [**Loo**]) to conclude that there exists a real analytic curve

$$\alpha : [0, \epsilon) \to \{\mathbf{0}\} \cup (\overset{\circ}{\mathbb{D}} \times \mathcal{U} - \Sigma f)$$

such that $\alpha(u) = \mathbf{0}$ if and only if $u = 0$ and such that

$$\lim_{u \to 0} T_{\alpha(u)} V(f - f(\alpha(u))) = \mathcal{T}.$$

As α is real analytic, it is trivial to show that, for all large j,

$$\lim_{u \to 0} \frac{\operatorname{grad}(f + z_0^j)_{|\alpha(u)}}{|\operatorname{grad}(f + z_0^j)_{|\alpha(u)}|} = \lim_{u \to 0} \frac{\operatorname{grad}(f)_{|\alpha(u)}}{|\operatorname{grad}(f)_{|\alpha(u)}|}.$$

Therefore, for all large j, the family $f_t + z_0^j$ also has $\mathcal{T}$ as a limit to level hypersurfaces, i.e., $\overset{\circ}{\mathbb{D}} \times \mathbf{0}$ does not satisfy the $a_{f+z_0^j}$ condition at the origin with respect to the ambient stratum.

However, $\lambda^0_{f_t,\mathbf{z}}(\mathbf{0})$ is independent of t and, applying Theorem 4.5, if $j \geqslant 2 + \lambda^0_{f_t,\mathbf{z}}(\mathbf{0})$, then the family $f_t + z_0^j$ has Lê numbers independent of t, and $f_0 + z_0^j$ has a critical locus of dimension $s - 1$. Thus, our inductive hypothesis contradicts the previous paragraph. □

Corollary 6.6. *Let $h : \mathcal{U} \to \mathbb{C}$ be an analytic function on an open subset of $\mathbb{C}^{n+1}$, let $\mathbf{z} = (z_0, \dots, z_n)$ be a linear choice of coordinates for $\mathbb{C}^{n+1}$, let M be an analytic submanifold of $V(h)$, let $\mathbf{q} \in M$, and let s denote $\dim_{\mathbf{q}} \Sigma h$.*

If, for each i such that $0 \leqslant i \leqslant s$, $\lambda^i_{h,\mathbf{z}}(\mathbf{p})$ is defined and is independent of $\mathbf{p}$, for all $\mathbf{p} \in M$ near $\mathbf{q}$, then M satisfies Thom's a_h condition at $\mathbf{q}$ with respect to the ambient stratum; that is, if $\mathbf{q}_i$ is a sequence of points in $\mathcal{U} - \Sigma h$ such that $\mathbf{q}_i \to \mathbf{q}$ and such that $T_{\mathbf{q}_i} V(h - h(\mathbf{q}_i))$ converges to some $\mathcal{T}$, then $T_{\mathbf{q}} M \subseteq \mathcal{T}$.

Proof. This follows from 6.5 by a fairly standard trick. Let $\mathbf{c}(t)$ be a smooth analytic path in M such that $\mathbf{c}(0) = \mathbf{q}$. If we can show that any limiting tangent plane, $\mathcal{T}$, contains the tangent to the image of $\mathbf{c}$ at $\mathbf{q}$, then we will be finished.

So, take such a $\mathbf{c}$, and suppose that we have a sequence of points, $\mathbf{q}_i$, in $\mathcal{U} - \Sigma h$ such that $\mathbf{q}_i \to \mathbf{q}$ and such that $T_{\mathbf{q}_i} V(h - h(\mathbf{q}_i)) \to \mathcal{T}$.

Define $f(t, \mathbf{z}) := h(\mathbf{z} + \mathbf{c}(t))$, and consider the sequence of points $(0, \mathbf{q}_i - \mathbf{q})$. If one now applies the theorem, the result is that $\mathbf{c}'(0) \subseteq \mathcal{T}$; we leave the details to the reader. □

Remark 6.7. It is important to note that, in 6.6, we only require that the coordinates are generic enough so that the Lê numbers are defined; we are not requiring that the coordinates are prepolar.

On the other hand, Corollary 6.6 tells us how we can obtain good stratifications: if we have an analytic stratification of $V(h)$ such that the Lê numbers are defined and constant along the strata, then the stratification is actually a good stratification. However, there is no guarantee that the coordinates used to define the Lê numbers are prepolar with respect to this good stratification.

Now we can prove the multi-parameter version of the result of Lê and Saito.

Theorem 6.8. *Let M be an open neighborhood of the origin in $\mathbb{C}^k$, let $\mathcal{U}$ be an open neighborhood of the origin in $\mathbb{C}^{n+1}$, and let $f : (M \times \mathcal{U}, M \times \mathbf{0}) \to (\mathbb{C}, 0)$ be an analytic function; we write $f_{\mathbf{t}}$ for the function defined by $f_{\mathbf{t}}(\mathbf{z}) := f(\mathbf{t}, \mathbf{z})$, where $\mathbf{t} \in M$ and $\mathbf{z} \in \mathcal{U}$.*

Suppose that $\dim_{\mathbf{0}} \Sigma f_{\mathbf{0}} = 0$ and that, for all $\mathbf{t}$ near the origin, the Milnor number of $f_{\mathbf{t}}$ at the origin is independent of $\mathbf{t}$. Then, $M \times \mathbf{0}$ satisfies Thom's a_f condition at the origin with respect to the ambient stratum, i.e., if $\mathbf{p}_i$ is a sequence of points in $M \times \mathcal{U} - \Sigma f$ such that $\mathbf{p}_i \to \mathbf{0}$ and such that $T_{\mathbf{p}_i} V(f - f(\mathbf{p}_i))$ converges to some $\mathcal{T}$, then $T_{\mathbf{0}}(M \times \mathbf{0}) \subseteq \mathcal{T}$.

Proof. If the constant value of the Milnor number is 0, then there is nothing to prove. Note, though, that if the constant value of the Milnor number is non-zero, then it follows from Sard's theorem that $M \times \mathbf{0} \subseteq \Sigma f$, i.e., the critical points of the $f_{\mathbf{t}}$ are not merely a result of critical points of the map $\mathbf{t}$ restricted to the smooth part of $V(f)$.

Let $\mathbf{a}$ be an element of M near the origin. The Milnor number of $f_{\mathbf{a}}$ at the origin satisfies the equality

$$\mu_{\mathbf{0}}(f_{\mathbf{a}}) = \left[V\left(t_0 - a_0, \dots, t_{k-1} - a_{k-1}, \frac{\partial f}{\partial z_0}, \dots, \frac{\partial f}{\partial z_n}\right)\right]_{(\mathbf{a},\mathbf{0})}.$$

In particular,

$$\dim_{(\mathbf{a},\mathbf{0})} V\left(t_0 - a_0, \dots, t_{k-1} - a_{k-1}, \frac{\partial f}{\partial z_0}, \dots, \frac{\partial f}{\partial z_n}\right) = 0.$$

This immediately implies that $\dim_{(\mathbf{a},\mathbf{0})}\Sigma f \leqslant k$. Hence,

$$\left[V\left(\frac{\partial f}{\partial z_0}, \dots, \frac{\partial f}{\partial z_n}\right)\right] = \Gamma^k_{f,(\mathbf{t},\mathbf{z})} + \Lambda^k_{f,(\mathbf{t},\mathbf{z})},$$

both $\gamma^k_{f,(\mathbf{t},\mathbf{z})}(\mathbf{a},\mathbf{0})$ and $\lambda^k_{f,(\mathbf{t},\mathbf{z})}(\mathbf{a},\mathbf{0})$ exist, and

$$\mu_{\mathbf{0}}(f_{\mathbf{a}}) = \gamma^k_{f,(\mathbf{t},\mathbf{z})}(\mathbf{a},\mathbf{0}) + \lambda^k_{f,(\mathbf{t},\mathbf{z})}(\mathbf{a},\mathbf{0}).$$

As $\mu_{\mathbf{0}}(f_{\mathbf{a}})$ is independent of $\mathbf{a}$, and both $\gamma^k_{f,(\mathbf{t},\mathbf{z})}(\mathbf{a},\mathbf{0})$ and $\lambda^k_{f,(\mathbf{t},\mathbf{z})}(\mathbf{a},\mathbf{0})$ are upper-semicontinuous as functions of $\mathbf{a}$, we conclude that both $\gamma^k_{f,(\mathbf{t},\mathbf{z})}(\mathbf{a},\mathbf{0})$ and $\lambda^k_{f,(\mathbf{t},\mathbf{z})}(\mathbf{a},\mathbf{0})$ are independent of $\mathbf{a}$.

This implies that $\gamma^k_{f,(\mathbf{t},\mathbf{z})}(\mathbf{a},\mathbf{0})$ is independent of $\mathbf{a}$ for $(\mathbf{a},\mathbf{0})$ in a k-dimensional component of Σf. But, $\Gamma^k_{f,(\mathbf{t},\mathbf{z})}$ cannot contain a component of Σf. Therefore, the constant value of $\gamma^k_{f,(\mathbf{t},\mathbf{z})}(\mathbf{a},\mathbf{0})$ for $\mathbf{a} \in M$ must be 0; that is, $\Gamma^k_{f,(\mathbf{t},\mathbf{z})}$ does not intersect $M \times \mathbf{0}$ near the origin.

But, all the lower-dimensional relative polar cycles are contained in $\Gamma^k_{f,(\mathbf{t},\mathbf{z})}$; thus, none of them hit $M \times \mathbf{0}$. This implies that $\lambda^i_{f,(\mathbf{t},\mathbf{z})}(\mathbf{a},\mathbf{0}) = 0$ for all $\mathbf{a} \in M$ and all i with $0 \leqslant i \leqslant k-1$. As we already saw that $\lambda^k_{f,(\mathbf{t},\mathbf{z})}(\mathbf{a},\mathbf{0})$ is independent of $\mathbf{a} \in M$, we see that all the Lê numbers of f are constant along $M \times \mathbf{0}$.

Now, apply Corollary 6.6. □

Finally, we have the multi-parameter generalization of the result of Lê and Saito, where the critical loci may have arbitrary dimension.

Theorem 6.9. *Let M be an open neighborhood of the origin in $\mathbb{C}^k$, let $\mathcal{U}$ be an open neighborhood of the origin in $\mathbb{C}^{n+1}$, and let $f : (M \times \mathcal{U}, M \times \mathbf{0}) \to (\mathbb{C}, 0)$ be an analytic function; we write $f_{\mathbf{t}}$ for the function defined by $f_{\mathbf{t}}(\mathbf{z}) := f(\mathbf{t},\mathbf{z})$, where $\mathbf{t} \in M$ and $\mathbf{z} \in \mathcal{U}$.*

Let $s = \dim_{\mathbf{0}}\Sigma f_{\mathbf{0}}$. Suppose that, for all small $\mathbf{t}$, for all i with $0 \leqslant i \leqslant s$, $\lambda^i_{f_{\mathbf{t}},\mathbf{z}}(\mathbf{0})$ is defined and is independent of $\mathbf{t}$. Then, $M \times \mathbf{0}$ satisfies Thom's a_f condition at the origin with respect to the ambient stratum, i.e., if $\mathbf{p}_i$ is a sequence of points in $M \times \mathcal{U} - \Sigma f$ such that $\mathbf{p}_i \to \mathbf{0}$ and such that $T_{\mathbf{p}_i} V(f - f(\mathbf{p}_i))$ converges to some $\mathcal{T}$, then $T_{\mathbf{0}}(M \times \mathbf{0}) \subseteq \mathcal{T}$.

Proof. The proof is by induction on s. To obtain 6.9 from 6.8, one follows word for word our derivation of 6.5 from 6.4. □

Chapter 7. ALIGNED SINGULARITIES

In this chapter, we once again consider analytic functions $h : \mathcal{U} \to \mathbb{C}$. We wish to investigate those h for which the critical locus, Σh, is of a particularly nice form – a form which generalizes isolated singularities, smooth one-dimensional singularities (line singularities), and the singularities found in hyperplane arrangements (see Chapter 5).

The obvious generalization of merely requiring the irreducible components of Σh to be smooth appears to be too general to yield nice results; what one would like is to put some restrictions on the subset of Σh where h fails to be "equisingular". For instance, in the case where Σh is smooth and 2-dimensional, any reasonable notion of equisingularity could fail on a subset of dimension at most one; a reasonable condition to impose is that this one-dimensional subset itself be smooth. Essentially this is what we require of an aligned singularity.

For convenience, throughout this section, we concentrate our attention on hypersurface germs at the origin.

Definition 7.1. If $h : (\mathcal{U}, \mathbf{0}) \to (\mathbb{C}, 0)$ is an analytic function, then an *aligned good stratification* for h at the origin is a good stratification for h at the origin in which the closure of each stratum of the singular set is smooth at the origin.

If such an aligned good stratification exists, we say that h has an *aligned singularity* at the origin.

If $\{S_\alpha\}$ is an aligned good stratification for h at the origin, then we say that a linear choice of coordinates, $\mathbf{z}$, is *an aligning set of coordinates* for $\{S_\alpha\}$ provided that for each i, $V(z_0, \dots, z_{i-1})$ transversely intersects the closure of each stratum of dimension at least i at the origin. Naturally, we say simply that a set of coordinates, $\mathbf{z}$, is *aligning for* h at the origin provided that there exists an aligned good stratification for h at the origin with respect to which $\mathbf{z}$ is aligning.

Note that, given an aligned singularity, aligning sets of coordinates are generic (in the IPZ-topology) and prepolar. It is important that, in fact, aligning coordinates are prepolar at **all** points in an entire neighborhood of the origin; the importance of this fact stems from the following result (in which we are **not** assuming that h has an aligned singularity).

Theorem 7.2. *If the coordinates* $\mathbf{z}$ *are prepolar at all points in a neighborhood* Ω *of a point* $\mathbf{p}$ *with respect to a good stratification* $\{S_\alpha\}$ *for* h *at* $\mathbf{p}$*, then, inside* Ω*, for all* i*,*

$$|\Lambda^i_{h,\mathbf{z}}| \subseteq \bigcup_{\dim_{\mathbf{0}} S_\alpha \leqslant i} S_\alpha .$$

Proof. When $i = 0$, we must show that if $\mathbf{p}$ is in $\Lambda^0_{h,\mathbf{z}}$, then $\mathbf{p}$ is also a stratum. Suppose not.

Then, $\mathbf{p}$ is in some stratum S of dimension at least 1 and, as $\mathbf{p} \in \Lambda^0_{h,\mathbf{z}}$, we must have $\mathbf{p} \in \Gamma^1_{h,\mathbf{z}}$. As our coordinates are prepolar, $V(z_0 - p_0)$ transversely intersects S at $\mathbf{p}$. This, however, is a contradiction, since $\mathbf{p} \in \Gamma^1_{h,\mathbf{z}}$ implies that there is a sequence of limiting tangent planes to level hypersurfaces which converges to $\{0\} \times \mathbb{C}^n$ at $\mathbf{p}$ and, hence, $T_{\mathbf{p}}S$ should be contained in $\{0\} \times \mathbb{C}^n$ since S is a good stratum.

Now, suppose that we have a point $\mathbf{q} \in \Omega$ such that

$$\mathbf{q} \in |\Lambda^i_{h,\mathbf{z}}| \text{ and } \mathbf{q} \notin \bigcup_{\dim_{\mathbf{0}} S_\alpha \leqslant i} S_\alpha.$$

Then, $\mathbf{q} \in |\Lambda^i_{h,\mathbf{z}}| \cap S_\beta$ for some good stratum S_β of dimension at least $i+1$. As $\mathbf{z}$ is prepolar at $\mathbf{q}$,

$$\mathcal{S}' := \{S_\alpha \cap V(z_0 - q_0, \dots, z_{i-1} - q_{i-1})\}_\alpha$$

is a good stratification for $h_{|_{V(z_0-p_0,\dots,z_{i-1}-p_{i-1})}}$ at $\mathbf{q}$; in addition, $V(z_0 - p_0, \dots, z_{i-1} - p_{i-1})$ transversely intersects S_β at $\mathbf{q}$ in a set of dimension at least 1. Thus, $\mathcal{S}'$ does not contain $\{\mathbf{q}\}$ as a stratum.

Now, let $\hat{\mathbf{z}} := (z_i, \dots, z_n)$. Then, $\hat{\mathbf{z}}$ is prepolar with respect to $\mathcal{S}'$ and so, we conclude from the $i = 0$ case (at the beginning of the proof) that

$$\mathbf{q} \notin \Lambda^0_{h_{|_{V(z_0-p_0,\dots,z_{i-1}-p_{i-1})}},\hat{\mathbf{z}}}.$$

By repeated applications of Theorem 1.26 and Proposition 1.18, it follows that $\mathbf{q} \notin \Lambda^i_{h,\mathbf{z}}$; this is a contradiction. □

Closely related to the notion of aligning sets of coordinates is

Definition 7.3. If $h : (\mathcal{U}, \mathbf{0}) \to (\mathbb{C}, 0)$ is an analytic function on an open subset of $\mathbb{C}^{n+1}$, then a linear choice of coordinates, $\mathbf{z}$, for $\mathbb{C}^{n+1}$ is *pre-aligning for h* at the origin provided that for each Lê cycle, $\Lambda^i_{h,\mathbf{z}}$, and for each irreducible component, C, of $\Lambda^i_{h,\mathbf{z}}$ passing through the origin, the following conditions are satisfied:

i) $\dim_{\mathbf{0}} C = i$;

ii) C is smooth at the origin;

iii) $V(z_0, z_1, \dots, z_{i-1})$ transversely intersects C at the origin.

Proposition 7.4. *If h has an aligned singularity at the origin, then for a generic linear reorganization of the coordinates $\mathbf{z}$, $\mathbf{z}$ is pre-aligning for h at the origin, and for all $\mathbf{p}$ near the origin, the reduced Euler characteristic of the Milnor fibre of h at $\mathbf{p}$ is given by*

$$\widetilde{\chi}(F_{h,\mathbf{p}}) = \sum_{i=0}^{s} (-1)^{n-i} \lambda^i_{h,\mathbf{p}}(\mathbf{0}).$$

Proof. One may simply choose $\mathbf{z}$ to be aligning. Theorem 7.2 then implies that $\mathbf{z}$ is pre-aligning. The Euler characteristic statement follows at once from Theorem 3.3. □

Remark 7.5. It is tempting to think that if $\mathbf{z}$ is a set of pre-aligning coordinates for h at the origin, then we can produce an aligned good stratification by considering the components of

$$\Lambda^i_{h,\mathbf{z}} - \bigcup_{j\leqslant i-1} \Lambda^j_{h,\mathbf{z}}.$$

This might seem reasonable in light of Corollary 6.6 and Remark 6.7. However, we see no reason for the higher Lê numbers to be constant along these proposed "strata".

We could define a more restricted class of singularities – *super aligned singularities* – by requiring the existence of an aligned good stratification in which, for every i with $0 \leqslant i \leqslant \dim_{\mathbf{0}}\Sigma h =: s$, there is at most one connected stratum, say S^i, of dimension i and

$$S^0 \subseteq \overline{S^1} \subseteq \cdots \subseteq \overline{S^{s-1}} \subseteq \overline{S^s}.$$

It is easy to see that this is equivalent to the existence of a set of pre-aligning coordinates, $\mathbf{z}$, for h at the origin such that each $\Lambda^i_{h,\mathbf{z}}$ has a single smooth component at the origin and, as germs of sets at the origin,

$$\Lambda^0_{h,\mathbf{z}} \subseteq \Lambda^1_{h,\mathbf{z}} \subseteq \cdots \subseteq \Lambda^s_{h,\mathbf{z}}.$$

For such super aligned singularities, we obtain a good stratification for h by taking the stratification

$$\{\Lambda^{j+1}_{h,\mathbf{z}} - \Lambda^j_{h,\mathbf{z}}\}.$$

Our main interest in aligned singularities is due to

Proposition 7.6. *Suppose that h has an aligned s-dimensional singularity at the origin and that the coordinates $\mathbf{z}$ are aligning. Then, the Lê cycles and Lê numbers can be characterized topologically in the following inductive manner:*

As a set, $\Lambda^s_{h,\mathbf{z}}$ equals the union of the s-dimensional components of the singular set of h. To determine the Lê cycle, to each s-dimensional component, C of Σh, we assign the multiplicity $m_C = (-1)^{n-s}\widetilde{\chi}(F_{h,\mathbf{p}})$ for generic $\mathbf{p} \in C$, where $F_{h,\mathbf{p}}$ denotes the Milnor fibre of h at $\mathbf{p}$ and $\widetilde{\chi}$ is the reduced Euler characteristic. Moreover, for all $\mathbf{p} \in |\Lambda^s_{h,\mathbf{z}}|$, $\lambda^s_{h,\mathbf{z}}(\mathbf{p}) = \sum_{\mathbf{p}\in C} m_C$.

Now, suppose that we have defined the Lê numbers, $\Lambda^i_{h,\mathbf{z}}(\mathbf{p})$ for all $i \geqslant k+1$ and for all $\mathbf{p}$ near the origin.

Then, as a set, $\Lambda^k_{h,\mathbf{z}}$ equals the closure of the k-dimensional components of the set of points $\mathbf{p} \in V(h)$ where

$$\widetilde{\chi}(F_{h,\mathbf{p}}) \neq \sum_{i=k+1}^{s} (-1)^{n-i}\lambda^i_{h,\mathbf{z}}(\mathbf{p}).$$

The Lê cycle is defined by assigning to each irreducible component C of this set the multiplicity

$$m_C = (-1)^{n-k}\left(\widetilde{\chi}(F_{h,\mathbf{p}}) - \sum_{i=k+1}^{s}(-1)^{n-i}\lambda^i_{h,\mathbf{z}}(\mathbf{p})\right),$$

for generic $\mathbf{p} \in C$. *Finally, for all* $\mathbf{p} \in |\Lambda^k_{h,\mathbf{z}}|$, *we have* $\lambda^k_{h,\mathbf{z}}(\mathbf{p}) = \sum_{\mathbf{p}\in C} m_C$.

Proof. As the aligning coordinates are prepolar at each point near the origin, this essentially follows from Theorem 3.3. However, in writing that

$$\lambda^k_{h,\mathbf{z}}(\mathbf{p}) = \sum_{\mathbf{p}\in C} m_C,$$

we are crucially using that the components of the Lê cycle are smooth and tranversely intersected by $V(z_0-p_0,\dots,z_{k-1}-p_{k-1})$. If this were not the case, the intersection multiplicities of $V(z_0-p_0,\dots,z_{k-1}-p_{k-1})$ with the components of the Lê cycles would enter the picture, and the characterization would no longer be purely topological. □

The following two corollaries are immediate:

Corollary 7.7. *If* h *has an aligned singularity at the origin, then all aligning coordinates* $\mathbf{z}$ *determine the same Lê cycles and Lê numbers.*

Corollary 7.8. *Let* f *and* g *be reduced, analytic germs with aligned singularities at the origin in* $\mathbb{C}^{n+1}$. *Let* $\mathbf{z}$ *and* $\tilde{\mathbf{z}}$ *be aligning sets of coordinates for* f *and* g, *respectively. If* H *is a local, ambient homeomorphism from the germ of* $V(f)$ *at the origin to the germ of* $V(g)$ *at the origin, then as germs of sets at the origin,*

$$H(\Lambda^i_{f,\mathbf{z}}) = \Lambda^i_{g,\tilde{\mathbf{z}}},$$

for all i, *and for all* $\mathbf{p}$ *near the origin in* $\mathbb{C}^{n+1}$,

$$\lambda^i_{f,\mathbf{z}}(\mathbf{p}) = \lambda^i_{g,\tilde{\mathbf{z}}}(H(\mathbf{p})),$$

for all i.

Now, we will give an amusing application of the results of this section. In [**Z**], Zariski conjectures that the multiplicity of a hypersurface at a point is an invariant of the local, ambient topological type of the hypersurface. A number of people have concentrated on the case of a one-parameter family of isolated singularities, but even this case has not been settled (however, for families of quasi-homogeneous isolated singularities, the proof of the conjecture has been given by Greuel [**Gr1**] and O'Shea [**O'S**]).

In our paper [**Mas13**], we prove a result which perhaps supplies a better place to look for counterexamples to the Zariski Multiplicity Conjecture; we prove, for families of hypersurfaces of dimension unequal to 2, that the Zariski Multiplicity Conjecture is true for families of hypersurfaces with isolated singularities if and only if it is true for families of hypersurfaces with smooth one-dimensional critical loci. The results of this section allow us to generalize this.

Theorem 7.9. *The following are equivalent:*

i) for all $n \geqslant 3$, the Zariski Multiplicity Conjecture is true for families of reduced analytic hypersurfaces $f_t : (\mathcal{U}, \mathbf{0}) \to (\mathbb{C}, 0)$, where $\mathcal{U}$ is an open subset of $\mathbb{C}^{n+1}$ and $\dim_{\mathbf{0}}\Sigma f_t = 0$;

ii) for all $n \geqslant 3$, there exists a k such that the Zariski Multiplicity Conjecture is true for families of reduced analytic hypersurfaces $f_t : (\mathcal{U}, \mathbf{0}) \to (\mathbb{C}, 0)$ with aligned singularities, where $\mathcal{U}$ is an open subset of $\mathbb{C}^{n+1}$ and $\dim_{\mathbf{0}}\Sigma f_t = k$;

iii) for all $n \geqslant 3$, for all k, the Zariski Multiplicity Conjecture is true for families of reduced analytic hypersurfaces $f_t : (\mathcal{U}, \mathbf{0}) \to (\mathbb{C}, 0)$ with aligned singularities, where $\mathcal{U}$ is an open subset of $\mathbb{C}^{n+1}$ and $\dim_{\mathbf{0}}\Sigma f_t = k$.

Proof. Certainly, iii) implies ii). We will show that ii) implies i) and that i) implies iii).

Suppose that ii) is true for some $k \geqslant 1$. Let $f_t : (\mathcal{U}, \mathbf{0}) \to (\mathbb{C}, 0)$ be an analytic family, where $\mathcal{U}$ is an open subset of $\mathbb{C}^{n+1}$, $n \geqslant 3$, $\dim_{\mathbf{0}}\Sigma f_t = 0$, and such that the local ambient topological type of the hypersurfaces $V(f_t)$ at the origin is independent of t. Then clearly, the family

$$\tilde{f}_t : (\mathcal{U} \times \mathbb{C}^k, \mathbf{0}) \to (\mathbb{C}, 0)$$

defined by $\tilde{f}_t(\mathbf{z}, \mathbf{w}) := f_t(\mathbf{z})$ is a family of aligned singularities of dimension k with constant topological type.

Hence, by ii), $\operatorname{mult}_{\mathbf{0}} \tilde{f}_t$ is independent of t. Now, as $\operatorname{mult}_{\mathbf{0}} f_t$ clearly equals $\operatorname{mult}_{\mathbf{0}} \tilde{f}_t$, we are finished with the implication that ii) implies i).

The interesting implication is, of course, that i) implies iii). Ideally, we would like to be able to select linear coordinates, $\mathbf{z}$, for $\mathbb{C}^{n+1}$ which are aligning for f_t at the origin for all small t; however, a proof that this is possible seems problematic. We will avoid needing such a result by being somewhat devious and applying the Baire Category Theorem.

Suppose that i) is true, and that we have a family of reduced analytic hypersurfaces $f_t : (\mathcal{U}, \mathbf{0}) \to (\mathbb{C}, 0)$ with aligned singularities, where $\mathcal{U}$ is an open subset of $\mathbb{C}^{n+1}$, $n \geqslant 3$, $\dim_{\mathbf{0}}\Sigma f_t = k$, and such that the local ambient topological type of the hypersurfaces $V(f_t)$ at the origin is independent of t.

Let t_m be an infinite sequence in $\mathbb{C}$ which approaches 0, e.g., $t_m = \frac{1}{m}$. For each t_m, there exists a generic subset of $PGL(\mathbb{C}^{n+1})$ representing aligned coordinates for f_{t_m}. We may apply the Baire Category Theorem to conclude that there exists a choice of coordinates, $\mathbf{z}$, which is aligning for f_0 and for f_{t_m} for all m. Let us fix such a choice of coordinates.

Then, by 7.8, the Lê numbers $\lambda^i_{f_0,\mathbf{z}}(\mathbf{0})$ are equal to the Lê numbers $\lambda^i_{f_{t_m},\mathbf{z}}(\mathbf{0})$ for all large m. By 4.6, if we take $0 \ll j_0 \ll j_1 \ll \cdots \ll j_{s-1}$, then $f_0 + z_0^{j_0} + z_1^{j_1} + \cdots + z_{s-1}^{j_{s-1}}$ has an isolated singularity at the origin; this implies that, for all small t, $f_t + z_0^{j_0} + z_1^{j_1} + \cdots + z_{s-1}^{j_{s-1}}$ has, at worst, an isolated singularity at the origin. In addition, 4.6 tells us that $f_0 + z_0^{j_0} + z_1^{j_1} + \cdots + z_{k-1}^{j_{k-1}}$ has the same Milnor number at the origin as $f_{t_m} + z_0^{j_0} + z_1^{j_1} + \cdots + z_{k-1}^{j_{k-1}}$ for all large m. As the Milnor number at the origin in the family $f_t + z_0^{j_0} + z_1^{j_1} + \cdots + z_{k-1}^{j_{k-1}}$ is upper-semicontinuous, it follows that, in fact, the Milnor number in this family is independent of t for all small t.

Hence, by [**L-R**], the local, ambient topological type is independent of t in the family $f_t + z_0^{j_0} + z_1^{j_1} + \cdots + z_{k-1}^{j_{k-1}}$. Since we are assuming i), this implies that the multiplicity is independent of t in the family $f_t + z_0^{j_0} + z_1^{j_1} + \cdots + z_{k-1}^{j_{k-1}}$. Finally, as the j_m's are arbitrarily large, this implies that the multiplicity is independent of t in the family f_t. $\square$

Chapter 8. SUSPENDING SINGULARITIES

In [**Ok**], [**Sak**], and [**Se-Th**], the general question is addressed of how the structure of the Milnor fibre of $f(\mathbf{z})+g(\mathbf{w})$ (where $\mathbf{z}$ and $\mathbf{w}$ are disjoint sets of variables) depends on the Milnor fibres of f and g. However, in each of these papers, f and g have isolated singularities or are quasi-homogeneous. Sakamoto remarks at the end of his paper that, by using Lê's notion of a good stratification, he can prove his main lemmas without the isolated singularity assumptions. As we crucially need this result in the special case of $f(\mathbf{z}) + w^j$, we will use our results in the appendix to indicate how one needs to modify Sakamoto's proof.

After we describe the homotopy-type of the Milnor fibre of $f(\mathbf{z}) + w^j$, we will use this description to give a new generalization of the formula of Lê and Iomdine [**Lê4**] – a different generalization than the formulas of Chapter 4. This new generalization appears in [**Mas12**].

Proposition 8.1. *If $j \geqslant 2$, then up to homotopy, the Milnor fibre of $\tilde{h}(w, \mathbf{z}) := h(\mathbf{z}) + w^j$ at the origin is the one-point union (wedge) of $j-1$ copies of the suspension of the Milnor fibre of h at the origin.*

Proof. By Proposition C.14, we may use neighborhoods of the form $\mathbb{D}_\omega \times B_\epsilon$, $0 < \omega \ll \epsilon$, to define the Milnor fibre of $\tilde{h}$ at the origin.

Now, for $0 < |\xi| \ll \omega \ll \epsilon$, consider the map

$$\big(\mathbb{D}_\omega \times B_\epsilon\big) \cap V(h + w^j - \xi) \xrightarrow{\ w\ } \mathbb{D}_\omega .$$

This map is a proper, stratified submersion above all points of $\mathbb{D}_\omega - V(w^j - \xi)$, i.e., except at the j roots of ξ. Thus, except above these j points, the fibre is the same as that above 0, which is clearly nothing more than the Milnor fibre of h at the origin. In addition, above each point of $V(w^j - \xi)$, the fibre is $B_\epsilon \cap V(h)$, which is contractible.

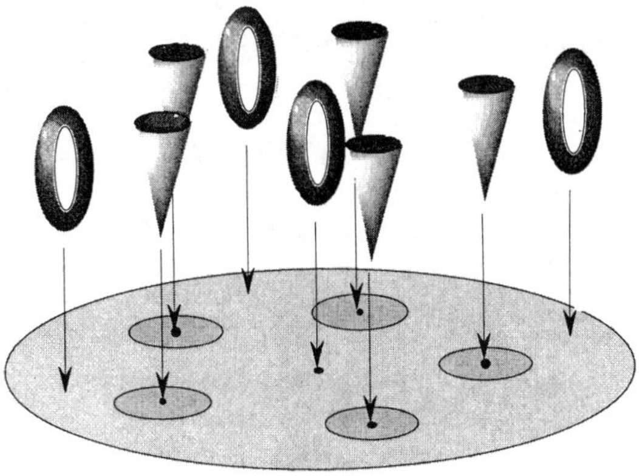

Figure 8.2. The fibres over the disc

In fact, around each point $\alpha_1, \dots, \alpha_j$ in $V(w^j - \xi) \subseteq \mathbb{D}_\omega$, there is an arbitrarily small disc $\mathbb{D}_{\alpha_i} \subseteq \mathbb{D}_\omega$ above which the total space is contractible. We choose the $\mathbb{D}_{\alpha_i}$ disjoint. Connect all of the $\mathbb{D}_{\alpha_i}$ to the origin by disjoint paths. Let P denote the subset of $\mathbb{D}_\omega$ consisting of the paths; so, P is a contractible set which has exactly one point in common with each of the $\mathbb{D}_{\alpha_i}$, and the fibre above each point of P has the homotopy-type of the Milnor fibre of h at the origin.

The result now follows easily. For the details, we refer the reader to Sakamoto [**Sak**] – the remainder of our proof now follows his exactly. □

We shall now use 8.1 to give our second generalization of the formula of Lê and Iomdine.

Let $\mathcal{U}$ be an open neighborhood of the origin in $\mathbb{C}^{n+1}$, let $h : (\mathcal{U}, \mathbf{0}) \to (\mathbb{C}, 0)$ be an analytic function, and suppose that the linear from $L : \mathbb{C}^{n+1} \to \mathbb{C}$ is prepolar with respect to h at the origin.

The formula of Lê and Iomdine says that, if $\dim_{\mathbf{0}} \Sigma h = 1$, then, for all large j, $h + L^j$ has an isolated singularity at the origin and

$$b_n(h + L^j) \; = \; \mu(h + L^j) \; = \; b_n(h) - b_{n-1}(h) + j \sum_{\nu} m_\nu \delta_\nu(h),$$

where $b_i()$ denotes the i-th Betti number of the Milnor fibre of a function at the origin, μ denotes the Milnor number of the isolated singularity at the origin, the summation is over all components, ν, of Σh, m_ν is the local degree of L restricted to ν at the origin, and $\delta_\nu(h)$ is the Milnor number of a generic hyperplane slice of h at a point $\mathbf{p} \in \nu - \mathbf{0}$ sufficiently close to the origin.

This formula has, at least, two possible generalizations. One generalization is in terms of Lê numbers, as given in Chapter 4. But, while there are Morse inequalities between the Lê numbers and the Betti numbers of the Milnor fibre, the Lê numbers are not themselves (generally) Betti numbers of the Milnor fibre. So, one might ask for a generalization of the formula of Lê and Iomdine which generalizes the Betti number information.

In remainder of this chapter, we prove that, if $\dim_{\mathbf{0}} \Sigma h = s \geqslant 1$, then, for all large j, $\dim_{\mathbf{0}} \Sigma(h + L^j) = s - 1$ and

$$b_n(h + L^j) \; = \; b_n(h) - b_{n-1}(h) + j\big(b_{n-1}(h_{|V(L)}) - \gamma^1_{h,L}(\mathbf{0})\big).$$

In the case where h has a one-dimensional critical locus at the origin, it is easy to show that this new formula reduces to that of Lê and Iomdine.

We consider this new Lê-Iomdine formula interesting because it implies that

$$b_{n-1}(h_{|V(L)}) \geqslant \gamma^1_{h,L}(\mathbf{0}).$$

In terms of deformations, this says that the top possible non-zero Betti number of the Milnor fibre of $h_{|V(L)}$ is greater than or equal to $\gamma^1_{h,L}(\mathbf{0})$ for all h which have $V(L)$ as a prepolar slice. Thus, if we define h to be a *prepolar deformation* of $h_{|V(L)}$ precisely when $V(L)$ is a prepolar slice of h, we obtain a class of deformations of $h_{|V(L)}$ which give lower bounds on the top Betti number of the Milnor fibre. This also suggests that it might be useful to study prepolar deformations for which $\gamma^1_{h,L}(\mathbf{0})$ obtains its maximum value.

We will need

Proposition 8.3. *If $j \geqslant 2$ and $\mathcal{S}$ is a good stratification for h at the origin, then*

$$\{V(h+w^j) - \Sigma(h+w^j)\} \ \cup \ \{0 \times S \mid S \text{ is a singular stratum of } \mathcal{S}\}$$

is a good stratification for $h + w^j$ at the origin.

Proof. As $j \geqslant 2$, $\Sigma(h+w^j) = \{0\} \times \Sigma h$.

Let $\mathbf{p} = (0, \mathbf{q}) \in \{0\} \times \Sigma h$, where $S \in \mathcal{S}$, and let $\mathbf{p}_i = (u_i, \mathbf{q}_i)$ be a sequence of points in $\mathbb{C} \times \mathcal{U} - \{0\} \times \Sigma h$ such that $\mathbf{p}_i \to \mathbf{p}$ and

$$T_{\mathbf{p}_i} V(h + w^j - (h+w^j)_{|\mathbf{p}_i}) \to \mathcal{T}.$$

We wish to show that $T_{\mathbf{p}}(\{0\} \times S) = \{0\} \times T_{\mathbf{q}} S \subseteq \mathcal{T}$.

If $\mathcal{T} = T_{\mathbf{p}} V(w) = \{0\} \times \mathbb{C}^{n+1}$, then we are finished. So, suppose otherwise. Then, by taking a subsequence, we may assume that $\mathbf{q}_i \notin \Sigma h$ and that

$$T_{\mathbf{q}_i} V(h - h(\mathbf{q}_i)) \to \eta.$$

As $\mathcal{T}$ transversely intersects $T_{\mathbf{p}} V(w)$, $T_{\mathbf{p}_i} V(h + w^j - (h+w^j)_{|\mathbf{p}_i})$ transversely intersects $T_{\mathbf{p}_i} V(w - w_i)$ for all $\mathbf{p}_i$ close to $\mathbf{p}$. Thus,

$$\mathcal{T} \cap (\{0\} \times \mathbb{C}^{n+1}) = \lim T_{\mathbf{p}_i} V(h + w^j - (h+w^j)_{|\mathbf{p}_i}) \cap T_{\mathbf{p}_i} V(w - w_i) =$$

$$\lim T_{\mathbf{p}_i} V(h + w^j - (h+w^j)_{|\mathbf{p}_i}, w - w_i) = \lim T_{\mathbf{p}_i} V(h - h(\mathbf{q}_i), w - w_i) = \{0\} \times \eta.$$

Now, as S is a good stratum for h, $T_{\mathbf{q}}(S) \subseteq \eta$ and the proposition follows. □

Corollary 8.4. *If $V(z_0)$ is a prepolar slice for h at $\mathbf{0}$ then, for all $j \geqslant 2 + \lambda^0_{h,\mathbf{z}}(\mathbf{0})$, $V(z_0 - w)$ is a prepolar slice for $h + w^j$ at $\mathbf{0}$.*

Proof. In light of the proposition, all that we must show is that, for all $j \geqslant 2 + \lambda^0_{h,\mathbf{z}}(\mathbf{0})$, $\Sigma(h+w^j) \cap V(z_0 - w) = \Sigma(h + w^j{}_{|V(z_0 - w)})$.

Now,

$$\Sigma(h+w^j) \cap V(z_0 - w) = (\{0\} \times \Sigma h) \cap V(z_0 - w) = \{0\} \times (\Sigma h \cap V(z_0)).$$

On the other hand,

$$\Sigma(h + w^j{}_{|V(z_0 - w)}) = (\mathbb{C} \times \Sigma(h + z_0^j)) \cap V(z_0 - w).$$

But, near the origin and for $j \geqslant 2 + \lambda^0_{h,\mathbf{z}}(\mathbf{0})$, $\Sigma(h + z_0^j) = \Sigma h \cap V(z_0)$ by 4.3.iii. The conclusion follows. □

Theorem 8.5. *If $V(z_0)$ is a prepolar slice of h at $\mathbf{0}$ then, for all $j \geqslant 2+\lambda^0_{h,\mathbf{z}}(\mathbf{0})$,*

$$b_n(h+z_0^j) \;=\; b_n(h)-b_{n-1}(h)+j\big(b_{n-1}(h_{|_{V(z_0)}})-\gamma^1_{h,z_0}(\mathbf{0})\big),$$

where $b_i()$ denotes the i-th Betti number of the Milnor fibre at the origin.

In particular, $b_{n-1}(h_{|_{V(z_0)}}) \geqslant \gamma^1_{h,z_0}(\mathbf{0})$.

Proof.

After applying Proposition 3.1 to $h+w^j$ and the slice $V(z_0-w)$, and considering the long exact sequence of the pair, we have

$$b_{n+1}(h+w^j)-b_n(h+w^j)+b_n(h+z_0^j)=\left(\Gamma^1_{h+w^j,z_0-w}\cdot V(h+w^j)\right)_{\mathbf{0}}$$

which, by 4.3.v, equals $j\lambda^0_{h,\mathbf{z}}(\mathbf{0})$.

Now, as the Milnor fibre of $h+w^j$ has the homotopy-type of the one-point union of $j-1$ copies of the suspension of the Milnor fibre of h, we obtain

$$(j-1)b_n(h)-(j-1)b_{n-1}(h)+b_n(h+z_0^j)=j\lambda^0_{h,\mathbf{z}}(\mathbf{0}).$$

Using 3.1 on h itself and rearranging, we get

$$b_n(h+z_0^j)=b_n(h)-b_{n-1}(h)+j\left[\lambda^0_{h,\mathbf{z}}(\mathbf{0})-\left(\left(\Gamma^1_{f,z_0}\cdot V(f)\right)_{\mathbf{0}}-b_{n-1}(h_{|_{V(z_0)}})\right)\right].$$

Finally, using the formula of Proposition 1.20 that

$$\left(\Gamma^1_{h,z_0}\cdot V(h)\right)_{\mathbf{0}}=\gamma^1_{h,z_0}(\mathbf{0})+\lambda^0_{h,z_0}(\mathbf{0}),$$

we obtain the desired result. □

The result of Theorem 8.5 is best thought of in terms of prepolar deformations: every prepolar deformation, h, of a fixed h_0 yields a lower bound on the top Betti number of the Milnor fibre of h_0.

One might hope that, by considering a prepolar deformation, h, for which $\gamma^1_{h,\mathbf{z}}(\mathbf{0})$ obtains its maximal value, one would actually obtain the top Betti number of the Milnor fibre of h_0. This seems unlikely however; certain singularities seem to be "rigid" with respect to prepolar deformations, in the weak sense that any prepolar deformation, h, has no polar curve at the origin.

Nonetheless, the lower bounds provided by prepolar deformations give new data which helps describe the Milnor fibre of a completely general affine hypersurface singularity; these data do not appear to follow from our Morse inequalities between the Betti numbers of the Milnor fibre and the Lê numbers of the hypersurface, as given in Theorem 3.3. As part of these Morse inequalities, we showed that $\lambda^0_{h_0,\tilde{z}}(\mathbf{0})$, provides an upper-bound on the top Betti number of the Milnor fibre of h_0. Also, it follows from 1.18 that if h is a prepolar deformation of h_0, then

$$\lambda^0_{h_0,\tilde{z}}(\mathbf{0})=\lambda^1_{h,\mathbf{z}}(\mathbf{0})+\gamma^1_{h,\mathbf{z}}(\mathbf{0}).$$

Thus, given a prepolar deformation, h, of h_0, we have bounded the top Betti number of the Milnor fibre of h_0:

$$\gamma^1_{h,\mathbf{z}}(\mathbf{0}) \leqslant b_{n-1}(h_0) \leqslant \lambda^1_{h,\mathbf{z}}(\mathbf{0}) + \gamma^1_{h,\mathbf{z}}(\mathbf{0}).$$

As $\lambda^0_{h_0,\tilde{z}}(\mathbf{0})$ is fixed, a prepolar deformation, h, with maximal $\gamma^1_{h,\mathbf{z}}(\mathbf{0})$ will have minimal $\lambda^1_{h,\mathbf{z}}(\mathbf{0})$. We prefer to call such a deformation a *minimal prepolar deformation*, instead of a maximal one. Note that a minimal prepolar deformation will not only have the maximal possible lower bound on the top Betti number of the Milnor fibre, it also provides the smallest difference between our general upper and lower bounds. One might hope that it is always possible to find a prepolar deformation, h, for which $\lambda^1_{h,\mathbf{z}}(\mathbf{0}) = 0$, for then we would have $b_{n-1}(h_0) = \gamma^1_{h,z_0}(\mathbf{0})$; unfortunately, Proposition 1.29 implies that it is usually impossible to find such a deformation.

Chapter 9. CONSTANCY OF THE MILNOR FIBRATIONS

In this chapter, we prove what is perhaps our most important result, and certainly the result which requires the most machinery – we generalize the result of Lê and Ramanujam [**L-R**] as stated in Theorem 0.10 in the introduction. Basically, we prove that if the Lê numbers are constant in a one-parameter family, then the Milnor fibrations are constant in the family, **regardless of the dimension of the critical loci**.

Unfortunately, we do not obtain the result that the local, ambient topological-type of the hypersurfaces remains constant in the family. It is an open question whether the constancy of the Lê numbers is strong enough to imply this topological constancy.

While the idea behind our proof of this generalized Lê-Ramanujam is simple, the technical details are horrendous. It is this chapter alone which is responsible for the existence of Appendix C of this book; we have relegated most of the technical details to the appendix. Before we prove the main result, there remain only two lemmas which we need (besides the results which appear in the appendix). Also, we will restate one of the results from the appendix in a form which is comprehensible without reading the entire appendix.

First, however, we wish to sketch the proof of the main theorem, so that the reader can see that the idea really is fairly easy. On the other hand, the proof is not straightforward – instead, it uses a trick which gives one very little insight as to why the result should be true.

Throughout this chapter, $\mathcal{U}$ will denote an open neighborhood of the origin in $\mathbb{C}^{n+1}$ and $f_t : (\mathcal{U}, \mathbf{0}) \to (\mathbb{C}, 0)$ will be an analytic family in the variables $\mathbf{z} = (z_0, \dots, z_n)$. Let $s = \dim_{\mathbf{0}} \Sigma f_0$.

A sketch of the proof is as follows:

The result of Proposition 8.1 is that the Milnor fibre of $f_t + w^j$ at the origin is homotopy-equivalent to the one-point union of $j-1$ copies of the suspension of the Milnor fibre of f_t at the origin. So, it certainly seems reasonable to expect that the Milnor fibrations are independent of t in the family f_t if and only if the Milnor fibrations are independent of t in the family $f_t + w^j$. But why should the family $f_t + w^j$ be any easier to study than the family f_t itself?

It is easier because we have very nice hyperplanes defined by $L = w - z_0$ such that, when we take the sections $(f_t + w^j)_{|_{V(L)}}$, we get the family $f_t + z_0^j$ which, for generic z_0 and for large j, is a family of singularities of one less dimension (this follows from the results of Chapter 4); that is, $\dim_{\mathbf{0}} \Sigma(f_0 + z_0^j) = s - 1$. Moreover, Theorem 0.9, Lê's attaching result, tells us how the Milnor fibre of $f_t + w^j$ is obtained from the Milnor fibre of a generic hyperplane section. The Milnor fibre of $f_t + w^j$ is obtained from the Milnor fibre of $f_t + z_0^j$ by attaching $\left(\Gamma^1_{f_t + w^j, w - z_0} \cdot V(f_t + w^j)\right)_{\mathbf{0}}$ $(n+1)$-handles.

By induction on s, we may require the Milnor fibrations of $f_t + z_0^j$ to be independent of t. If we also require the number of attached $(n+1)$-handles to be independent of t, it seems reasonable to expect that the Milnor fibrations of the family $f_t + w^j$ should be independent of t and, thus, that the Milnor fibrations of f_t are independent of t.

The Lê numbers enter the picture because Lemma 4.3 says that, for large j, the intersection number $\left(\Gamma^1_{f_t+w^j, w-z_0} \cdot V(f_t+w^j)\right)_{\mathbf{0}} = j\lambda^0_{f_t,\mathbf{z}}(\mathbf{0})$. Combining this with the Lê-Iomdine formulas of Theorem 4.5, we find that the inductive requirement that the Milnor fibrations of $f_t + z_0^j$ are independent of t amounts to requiring all the Lê numbers of f_t to be independent of t.

We first wish to prove a result which will tell us that the main theorem of this chapter is not vacuously true.

Lemma 9.1. *For all i with $0 \leqslant i \leqslant n$, for a generic linear reorganization of the coordinates $(z_0, \dots, z_i)$, $(z_0, \dots, z_i)$ is prepolar at the origin for f_t for all small t.*

Proof. Fix a good stratification, $\mathfrak{G}$, for f in a neighborhood, $\mathcal{V}$, of the origin. By refining if necessary, we may also assume that $\mathfrak{G}$ satisfies Whitney's condition a). We will also assume that

$$S := \mathcal{V} \cap \{t\text{-axis}\} - \{\mathbf{0}\} = \mathcal{V} \cap (\mathbb{C} \times \{\mathbf{0}\}) - \{\mathbf{0}\}$$

and $\{\mathbf{0}\}$ are strata of $\mathfrak{G}$. As the function t has isolated stratified critical values, $V(t - t_0)$ transversely intersects all strata of $\mathfrak{G}$ near $(t_0, \mathbf{0})$; hence, $V(t - t_0) \cap \mathfrak{G}$ provides a good stratification for f_{t_0} at $\mathbf{0}$ which still satisfies Whitney's condition a).

By induction on i, we will prove that: for all i with $0 \leqslant i \leqslant n$, for a generic linear reorganization of the coordinates $(z_0, \dots, z_i)$, $(z_0, \dots, z_i)$ is prepolar at the origin for f_{t_0} with respect to the good stratification $V(t - t_0) \cap \mathfrak{G}$ for all small, non-zero t_0.

$i = 0$: Using the terminology of Goresky and MacPherson [**G-M2**], the set of degenerate conormal covectors to S is a complex analytic subvariety of codimension at least 1 inside the total space of the conormal bundle of S inside $\mathbb{C}^{n+2}$ (see [**G-M2**, Prop. 1.8, p.44]). Projectivizing and dualizing, this says that the set Ω, defined by

$$\{(\mathbf{p}, H) \in S \times G_n(\mathbb{C}^{n+1}) \mid T_{\mathbf{p}}(\mathbb{C} \times H) \text{ contains a generalized tangent plane of } \mathfrak{G} \text{ at } \mathbf{p}\},$$

has dimension at most n. Hence, $\overline{\Omega} \subseteq (\mathcal{V} \cap (\mathbb{C} \times \mathbf{0})) \times G_n(\mathbb{C}^{n+1})$ has dimension at most n and the fibre over $\mathbf{0}$, call it Ψ, must therefore have dimension at most $n-1$. Thus, $G_n(\mathbb{C}^{n+1}) - \Psi$ is open and dense in $G_n(\mathbb{C}^{n+1})$. We claim that, for all H in $G_n(\mathbb{C}^{n+1}) - \Psi$, H is a prepolar slice for f_t at $\mathbf{0}$ for all small, non-zero t.

Certainly, if $H \notin \Psi$, then, for all small, non-zero t_0, $T_{(t_0,\mathbf{0})}(\mathbb{C} \times H)$ contains no generalized tangent plane from $\mathfrak{G}$ at $(t_0, \mathbf{0})$. Also, Whitney's condition a) guarantees that all limiting tangent planes from strata of $\mathfrak{G}$ at $(t_0, \mathbf{0})$ actually contains $T_{(t_0,\mathbf{0})}S = \mathbb{C} \times \mathbf{0}$. Combining these two facts, it follows easily that $T_{\mathbf{0}}H$ contains no generalized tangent plane from $V(t - t_0) \cap \mathfrak{G}$ at $\mathbf{0}$. Thus, H is prepolar for f_{t_0} at $\mathbf{0}$ with respect to the good stratification $V(t - t_0) \cap \mathfrak{G}$. (Actually, this implies much more – it implies that H is *polar*, as defined in [**Mas8**].)

$i \geqslant 1$: Now, assume that we have already chosen $(z_0, \dots, z_{i-1})$ generically (in the IPZ-topology) so that $(z_0, \dots, z_{i-1})$ is prepolar at the origin for f_{t_0} with respect to the good stratification $V(t-t_0)\cap\mathfrak{G}$ for all small, non-zero t_0.

Then, there exists a good stratification, $\mathfrak{G}'$, for $f_{|_{V(z_0,\dots,z_{i-1})}}$ at the origin which satisfies Whitney's condition a). Though $\mathfrak{G}'$ may not necessarily be chosen to equal

$$\mathfrak{G}\cap V(z_0,\dots,z_{i-1}),$$

after refining $\mathfrak{G}'$ using Proposition C.2, we may certainly assume that each stratum of $\mathfrak{G}'$ is contained in a stratum of $\mathfrak{G}$ and that, in some neighborhood of the origin, $(\mathbb{C}\times\mathbf{0})-\mathbf{0}$ and $\mathbf{0}$ are strata of $\mathfrak{G}'$.

By the $i=0$ case, for a generic choice of z_i, $V(z_i)$ is a prepolar slice for $f_{t_0|_{V(z_0,\dots,z_{i-1})}}$ at the origin with respect to $V(t-t_0)\cap\mathfrak{G}'$ for all small non-zero t_0. But, as each stratum of $\mathfrak{G}'$ is contained in a stratum of $\mathfrak{G}$, this last statement is stronger than saying that $V(z_i)$ is prepolar with respect to $V(t-t_0)\cap\mathfrak{G}$. This concludes the induction.

Finally, to finish the proof, one simply chooses coordinates as generic as given above and generically enough so that the coordinates are prepolar for f_0 at the origin. □

We shall also need the following uniform version of Proposition 4.19:

Proposition 9.2. *If $(z_0,\dots,z_i)$ is prepolar at the origin for f_t for all small t, then, for all large j, $(z_1,\dots,z_i)$ is prepolar for $f_t+z_0^j$ for all small t.*

Proof. In light of Corollary 4.18, what we need to show is that, for all large j, for all k with $0\leqslant k\leqslant i$ and for all small t_0,

$$(*)\qquad \dim_{\mathbf{0}}\Gamma^{k+1}_{f_{t_0},\mathbf{z}}\cap V\left(t-t_0,\frac{\partial f}{\partial z_0}+jz_0^{j-1}\right)\cap V(z_1,\dots,z_k)\leqslant 0.$$

In fact, we only have to show that $(*)$ holds for small **non-zero** t_0, for then – by 4.19 – we may impose the extra largeness condition on j so that $(*)$ also holds for $t_0=0$.

By Lemma 4.14, for all small non-zero t_0, $\Gamma^{k+1}_{f_{t_0},\mathbf{z}}=\Gamma^{k+2}_{f,(t,\mathbf{z})}\cap V(t-t_0)$ as sets, in a neighborhood of the origin. In addition, by Theorem 1.26, $\gamma^{k+1}_{f_{t_0},\mathbf{z}}(\mathbf{0})$ exists for all small t_0; thus, $\dim_{\mathbf{0}}\Gamma^{k+1}_{f_{t_0},\mathbf{z}}\cap V(z_0,z_1,\dots,z_k)\leqslant 0$. Putting these two facts together, we find that, for all small non-zero t_0,

$$\dim_{\mathbf{0}}V(t-t_0)\cap\Gamma^{k+2}_{f,(t,\mathbf{z})}\cap V(z_0,z_1,\dots,z_k)\leqslant 0.$$

Let W denote the union of those irreducible analytic components, C, of

$$\Gamma^{k+2}_{f,(t,\mathbf{z})}\cap V(z_1,\dots,z_k),$$

at the origin, which satisfy the property that the t-axis is contained in $C\cap V(z_0)$. By the above, $W\cap V(z_0)$ is at most one-dimensional at the origin and, for all small non-zero t_0, $\Gamma^{k+1}_{f_{t_0},\mathbf{z}}\cap V(z_1,\dots,z_k)=W\cap V(t-t_0)$ as germs of sets at $(t_0,\mathbf{0})$.

It follows that each irreducible component of W at the origin is at most 2-dimensional. We would like to show that, for all large j, $W \cap V\left(\frac{\partial f}{\partial z_0} + jz_0^{j-1}\right)$ is at most 1-dimensional at the origin, for then – by intersecting with $V(t - t_0)$ – we obtain $(*)$.

This is easy. For if a 2-dimensional component C of W is contained in $V\left(\frac{\partial f}{\partial z_0} + j_0 z_0^{j_0-1}\right)$ and $V\left(\frac{\partial f}{\partial z_0} + j_1 z_0^{j_1-1}\right)$ for $j_0 \neq j_1$, then $C \subseteq V(z_0)$. However, we know that $W \cap V(z_0)$ is at most 1-dimensional. Hence, there are only a finite number of "bad" values for j. □

Proposition 9.3 *Suppose that $V(z_0)$ is a prepolar slice for f_t at the origin for all small t, and that $V(t)$ does not occur as the limit of tangent spaces to level hypersurfaces of f or $f_{|V(z_0)}$ at the origin. Then, for all small non-zero t_0, there is a natural inclusion of pairs of Milnor fibres $(F_{f_{t_0},\mathbf{0}}, F_{f_{t_0},\mathbf{0}} \cap V(z_0)) \hookrightarrow (F_{f_0,\mathbf{0}}, F_{f_0,\mathbf{0}} \cap V(z_0))$.*

If we also assume that the intersection number $\left(\Gamma^1_{f_t,z_0} \cdot V(f_t)\right)_\mathbf{0}$ is independent of t for all small t, then we have the following three results:

i) if the inclusion $F_{f_{t_0},\mathbf{0}} \cap V(z_0) \hookrightarrow F_{f_0,\mathbf{0}} \cap V(z_0)$ induces isomorphisms on integral homology groups, then so does the inclusion $F_{f_{t_0},\mathbf{0}} \hookrightarrow F_{f_0,\mathbf{0}}$;

ii) if $s \leqslant n-2$ and the inclusion $F_{f_{t_0},\mathbf{0}} \cap V(z_0) \hookrightarrow F_{f_0,\mathbf{0}} \cap V(z_0)$ is a homotopy-equivalence, then so is the inclusion $F_{f_{t_0},\mathbf{0}} \hookrightarrow F_{f_0,\mathbf{0}}$, and the fibre homotopy-type of the Milnor fibrations is independent of t for all small t; and

iii) if $s \leqslant n-3$ and the inclusion $F_{f_{t_0},\mathbf{0}} \cap V(z_0) \hookrightarrow F_{f_0,\mathbf{0}} \cap V(z_0)$ is a homotopy-equivalence, then the inclusion $F_{f_{t_0},\mathbf{0}} \hookrightarrow F_{f_0,\mathbf{0}}$ is a diffeomorphism and, moreover, the diffeomorphism-type of the Milnor fibrations is independent of t for all small t.

Proof. This is primarily a restatement of Theorem C.13 from the appendix.

The condition on $V(t)$ is that $V(t)$ is not in the Thom set of either f or $f_{|V(z_0)}$ at the origin (see Definition C.8). Therefore, by Proposition C.9, the families f_t and $f_{t|V(z_0)}$ satisfy the universal conormal condition at the origin (see Definition C.10) and this produces the inclusion of pairs of Milnor fibres. Now, i) and ii) follow from Theorem C.13, and iii) follows from ii) together with Proposition C.12. □

We are now able to prove the main result of this chapter: our generalization of part of the result of Lê and Ramanujam. Essentially, we prove that the constancy of the Lê numbers in a family implies the constancy of the Milnor fibrations in the family.

Theorem 9.4. *Let $s := \dim_\mathbf{0} \Sigma f_0$. Suppose that, for all t small, $(z_0, \dots, z_{s-1})$ is prepolar for f_t at $\mathbf{0}$ and that the Lê numbers, $\lambda^i_{f_t,\mathbf{z}}(\mathbf{0})$, are independent of t for each i with $0 \leqslant i \leqslant s$. Then,*

i) the homology of the Milnor fibre of f_t at the origin is independent of t for all t small;

if $s \leqslant n-2$,

ii) the fibre homotopy-type of the Milnor fibrations of f_t *at the origin is independent of* t *for all* t *small;*

and, if $s \leqslant n-3$,

iii) the diffeomorphism-type of the Milnor fibrations of f_t *at the origin is independent of* t *for all* t *small.*

Proof. By induction on s.

For $s=0$, this is the result of Lê and Ramanujam [**L-R**]. Now, assume that $s \geqslant 1$ and that we know the result for families of hypersurfaces with critical loci of dimension $\leqslant s-1$.

Let j be so large that the uniform Lê-Iomdine formulas of Theorem 4.15 hold and so large that Proposition 9.2 holds. Finally, using that $\lambda^0_{f_t,\mathbf{z}}(\mathbf{0})$ is independent of t, let j be so large that $j \geqslant 2+\lambda^0_{f_t,\mathbf{z}}(\mathbf{0})$ for all small t, so that we may apply Lemma 4.3 and Corollary 8.4.

Consider the family f_t+w^j, where w is a variable disjoint from the z's. The dimension of the critical locus at $t=0$ is still equal to s. As the Lê numbers of f_t are independent of t, we may apply Theorem 6.5 to conclude that $V(t)$ is not the limit of tangent spaces to level hypersurfaces of f at the origin. It follows trivially that $V(t)$ is not the limit of tangent spaces to level hypersurfaces of $f+w^j$ at the origin. Moreover, using the uniform Lê-Iomdines formulas of Theorem 4.15, we find that the Lê numbers of $(f_t+w^j)_{|V(z_0-w)} = f_t+z_0^j$ at $\mathbf{0}$ with respect to $(z_1,\dots,z_{s-1})$ are independent of t, and that $\dim_{\mathbf{0}}\Sigma(f_0+z_0^j)=s-1$.

Therefore, by our inductive hypothesis, we already have the constancy results for the family $(f_t+w^j)_{|V(z_0-w)}$. In addition, we may use Theorem 6.5 again to conclude that $V(t)$ is not the limit of tangent spaces to level hypersurfaces of $(f+w^j)_{|V(z_0-w)}$ at the origin. By 8.4, $V(z_0-w)$ is a prepolar slice for f_t+w^j at the origin for all small t. Also, by Lemma 4.3.v, $\left(\Gamma^1_{f_t+w^j,w-z_0}\cdot V(f_t+w^j)\right)_{\mathbf{0}} = j\lambda^0_{f_t,\mathbf{z}}(\mathbf{0})$, which is independent of t for all small t. Thus, we are in a position to apply Proposition 9.3 to the family f_t+w^j and we conclude the desired constancy results for this family.

Finally, now that we know the results for f_t+w^j, we apply Proposition C.16 to conclude that the result actually holds for f_t itself. □

Remark 9.5. While we are not very fond of discussing it, as we mentioned in Remark 1.27, for a fixed function h and a fixed point $\mathbf{p}$, there exist *generic Lê numbers* of h at $\mathbf{p}$ – that is, as one varies the linear choice of coordinates, $\mathbf{z}$, through coordinates for which the Lê numbers are defined, one finds a generic value for each of the Lê numbers, $\lambda^i_{h,\mathbf{z}}(\mathbf{p})$; let us denote this generic value simply by $\lambda^i_h(\mathbf{p})$.

We are not fond of discussing these generic Lê numbers because the Lê numbers are intended to be effectively calculable, and we know of no effective way of knowing when a coordinate choice is sufficiently generic to give $\lambda^i_h(\mathbf{p})$. We do know, by Corollary 4.16, that the tuple of generic Lê numbers $(\lambda^s_h(\mathbf{p}),\dots,\lambda^0_h(\mathbf{p}))$ is minimal with respect to the lexigraphic ordering.

We mention all this here because the tuple $(\lambda_h^s(\mathbf{p}), \dots, \lambda_h^0(\mathbf{p}))$ is an analytic invariant (this follows from the relationship between the Lê numbers and the polar multiplicities; see [**Mas11**] or Part IV, Theorems 1.10 and 3.2) , and so the reader may wonder whether Theorem 9.4 stills holds under the assumption that the generic Lê numbers of f_t are independent of t at the origin. The answer to this question is easily seen to be: yes. This follows quickly from 9.4 itself.

Suppose that the generic Lê numbers of f_t at the origin are independent of t. Choose coordinates $\mathbf{z}$ that are so generic that $\mathbf{z}$ is prepolar for f_t at the origin for all small t and so that $\mathbf{z}$ is generic enough to give the generic Lê numbers of f_0 at the origin. Then, for all small t, we have

$$(\lambda_{f_t}^s(\mathbf{0}), \dots, \lambda_{f_t}^0(\mathbf{0})) \leqslant (\lambda_{f_t,\mathbf{z}}^s(\mathbf{0}), \dots, \lambda_{f_t,\mathbf{z}}^0(\mathbf{0}))$$

$$\leqslant (\lambda_{f_0,\mathbf{z}}^s(\mathbf{0}), \dots, \lambda_{f_0,\mathbf{z}}^0(\mathbf{0})) = (\lambda_{f_0}^s(\mathbf{0}), \dots, \lambda_{f_0}^0(\mathbf{0})),$$

where we are using the lexigraphic ordering. Since

$$(\lambda_{f_t}^s(\mathbf{0}), \dots, \lambda_{f_t}^0(\mathbf{0})) = (\lambda_{f_0}^s(\mathbf{0}), \dots, \lambda_{f_0}^0(\mathbf{0})),$$

it follows that $(\lambda_{f_t,\mathbf{z}}^s(\mathbf{0}), \dots, \lambda_{f_t,\mathbf{z}}^0(\mathbf{0}))$ equals the tuple at $t = 0$. Therefore, we may apply Theorem 9.4.

Chapter 10. ANOTHER CHARACTERIZATION OF THE LÊ CYCLES

In this chapter, we give an alternative characterization of the Lê cycles and Lê numbers of a hypersurface singularity. This alternative characterization is generalized in Part IV, Chapter 3, where we see that the case of the Lê numbers of a function f is just the case where the underlying complex of sheaves is the sheaf of vanishing cycles along f.

As a consequence of Theorem 3.3, the Lê cycles can be characterized formally by requiring that the alternating sum of the Lê numbers yields the reduced Euler characteristic of the Milnor fibre at each point – provided that we know that a fixed choice of coordinates is prepolar at **every** point. We will show this now.

Lemma 10.1. *Let X be an analytic subset containing the origin in some $\mathbb{C}^N$ and let $\{S_\alpha\}$ be an analytic stratification of X with connected strata. Let $p : \mathbb{C}^N \to \mathbb{C}^k$ be a linear map such that $p_{|S_\alpha}$ is a submersion if $\dim S_\alpha \geqslant k$ and $\dim_{\mathbf{0}} \left(p^{-1}(\mathbf{0}) \cap \overline{S}_\alpha\right) = 0$ if $\dim S_\alpha \leqslant k-1$.*

Then, for generic linear $\pi : \mathbb{C}^k \to \mathbb{C}^{k-1}$, there exists a refinement $\{R_\beta\}$ of $\{S_\alpha\}$ in some neighborhood of the origin which preserves the strata of dimension greater than or equal to k and such that $\pi \circ p_{|R_\beta}$ is a submersion if $\dim R_\beta \geqslant k-1$ and $\dim_{\mathbf{0}} \left((\pi \circ p)^{-1}(\mathbf{0}) \cap \overline{R}_\beta\right) = 0$ if $\dim R_\beta \leqslant k-2$.

Proof. Let X' denote the union of those strata S_α such that $\dim S_\alpha$ is less than k. Then, as $\dim_{\mathbf{0}}(p^{-1}(\mathbf{0}) \cap X') = 0$, there exists an open neighborhood, $\mathcal{U}$, of the origin in $\mathbb{C}^N$ and an open neighborhood of the origin, $\mathcal{V}$, in $\mathbb{C}^k$ such that the restriction of p to a map from $\mathcal{U} \cap X'$ to $\mathcal{V}$ is finite. Therefore, as the conclusion of the lemma is purely local in nature, we may reduce ourselves to considering only the case where $p_{|X'}$ is a finite map.

Thus, $p(X')$ is an analytic subset of $\mathbb{C}^k$ of dimension at most $k-1$ and so, for generic lines, L, through $\mathbf{0}$ in $\mathbb{C}^k$, we have that $L \cap p(X') = \{\mathbf{0}\}$ near the origin. Hence, for generic linear $\pi : \mathbb{C}^k \to \mathbb{C}^{k-1}$, we must have that $\pi^{-1}(\mathbf{0}) \cap p(X') = \{\mathbf{0}\}$, and so

$$(\pi \circ p)^{-1}(\mathbf{0}) \cap X' = p^{-1}\left(\pi^{-1}(\mathbf{0}) \cap p(X')\right) \cap X' = p^{-1}(\mathbf{0}) \cap X'.$$

But, by hypothesis, $p^{-1}(\mathbf{0}) \cap X' = \{\mathbf{0}\}$, and thus – for such a generic π – any refinement of X' will give a refinement of $\{S_\alpha\}$ which preserves the strata of dimension greater than or equal to k and realizes the desired intersection condition.

Now, we have the restriction $\pi \circ p : X' \to \mathbb{C}^{k-1}$. By the above, if we once again take sufficiently small neighborhoods around the origins, $\pi \circ p$ restricts to a finite map on X'. Assume then that $(\pi \circ p)_{|X'}$ is finite. If $\dim X' = k-1$, then $\pi \circ p$ is a submersion on $X' - W$, where W is a closed subset of X' of dimension at most $k-2$. Thus, we may refine the stratification so that $X' \cap (\pi \circ p)^{-1}(\pi \circ p)(W)$ is a union of strata, where the strata have dimension at most $k-2$ since $\pi \circ p$ restricted to X' is a finite map. This yields the desired refinement $\{R_\beta\}$. $\square$

Proposition 10.2. *Let $\mathcal{U}$ be an open subset of $\mathbb{C}^{n+1}$ containing the origin, and let $h : (\mathcal{U}, \mathbf{0}) \to (\mathbb{C}, \mathbf{0})$ be an analytic map. Then, for a generic linear reorganization, $\mathbf{z}$, of coordinate systems for $\mathbb{C}^{n+1}$, there exists an open neighborhood, $\mathcal{U}'$, of the origin such that $\mathbf{z}$ is prepolar for $h_{|_{\mathcal{U}'}}$, i.e., for all $\mathbf{p} \in \mathcal{U}' \cap V(h)$, if we let s denote $\dim_{\mathbf{p}}\Sigma h$, then $(z_0, \dots, z_{s-1})$ is prepolar for h at $\mathbf{p}$.*

Proof. Begin by applying the lemma to $X = V(h)$, endowed with some good stratification, and $p = \mathrm{id} : \mathbb{C}^{n+1} \to \mathbb{C}^{n+1}$. By an inductive application of the lemma, for a generic linear reorganization of coordinates, $\mathbf{z}$, for $\mathbb{C}^{n+1}$, we arrive at a stratification $\{S_\alpha\}$, which is a refinement of the original good stratification (in a possibly smaller neighborhood of the origin), such that for each i, the map $(z_0, \dots, z_i)$ is a submersion when restricted to a stratum of dimension greater than or equal to $i + 1$. This would say precisely that $(z_0, \dots, z_n)$ is prepolar at each point of $V(h)$ with respect to the stratification $\{S_\alpha\}$ – provided that $\{S_\alpha\}$ is actually a good stratification for h.

But, any refinement of a good stratification is another good stratification – except for the fact that we may have refined the smooth part of $V(h)$, which we required to be a stratum of any good stratification. However,

$$\{V(h) - \Sigma h\} \cup \{S_\alpha \mid S_\alpha \subseteq \Sigma h\}$$

is certainly a good stratification for h, and what remains for us to show is that, for generically reorganized coordinates, $\mathbf{z}$, for each $\mathbf{p} \in \Sigma h$, $V(z_0 - p_0, \dots, z_{s-1} - p_{s-1})$ transversely intersects the smooth part of $V(h)$ near $\mathbf{p}$, where again $s = \dim_{\mathbf{p}}\Sigma h$.

Assume then that we have chosen the coordinates $\mathbf{z}$ generic as above and also generic enough so that $(z_0, \dots, z_{s-1})$ is prepolar for h at $\mathbf{0}$, where $s = \dim_{\mathbf{0}}\Sigma h$. By Theorem 1.26, $\gamma^s_{h,\mathbf{z}}(\mathbf{0})$ exists, and so $\gamma^s_{h,\mathbf{z}}(\mathbf{p})$ exists for all $\mathbf{p}$ near $\mathbf{0}$. But, by Remark 1.6, this implies that if $\mathbf{p} \in \Sigma h$, then

$$\Sigma(h_{|_{V(z_0-p_0,\dots,z_{k-1}-p_{k-1})}}) = V(z_0 - p_0, \dots, z_{k-1} - p_{k-1}) \cap \Sigma h$$

at $\mathbf{p}$, i.e., $V(z_0 - p_0, \dots, z_{s-1} - p_{s-1})$ transversely intersects the smooth part of $V(h)$ near $\mathbf{p}$ (and, of course, $s = \dim_{\mathbf{0}}\Sigma h \geqslant \dim_{\mathbf{p}}\Sigma h$ for all $\mathbf{p}$ near $\mathbf{0}$). □

In Part IV, Chapter 3, we will generalize the result below to the case where the underlying space is arbitrary.

Theorem 10.3. *For a generic linear reorganization of coordinates, $\mathbf{z}$, for $\mathbb{C}^{n+1}$, the Lê cycles are a collection of analytic cycle germs, $\Lambda^i_{h,\mathbf{z}}$, in Σh at the origin such that each $\Lambda^i_{h,\mathbf{z}}$ is purely i-dimensional and properly intersects $V(z_0, \dots, z_{i-1})$ at the origin, and*

$$\widetilde{\chi}(F_{h,\mathbf{p}}) = \sum_{i=0}^{s}(-1)^{n-i}\left(\Lambda^i_{h,\mathbf{z}} \cdot V(z_0 - p_0, \dots, z_{i-1} - p_{i-1})\right)_{\mathbf{p}},$$

for all $\mathbf{p} \in \Sigma h$ *near* $\mathbf{0}$, *where* $\widetilde{\chi}(F_{h,\mathbf{p}})$ *is the reduced Euler characteristic of the Milnor fibre of* h *at* $\mathbf{p}$*; specifically, this is the case if* $\mathbf{z}$ *is prepolar for* h *in a neighborhood of the origin.*

Moreover, if $\mathbf{z}$ *is any linear coordinate system such that such cycles exist, then they are unique.*

Proof. The first statement follows immediately from the previous proposition and Theorem 3.3. As for the uniqueness assertion, this is a fairly standard argument for constructible functions.

Suppose that we had two such collections, $\Lambda^i_{h,\mathbf{z}}$ and $\Omega^i_{h,\mathbf{z}}$. Let s denote $\dim_{\mathbf{0}}\Sigma h$. We will show that $\Lambda^i_{h,\mathbf{z}}$ and $\Omega^i_{h,\mathbf{z}}$ agree by downward induction on i.

For a generic point, $\mathbf{p}$, in an s-dimensional component, ν, of the support of $\Lambda^s_{h,\mathbf{z}}$, $\mathbf{p}$ will be a smooth point of ν, $V(z_0 - p_0, \ldots, z_{s-1} - p_{s-1})$ will transversely intersect ν at $\mathbf{p}$, and $\mathbf{p}$ will not be in the support of any of the lower-dimensional $\Lambda^i_{h,\mathbf{z}}$ or $\Omega^i_{h,\mathbf{z}}$. Thus, at such a $\mathbf{p}$,

$$\left(\Omega^s_{h,\mathbf{z}} \cdot V(z_0 - p_0, \ldots, z_{s-1} - p_{i-1})\right)_{\mathbf{p}} = \left(\Lambda^s_{h,\mathbf{z}} \cdot V(z_0 - p_0, \ldots, z_{s-1} - p_{i-1})\right)_{\mathbf{p}}.$$

Of course, the same conclusion would have followed it we had chosen a generic point, $\mathbf{p}$, in an s-dimensional component, ν, of the support of $\Omega^s_{h,\mathbf{z}}$, It follows that $\Lambda^s_{h,\mathbf{z}} = \Omega^s_{h,\mathbf{z}}$.

Now, suppose that we have shown that $\Lambda^i_{h,\mathbf{z}} = \Omega^i_{h,\mathbf{z}}$ for all i greater than some k. Then,

$$\sum_{i=0}^{k}(-1)^{n-i}\left(\Lambda^i_{h,\mathbf{z}} \cdot V(z_0 - p_0, \ldots, z_{i-1} - p_{i-1})\right)_{\mathbf{p}} =$$

$$\sum_{i=0}^{k}(-1)^{n-i}\left(\Omega^i_{h,\mathbf{z}} \cdot V(z_0 - p_0, \ldots, z_{i-1} - p_{i-1})\right)_{\mathbf{p}},$$

and we repeat the argument above with k in place of s. The conclusion follows. □

Theorem 10.3 leaves us with a very strange set of affairs; it tells us that, for generic $\mathbf{z}$, we could have defined the Lê cycles by their characterization in the theorem. This means that the Lê cycles and hence, the Lê numbers, are determined by the choice of $\mathbf{z}$ and the data of the Euler characteristic of the Milnor fibre at each point. But, had we defined the Lê cycles and numbers this way, then we would have produced the Morse inequalities on the Betti numbers of the Milnor fibres in Theorem 3.3 from seemingly much less data. This phenomenon occurs in a much more general setting; the explanation appears in [**Mas11**].

Part III. ISOLATED CRITICAL POINTS OF FUNCTIONS ON SINGULAR SPACES

Chapter 0. INTRODUCTION

In the introduction to Part II, we discussed known results for functions, or families of functions, with isolated critical points. The remainder of Part II was devoted to the development of the Lê cycles and Lê numbers – a generalization to functions with non-isolated critical loci of the data provided by the Milnor number of an isolated critical point.

All of the functions considered in Part II had open subsets of affine space as their domains. Here, in Part III, we will begin our generalization of the Lê numbers to the case where the functions considered have arbitrary analytic spaces as their domains. However, since the Lê numbers generalize the Milnor number of an isolated critical point (on affine space), if we want to have Lê numbers for functions with arbitrary domains, then we must first develop some sort of "Milnor number" theory for functions with "isolated critical points" with arbitrary domains.

This immediately leads us to two fundamental questions for functions with arbitrary domains:

- What is the proper notion of the "critical locus" of such a function?
- What is the proper notion of the "Milnor number" of such a function when its "critical locus"
consists of an isolated point?

In the remainder of this introduction, we will discuss possible definitions of the "critical locus" for functions; we will first discuss the non-controversial case of functions on affine space, and then move on to the much more complicated general case. There is a slight bit of overlap here with some of the material presented in the introduction to Part II, but we feel that this minor repetition aids the exposition.

Let $\mathcal{U}$ be an open subset of $\mathbb{C}^{n+1}$, let $\mathbf{z} := (z_0, z_1, \dots, z_n)$ be coordinates for $\mathbb{C}^{n+1}$, and suppose that $\tilde{f} : \mathcal{U} \to \mathbb{C}$ is an analytic function. Then, all conceivable definitions of the *critical locus*, $\Sigma\tilde{f}$, of $\tilde{f}$ agree: one can consider the points, $\mathbf{x}$, where the derivative vanishes, i.e., $d_{\mathbf{x}}\tilde{f} = 0$, or one can consider the points, $\mathbf{x}$, where the Taylor series of $\tilde{f}$ at $\mathbf{x}$ has no linear term, i.e., $\tilde{f} - \tilde{f}(\mathbf{x}) \in \mathfrak{m}^2_{\mathcal{U},\mathbf{x}}$ (where $\mathfrak{m}_{\mathcal{U},\mathbf{x}}$ is the maximal ideal in the coordinate ring of $\mathcal{U}$ at $\mathbf{x}$), or one can consider the points, $\mathbf{x}$, where the Milnor fibre of $\tilde{f}$ at $\mathbf{x}$, $F_{\tilde{f},\mathbf{x}}$, is not trivial (where, here, "trivial" could mean even up to analytic isomorphism).

Now, suppose that X is an analytic subset of $\mathcal{U}$, and let $f := \tilde{f}_{|X}$. Then, what should be meant by "the critical locus of f"? It is not clear what the relationship is between points, $\mathbf{x}$, where $f - f(\mathbf{x}) \in \mathfrak{m}^2_{X,\mathbf{x}}$ and points where the Milnor fibre, $F_{f,\mathbf{x}}$, is not trivial (with any definition of trivial); moreover, the derivative $d_{\mathbf{x}}f$ does not even exist.

We are guided by the successes of Morse Theory and stratified Morse Theory to choosing the Milnor fibre definition as our primary notion of critical locus, for we believe that critical points should coincide with changes in the topology of the level hypersurfaces of f. Therefore, we make the following definition:

Definition 0.1. The $\mathbb{C}$-*critical locus of* f, $\Sigma_{\mathbb{C}} f$, is given by

$$\Sigma_{\mathbb{C}} f := \{\mathbf{x} \in X \mid H^*(F_{f,\mathbf{x}};\ \mathbb{C}) \neq H^*(point;\ \mathbb{C})\}.$$

(The reasons for using field coefficients, rather than $\mathbb{Z}$, are technical: we want Lemma 4.1 to be true.)

In Chapter 1, we will compare and contrast the $\mathbb{C}$-critical locus with other possible notions of critical locus, including the ones mentioned above and the stratified critical locus.

After Chapter 1, the remainder of Part III is dedicated to showing that Definition 0.1 really yields a useful, calculable definition of the critical locus. We show this by looking at the case of a generalized isolated singularity, i.e., an isolated point of $\Sigma_{\mathbb{C}} f$, and showing that, at such a point, there is a workable definition of the Milnor number(s) of f; we show that the Betti numbers of the Milnor fibre can be calculated (3.7.ii), and we give a generalization of the result of Lê and Saito [**L-S**] that constant Milnor number throughout a family implies Thom's a_f condition holds. Specifically, in Corollary 6.14, we prove (with slightly weaker hypotheses) that:

Theorem 0.2. *Let W be a (not necessarily purely) d-dimensional analytic subset of an open subset of $\mathbb{C}^n$. Let Z be a d-dimensional irreducible component of W. Let $X := \overset{\circ}{\mathbb{D}} \times W$ be the product of an open disk about the origin with W, and let $Y := \overset{\circ}{\mathbb{D}} \times Z$.*

Let $f : (X, \overset{\circ}{\mathbb{D}} \times \{\mathbf{0}\}) \to (\mathbb{C}, 0)$ be an analytic function, and let $f_t(\mathbf{z}) := f(t, \mathbf{z})$. Suppose that f_0 is in the square of the maximal ideal of Z at $\mathbf{0}$.

Suppose that $\mathbf{0}$ is an isolated point of $\Sigma_{\mathbb{C}}(f_0)$ and that, for all small a, the reduced Betti number $\tilde{b}_{d-1}(F_{f_a,(a,\mathbf{0})})$ is independent of a .

Then, $\tilde{b}_{d-1}(F_{f_a,(a,\mathbf{0})}) \neq 0$ and, near $\mathbf{0}$, $\Sigma(f_{|_{Y_{\mathrm{reg}}}}) \subseteq \overset{\circ}{\mathbb{D}} \times \{\mathbf{0}\}$ and the pair

$$(Y_{\mathrm{reg}} - \Sigma(f_{|_{Y_{\mathrm{reg}}}}),\ \overset{\circ}{\mathbb{D}} \times \{\mathbf{0}\})$$

satisfies Thom's a_f condition at $\mathbf{0}$.

Thom's a_f is important for several reasons, but perhaps the best reason is because it is an hypothesis of Thom's Second Isotopy Lemma. General results on the a_f condition have been proved by many researchers: Hironaka, Lê, Saito, Henry, Merle, Sabbah, Briançon, Maisonobe, Parusiński, etc., and the above theorem is closely related to the recent results contained in [**BMM**] and [**P2**]. However, the reader should contrast the hypotheses of Theorem 0.2 with those of the main theorem of [**BMM**] (Theorem 5.2.1); our main hypothesis is that a single number is constant throughout the family, while the main hypothesis of Theorem 4.2.1 of [**BMM**] is a condition which requires one to check an infinite amount of data: the property of local stratified triviality. Moreover, the Betti numbers that we require to be constant are actually calculable.

While much of this part is fairly technical in nature, there are three new, key ideas that guide us throughout.

The first of these fundamental precepts is: controlling the **vanishing cycles** in a family of functions is enough to control Thom's a_f condition and, perhaps, the topology throughout the family. While this may seem like an obvious principle – given the results of Lê and Saito in [**L-S**] and of Lê and Ramanujam in [**L-R**] – in fact, in the general setting, most of the known results seem to require the constancy of much stronger data, e.g., the constancy of the polar multiplicities [**Te6**] or that one has the local stratified triviality property [**BMM**]. In a very precise sense, controlling the polar multiplicities corresponds to controlling the **nearby cycles** of the family of functions, instead of merely controlling the vanishing cycles. As we show in Corollary 5.4, controlling the characteristic cycle of the vanishing cycles is sufficient for obtaining the a_f condition.

Our second fundamental idea is: the correct setting for all of our cohomological results is where perverse sheaves are used as coefficients. While papers on intersection cohomology abound, and while perverse sheaves are occasionally used as a tool (e.g., [**BMM**, 4.2.1]), we are not aware of any other work on general singularities in which arbitrary perverse sheaves of coefficients are used in an integral fashion throughout. The importance of perverse sheaves in Part III begins with Theorem 4.2, where we give a description of the critical locus of a function with respect to a perverse sheaf.

The third new feature of Part III is the recurrent use of the perverse cohomology of a complex of sheaves. This device allows us to take our general results about perverse sheaves and translate them into statements about the constant sheaf. The reason that we use perverse cohomology, instead of intersection cohomology, is because perverse cohomology has such nice functorial properties: it commutes with Verdier dualizing, and with taking nearby and vanishing cycles (shifted by $[-1]$). If we were only interested in proving results for local complete intersections (l.c.i.'s), we would never need the perverse cohomology; however, we want to prove completely general results. The perverse cohomology seems to be a hitherto unused tool for accomplishing this goal.

Part III is organized as follows:

In Chapter 1, we discuss seven different notions of the "critical locus" of a function. We give examples to show that, in general, all of these notions are different.

In Chapter 2, we discuss the relative polar curve of Lê and Teissier. We need to relate the intrinsic definition to a conormal characterization in terms of gap sheaves.

Chapter 3 is devoted to proving an "index theorem", Theorem 3.10, which provides the main link between the topological data of the Milnor fibre and the algebraic data obtained by blowing-up the image of $d\tilde{f}$ inside the appropriate space. This theorem is presented with coefficients in a bounded, constructible complex of sheaves; this level of generality is absolutely necessary in order to obtain the results in the remainder of Part III.

Chapter 4 uses the index theorem of Chapter 3 to show that $\Sigma_{\mathbb{C}} f$ and the Betti numbers of the Milnor fibre really are fairly well-behaved. This is accomplished by applying Theorem 3.10 in the case where the complex of sheaves is taken to be the perverse cohomology of the shifted constant sheaf. Perverse cohomology essentially gives us the "closest" perverse sheaf to the constant sheaf. Many of the results of Chapter 4 are stated for arbitrary perverse sheaves, for this seems to be the most natural setting.

Chapter 5 contains the necessary results from conormal geometry that we will need in order to conclude that topological data implies that Thom's a_f condition holds. The primary result of this chapter is Corollary 5.4, which once again relies on the index theorem from Chapter 3.

Chapter 6 begins with a discussion of "continuous families of constructible complexes of sheaves". We then prove in Theorem 6.7 that additivity of Milnor numbers occurs in continuous families of perverse sheaves, and we use this to conclude additivity of the Betti numbers of the Milnor fibres, by once again resorting to the perverse cohomology of the shifted constant sheaf. Finally, in Corollaries 6.11 and 6.12, we prove that the constancy of the Milnor/Betti number(s) throughout a family implies that the a_f condition holds – we prove this first in the setting of arbitrary perverse sheaves, and then for perverse cohomology of the shifted constant sheaf. By translating our hypotheses from the language of the derived category back into more down-to-Earth terms, we obtain Corollary 6.12, which leads to Theorem 0.2 above.

Chapter 1. CRITICAL AVATARS.

We continue with $\mathcal{U}$, $\mathbf{z}$, $\tilde{f}$, X, and f as in the introduction.

In this chapter, we will investigate seven possible notions of the "critical locus" of a function on a singular space, one of which is the $\mathbb{C}$-critical locus already defined in 0.1.

Definition 1.1. The *algebraic critical locus of* f, $\Sigma_{\rm alg} f$, is defined by

$$\Sigma_{\rm alg} f := \{\mathbf{x} \in X \mid f - f(\mathbf{x}) \in \mathfrak{m}^2_{X,\mathbf{x}}\}.$$

Remark 1.2. It is a trivial exercise to verify that

$$\Sigma_{\rm alg} f = \{\mathbf{x} \in X \mid \text{there exists a local extension, } \hat{f}, \text{ of } f \text{ to } \mathcal{U} \text{ such that } d_{\mathbf{x}}\hat{f} = 0\}.$$

Note that $\mathbf{x}$ being in $\Sigma_{\rm alg} f$ does **not** imply that **every** local extension of f has zero for its derivative at $\mathbf{x}$.

One might expect that $\Sigma_{\rm alg} f$ is always a closed set; in fact, it need not be. Consider the example where $X := V(xy) \subseteq \mathbb{C}^2$, and $f = y_{|_X}$. We leave it as an exercise for the reader to verify that $\Sigma_{\rm alg} f = V(y) - \{\mathbf{0}\}$.

There are five more variants of the critical locus of f that we will consider. We let $X_{\rm reg}$ denote the regular (or smooth) part of X and, if M is an analytic submanifold of $\mathcal{U}$, we let $T^*_M\mathcal{U}$ denote the conormal space to M in $\mathcal{U}$ (that is, the elements $(\mathbf{x}, \eta)$ of the cotangent space to $\mathcal{U}$ such that $\mathbf{x} \in M$ and η annihilates the tangent space to M at $\mathbf{x}$). We let $N(X)$ denote the Nash modification of X, so that the fibre $N_{\mathbf{x}}(X)$ at $\mathbf{x}$ consists of limits of tangent planes from the regular part of X.

We also remind the reader that complex analytic spaces possess canonical Whitney stratifications (see [**Te6**]).

Definition 1.3. We define the *regular critical locus of* f, $\Sigma_{\rm reg} f$, to be the critical locus of the restriction of f to $X_{\rm reg}$, i.e., $\Sigma_{\rm reg} f = \Sigma(f_{|_{X_{\rm reg}}})$.

We define the *Nash critical locus of* f, $\Sigma_{\rm Nash} f$, to be the set of those $\mathbf{x} \in X$ such that there exists a local extension, $\hat{f}$, of f to $\mathcal{U}$ such that, for all $T \in N_{\mathbf{x}}(X)$, $d_{\mathbf{x}}\hat{f}(T) \equiv 0$.

We define the *conormal-regular critical locus of* f, $\Sigma_{\rm cnr} f$, to be the set of those $\mathbf{x} \in X$ such that there exists a local extension, $\hat{f}$, of f to $\mathcal{U}$ such that $(\mathbf{x}, d_{\mathbf{x}}\hat{f}) \in \overline{T^*_{X_{\rm reg}}\mathcal{U}}$. It is trivial to see that this set is equal to the set of those $\mathbf{x} \in X$ such that there exists a local extension, $\hat{f}$, of f to $\mathcal{U}$ such that **there exists** $T \in N_{\mathbf{x}}(X)$ such that $d_{\mathbf{x}}\hat{f}(T) \equiv 0$.

Let $\mathcal{S} = \{S_\alpha\}$ be a (complex analytic) Whitney stratification of X. We define the *$\mathcal{S}$-stratified critical locus of f*, $\Sigma_{\mathcal{S}} f$, to be $\bigcup_\alpha \Sigma(f_{|S_\alpha})$. If $\mathcal{S}$ is clear, we simply call $\Sigma_{\mathcal{S}} f$ the stratified critical locus.

If $\mathcal{S}$ is, in fact, the canonical Whitney stratification of X, then we write $\Sigma_{\mathrm{can}} f$ in place of $\Sigma_{\mathcal{S}} f$, and call it the *canonical stratified critical locus.*

We define the *relative differential critical locus of f*, $\Sigma_{\mathrm{rdf}} f$, to be the union of the singular set of X and $\Sigma_{\mathrm{reg}} f$.

If $\mathbf{x} \in X$ and $h_1, \dots, h_j$ are equations whose zero-locus defines X near $\mathbf{x}$, then $\mathbf{x} \in \Sigma_{\mathrm{rdf}} f$ if and only if the rank of the Jacobian map of $(\tilde{f}, h_1, \dots, h_j)$ at $\mathbf{x}$ is not maximal among all points of X near $\mathbf{x}$. By using this Jacobian, we could (but will not) endow $\Sigma_{\mathrm{rdf}} f$ with a scheme structure (the *critical space*) which is independent of the choice of the extension $\tilde{f}$ and the defining functions $h_1, \dots, h_n$ (see [**Loo**, 4.A]). The proof of the independence uses relative differentials; this is the reason for our terminology.

Remark 1.4. In terms of conormal geometry, $\Sigma_{\mathcal{S}} f = \left\{\mathbf{x} \in X \mid (\mathbf{x}, d_{\mathbf{x}}\tilde{f}) \in \bigcup_\alpha T^*_{S_\alpha}\mathcal{U}\right\}$ or, using Whitney's condition a) again, $\Sigma_{\mathcal{S}} f = \left\{\mathbf{x} \in X \mid (\mathbf{x}, d_{\mathbf{x}}\tilde{f}) \in \bigcup_\alpha \overline{T^*_{S_\alpha}\mathcal{U}}\right\}$.

Clearly, $\Sigma_{\mathrm{rdf}} f$ is closed, and it is an easy exercise to show that Whitney's condition a) implies that $\Sigma_{\mathcal{S}} f$ is closed. On the other hand, $\Sigma_{\mathrm{reg}} f$ is, in general, not closed and, in order to have any information at singular points of X, we will normally look at its closure $\overline{\Sigma_{\mathrm{reg}} f}$.

Looking at the definition of $\Sigma_{\mathrm{cnr}} f$, one might expect that $\overline{\Sigma_{\mathrm{reg}} f} = \Sigma_{\mathrm{cnr}} f$. In fact, we shall see in Example 1.8 that this is false. That $\Sigma_{\mathrm{cnr}} f$ is, itself, closed is part of the following proposition. (Recall that $\tilde{f}$ is our fixed extension of f to all of $\mathcal{U}$.)

In the following proposition, we show that, in the definitions of the Nash and conormal-regular critical loci, we could have used "for all" in place of "there exists" for the local extensions; in particular, this implies that we can use the fixed extension $\tilde{f}$. Finally, we show that the conormal-regular critical locus is closed.

Proposition 1.5. *The Nash critical locus of f is equal to*

$$\{\mathbf{x} \in X \mid \textit{for all local extensions, } \hat{f}, \textit{ of } f \textit{ to } \mathcal{U},\ d_{\mathbf{x}}\hat{f}(T) \equiv 0, \textit{ for all } T \in N_{\mathbf{x}}(X)\} =$$

$$\{\mathbf{x} \in X \mid d_{\mathbf{x}}\tilde{f}(T) \equiv 0, \ \textit{for all } T \in N_{\mathbf{x}}(X)\}.$$

The conormal-regular critical locus of f is equal to

$$\{\mathbf{x} \in X \mid \textit{for all local extensions, } \hat{f}, \textit{ of } f \textit{ to } \mathcal{U},\ (\mathbf{x}, d_{\mathbf{x}}\hat{f}) \in \overline{T^*_{X_{\mathrm{reg}}}\mathcal{U}}\ \}$$

$$= \left\{\mathbf{x} \in X \mid (\mathbf{x}, d_{\mathbf{x}}\tilde{f}) \in \overline{T^*_{X_{\mathrm{reg}}}\mathcal{U}}\right\}.$$

In addition, $\Sigma_{\mathrm{cnr}} f$ is closed.

Proof. Let

$$Z := \{\mathbf{x} \in X \mid \text{for all local extensions, } \hat{f}, \text{ of } f \text{ to } \mathcal{U},\ d_{\mathbf{x}}\hat{f}(T) \equiv 0, \text{ for all } T \in N_{\mathbf{x}}(X)\}.$$

Clearly, we have $Z \subseteq \Sigma_{\mathrm{Nash}} f$.

Suppose now that $\mathbf{x} \in \Sigma_{\mathrm{Nash}} f$. Then, there exists a local extension, $\hat{f}$, of f to $\mathcal{U}$ such that $d_{\mathbf{x}}\hat{f}(T) \equiv 0$, for all $T \in N_{\mathbf{x}}(X)$. Let $\check{f}$ be another local extension of f to $\mathcal{U}$ and let $T_\infty \in N_{\mathbf{x}}(X)$; to show that $\mathbf{x} \in Z$, what we must show is that $d_{\mathbf{x}}\check{f}(T_\infty) \equiv 0$.

Suppose not. Then, there exists $\mathbf{v} \in T_\infty$ such that $d_{\mathbf{x}}\check{f}(\mathbf{v}) \neq 0$, but $d_{\mathbf{x}}\hat{f}(\mathbf{v}) = 0$. Therefore, there exist $\mathbf{x}_i \in X_{\mathrm{reg}}$ and $\mathbf{v}_i \in T_{\mathbf{x}_i}X_{\mathrm{reg}}$ such that $\mathbf{x}_i \to \mathbf{x}$, $T_{\mathbf{x}_i}X_{\mathrm{reg}} \to T_\infty$, and $\mathbf{v}_i \to \mathbf{v}$.

Let $\mathcal{V}$ be an open neighborhood of x in $\mathcal{U}$ which in $\hat{f}$ and $\check{f}$ are both defined. Let $\Phi : \mathcal{V} \cap \overline{TX_{\mathrm{reg}}} \to \mathbb{C}$ be defined by $\Phi(\mathbf{p}, \mathbf{w}) = d_{\mathbf{p}}(\hat{f} - \check{f})(\mathbf{w})$. Then, Φ is continuous, and so $\Phi^{-1}(0)$ is closed. As $(\hat{f} - \check{f})_{|X \cap \mathcal{V}} \equiv 0$, $(\mathbf{x}_i, \mathbf{v}_i) \in \Phi^{-1}(0)$, and thus $(\mathbf{x}, \mathbf{v}) \in \Phi^{-1}(0)$ – a contradiction. Therefore, $Z = \Sigma_{\mathrm{Nash}} f$.

It follows immediately that $\Sigma_{\mathrm{Nash}} f = \{\mathbf{x} \in X \mid d_{\mathbf{x}}\tilde{f}(T) \equiv 0, \text{ for all } T \in N_{\mathbf{x}}(X)\}$.

Now, let $W := \{\mathbf{x} \in X \mid \text{for all local extensions, } \hat{f}, \text{ of } f \text{ to } \mathcal{U},\ (\mathbf{x}, d_{\mathbf{x}}\hat{f}) \in \overline{T^*_{X_{\mathrm{reg}}}\mathcal{U}}\}$. Then, $W \subseteq \Sigma_{\mathrm{cnr}} f$.

Suppose now that $\mathbf{x} \in \Sigma_{\mathrm{cnr}} f$. Then, there exists a local extension, $\hat{f}$, of f to $\mathcal{U}$ such that $(\mathbf{x}, d_{\mathbf{x}}\hat{f}) \in \overline{T^*_{X_{\mathrm{reg}}}\mathcal{U}}$. Let $(\mathbf{x}_i, \eta_i) \in T^*_{X_{\mathrm{reg}}}\mathcal{U}$ be such that $(\mathbf{x}_i, \eta_i) \to (\mathbf{x}, d_{\mathbf{x}}\hat{f})$. Let $\check{f}$ be another local extension of f to $\mathcal{U}$; to show that $\mathbf{x} \in W$, what we must show is that $(\mathbf{x}, d_{\mathbf{x}}\check{f}) \in \overline{T^*_{X_{\mathrm{reg}}}\mathcal{U}}$.

Since $(\check{f} - \hat{f})_{|X \cap \mathcal{V}} \equiv 0$, for all $\mathbf{q} \in X_{\mathrm{reg}}$, $(\mathbf{q}, d_{\mathbf{q}}(\check{f} - \hat{f})) \in T^*_{X_{\mathrm{reg}}}\mathcal{U}$; in particular,

$$(\mathbf{x}_i, d_{\mathbf{x}_i}(\check{f} - \hat{f})) \in T^*_{X_{\mathrm{reg}}}\mathcal{U}.$$

Thus, $(\mathbf{x}_i, \eta_i + d_{\mathbf{x}_i}(\check{f} - \hat{f})) \in T^*_{X_{\mathrm{reg}}}\mathcal{U}$, and $(\mathbf{x}_i, \eta_i + d_{\mathbf{x}_i}(\check{f} - \hat{f})) \to (\mathbf{x}, d_{\mathbf{x}}\check{f})$. Therefore, $(\mathbf{x}, d_{\mathbf{x}}\check{f}) \in \overline{T^*_{X_{\mathrm{reg}}}\mathcal{U}}$, and $W = \Sigma_{\mathrm{cnr}} f$.

It follows immediately that $\Sigma_{\mathrm{cnr}} f = \{\mathbf{x} \in X \mid (\mathbf{x}, d_{\mathbf{x}}\tilde{f}) \in \overline{T^*_{X_{\mathrm{reg}}}\mathcal{U}}\}$.

Finally, we need to show that $\Sigma_{\mathrm{cnr}} f$ is closed. Let $\Psi : X \to T^*\mathcal{U}$ be given by $\Psi(\mathbf{x}) = (\mathbf{x}, d_{\mathbf{x}}\tilde{f})$. Then, Ψ is a continuous map and, by the above, $\Sigma_{\mathrm{cnr}} f = \Psi^{-1}(\overline{T^*_{X_{\mathrm{reg}}}\mathcal{U}})$. □

Proposition 1.6. *There are inclusions*

$$\overline{\Sigma_{\mathrm{reg}} f} \subseteq \overline{\Sigma_{\mathrm{alg}} f} \subseteq \overline{\Sigma_{\mathrm{Nash}} f} \subseteq \Sigma_{\mathrm{cnr}} f \subseteq \overline{\Sigma_{\mathrm{C}} f} \subseteq \Sigma_{\mathrm{can}} f \subseteq \Sigma_{\mathrm{rdf}} f.$$

In addition, if $\mathcal{S}$ is a Whitney stratification of X, then $\Sigma_{\mathrm{can}} f \subseteq \Sigma_{\mathcal{S}} f$.

Proof. Clearly, $\Sigma_{\mathrm{reg}} f \subseteq \Sigma_{\mathrm{alg}} f \subseteq \Sigma_{\mathrm{Nash}} f \subseteq \Sigma_{\mathrm{cnr}} f$, and so the containments for their closures follows (recall, also, that $\Sigma_{\mathrm{cnr}} f$ is closed). It is also obvious that $\Sigma_{\mathrm{can}} f \subseteq \Sigma_{\mathrm{rdf}} f$ and $\Sigma_{\mathrm{can}} f \subseteq \Sigma_{\mathcal{S}} f$.

That $\Sigma_{\mathbb{Z}} f \subseteq \Sigma_{\mathrm{can}} f$ follows from Stratified Morse Theory [**Go-Mac1**], and so, since $\Sigma_{\mathrm{can}} f$ is closed, $\overline{\Sigma_{\mathbb{C}} f} \subseteq \Sigma_{\mathrm{can}} f$.

It remains for us to show that $\Sigma_{\mathrm{cnr}} f \subseteq \overline{\Sigma_{\mathbb{C}} f}$. Unfortunately, to reach this conclusion, we must refer ahead to Theorem 4.6, from which it follows immediately. (However, that $\overline{\Sigma_{\mathrm{alg}} f} \subseteq \overline{\Sigma_{\mathbb{C}} f}$ follows from A'Campo's Theorem [**A'C**].) □

Remark 1.7. For a fixed stratification $\mathcal{S}$, for all $\mathbf{x} \in X$, there exists a neighborhood $\mathcal{W}$ of $\mathbf{x}$ in X such that $\mathcal{W} \cap \Sigma_{\mathcal{S}} f \subseteq f^{-1} f(\mathbf{x})$. This is easy to show: the level hypersurfaces of f close to $V(f - f(\mathbf{x}))$ will be transverse to all of the strata of $\mathcal{S}$ near $\mathbf{x}$. All of our other critical loci which are contained in $\Sigma_{\mathcal{S}} f$ (i.e., all of them except $\Sigma_{\mathrm{rdf}} f$) also satisfy this local isolated critical value property.

Example 1.8. In this example, we wish to look at the containments given in Proposition 1.6, and investigate whether the containments are proper, and also investigate what would happen if we did not take closures in the four cases where we do.

The same example that we used in Remark 1.2 shows that none of $\Sigma_{\mathrm{reg}} f$, $\Sigma_{\mathrm{alg}} f$, $\Sigma_{\mathrm{Nash}} f$, or $\Sigma_{\mathbb{C}} f$ are necessarily closed; if $X := V(xy) \subseteq \mathbb{C}^2$, and $f = y_{|_X}$, then all four critical sets are precisely $V(y) - \{\mathbf{0}\}$. Additionally, since $\Sigma_{\mathrm{cnr}} f = V(y)$, this example also shows that, in general, $\Sigma_{\mathrm{cnr}} f \not\subseteq \Sigma_{\mathbb{C}} f$.

If we continue with $X = V(xy)$ and let $g := (x + y)^2_{|_X}$, then $\overline{\Sigma_{\mathrm{alg}} g} = \{\mathbf{0}\}$ and $\overline{\Sigma_{\mathrm{reg}} g} = \emptyset$; thus, in general, $\overline{\Sigma_{\mathrm{reg}} f} \neq \overline{\Sigma_{\mathrm{alg}} f}$.

While it is easy to produce examples where $\Sigma_{\mathrm{Nash}} f$ is not equal to $\Sigma_{\mathrm{alg}} f$ and examples where $\Sigma_{\mathrm{Nash}} f$ is not equal to $\Sigma_{\mathrm{cnr}} f$, it is not quite so easy to come up with examples where all three of these sets are distinct. We give such an example here.

Let $Z := V((y - zx)(y^2 - x^3)) \subseteq \mathbb{C}^3$ and $L := y_{|_Z}$. Then, one easily verifies that $\Sigma_{\mathrm{alg}} L = \emptyset$, $\Sigma_{\mathrm{Nash}} L = \{\mathbf{0}\}$, and $\Sigma_{\mathrm{cnr}} L = \mathbb{C} \times \{\mathbf{0}\}$.

If $X = V(xy)$ and $h := (x + y)_{|_X}$, then $\Sigma_{\mathbb{C}} h = \{\mathbf{0}\}$ and $\Sigma_{\mathrm{cnr}} h = \emptyset$; thus, in general, $\Sigma_{\mathrm{cnr}} f \neq \overline{\Sigma_{\mathbb{C}} f}$.

Let $W := V(z^5 + ty^6 z + y^7 x + x^{15}) \subseteq \mathbb{C}^4$; this is the example of Briançon and Speder [**B-S**] in which the topology along the t-axis is constant, despite the fact that the origin is a point-stratum in the canonical Whitney stratification of W. Hence, if we let r denote the restriction of t to W, then, for values of r close to 0, $\mathbf{0}$ is the only point in $\Sigma_{\mathrm{can}} r$ and $\mathbf{0} \notin \Sigma_{\mathbb{C}} r$. Therefore, $\mathbf{0} \in \Sigma_{\mathrm{can}} r - \overline{\Sigma_{\mathbb{C}} r}$, and so, in general, $\overline{\Sigma_{\mathbb{C}} f} \neq \Sigma_{\mathrm{can}} f$.

Using the coordinates (x, y, z) on $\mathbb{C}^3$, consider the cross-product $Y := V(y^2 - x^3) \subseteq \mathbb{C}^3$. The canonical Whitney stratification of Y is given by $\{Y - \{\mathbf{0}\} \times \mathbb{C},\ \{\mathbf{0}\} \times \mathbb{C}\}$. Let $\pi := z_{|_Y}$. Then, $\Sigma_{\mathrm{can}} \pi = \emptyset$, while $\Sigma_{\mathrm{rdf}} \pi = \{\mathbf{0}\} \times \mathbb{C}$. Thus, in general, $\Sigma_{\mathrm{can}} f \neq \Sigma_{\mathrm{rdf}} f$.

It is, of course, easy to throw extra, non-canonical, Whitney strata into almost any example in order to see that, in general, $\Sigma_{\mathrm{can}} f \neq \Sigma_{\mathcal{S}} f$.

To summarize the contents of this example and Proposition 1.6: we have seven seemingly reasonable definitions of "critical locus" for complex analytic functions on singular spaces (we are not counting $\Sigma_{\mathcal{S}} f$, since it is not intrinsically defined). All of our critical locus avatars agree for manifolds. The sets $\Sigma_{\mathrm{reg}} f$, $\Sigma_{\mathrm{alg}} f$, $\Sigma_{\mathrm{Nash}} f$, and $\Sigma_{\mathbb{C}} f$ need not be closed. There is a chain of containments among the closures of these critical loci, but – in general – none of the sets are equal.

However, we consider the sets $\overline{\Sigma_{\mathrm{reg}} f}$, $\overline{\Sigma_{\mathrm{alg}} f}$, $\overline{\Sigma_{\mathrm{Nash}} f}$, and $\Sigma_{\mathrm{cnr}} f$ to be too small; these "critical loci" do not detect the change in topology at the level hypersurface $h = 0$ in the simple example $X = V(xy)$ and $h = (x+y)_{|X}$ (from Example 1.8).

Despite the fact that the Stratified Morse Theory of [**Go-Mac1**] yields nice results and requires one to consider the stratified critical locus, we also will not use $\Sigma_{\mathrm{can}} f$ (or any other $\Sigma_{\mathcal{S}} f$) as our primary notion of critical locus; $\Sigma_{\mathrm{can}} f$ is often too big. As we saw in the Briançon-Speder example in Example 1.8, the stratified critical locus sometimes forces one to consider "critical points" which do not correspond to changes in topology.

Certainly, $\Sigma_{\mathrm{rdf}} f$ is far too large, if we want critical points to have **any** relation to changes in the topology of level hypersurfaces: if X has a singular set ΣX, then the critical space of the projection $\pi : X \times \mathbb{C} \to \mathbb{C}$ would consist of $\Sigma X \times \mathbb{C}$, despite the obvious triviality of the family of level hypersurfaces defined by π.

Therefore, we choose to concentrate our attention on the $\mathbb{C}$-critical locus, and we will justify this choice with the results in the remainder of Part III.

Note that we consider $\Sigma_{\mathbb{C}} f$, not its closure, to be the correct notion of critical locus; we think that this is the more natural definition, and we consider the question of when $\Sigma_{\mathbb{C}} f$ is closed to be an interesting one. It is true, however, that all of our results refer to $\overline{\Sigma_{\mathbb{C}} f}$. We should mention here that, while $\Sigma_{\mathbb{C}} f$ need not be closed, the existence of Thom stratifications [**Hi**] implies that $\Sigma_{\mathbb{C}} f$ is at least analytically constructible; hence, $\overline{\Sigma_{\mathbb{C}} f}$ is an analytic subset of X.

Before we leave this chapter, in which we have already looked at seven definitions of "critical locus", we need to look at one last variant. As we mentioned at the end of the introduction, even though we wish to investigate the Milnor fibre with coefficients in $\mathbb{C}$, the fact that the shifted constant sheaf on a non-l.c.i. need not be perverse requires us to take the perverse cohomology of the constant sheaf. This means that we need to consider the hypercohomology of Milnor fibres with coefficients in an arbitrary bounded, constructible complex of sheaves of modules. As we wish to discuss Euler characteristics, we need for the rank of a finitely-generated module to be defined and additive over exact sequences; thus, we must choose our base ring to be a p.i.d. However, since the rank of a module over a p.i.d. equals the dimension of the associated vector space over the quotient field, we may as well restrict ourselves to the case where the base ring is, in fact, a field.

The $\mathbb{C}$−critical locus is nicely described in terms of vanishing cycles (see [**K-S**] for general properties of vanishing cycles, but be aware that we use the more traditional shift):

$$\Sigma_{\mathbb{C}} f = \{\mathbf{x} \in X \mid H^*(\phi_{f-f(\mathbf{x})} \mathbb{C}^\bullet_X)_{\mathbf{x}} \neq 0\}.$$

This definition generalizes easily to yield a definition of the critical loci of f with respect to arbitrary bounded, constructible complexes of sheaves on X.

Let R be a p.i.d. Let $\mathcal{S} := \{S_\alpha\}$ be a Whitney stratification of X, and let $\mathbf{F}^\bullet$ be a bounded complex of sheaves of R-modules which is constructible with respect to $\mathcal{S}$.

Definition 1.9. The $\mathbf{F}^\bullet$*-critical locus of* f , $\Sigma_{\mathbf{F}^\bullet} f$, is defined by

$$\Sigma_{\mathbf{F}^\bullet} f := \{\mathbf{x} \in X \mid H^*(\phi_{f-f(\mathbf{x})}\mathbf{F}^\bullet)_{\mathbf{x}} \neq 0\}.$$

Remark 1.10. Stratified Morse Theory (see [**Go-Mac1**]) implies that $\Sigma_{\mathbf{F}^\bullet} f \subseteq \Sigma_{\mathcal{S}} f$ (alternatively, this follows from 8.4.1 and 8.6.12 of [**K-S**], combined with the facts that complex analytic Whitney stratifications are w-stratifications, and w-stratifications are μ-stratifications.)

We could discuss three more notions of the critical locus of a function – two of which are obtained by picking specific complexes for $\mathbf{F}^\bullet$ in Definition 1.9. However, we will defer the introduction of these new critical loci until Chapter 4; at that point, we will have developed the tools necessary to say something interesting about these three new definitions.

Chapter 2. THE RELATIVE POLAR CURVE

In this chapter, we discuss the relative polar curve of Lê and Teissier ([**L-T2**],[**Te5**], [**Te6**], [**Te7**], etc.). This object is now a standard part of singularity theory, and the reader is most likely familiar with some of the results that appear here. However, we shall use our early results on gap varieties and cycles to present the theory in the conormal form in which we will use it in the next chapter. Our treatment of the higher-dimensional relative polar varieties, when the underlying space is singular, does not appear here; it is a major portion of Part IV.

We let $\mathcal{U}$ be an open subset of $\mathbb{C}^{n+1}$, and let X be a reduced analytic subspace of $\mathcal{U}$ with analytic components $\{X_j\}$. Let d_j denote the dimension of X_j, let $c_j := n+1-d_j$ denote its codimension, and let d denote the global dimension of X. Let $\tilde{f} : \mathcal{U} \to \mathbb{C}$ be an analytic function, and let $f := \tilde{f}_{|X}$.

For $g_1, \dots, g_j \in \mathcal{O}_\mathcal{U}$, let $\mathrm{Jac}(g_1, \dots, g_j)$ denote the Jacobian matrix of $(g_1, \dots, g_j)$, which has the partial derivatives of g_i in its i-th row. For any matrix A of functions, we let $\mathrm{Min}_i(A)$ denote the sheaf of ideals generated by the determinants of the $i \times i$ minors of A. Let $J_i(g_1, \dots, g_j)$ denote $\mathrm{Min}_i(\mathrm{Jac}(g_1, \dots, g_j))$.

We use $\mathbf{z} := (z_0, \dots, z_n)$ as coordinates on $\mathcal{U}$. We let $\eta : T^*\mathcal{U} \to \mathcal{U}$ denote the cotangent bundle, and we identify the cotangent space $T^*\mathcal{U}$ with $\mathcal{U} \times \mathbb{C}^{n+1}$ by using $dz_0, \dots, dz_n$ as a basis. We use $\mathbf{w} := (w_0, \dots, w_n)$ as coordinates for the cotangent vectors, i.e., a cotangent vector is $w_0 dz_0 + \dots + w_n dz_n$. If $\hat{\mathbf{z}}$ is a linear change of coordinates applied to $\mathbf{z}$, then we let $\widehat{\mathbf{w}}$ denote cotangent coordinates with respect to the new basis $d\hat{z}_0, \dots, d\hat{z}_n$, i.e., if A is in $\mathrm{Gl}_{n+1}(\mathbb{C})$ and $\hat{\mathbf{z}} := A\mathbf{z}$, then $\mathbf{w} = A^t \widehat{\mathbf{w}}$.

Definition 2.1. The *relative polar curve of X with respect to f and z_0*, $\Gamma^1_{f,z_0}(X)$, is defined as a set by

$$\Gamma^1_{f,z_0} := \overline{\Sigma\big((f, z_0)_{|X_{\mathrm{reg}} - \Sigma(f_{|X_{\mathrm{reg}}})}\big)}.$$

Let $\mathcal{V}$ be an open subset of $\mathcal{U}$, and suppose that $\mathbf{h} := (h_1, \dots, h_l)$ defines X in $\mathcal{V}$. Then, Γ^1_{f,z_0} is defined as a scheme on $\mathcal{V}$ by

$$\bigcup_j \Big(X_j \cap V\big(J_{c_j+2}(\mathbf{h}, \tilde{f}, z_0)\big)\Big) \neg \big(\Sigma X \cup \overline{\Sigma(f_{|X_{\mathrm{reg}}})}\big).$$

(We remind the reader that the union is given a scheme structure by using the intersection of the underlying ideal sheaves.)

These definitions are independent of the defining equations $\mathbf{h}$ and the choice of the extension, $\tilde{f}$, of f.

Lê and Teissier prove:

Proposition 2.2. *Let* $\mathbf{p} \in X$. *Then, for a generic choice of the linear form,* z_0, *in a neighborhood of* $\mathbf{p}$, Γ^1_{f,z_0} *is reduced, is purely* 1*-dimensional, and the restriction of either of the maps* f *or* z_0 *to this scheme is finite.*

Moreover, if X *is irreducible at* $\mathbf{p}$, *and* Y *is a proper analytic subset of* X, *then for a generic choice of* z_0, Γ^1_{f,z_0} *has no component contained in* Y *at* $\mathbf{p}$.

Proof. The first paragraph is well-known and is proved in many places; see, for instance, [**L-T2**], 4.2.1 and [**Te6**], 4.1.3.2. However, for lack of a convenient reference, we will prove the second statement.

Suppose that X is irreducible. If f is constant on X, then Γ^1_{f,z_0} is empty, and we are finished; so, assume that f is not constant. It clearly suffices to prove the result when Y is irreducible; so, we assume that it is. Also, assume that $Y \not\subseteq \Sigma X \cup \Sigma(f_{|_{X_{\rm reg}}})$, for otherwise there is nothing to prove. Finally, it suffices to prove that Γ^1_{f,z_0} has no component contained in $Y_{\rm reg}$ at $\mathbf{p}$; for if Γ^1_{f,z_0} has a component contained in ΣY, we replace Y by ΣY and induct on the dimension of Y.

Let $\overset{\circ}{X} := X_{\rm reg} - \Sigma(f_{|_{X_{\rm reg}}})$, and consider the relative conormal variety

$$T^*_{f_{|_{\overset{\circ}{X}}}}\mathcal{U} := \{(\mathbf{x}, \eta) \in T^*\mathcal{U} \mid \mathbf{x} \in \overset{\circ}{X},\ \eta(T_{\mathbf{x}}\overset{\circ}{X} \cap \ker d_{\mathbf{x}}\tilde{f}) = 0\}.$$

The dimension of the fibre over a point $\mathbf{x} \in \overset{\circ}{X}$ is precisely $n + 2 - \dim X$, and $(\mathbf{x}, d_{\mathbf{x}}z_0) \in T^*_{f_{|_{\overset{\circ}{X}}}}\mathcal{U}$ if and only if $\mathbf{x} \in \Gamma^1_{f,z_0}$. It follows that $\overline{\eta^{-1}(Y_{\rm reg}) \cap T^*_{f_{|_{\overset{\circ}{X}}}}\mathcal{U}}$ is irreducible of dimension at most $n+2-\dim X+\dim Y \leqslant n+1$. Therefore, the fibre of this space over $\mathbf{p}$ is conic and of dimension at most n, and so does not contain $d_{\mathbf{p}}z_0$ for generic z_0. The result follows. □

In general, we do not care is z_0 is chosen so that Γ^1_{f,z_0} is reduced, but we do want that the restrictions of f and z_0 are finite.

We now wish to characterize the relative polar curve in terms of conormal spaces and gap sheaves.

We remind the reader that $\overline{T^*_{X_{\rm reg}}\mathcal{U}}$ is purely $(n+1)$-dimensional. Also, if $X = V(\mathbf{h})$ in an open subset $\mathcal{V}$ of $\mathcal{U}$, then it is trivial to see that, over $\mathcal{V}$, the reduced space $\overline{T^*_{X_{\rm reg}}\mathcal{U}}$ is given by

$$\bigcup_j \left((X_j \times \mathbb{C}^{n+1}) \cap \operatorname{Min}_{c_j+1}\left(\begin{matrix} \mathbf{w} \\ \operatorname{Jac}(\mathbf{h}) \end{matrix} \right) \right) \neg (\Sigma X \times \mathbb{C}^{n+1}).$$

Note that a generic linear reorganization of $\mathbf{z}$ (and the corresponding reorganization of $\mathbf{w}$) produces a generic linear reorganization of $\left(w_0 - \frac{\partial \tilde{f}}{\partial z_0}, \dots, w_n - \frac{\partial \tilde{f}}{\partial z_n}\right)$.

Definition 2.3. The *relative conormal polar curve of X with respect to $\tilde{f}$ and $\mathbf{z}$*, $\Gamma^1_{\tilde{f},\mathbf{z}}(\overline{T^*_{X_{\mathrm{reg}}}\mathcal{U}})$, is defined to be the 1-th gap variety of $\mathbf{w} - \frac{\partial \tilde{f}}{\partial \mathbf{z}} := \left(w_0 - \frac{\partial \tilde{f}}{\partial z_0}, \dots, w_n - \frac{\partial \tilde{f}}{\partial z_n}\right)$ restricted to $\overline{T^*_{X_{\mathrm{reg}}}\mathcal{U}}$, i.e.,

$$\Gamma^1_{\tilde{f},\mathbf{z}}(\overline{T^*_{X_{\mathrm{reg}}}\mathcal{U}}) := \left(\overline{T^*_{X_{\mathrm{reg}}}\mathcal{U}} \cap V\left(w_1 - \frac{\partial \tilde{f}}{\partial z_1}, \dots, w_n - \frac{\partial \tilde{f}}{\partial z_n}\right)\right) \neg V\left(\mathbf{w} - \frac{\partial \tilde{f}}{\partial \mathbf{z}}\right).$$

Note that we are not claiming that the relative conormal polar curve is independent of the extension $\tilde{f}$.

We need to investigate the relation between the relative conormal polar curve and the ordinary polar curve. The following lemma tells us that they agree over the regular part of X, regardless of whether the two varieties are even really curves.

Lemma 2.4. *Let $\mathbf{p} \in X$ be a regular point and suppose that $d_{\mathbf{p}} z_0 \notin (\overline{T^*_{X_{\mathrm{reg}}}\mathcal{U}})_{\mathbf{p}}$. Then, in a neighborhood of $\mathbf{p}$, the projection map η restricted to $\Gamma^1_{\tilde{f},\mathbf{z}}(\overline{T^*_{X_{\mathrm{reg}}}\mathcal{U}})$ induces an isomorphism onto Γ^1_{f,z_0}.*

Proof. Because X is smooth at $\mathbf{p}$ and $d_{\mathbf{p}} z_0 \notin (\overline{T^*_{X_{\mathrm{reg}}}\mathcal{U}})_{\mathbf{p}}$, we may use an analytic change of coordinates at $\mathbf{p}$ to reduce ourselves to the case where $X = \mathcal{V}$, where $\mathcal{V}$ is an open subset of $\mathbb{C}^d \times \{\mathbf{0}\}$ and $\overline{T^*_{X_{\mathrm{reg}}}\mathcal{U}} = \mathcal{V} \times (\{\mathbf{0}\} \times \mathbb{C}^{n+1-d})$.

Now, one sees easily that $\Gamma^1_{f,z_0} = \left(V\left(\frac{\partial f}{\partial z_1}, \dots, \frac{\partial f}{\partial z_{d-1}}\right) \neg V\left(\frac{\partial f}{\partial z_0}\right)\right) \times \{\mathbf{0}\}$ and that

$$\Gamma^1_{\tilde{f},\mathbf{z}}(\overline{T^*_{X_{\mathrm{reg}}}\mathcal{U}}) =$$

$$\left((\mathcal{V} \times (\{\mathbf{0}\} \times \mathbb{C}^{n+1-d})) \cap V\left(w_1 - \frac{\partial \tilde{f}}{\partial z_1}, \dots, w_n - \frac{\partial \tilde{f}}{\partial z_n}\right)\right) \neg V\left(w_0 - \frac{\partial \tilde{f}}{\partial z_0}\right) =$$

$$\left((\mathcal{V} \times (\{\mathbf{0}\} \times \mathbb{C}^{n+1-d})) \cap V\left(\frac{\partial \tilde{f}}{\partial z_1}, \dots, \frac{\partial \tilde{f}}{\partial z_{d-1}}, w_d - \frac{\partial \tilde{f}}{\partial z_d}, \dots, w_n - \frac{\partial \tilde{f}}{\partial z_n}\right)\right) \neg V\left(\frac{\partial \tilde{f}}{\partial z_0}\right) =$$

$$\left(\Gamma^1_{f,z_0} \times \mathbb{C}^{n+1}\right) \cap V\left(w_d - \frac{\partial \tilde{f}}{\partial z_d}, \dots, w_n - \frac{\partial \tilde{f}}{\partial z_n}\right).$$

Thus, η restricted to $\Gamma^1_{\tilde{f},\mathbf{z}}(\overline{T^*_{X_{\mathrm{reg}}}\mathcal{U}})$ (and to its image) has as its inverse the map

$$\tau : \Gamma^1_{f,z_0} \to \Gamma^1_{\tilde{f},\mathbf{z}}(\overline{T^*_{X_{\mathrm{reg}}}\mathcal{U}})$$

given by $\tau(\mathbf{x}) = \left(\mathbf{x}, \mathbf{0}, \frac{\partial \tilde{f}}{\partial z_d}, \dots, \frac{\partial \tilde{f}}{\partial z_n}\right)$. □

We now prove the fundamental result for polar curves on general spaces.

Theorem 2.5. *Let* $\mathbf{p} \in X$. *For a generic linear reorganization,* $\mathbf{z}$,

0) $d_{\mathbf{p}} z_0 \notin \left(\overline{T^*_{X_{\text{reg}}}\mathcal{U}}\right)_{\mathbf{p}}$;

i) Γ^1_{f,z_0} *is purely* 1*-dimensional at* $\mathbf{p}$;

ii) $f_{|_{\Gamma^1_{f,z_0}}}$ *is finite at* $\mathbf{p}$; *and*

iii) $\Gamma^1_{\tilde{f},\mathbf{z}}\left(\overline{T^*_{X_{\text{reg}}}\mathcal{U}}\right)$ *has no components contained in* $\eta^{-1}(\Sigma X)$ *at* $(\mathbf{p}, d_{\mathbf{p}}\tilde{f})$.

In addition, whenever 0)-iii) hold, then we also have

iv) $z_0{}_{|_{\Gamma^1_{f,z_0}}}$ *is finite at* $\mathbf{p}$;

v) $\Gamma^1_{\tilde{f},\mathbf{z}}\left(\overline{T^*_{X_{\text{reg}}}\mathcal{U}}\right)$ *is purely* 1*-dimensional at* $(\mathbf{p}, d_{\mathbf{p}}\tilde{f})$;

vi) $(f \circ \eta)_{|_{\Gamma^1_{\tilde{f},\mathbf{z}}(\overline{T^*_{X_{\text{reg}}}\mathcal{U}})}}$ *is finite at* $(\mathbf{p}, d_{\mathbf{p}}\tilde{f})$;

vii) $(z_0 \circ \eta)_{|_{\Gamma^1_{\tilde{f},\mathbf{z}}(\overline{T^*_{X_{\text{reg}}}\mathcal{U}})}}$ *is finite at* $(\mathbf{p}, d_{\mathbf{p}}\tilde{f})$; *and*

viii) $\left(\Gamma^1_{\tilde{f},\mathbf{z}}\left(\overline{T^*_{X_{\text{reg}}}\mathcal{U}}\right)\cdot V\left(w_0 - \frac{\partial \tilde{f}}{\partial z_0}\right)\right)_{(\mathbf{p},d_{\mathbf{p}}\tilde{f})} = \left(\Gamma^1_{f,z_0}\cdot V(f - f(\mathbf{p}))\right)_{\mathbf{p}} - \left(\Gamma^1_{f,z_0}\cdot V(z_0 - z_0(\mathbf{p}))\right)_{\mathbf{p}}$.

Proof. For notational convenience, we shall assume that $f(\mathbf{p}) = z_0(\mathbf{p}) = 0$.

That 0) holds generically is trivial. That i) and ii) hold generically follows from 2.2. As $\eta^{-1}(\Sigma X) \cap \overline{T^*_{X_{\text{reg}}}\mathcal{U}}$ has dimension at most n, I.2.11 implies that for a generic linear reorganization

$$\left(\eta^{-1}(\Sigma X) \cap \overline{T^*_{X_{\text{reg}}}\mathcal{U}} \cap V\left(w_1 - \frac{\partial \tilde{f}}{\partial z_1}, \dots, w_n - \frac{\partial \tilde{f}}{\partial z_n}\right)\right) \neg\, V\left(\mathbf{w} - \frac{\partial \tilde{f}}{\partial \mathbf{z}}\right)$$

is purely 0-dimensional at $(\mathbf{p}, d_{\mathbf{p}}\tilde{f})$. As every component of $\Gamma^1_{\tilde{f},\mathbf{z}}\left(\overline{T^*_{X_{\text{reg}}}\mathcal{U}}\right)$ has dimension at least 1 (by I.2.2), iii) holds for generic $\mathbf{z}$.

Now suppose that 0)-iii) hold.

By the lemma, η yields a local isomorphism between the schemes $\Gamma^1_{\tilde{f},\mathbf{z}}\big(\overline{T^*_{X_{\rm reg}}\mathcal{U}}\big)-\eta^{-1}(\Sigma X)$ and $\Gamma^1_{f,z_0}-\Sigma X$; combining this with iii), we conclude that i) implies v), ii) implies vi), and iv) implies vii). It remains for us to show that iv) and viii) hold.

Let C be an irreducible component of Γ^1_{f,z_0} and suppose that $\gamma(t)$ is an analytic parameterization of C near $\mathbf{p}$ such that $\gamma(0)=\mathbf{p}$. For $t\neq 0$, $\gamma(t)\in X_{\rm reg}$, $d_{\gamma(t)}\tilde{f}\notin T^*_{X_{\rm reg}}\mathcal{U}$, and there must exist complex numbers $a(t)$ and $b(t)$, not both zero, such that $a(t)d_{\gamma(t)}\tilde{f}-b(t)d_{\gamma(t)}z_0\in T^*_{X_{\rm reg}}\mathcal{U}$. If $a(t)$ were zero for an infinite number of t, then, since $T^*_{X_{\rm reg}}\mathcal{U}$ is conic, $d_{\gamma(t)}z_0$ would be in $T^*_{X_{\rm reg}}\mathcal{U}$ for an infinite number of t; this would contradict 0). Thus, for small $t\neq 0$, there exists $c(t)$ such that $d_{\gamma(t)}\tilde{f}-c(t)d_{\gamma(t)}z_0\in T^*_{X_{\rm reg}}\mathcal{U}$. By evaluating this form on the tangent vector $\gamma'(t)$, we conclude that $\big(f(\gamma(t))\big)'-c(t)\big(z_0(\gamma(t))\big)'\equiv 0$.

If iv) were false, then $z_0(\gamma(t))$ would be zero (for some component C), and, hence, we would have that $\big(f(\gamma(t))\big)'\equiv 0$; but, this would imply that $f(\gamma(t))\equiv 0$, in contradiction of ii). Therefore, we have shown iv).

Note that $c(t)$ is uniquely determined by

$$c(t)=\frac{\big(f(\gamma(t))\big)'}{\big(z_0(\gamma(t))\big)'},$$

where $z_0(\gamma(t))\not\equiv 0$ by iv). If $|c(t)|\to\infty$ as $t\to 0$, then since $\frac{1}{c(t)}d_{\gamma(t)}\tilde{f}-d_{\gamma(t)}z_0\in T^*_{X_{\rm reg}}\mathcal{U}$, we would once again conclude that $d_{\mathbf{p}}z_0\in\big(\overline{T^*_{X_{\rm reg}}\mathcal{U}}\big)_{\mathbf{p}}$, in contradiction to 0). It follows that $c(t)$ is analytic at 0, and thus that $c(t)$ approaches some finite value q_C as $t\to 0$. Therefore, the component C corresponds to a component, $\widehat{C}$, of $\Gamma^1_{\tilde{f},\mathbf{z}}\big(\overline{T^*_{X_{\rm reg}}\mathcal{U}}\big)$ through $(\mathbf{p},d_{\mathbf{p}}\tilde{f})$ if and only if $q_C=0$; for $\widehat{C}$ is parameterized by $\hat{\gamma}(t)=\big(\gamma(t),\ d_{\gamma(t)}\tilde{f}-c(t)d_{\gamma(t)}z_0\big)$.

Now suppose that, at $(\mathbf{p},d_{\mathbf{p}}\tilde{f})$, the cycle $\Gamma^1_{\tilde{f},\mathbf{z}}\big(\overline{T^*_{X_{\rm reg}}\mathcal{U}}\big)$ equals $\Sigma n_V[V]$. As the restriction of η to $\Gamma^1_{\tilde{f},\mathbf{z}}\big(\overline{T^*_{X_{\rm reg}}\mathcal{U}}\big)$ is a local isomorphism onto Γ^1_{f,z_0} over smooth points of X, we conclude that the cycle Γ^1_{f,z_0} can be written as

$$\Gamma^1_{f,z_0}=\sum_V n_V[\eta(V)]\ +\ \sum_{q_C\neq 0} n_C[C].$$

Therefore, to demonstrate viii), we need to show two things:

a) if $q_C\neq 0$, then $(C\cdot V(f))_{\mathbf{p}}-(C\cdot V(z_0))_{\mathbf{p}}=0$, and

b) if $q_C=0$, then $\big(\widehat{C}\cdot V\big(w_0-\frac{\partial\tilde{f}}{\partial z_0}\big)\big)_{(\mathbf{p},d_{\mathbf{p}}\tilde{f})}=(C\cdot V(f))_{\mathbf{p}}-(C\cdot V(z_0))_{\mathbf{p}}$.

We show a) and b) by calculating intersection numbers via parameterizations (see Appendix A.9).

If $q_C\neq 0$, then $\big(f(\gamma(t))\big)'$ and $\big(z_0(\gamma(t))\big)'$ must have the same t-multiplicity. Thus, $f(\gamma(t))$ and $z_0(\gamma(t))$ have the same t-multiplicity, and so

$$(C\cdot V(f))_{\mathbf{p}}-(C\cdot V(z_0))_{\mathbf{p}}\ =\ \mathrm{mult}_t f(\gamma(t))-\mathrm{mult}_t z_0(\gamma(t))\ =\ 0.$$

If $q_C = 0$, then

$$\Big(\widehat{C}\cdot V\Big(w_0 - \frac{\partial \tilde{f}}{\partial z_0}\Big)\Big)_{(\mathbf{p}, d_{\mathbf{p}}\tilde{f})} = \operatorname{mult}_t\Big(w_0(\hat{\gamma}(t)) - \frac{\partial \tilde{f}}{\partial z_0}\Big|_{\hat{\gamma}(t)}\Big) =$$

$$\operatorname{mult}_t\Big(\frac{\partial \tilde{f}}{\partial z_0}\Big|_{\hat{\gamma}(t)} - c(t) - \frac{\partial \tilde{f}}{\partial z_0}\Big|_{\hat{\gamma}(t)}\Big) = \operatorname{mult}_t(c(t)) = \operatorname{mult}_t\big(f(\gamma(t))\big)' - \operatorname{mult}_t\big(z_0(\gamma(t))\big)' =$$

$$\operatorname{mult}_t\big(f(\gamma(t))\big) - \operatorname{mult}_t\big(z_0(\gamma(t))\big) = (C\cdot V(f))_{\mathbf{p}} - (C\cdot V(z_0))_{\mathbf{p}}. \quad \square$$

Remark 2.6. The point of 2.5.viii is that, for generic $\mathbf{z}$, the quantity

$$\Big(\Gamma^1_{\tilde{f},\mathbf{z}}\big(\overline{T^*_{X_{\mathrm{reg}}}\mathcal{U}}\big)\cdot V\Big(w_0 - \frac{\partial \tilde{f}}{\partial z_0}\Big)\Big)_{(\mathbf{p}, d_{\mathbf{p}}\tilde{f})}$$

is, in fact, the multiplicity of the 0-th Vogel cycle of $(\mathbf{w} - \frac{\partial \tilde{f}}{\partial \mathbf{z}})\big|_{\overline{T^*_{X_{\mathrm{reg}}}\mathcal{U}}}$ at $(\mathbf{p}, d_{\mathbf{p}}\tilde{f})$ (recall I.2.14). This enables us to apply results from Part I.

Note, also, in the affine case where $X = \mathcal{U}$, the equality of 2.5.viii reduces to the well-known formula from Proposition II.1.20:

$$\Big(\Gamma^1_{f,z_0}\cdot V\Big(\frac{\partial f}{\partial z_0}\Big)\Big)_{\mathbf{p}} = \big(\Gamma^1_{f,z_0}\cdot V(f - f(\mathbf{p}))\big)_{\mathbf{p}} - \big(\Gamma^1_{f,z_0}\cdot V(z_0 - z_0(\mathbf{p}))\big)_{\mathbf{p}}.$$

Chapter 3. THE LINK BETWEEN THE ALGEBRAIC AND TOPOLOGICAL POINTS OF VIEW.

We continue with our previous notation: X is a d-dimensional complex analytic space contained in some open subset $\mathcal{U}$ of some $\mathbb{C}^{n+1}$, $\tilde{f} : \mathcal{U} \to \mathbb{C}$ is a complex analytic function, $f = \tilde{f}_{|_X}$, $\mathcal{S} = \{S_\alpha\}$ is a Whitney stratification of X with connected strata, the base ring R is a p.i.d., and $\mathbf{F}^\bullet$ is a bounded complex of sheaves of R-modules which is constructible with respect to $\mathcal{S}$. In addition, $\mathbb{N}_\alpha$ and $\mathbb{L}_\alpha$ are, respectively, the normal slice and complex link of the d_α-dimensional stratum S_α (see [**Go-Mac1**]).

In this chapter, we are going to prove a general result which describes the characteristic cycle of $\phi_f\mathbf{F}^\bullet$ in terms of blowing-up the image of $d\tilde{f}$ inside the conormal spaces to strata. We will have to wait until the next chapter (on results for perverse sheaves) to actually show how this provides a relationship between $\Sigma_{\mathbf{F}^\bullet} f$ and $\Sigma_{\mathcal{S}} f$ in the case where $\mathbf{F}^\bullet$ is perverse.

Because d is the global dimension of X, and we are not assuming that X is pure-dimensional, or that f is not constant on a d-dimensional component of X, if $v \in \mathbb{C}$, then the dimension of $V(f - v)$ could be anything between 0 and d. Hence, we let $\hat{d}_v := 1 + \dim V(f - v)$, and will usually denote $\hat{d}_0$ by simply $\hat{d}$. Of course, if we work locally, or assume that X is pure-dimensional, and require f not to vanish on a component of X, then $\hat{d}$ will have attain its "expected" value of d.

Definition 3.1. Recall that the *characteristic cycle,* $\mathrm{Ch}(\mathbf{F}^\bullet)$, *of* $\mathbf{F}^\bullet$ in $T^*\mathcal{U}$ is the linear combination $\sum_\alpha m_\alpha(\mathbf{F}^\bullet)\left[\overline{T^*_{S_\alpha}\mathcal{U}}\right]$, where the $m_\alpha(\mathbf{F}^\bullet)$ are integers given by

$$m_\alpha(\mathbf{F}^\bullet) \;:=\; (-1)^d\chi(\phi_{L_{|_X}}[-1]\mathbf{F}^\bullet)_\mathbf{x} \;=\; (-1)^d\chi(\phi_{L_{|_{\mathbb{N}_\alpha}}}[-1]\mathbf{F}^\bullet{}_{|_{\mathbb{N}_\alpha}}[-d_\alpha])_\mathbf{x} \;=\; (-1)^{d-d_\alpha}\chi\big(\mathbb{H}^*(\mathbb{N}_\alpha, \mathbb{L}_\alpha;\ \mathbf{F}^\bullet)\big)$$

for any point $\mathbf{x}$ in S_α, with normal slice $\mathbb{N}_\alpha$ at $\mathbf{x}$, and any $L : (\mathcal{U}, x) \to (\mathbb{C}, 0)$ such that $d_\mathbf{x}L$ is a non-degenerate covector at $\mathbf{x}$ (with respect to our fixed stratification; see [**Go-Mac1**]) and $L_{|_{S_\alpha}}$ has a Morse singularity at $\mathbf{x}$. This cycle is independent of all the choices made (see, for instance, [**K-S**, Chapter IX]).

We need a number of preliminary results before we can prove the main theorem (Theorem 3.10) of this section.

Definition 3.2. Recall that, if M is an analytic submanifold of $\mathcal{U}$ and $M \subseteq X$, then the *relative conormal space (of M with respect to f in $\mathcal{U}$),* $T^*_{f_{|_M}}\mathcal{U}$, is given by

$$T^*_{f_{|_M}}\mathcal{U} := \{(\mathbf{x}, \eta) \in T^*\mathcal{U} \mid \mathbf{x} \in M,\ \eta\big(\ker d_\mathbf{x}(f_{|_M})\big) = 0\} = \{(\mathbf{x}, \eta) \in T^*\mathcal{U} \mid \mathbf{x} \in M,\ \eta\big(T_\mathbf{x}M \cap \ker d_\mathbf{x}\tilde{f}\big) = 0\}.$$

We define the *total relative conormal cycle,* $T^*_{f,\mathbf{F}^\bullet}\mathcal{U}$, by

$$T^*_{f,\mathbf{F}^\bullet}\mathcal{U} := \sum_{S_\alpha \not\subseteq f^{-1}(0)} m_\alpha(\mathbf{F}^\bullet)\left[\overline{T^*_{f_{|_{S_\alpha}}}\mathcal{U}}\right].$$

From this point, through Lemma 3.9, it will be convenient to assume that we have refined our stratification $\mathcal{S} = \{S_\alpha\}$ so that $V(f)$ is a union of strata. By Remark 1.7, this implies that, in a neighborhood of $V(f)$, if $S_\alpha \not\subseteq V(f)$, then $\Sigma(f_{|_{S_\alpha}}) = \emptyset$.

We first stated Theorems 3.3 and 3.4 below in our earlier works [**Mas2**] and [**Mas5**]. In those papers, we were mainly concerned with local questions, and we also tacitly assumed that f was not constant on any irreducible component of X. We did **not** correctly adjust the sign for degenerate cases. We correct this error in the statements below – the proofs remain the same.

We shall need the following important result from [**BMM**, 3.4.2].

Theorem 3.3. ([**BMM**]) *The shifted characteristic cycle of the sheaf of nearby cycles of* $\mathbf{F}^\bullet$ *along* f, $(-1)^{d-\hat{d}}\operatorname{Ch}(\psi_f\mathbf{F}^\bullet)$, *is isomorphic to the intersection product* $T^*_{f,\mathbf{F}^\bullet}\mathcal{U}\cdot(V(f)\times\mathbb{C}^{n+1})$ *in* $\mathcal{U}\times\mathbb{C}^{n+1}$.

We should note here that the context of [**BMM**] is that of $\mathcal{D}$-modules and, hence, in that work, the complex of sheaves was a complex of $\mathbb{C}$-vector spaces. However, Theorem 3.3 can easily be recovered by combining the first formula of Theorem 3.4 with Lemma 3.5 (see below), keeping in mind that Lemma 3.5 relies on the result of Theorem 3.3 with $\mathbb{C}$-complexes only.

Let $\Gamma^1_{f,L}(S_\alpha)$ denote the relative polar curve of $f_{|_{\overline{S_\alpha}}}$ with respect to a generic linear form L (see Chapter 2 and [**M1**] and [**M3**]). It is important to note that the second part of 2.2 implies that $\Gamma^1_{f,L}(S_\alpha)$ has no components contained in any strata $S_\beta \subseteq \overline{S_\alpha}$ such that $S_\beta \neq S_\alpha$.

It is convenient to have a specific point in X at which to work. Below, we concentrate our attention at the origin; of course, if the origin is not in X (or, if the origin is not in $V(f)$), then we obtain zeroes for all the terms below. For any bounded, constructible complex $\mathbf{A}^\bullet$ on a subspace of $\mathcal{U}$, let $m_\mathbf{0}(\mathbf{A}^\bullet)$ equal the coefficient of $\left[T^*_{\{\mathbf{0}\}}\mathcal{U}\right]$ in the characteristic cycle of $\mathbf{A}^\bullet$.

We need to state one further result without proof – this result can be obtained from [**BMM**], but we give the result as stated in [**Mas5**, 4.6], with the added corrections of $(-1)^{d-\hat{d}}$ in various places.

Theorem 3.4. *For generic linear forms* L, *we have the following formulas:*

$$(-1)^{d-\hat{d}} m_\mathbf{0}(\psi_f\mathbf{F}^\bullet) = \sum_{S_\alpha \not\subseteq V(f)} m_\alpha(\mathbf{F}^\bullet)\left(\Gamma^1_{f,L}(S_\alpha)\cdot V(f)\right)_\mathbf{0};$$

$$m_{\mathbf{0}}(\mathbf{F}^\bullet)+(-1)^{d-\hat{d}}m_{\mathbf{0}}(\mathbf{F}^\bullet_{|V(f)}) = \sum_{S_\alpha\not\subseteq V(f)} m_\alpha(\mathbf{F}^\bullet)\big(\Gamma^1_{f,L}(S_\alpha)\cdot V(L)\big)_{\mathbf{0}};\ \textit{and}$$

$$(-1)^{d-\hat{d}}m_{\mathbf{0}}(\phi_f\mathbf{F}^\bullet) = m_{\mathbf{0}}(\mathbf{F}^\bullet)+\sum_{S_\alpha\not\subseteq V(f)} m_\alpha(\mathbf{F}^\bullet)\Big(\big(\Gamma^1_{f,L}(S_\alpha)\cdot V(f)\big)_{\mathbf{0}} - \big(\Gamma^1_{f,L}(S_\alpha)\cdot V(L)\big)_{\mathbf{0}}\Big).$$

Lemma 3.5. *If $S_\alpha \not\subseteq f^{-1}(0)$, then the coefficient of $\left[\mathbb{P}(T^*_{\{\mathbf{0}\}}\mathcal{U})\right]$ in $\mathbb{P}(\overline{T^*_{f_{|S_\alpha}}\mathcal{U}})\cdot(V(f)\times\mathbb{P}^n)$ is given by $\big(\Gamma^1_{f,L}(S_\alpha)\cdot V(f)\big)_{\mathbf{0}}$.*

Proof. Take a complex of sheaves, $\mathbf{F}^\bullet$, which has a characteristic cycle consisting only of $\left[\overline{T^*_{S_\alpha}\mathcal{U}}\right]$ (see, for instance, [**M1**]). Now, apply the formula for $m_{\mathbf{0}}(\psi_f\mathbf{F}^\bullet)$ from Theorem 3.4 together with Theorem 3.3. □

We need to establish some notation that we shall use throughout the remainder of this section.

Using the isomorphism, $T^*\mathcal{U}\cong\mathcal{U}\times\mathbb{C}^{n+1}$, we consider $\mathrm{Ch}(\mathbf{F}^\bullet)$ as a cycle in $X\times\mathbb{C}^{n+1}$; we use $\mathbf{z} := (z_0,\dots,z_n)$ as coordinates on $\mathcal{U}$ and $\mathbf{w} := (w_0,\dots,w_n)$ as the cotangent coordinates.

Let I denote the sheaf of ideals on $\mathcal{U}$ given by the image of $d\tilde{f}$, i.e.,

$$I = \big\langle w_0 - \frac{\partial\tilde{f}}{\partial z_0},\dots,w_n-\frac{\partial\tilde{f}}{\partial z_n}\big\rangle.$$

For all α, let $B_\alpha = \mathrm{Bl}_{\mathrm{im}\,d\tilde{f}}\overline{T^*_{S_\alpha}\mathcal{U}}$ denote the blow-up of $\overline{T^*_{S_\alpha}\mathcal{U}}$ along the image of I in $\overline{T^*_{S_\alpha}\mathcal{U}}$, and let E_α denote the corresponding exceptional divisor. For all α, we have

$$E_\alpha\subseteq B_\alpha\subseteq X\times\mathbb{C}^{n+1}\times\mathbb{P}^n.$$

Let $\pi : X\times\mathbb{C}^{n+1}\times\mathbb{P}^n\to X\times\mathbb{P}^n$ denote the projection. Note that, if $(\mathbf{x},\mathbf{w},[\eta])\in E_\alpha$, then $\mathbf{w}=d_{\mathbf{x}}\tilde{f}$ and so, for all α, π induces an isomorphism from E_α to $\pi(E_\alpha)$. We refer to $E:=\sum_\alpha m_\alpha E_\alpha$ as the *total exceptional divisor* inside the total blow-up $\mathrm{Bl}_{\mathrm{im}\,d\tilde{f}}\mathrm{Ch}(\mathbf{F}^\bullet) := \sum_\alpha m_\alpha \mathrm{Bl}_{\mathrm{im}\,d\tilde{f}}\left[\overline{T^*_{S_\alpha}\mathcal{U}}\right]$.

Lemma 3.6. *For all S_α, there is an inclusion $\pi\left(\mathrm{Bl}_{\mathrm{im}\,d\tilde{f}}\overline{T^*_{S_\alpha}\mathcal{U}}\right)\subseteq\mathbb{P}(\overline{T^*_{f_{|S_\alpha}}\mathcal{U}})$.*

Proof. This is entirely straightforward. Suppose that

$$(\mathbf{x}, \mathbf{w}, [\eta]) \in \mathrm{Bl}_{\mathrm{im}\, d\tilde{f}}\, \overline{T^*_{S_\alpha}\mathcal{U}} = \overline{\mathrm{Bl}_{\mathrm{im}\, d\tilde{f}}\, T^*_{S_\alpha}\mathcal{U}}.$$

Then, we have a sequence $(\mathbf{x}_i, \mathbf{w}_i, [\eta_i]) \in \mathrm{Bl}_{\mathrm{im}\, d\tilde{f}}\, T^*_{S_\alpha}\mathcal{U}$ such that $(\mathbf{x}_i, \mathbf{w}_i, [\eta_i]) \to (\mathbf{x}, \mathbf{w}, [\eta])$.

By definition of the blow-up, for each $(\mathbf{x}_i, \mathbf{w}_i, [\eta_i])$, there exists a sequence $(\mathbf{x}_i^j, \mathbf{w}_i^j) \in T^*_{S_\alpha}\mathcal{U} - \mathrm{im}\, d\tilde{f}$ such that $(\mathbf{x}_i^j, \mathbf{w}_i^j, [\mathbf{w}_i^j - d_{\mathbf{x}_i^j}\tilde{f}]) \to (\mathbf{x}_i, \mathbf{w}_i, [\eta_i])$. Now, $(\mathbf{x}_i^j, [\mathbf{w}_i^j - d_{\mathbf{x}_i^j}\tilde{f}])$ is clearly in $\mathbb{P}(T^*_{f_{|S_\alpha}}\mathcal{U})$, and so each $(\mathbf{x}_i, [\eta_i])$ is in $\mathbb{P}(\overline{T^*_{f_{|S_\alpha}}\mathcal{U}})$. Therefore, $(\mathbf{x}, [\eta]) \in \mathbb{P}(\overline{T^*_{f_{|S_\alpha}}\mathcal{U}})$. □

Lemma 3.7. *If $S_\alpha \not\subseteq f^{-1}(0)$, then the coefficient of $\left[\mathbb{P}(T^*_{\{\mathbf{0}\}}\mathcal{U})\right] = \{\mathbf{0}\} \times \mathbb{P}^n$ in $\pi_*(E_\alpha)$ equals $\left(\Gamma^1_{f,L}(S_\alpha) \cdot V(f)\right)_{\mathbf{0}} - \left(\Gamma^1_{f,L}(S_\alpha) \cdot V(L)\right)_{\mathbf{0}}$.*

Proof. By I.2.23, the multiplicity of $\{\mathbf{0}\} \times \mathbb{P}^n$ in $\pi_*(E_\alpha)$ equals $\left(\Delta^0_{\left(\mathbf{w} - \frac{\partial \tilde{f}}{\partial \mathbf{z}}\right)_{|\overline{T^*_{S_\alpha}\mathcal{U}}}}\right)_{(\mathbf{0}, d_{\mathbf{0}}\tilde{f})}$, for a generic choice of $\mathbf{z}$. By 2.5.viii and 2.6, this is equal to $\left(\Gamma^1_{f,L}(S_\alpha) \cdot V(f)\right)_{\mathbf{0}} - \left(\Gamma^1_{f,L}(S_\alpha) \cdot V(L)\right)_{\mathbf{0}}$ for a generic linear choice of L. □

Lemma 3.8. *For all α such that $S_\alpha \subseteq V(f)$, there is an inclusion of the exceptional divisor*

$$E_\alpha \cong \pi(E_\alpha) \subseteq \mathbb{P}(\overline{T^*_{f_{|S_\alpha}}\mathcal{U}}) \cap \left(V(f) \times \mathbb{P}^n\right).$$

Proof. That π is an isomorphism when restricted to the exceptional divisor is trivial: $(\mathbf{x}, \mathbf{w}, [\eta]) \in E_\alpha$ implies that $\mathbf{w} = d_{\mathbf{x}}\tilde{f}$. From Lemma 3.6, $\pi(E_\alpha) \subseteq \pi\left(\mathrm{Bl}_{\mathrm{im}\, d\tilde{f}}\, \overline{T^*_{S_\alpha}\mathcal{U}}\right) \subseteq \mathbb{P}(\overline{T^*_{f_{|S_\alpha}}\mathcal{U}})$. The result follows. □

Lemma 3.9. *If $S_\alpha \subseteq f^{-1}(0)$, then $E_\alpha \cong \pi(E_\alpha) = \mathbb{P}(\overline{T^*_{S_\alpha}\mathcal{U}})$.*

Proof. If $S_\alpha \subseteq f^{-1}(0)$, then $\mathbb{P}(T^*_{f_{|S_\alpha}}\mathcal{U}) = \mathbb{P}(T^*_{S_\alpha}\mathcal{U})$, and so, by 3.8, $\pi(E_\alpha) \subseteq \mathbb{P}(\overline{T^*_{S_\alpha}\mathcal{U}})$. We will demonstrate the reverse inclusion.

Suppose that we have $(\mathbf{x}, [\eta]) \in \mathbb{P}(\overline{T^*_{S_\alpha}\mathcal{U}})$. Then, there exists a sequence $(\mathbf{x}_i, \eta_i) \in T^*_{S_\alpha}\mathcal{U}$ such that $(\mathbf{x}_i, \eta_i) \to (\mathbf{x}, \eta)$. Hence, $\left(\mathbf{x}_i,\ \frac{1}{i}\eta_i + d_{\mathbf{x}_i}\tilde{f}\right) \in \overline{T^*_{S_\alpha}\mathcal{U}} - \mathrm{im}\, d\tilde{f}$ and

$$\left(\mathbf{x}_i,\ \frac{1}{i}\eta_i + d_{\mathbf{x}_i}\tilde{f},\ \left[\left(\frac{1}{i}\eta_i + d_{\mathbf{x}_i}\tilde{f}\right) - d_{\mathbf{x}_i}\tilde{f}\right]\right) \to (\mathbf{x}, d_{\mathbf{x}}\tilde{f}, [\eta]) \in E_\alpha. \quad \square$$

We come now to the main theorem of this section. This theorem relates the topological data provided by the vanishing cycles of a function f to the algebraic data given by blowing-up the image of the differential of an extension of f.

Theorem 3.10. *The projection π induces an isomorphism between the total exceptional divisor $E \subseteq \mathrm{Bl}_{\mathrm{im}\, d\tilde{f}}\, \mathrm{Ch}(\mathbf{F}^\bullet)$ and the sum over all $v \in \mathbb{C}$ of the projectivized characteristic cycles of the sheaves of vanishing cycles of $\mathbf{F}^\bullet$ along $f - v$, i.e.,*

$$E \cong \pi_*(E) = \sum_{v \in \mathbb{C}} (-1)^{d - \hat{d}_v} \mathbb{P}(\mathrm{Ch}(\phi_{f-v}\mathbf{F}^\bullet)).$$

Proof. Remarks 1.7 and 1.10 imply that, locally, $\operatorname{supp} \phi_{f-v}\mathbf{F}^\bullet \subseteq f^{-1}(v)$. As the various $\mathbb{P}(\mathrm{Ch}(\phi_{f-v}\mathbf{F}^\bullet))$ are disjoint for different values of v, we may immediately reduce ourselves to the case where we are working near $\mathbf{0} \in X$ and where $f(\mathbf{0}) = 0$. We refine our stratification so that, for all α, $\Sigma(f_{|S_\alpha}) = \emptyset$ unless $S_\alpha \subseteq V(f)$. As any newly introduced stratum will appear with a coefficient of zero in the characteristic cycle, the total exceptional divisor will not change. We need to show that $E \cong \pi(E) = \mathbb{P}(\mathrm{Ch}(\phi_f\mathbf{F}^\bullet))$.

Now, we will first show that $\pi(E)$ is Lagrangian.

If $S_\alpha \subseteq f^{-1}(0)$, then $\pi(E_\alpha) = \mathbb{P}(\overline{T^*_{S_\alpha}\mathcal{U}})$ by 3.9. If $S_\alpha \not\subseteq f^{-1}(0)$, then, by Theorem 3.3, $\mathbb{P}(\overline{T^*_{f_{|S_\alpha}}\mathcal{U}}) \cap \big(V(f) \times \mathbb{P}^n\big)$ is Lagrangian and, in particular, is purely n-dimensional. By Lemma 3.8, $\pi(E_\alpha)$ is a purely n-dimensional analytic set contained in

$$\mathbb{P}(\overline{T^*_{f_{|S_\alpha}}\mathcal{U}}) \cap \big(V(f) \times \mathbb{P}^n\big).$$

We need to show that $\pi(E_\alpha)$ is closed.

Suppose we have a sequence $(\mathbf{x}_i, [\eta_i]) \in \pi(E_\alpha)$ and $(\mathbf{x}_i, [\eta_i]) \to (\mathbf{x}, [\eta])$ in $\mathcal{U} \times \mathbb{P}^n$. Then, there exists a sequence $\mathbf{w}_i$ so that $(\mathbf{x}_i, \mathbf{w}_i, [\eta_i]) \in E_\alpha$; by definition of the exceptional divisor, this implies $\mathbf{w}_i = d_{\mathbf{x}_i}\tilde{f}$. Therefore, $(\mathbf{x}_i, \mathbf{w}_i, [\eta_i]) \to (\mathbf{x}, d_{\mathbf{x}}\tilde{f}, [\eta])$, which is contained in E_α since E_α is closed in $\mathcal{U} \times \mathbb{C}^{n+1} \times \mathbb{P}^n$. Thus, $(\mathbf{x}, [\eta]) \in \pi(E_\alpha)$, and so $\pi(E_\alpha)$ is closed and, hence, Lagrangian.

Now, $\pi(E)$ and $\mathbb{P}(\mathrm{Ch}(\phi_f\mathbf{F}^\bullet))$ are both supported over $\Sigma_{\mathcal{S}} f$ and, by taking normal slices to strata, we are reduced to the point-stratum case. Thus, what we need to show is: the coefficient of $\big[\mathbb{P}(T^*_{\{\mathbf{0}\}}\mathcal{U})\big]$ in E equals the coefficient of $\big[\mathbb{P}(T^*_{\{\mathbf{0}\}}\mathcal{U})\big]$ in $(-1)^{d-\hat{d}}\mathbb{P}(\mathrm{Ch}(\phi_f\mathbf{F}^\bullet))$. Using 3.4, this is equivalent to showing that the coefficient of $\big[\mathbb{P}(T^*_{\{\mathbf{0}\}}\mathcal{U})\big]$ in E equals

$$m_{\mathbf{0}}(\mathbf{F}^\bullet) + \sum_{S_\alpha \not\subseteq V(f)} m_\alpha \Big(\big(\Gamma^1_{f,L}(S_\alpha) \cdot V(f)\big)_{\mathbf{0}} - \big(\Gamma^1_{f,L}(S_\alpha) \cdot V(L)\big)_{\mathbf{0}} \Big)$$

for a generic linear form L.

But, by 3.9,

$$E = \sum_{\alpha} m_\alpha E_\alpha = \sum_{S_\alpha \subseteq V(f)} m_\alpha \left[\mathbb{P}(\overline{T^*_{S_\alpha}\mathcal{U}})\right] + \sum_{S_\alpha \not\subseteq V(f)} m_\alpha E_\alpha$$

and the coefficient of $\left[\mathbb{P}(T^*_{\{\mathbf{0}\}}\mathcal{U})\right]$ in $\sum_{S_\alpha \subseteq V(f)} m_\alpha \left[\mathbb{P}(\overline{T^*_{S_\alpha}\mathcal{U}})\right]$ is precisely $m_{\mathbf{0}}(\mathbf{F}^\bullet)$.

Therefore, we will be finished if we can show that the coefficient of $\left[\mathbb{P}(T^*_{\{\mathbf{0}\}}\mathcal{U})\right]$ in E_α equals $\left(\Gamma^1_{f,L}(S_\alpha) \cdot V(f)\right)_{\mathbf{0}} - \left(\Gamma^1_{f,L}(S_\alpha) \cdot V(L)\right)_{\mathbf{0}}$ if $S_\alpha \not\subseteq V(f)$. However, this is exactly the content of Lemma 3.7. □

Remark 3.11. In special cases, Theorem 3.10 was already known.

Consider the case where $X = \mathcal{U}$ and $\mathbf{F}^\bullet$ is the constant sheaf. Then, $\mathrm{Ch}(\mathbf{F}^\bullet) = \mathcal{U} \times \{\mathbf{0}\}$, and the image of $d\tilde{f}$ in $\mathcal{U} \times \{\mathbf{0}\}$ is simply defined by the Jacobian ideal of f. Hence, our result reduces to the result obtained from the work of Kashiwara in [**K**] and Lê-Mebkhout in [**L-M**] – namely, that the projectivized characteristic cycle of the sheaf of vanishing cycles is isomorphic to the exceptional divisor of the blow-up of the Jacobian ideal in affine space.

As a second special case, suppose that X and $\mathbf{F}^\bullet$ are completely general, but that $\mathbf{x}$ is an isolated point in the image of $\mathrm{Ch}(\phi_f\mathbf{F}^\bullet)$ in X (for instance, $\mathbf{x}$ might be an isolated point in $\mathrm{supp}\,\phi_f\mathbf{F}^\bullet$). Then, for every stratum for which $m_\alpha \neq 0$, $(\mathbf{x}, d_{\mathbf{x}}\tilde{f})$ is an isolated point of $\mathrm{im}\,d\tilde{f} \cap \overline{T^*_{S_\alpha}\mathcal{U}}$ or is not contained in the intersection at all.

Therefore, the last part of I.2.23 implies that the exceptional divisor of the blow-up of $\mathrm{im}\,d\tilde{f}$ in $\overline{T^*_{S_\alpha}\mathcal{U}}$ has one component over $(\mathbf{x}, d_{\mathbf{x}}\tilde{f})$ and that that component occurs with multiplicity precisely equal to the intersection multiplicity $\left(\mathrm{im}\,d\tilde{f} \cdot \overline{T^*_{S_\alpha}\mathcal{U}}\right)_{(\mathbf{x}, d_{\mathbf{x}}\tilde{f})}$ in $T^*\mathcal{U}$. Thus, we recover the results of three independent works appearing in [**G**], [**Lê**], and [**S**] – that the coefficient of $\{\mathbf{x}\} \times \mathbb{C}^{n+1}$ in $(-1)^{d-\hat{d}}\,\mathrm{Ch}(\phi_f\mathbf{F}^\bullet)$ is given by $\left(\mathrm{im}\,d\tilde{f} \cdot \mathrm{Ch}(\mathbf{F}^\bullet)\right)_{(\mathbf{x}, d_{\mathbf{x}}\tilde{f})}$. This result is usually stated in terms of the Euler characteristic: if $\mathbf{x}$ is an isolated point in $\mathrm{supp}\,\phi_f\mathbf{F}^\bullet$), then

$$\chi(\phi[-1]\mathbf{F}^\bullet)_{\mathbf{x}} = (-1)^d \left(\mathrm{im}\,d\tilde{f} \cdot \mathrm{Ch}(\mathbf{F}^\bullet)\right)_{(\mathbf{x}, d_{\mathbf{x}}\tilde{f})}.$$

In addition to generalizing the above results, Theorem 3.10 fits in well with Theorem 3.4.2 of [**BMM**]; that theorem contains a nice description of the characteristic cycles of the nearby cycles and of the restriction of a complex to a hypersurface. However, [**BMM**] does not contain a nice description of the vanishing cycles, nor does our Theorem 3.10 seem to follow easily from the results of [**BMM**]; in fact, Example 3.4.3 of [**BMM**] makes it clear that the general result contained in our Theorem 3.10 was unknown – for Briançon, Maisonobe, and Merle only derive the vanishing cycle result from their nearby cycle result in the easy, known case where the vanishing cycles are supported on an isolated point and, even then, they must make half a page of argument.

Corollary 3.12. *For each extension $\tilde{f}$ of f, let $E_{\tilde{f}}$ denote the exceptional divisor in* $\mathrm{Bl}_{\mathrm{im}\, d\tilde{f}}\overline{T^*_{X_{\mathrm{reg}}}\mathcal{U}}$. *Then, $\pi\big(E_{\tilde{f}}\big)$ is independent of $\tilde{f}$.*

Proof. We apply Theorem 3.10 to a complex of sheaves $\mathbf{F}^\bullet$ such that $m_\alpha = 1$ for each smooth component of X_{reg} and $m_\alpha = 0$ for every other stratum in some Whitney stratification of X (it is easy to produce such an $\mathbf{F}^\bullet$ – see, for instance, Lemma 3.1 of [**M1**]). The corollary follows from the fact that $\mathbb{P}(\mathrm{Ch}(\phi_f\mathbf{F}^\bullet))$ does not depend on the extension. □

Chapter 4. THE SPECIAL CASE OF PERVERSE SHEAVES.

We continue with our previous notation, **except that in this chapter we must assume that our base ring is a field.**

For the purposes of Part III, perverse sheaves are important because the vanishing cycles functor (shifted by -1) applied to a perverse sheaf once again yields a perverse sheaf and because of the following lemma.

Lemma 4.1. *If $\mathbf{P}^\bullet$ is a perverse sheaf on X, then $\operatorname{Ch}(\mathbf{P}^\bullet) = \sum_\alpha m_\alpha \left[\overline{T^*_{S_\alpha}\mathcal{U}}\right]$, where*

$$m_\alpha = (-1)^d \dim H^0(\mathbb{N}_\alpha, \mathbb{L}_\alpha;\ \mathbf{P}^\bullet_{|_{\mathbb{N}_\alpha}}[-d_\alpha]);$$

in particular, $(-1)^d \operatorname{Ch}(\mathbf{P}^\bullet)$ is a non-negative cycle.

If $\mathbf{P}^\bullet$ is perverse on X (or, even, perverse up to a shift), then $\operatorname{supp}\mathbf{P}^\bullet$ *equals the image in X of the characteristic cycle of $\mathbf{P}^\bullet$.*

Proof. The first statement follows from the definition of the characteristic cycle, together with the fact that a perverse sheaf supported on a point has non-zero cohomology only in degree zero.

The second statement follows at once from the fact that if $\mathbf{P}^\bullet$ is perverse up to a shift, then so is the restriction of $\mathbf{P}^\bullet$ to its support. Hence, by the support condition on perverse sheaves, there is an open dense set of the support, Ω, such that, for all $\mathbf{x} \in \Omega$, $H^*(\mathbf{P}^\bullet)_\mathbf{x}$ is non-zero in a single degree. The conclusion follows. □

The fact that the above lemma refers to the support of $\mathbf{P}^\bullet$, which is the closure of the set of points with non-zero stalk cohomology, means that we can use it to conclude something about the closure of the $\mathbf{P}^\bullet$-critical locus (recall Definition 1.9).

Theorem 4.2. *Let $\mathbf{P}^\bullet$ be a perverse sheaf on X, and suppose that the characteristic cycle of $\mathbf{P}^\bullet$ in $\mathcal{U}$ is given by $\operatorname{Ch}(\mathbf{P}^\bullet) = \sum_\alpha m_\alpha \left[\overline{T^*_{S_\alpha}\mathcal{U}}\right]$.*

Then, the closure of the $\mathbf{P}^\bullet$-critical locus of f is given by

$$\overline{\Sigma_{\mathbf{P}^\bullet} f} = \left\{\mathbf{x} \in X \mid (\mathbf{x}, d_\mathbf{x}\tilde{f}) \in |\operatorname{Ch}(\mathbf{P}^\bullet)|\right\} = \bigcup_{m_\alpha \neq 0} \Sigma_{\mathrm{cnr}}(f_{|_{\overline{S_\alpha}}}).$$

Proof. Let $\mathbf{q} \in X$, and let $v = f(\mathbf{q})$. Let $\mathcal{W}$ be an open neighborhood of $\mathbf{q}$ in X such that $\mathcal{W} \cap \Sigma_{\mathbf{P}^\bullet} f \subseteq V(f-v)$ (see the end of Remark 1.7). Then, $\mathcal{W} \cap \overline{\Sigma_{\mathbf{P}^\bullet} f} = \mathcal{W} \cap \operatorname{supp}\phi_{f-v}\mathbf{P}^\bullet$. As $\phi_{f-v}\mathbf{P}^\bullet[-1]$ is perverse, Lemma 4.1 tells us that $\operatorname{supp}\phi_{f-v}\mathbf{P}^\bullet$ equals the image in X of $\operatorname{Ch}(\phi_{f-v}\mathbf{P}^\bullet)$. Now, Theorem 3.10 tells us that this image is precisely

$$\bigcup_{m_\alpha \neq 0} \left\{\mathbf{x} \in \overline{S_\alpha} \mid (\mathbf{x}, d_\mathbf{x}\tilde{f}) \in \overline{T^*_{S_\alpha}\mathcal{U}}\right\},$$

since there can be no cancellation as all the non-zero m_α have the same sign.

Therefore, we have the desired equality of sets in an open neighborhood of every point; the theorem follows. □

We will use the *perverse cohomology* of the shifted constant sheaf, $\mathbb{C}^\bullet_X[k]$, in order to deal with non-l.c.i.'s; this perverse cohomology is denoted by ${}^\mu H^0(\mathbb{C}^\bullet_X[k]) = {}^\mu H^k(\mathbb{C}^\bullet_X)$ (see [**BBD**], [**K-S**], or Appendix B). Like the intersection cohomology complex, this sheaf has the property that it is the shifted constant sheaf on the smooth part of any component of X with dimension equal to $d = \dim X$.

We now list some properties of the perverse cohomology and of vanishing cycles that we will need later. For further properties, see Appendix B.

The perverse cohomology functor on X, ${}^\mu H^0$, is a functor from the derived category of bounded, constructible complexes on X to the Abelian category of perverse sheaves on X.

If $\mathbf{F}^\bullet$ is constructible with respect to $\mathcal{S}$, then ${}^\mu H^0(\mathbf{F}^\bullet)$ is also constructible with respect to $\mathcal{S}$, and $\left({}^\mu H^0(\mathbf{F}^\bullet)\right)_{|_{\mathbb{N}_\alpha}}[-d_\alpha]$ is naturally isomorphic to ${}^\mu H^0(\mathbf{F}^\bullet_{|_{\mathbb{N}_\alpha}}[-d_\alpha])$.

The functor ${}^\mu H^0$, applied to a perverse sheaf $\mathbf{P}^\bullet$ is canonically isomorphic to $\mathbf{P}^\bullet$. In addition, a bounded, constructible complex of sheaves $\mathbf{F}^\bullet$ is perverse if and only ${}^\mu H^0(\mathbf{F}^\bullet[k]) = 0$ for all $k \neq 0$. In particular, if X is an l.c.i., then ${}^\mu H^0(\mathbb{C}^\bullet_X[d]) \cong \mathbb{C}^\bullet_X[d]$ and ${}^\mu H^0(\mathbb{C}^\bullet_X[k]) = 0$ if $k \neq d$.

The functor ${}^\mu H^0$ commutes with vanishing cycles with a shift of -1, nearby cycles with a shift of -1, and Verdier dualizing. That is, there are natural isomorphisms

$${}^\mu H^0 \circ \phi_f[-1] \cong \phi_f[-1] \circ {}^\mu H^0, \qquad {}^\mu H^0 \circ \psi_f[-1] \cong \psi_f[-1] \circ {}^\mu H^0, \quad \text{and } \mathcal{D} \circ {}^\mu H^0 \cong {}^\mu H^0 \circ \mathcal{D}.$$

Let $\mathbf{F}^\bullet$ be a bounded complex of sheaves on X which is constructible with respect to a connected Whitney stratification $\{S_\alpha\}$ of X. Let $S_{\max}$ be a maximal stratum contained in the support of $\mathbf{F}^\bullet$, and let $m = \dim S_{\max}$. Then, $\left({}^\mu H^0(\mathbf{F}^\bullet)\right)_{|_{S_{\max}}}$ is isomorphic (in the derived category) to the complex which has $(\mathbf{H}^{-m}(\mathbf{F}^\bullet))_{|_{S_{\max}}}$ in degree $-m$ and zero in all other degrees.

In particular, $\operatorname{supp} \mathbf{F}^\bullet = \bigcup_i \operatorname{supp} {}^\mu H^0(\mathbf{F}^\bullet[i])$, and if $\mathbf{F}^\bullet$ is supported on an isolated point, $\mathbf{q}$, then $H^0({}^\mu H^0(\mathbf{F}^\bullet))_\mathbf{q} \cong H^0(\mathbf{F}^\bullet)_\mathbf{q}$.

Throughout the remainder of Part III, we let ${}^k\mathbf{P}^\bullet := {}^\mu H^0(\mathbb{C}^\bullet_X[k+1])$; it will be useful later to have a nice characterization of the characteristic cycle of ${}^k\mathbf{P}^\bullet$.

Proposition 4.3. *The complex ${}^k\mathbf{P}^\bullet$ is a perverse sheaf on X which is constructible with respect to $\mathcal{S}$ and the characteristic cycle $\operatorname{Ch}({}^k\mathbf{P}^\bullet)$ is equal to*

$$(-1)^d \sum_\alpha b_{k+1-d_\alpha}(\mathbb{N}_\alpha, \mathbb{L}_\alpha)\left[\overline{T^*_{S_\alpha}\mathcal{U}}\right],$$

where b_j denotes the j-th (relative) Betti number.

In particular, $H^*(\mathbb{L}_\alpha; \mathbb{C}) \cong H^*(point; \mathbb{C})$ if and only if $m_\alpha({}^k\mathbf{P}^\bullet) = 0$ for all k.

Proof. The constructibility claim follows from the fact that the constant sheaf itself is clearly constructible with respect to any Whitney stratification. The remainder follows trivially from the definition of the characteristic cycle, combined with two properties of ${}^{\mu}H^0$; namely, ${}^{\mu}H^0$ commutes with $\phi_f[-1]$, and ${}^{\mu}H^0$ applied to a complex which is supported at a point simply gives ordinary cohomology in degree zero and zeroes in all other degrees. See [**K-S**, 10.3]. □

Remark 4.4. As $\mathbb{N}_\alpha$ is contractible, it is possible to give a characterization of $b_{k+1-d_\alpha}(\mathbb{N}_\alpha, \mathbb{L}_\alpha)$ without referring to $\mathbb{N}_\alpha$; the statement gets a little complicated, however, since we have to worry about what happens near degree zero and because the link of a maximal stratum is empty. However, if we slightly modify the usual definitions of reduced cohomology and the corresponding reduced Betti numbers, then the statement becomes quite easy.

What we want is for the "reduced" cohomology $\widetilde{H}^k(A;\mathbb{C})$ to be the relative cohomology vector space $H^{k+1}(B, A;\mathbb{C})$, where B is a contractible set containing A, and we want $\tilde{b}_*()$ to be the Betti numbers of this "reduced" cohomology. Therefore, letting $b_k()$ denote the usual k-th Betti number, we define $\tilde{b}_*()$ by

$$\tilde{b}_k(A) = \begin{cases} b_k(A), & \text{if } k \neq 0 \text{ and } A \neq \emptyset \\ b_0(A) - 1, & \text{if } k = 0 \text{ and } A \neq \emptyset \\ 0, & \text{if } k \neq -1 \text{ and } A = \emptyset \\ 1, & \text{if } k = -1 \text{ and } A = \emptyset. \end{cases}$$

Thus, $\tilde{b}_k(A)$ is the k-th Betti number of the reduced cohomology, provided that A is not the empty set.

The special definition of $\tilde{b}_k()$ for the empty set implies that if S_α is maximal, so that $\mathbb{N}_\alpha = \textit{point}$ and $\mathbb{L}_\alpha = \emptyset$, then

$$b_{k+1-d_\alpha}(\mathbb{N}_\alpha, \mathbb{L}_\alpha) = \begin{cases} 0, & \text{if } k+1 \neq d_\alpha \\ 1, & \text{if } k+1 = d_\alpha \end{cases} = \tilde{b}_{k-d_\alpha}(\mathbb{L}_\alpha).$$

Thus, with this new notation,

$$\operatorname{Ch}({}^k\mathbf{P}^\bullet) = (-1)^d \sum_\alpha \tilde{b}_{k-d_\alpha}(\mathbb{L}_\alpha)\left[\overline{T^*_{S_\alpha}\mathcal{U}}\right].$$

By combining 4.2 with 4.3 and 4.4, we can now give a result about $\Sigma_{\mathbb{C}} f$. First, though, it will be useful to adopt the following terminology.

Definition 4.5. We say that the stratum S_α is *visible* (or, $\mathbb{C}$-*visible*) if $H^*(\mathbb{L}_\alpha;\mathbb{C}) \ncong H^*(\textit{point};\mathbb{C})$ (or, equivalently, if $H^*(\mathbb{N}_\alpha, \mathbb{L}_\alpha;\mathbb{C}) \neq 0$). Otherwise, the stratum is *invisible*.

The final line of Proposition 4.3 tells us that a stratum is visible if and only if there exists an integer k such that $\left[\overline{T^*_{S_\alpha}\mathcal{U}}\right]$ appears with a non-zero coefficient in $\operatorname{Ch}\left({}^k\mathbf{P}^\bullet\right)$.

Note that if S_α has an empty complex link (i.e., the stratum is maximal), then S_α is **visible**.

Theorem 4.6. *Then,*

$$\overline{\Sigma_{\mathbb{C}} f} = \bigcup_{k=-1}^{d-1} \overline{\Sigma_{{}^k\mathbf{P}^\bullet} f} = \bigcup_{\text{visible } S_\alpha} \left\{ \mathbf{x} \in \overline{S_\alpha} \mid (\mathbf{x}, d_{\mathbf{x}}\tilde{f}) \in \overline{T^*_{S_\alpha}\mathcal{U}} \right\} = \bigcup_{\text{visible } S_\alpha} \Sigma_{\text{cnr}}(f_{|_{\overline{S_\alpha}}}).$$

In particular, since all maximal strata are visible, $\Sigma_{\text{cnr}} f \subseteq \overline{\Sigma_{\mathbb{C}} f}$ (as stated in Proposition 1.6). Moreover, if $\mathbf{x}$ is an isolated point of $\Sigma_{\mathbb{C}} f$,then, for all Whitney stratifications, $\{R_\beta\}$, of X, the only possibly visible stratum which can be contained in $f^{-1}f(\mathbf{x})$ is $\{\mathbf{x}\}$.

Proof. Recall that, for any complex $\mathbf{F}^\bullet$, $\operatorname{supp} \mathbf{F}^\bullet = \bigcup_k \operatorname{supp} {}^\mu\!H^0(\mathbf{F}^\bullet[k])$. In addition, we claim that ${}^k\mathbf{P}^\bullet = 0$ unless $-1 \leqslant k \leqslant d-1$. By Lemma 4.1, ${}^k\mathbf{P}^\bullet = 0$ is equivalent to $\operatorname{Ch}({}^k\mathbf{P}^\bullet) = 0$; if k is not between -1 and $d-1$, then, using Proposition 4.3, $\operatorname{Ch}({}^k\mathbf{P}^\bullet) = 0$ follows from the fact that the complex link of a stratum has the homotopy-type of a finite CW complex of dimension no more than the complex dimension of the link (see [**Go-Mac1**]).

Now, in an open neighborhood of any point $\mathbf{q}$ with $v := f(\mathbf{q})$, we have

$$\overline{\Sigma_{\mathbb{C}} f} = \operatorname{supp} \phi_{f-v}\mathbb{C}^\bullet = \bigcup_k \operatorname{supp} {}^\mu\!H^0(\phi_{f-v}\mathbb{C}^\bullet_X[k]) =$$

$$\bigcup_k \operatorname{supp} \phi_{f-v}[-1]({}^\mu\!H^0(\mathbb{C}^\bullet_X[k+1])) = \bigcup_k \overline{\Sigma_{{}^k\mathbf{P}^\bullet} f}.$$

Now, applying Theorem 4.2, we have

$$\overline{\Sigma_{\mathbb{C}} f} = \bigcup_k \bigcup_{m_\alpha({}^k\mathbf{P}^\bullet) \neq 0} \left\{ \mathbf{x} \in \overline{S_\alpha} \mid (\mathbf{x}, d_{\mathbf{x}}\tilde{f}) \in \overline{T^*_{S_\alpha}\mathcal{U}} \right\}.$$

The desired conclusion follows. □

Remark 4.7. Those familiar with stratified Morse theory should find the result of Theorem 4.6 very un-surprising – it looks like it results from some break-down of the $\mathbb{C}$-critical locus into normal and tangential data, and naturally one gets no contributions from strata with trivial normal data. This is the approach that we took in Theorem 3.2 of [**Ma1**]. There is a slightly subtle, technical point which prevents us from taking this approach in our current setting: by taking normal slices at points in an open, dense subset of $\operatorname{supp} \phi_{f-v}\mathbb{C}^\bullet_X$, we could reduce ourselves to the case where $\Sigma_{\mathbb{C}} f$ consists of a single point, but we would **not** know that the point was a **stratified** isolated critical point. In particular, the case where $\operatorname{supp} \phi_{f-v}\mathbb{C}^\bullet_X$ consists of a single point, but where f has a non-isolated (stratified) critical locus coming from an invisible stratum causes difficulties with the obvious Morse Theory approach.

Remark 4.8. At this point, we wish to add to our hierarchy of critical loci from Proposition 1.6. Theorem 4.6 tells us that $\overline{\Sigma_{{}^k\mathbf{P}^\bullet} f} \subseteq \overline{\Sigma_{\mathbb{C}} f}$ for all k. If X is purely $(m+1)$-dimensional, then 4.2 implies that $\Sigma_{\text{cnr}} f \subseteq \overline{\Sigma_{{}^m\mathbf{P}^\bullet} f}$.

Now, suppose that X is irreducible of dimension $m+1$. Let $\mathbf{IC}^\bullet$ be the intersection cohomology sheaf (with constant coefficients) on X (see [**Go-Mac2**]); $\mathbf{IC}^\bullet$ is a simple object in the category of perverse sheaves. As the category of perverse sheaves on X is (locally) Artinian, and since ${}^m\mathbf{P}^\bullet$ is a perverse sheaf which is the shifted constant sheaf on the smooth part of X, it follows that $\mathbf{IC}^\bullet$ appears as a simple subquotient in any composition series for ${}^m\mathbf{P}^\bullet$. Consequently, $|\operatorname{Ch}(\mathbf{IC}^\bullet)| \subseteq |\operatorname{Ch}({}^m\mathbf{P}^\bullet)|$, and so 4.2 implies that $\overline{\Sigma_{\mathbf{IC}^\bullet} f} \subseteq \overline{\Sigma_{{}^m\mathbf{P}^\bullet} f}$. Moreover, 4.2 also implies that $\Sigma_{\text{cnr}} f \subseteq \overline{\Sigma_{\mathbf{IC}^\bullet} f}$. Therefore, we can extend our sequence of inclusions from Proposition 1.6 to:

$$\overline{\Sigma_{\text{reg}} f} \subseteq \overline{\Sigma_{\text{alg}} f} \subseteq \overline{\Sigma_{\text{Nash}} f} \subseteq \Sigma_{\text{cnr}} f \subseteq \overline{\Sigma_{\mathbf{IC}^\bullet} f} \subseteq \overline{\Sigma_{{}^m\mathbf{P}^\bullet} f} \subseteq \overline{\Sigma_{\mathbb{C}} f} \subseteq \Sigma_{\text{can}} f \subseteq \Sigma_{\text{rdf}} f.$$

Why not use one of these new critical loci as our most fundamental notion of the critical locus of f? Both $\Sigma_{\mathbf{IC}^\bullet} f$ and $\Sigma_{{}^m\mathbf{P}^\bullet} f$ are topological in nature, and easy examples show that they can be distinct from $\Sigma_{\mathbb{C}} f$. However, 4.6 tells us that $\Sigma_{{}^m\mathbf{P}^\bullet} f$ is merely one piece that goes into making up $\Sigma_{\mathbb{C}} f$ – we should include the other shifted perverse cohomologies. On the other hand, given the importance of intersection cohomology throughout mathematics, one should wonder why we do not use $\Sigma_{\mathbf{IC}^\bullet} f$ as our most basic notion.

Consider the node $X := V(y^2 - x^3 - x^2) \subseteq \mathbb{C}^2$ and the function $f := y_{|X}$. The node has a *small resolution of singularities* (see [**Go-Mac2**]) given by simply pulling the branches apart. As a result, the intersection cohomology sheaf on X is the constant sheaf shifted by one on $X - \{\mathbf{0}\}$, and the stalk cohomology at $\mathbf{0}$ is a copy of $\mathbb{C}^2$ concentrated in degree -1. Therefore, one can easily show that $\mathbf{0} \notin \overline{\Sigma_{\mathbf{IC}^\bullet} f}$.

As $\overline{\Sigma_{\mathbf{IC}^\bullet} f}$ fails to detect the simple change in topology of the level hypersurfaces of f as they go from being two points to being a single point, we do not wish to use $\Sigma_{\mathbf{IC}^\bullet} f$ as our basic type of critical locus. That is not to say that $\Sigma_{\mathbf{IC}^\bullet} f$ is not interesting in its own right; it is integrally tied to resolutions of singularities. For instance, it is easy to show (using the Decomposition Theorem [**BBD**]) that if $\widetilde{X} \xrightarrow{\pi} X$ is a resolution of singularities, then $\Sigma_{\mathbf{IC}^\bullet} f \subseteq \pi(\Sigma(f \circ \pi))$.

Now that we can "calculate" $\Sigma_{\mathbb{C}} f$ using Theorem 4.6, we are ready to generalize the Milnor number of a function with an isolated critical point.

Definition 4.9. If $\mathbf{P}^\bullet$ is a perverse sheaf on X, and $\mathbf{x}$ is an isolated point in $\Sigma_{\mathbf{P}^\bullet} f$ (or, if $\mathbf{x} \notin \Sigma_{\mathbf{P}^\bullet} f$), then we call $\dim_{\mathbb{C}} H^0(\phi_{f-f(\mathbf{x})}[-1]\mathbf{P}^\bullet)_{\mathbf{x}}$ the *Milnor number of f at $\mathbf{x}$ with coefficients in $\mathbf{P}^\bullet$* and we denote it by $\mu_{\mathbf{x}}(f; \mathbf{P}^\bullet)$.

This definition is reasonable for, in this case, $\phi_{f-f(\mathbf{x})}[-1]\mathbf{P}^\bullet$ is a perverse sheaf supported at the isolated point $\mathbf{x}$. Hence, the stalk cohomology of $\phi_{f-f(\mathbf{x})}[-1]\mathbf{P}^\bullet$ at $\mathbf{x}$ is possibly non-zero only in degree zero. Normally, we summarize that $\mathbf{x}$ is an isolated point in $\Sigma_{\mathbf{P}^\bullet} f$ or that $\mathbf{x} \notin \Sigma_{\mathbf{P}^\bullet} f$ by writing $\dim_{\mathbf{x}} \Sigma_{\mathbf{P}^\bullet} f \leqslant 0$ (we consider the dimension of the empty set to be $-\infty$).

Before we state the next proposition, note that it is always the case that

$$\left(\operatorname{im} d\tilde{f} \cdot T^*_{\{\mathbf{0}\}}\mathcal{U}\right)_{(\mathbf{0}, d_{\mathbf{0}}\tilde{f})} = 1.$$

Proposition 4.10. *For notational convenience, we assume that* $\mathbf{0} \in X$ *and that* $f(\mathbf{0}) = 0$.

Then, $\dim_{\mathbf{0}} \Sigma_{\mathbb{C}} f \leqslant 0$ *if and only if, for all* k, $\dim_{\mathbf{0}} \Sigma_{{}^k\mathbf{P}^\bullet} f \leqslant 0$. *Moreover, if* $\dim_{\mathbf{0}} \Sigma_{\mathbb{C}} f \leqslant 0$, *then,*

i) for all visible strata, S_α, *such that* $\dim S_\alpha \geqslant 1$, *the intersection of* $\operatorname{im} d\tilde{f}$ *and* $\overline{T^*_{S_\alpha}\mathcal{U}}$ *is, at most, 0-dimensional at* $(\mathbf{0}, d_{\mathbf{0}}\tilde{f})$,

and

$$\left(\operatorname{im} d\tilde{f} \cdot \overline{T^*_{S_\alpha}\mathcal{U}}\right)_{(\mathbf{0}, d_{\mathbf{0}}\tilde{f})} = \left(\Gamma^1_{f,L}(S_\alpha)\cdot V(f)\right)_{\mathbf{0}} - \left(\Gamma^1_{f,L}(S_\alpha)\cdot V(L)\right)_{\mathbf{0}},$$

where L *is a generic linear form, and*

ii) for all k,

$$\mu_{\mathbf{0}}(f; {}^k\mathbf{P}^\bullet) = \tilde{b}_k(F_{f,\mathbf{0}}) = (-1)^{\dim X}\left(\operatorname{im} d\tilde{f} \cdot \operatorname{Ch}({}^k\mathbf{P}^\bullet)\right)_{(\mathbf{0}, d_{\mathbf{0}}\tilde{f})} =$$

$$\sum_{\text{visible } S_\alpha} \tilde{b}_{k-d_\alpha}(\mathbb{L}_\alpha)\left(\operatorname{im} d\tilde{f} \cdot \overline{T^*_{S_\alpha}\mathcal{U}}\right)_{(\mathbf{0}, d_{\mathbf{0}}\tilde{f})} =$$

$$\sum_{\substack{\text{visible } S_\alpha \\ S_\alpha \text{ not maximal}}} \tilde{b}_{k-d_\alpha}(\mathbb{L}_\alpha)\left(\operatorname{im} d\tilde{f} \cdot \overline{T^*_{S_\alpha}\mathcal{U}}\right)_{(\mathbf{0}, d_{\mathbf{0}}\tilde{f})} + \sum_{\substack{S_\alpha \text{ maximal} \\ \dim S_\alpha = k+1}} \left(\operatorname{im} d\tilde{f} \cdot \overline{T^*_{S_\alpha}\mathcal{U}}\right)_{(\mathbf{0}, d_{\mathbf{0}}\tilde{f})}.$$

Proof. It follows at once from 4.6 that $\dim_{\mathbf{0}} \Sigma_{\mathbb{C}} f \leqslant 0$ if and only if, for all k, $\dim_{\mathbf{0}} \Sigma_{{}^k\mathbf{P}^\bullet} f \leqslant 0$.

i) follows immediately from Lemma 3.7 (combined with Remark 3.11).

It remains for us to prove ii). As in the proof of 4.6, we have

$${}^\mu H^0(\phi_f \mathbb{C}^\bullet_X[k]) = \phi_f[-1]\left({}^\mu H^0(\mathbb{C}^\bullet_X[k+1])\right) = \phi_f[-1]{}^k\mathbf{P}^\bullet.$$

It follows that

$$\mu_{\mathbf{0}}(f; {}^k\mathbf{P}^\bullet) = \dim_{\mathbb{C}} H^0(\phi_f[-1]{}^k\mathbf{P}^\bullet)_{\mathbf{0}} = \dim_{\mathbb{C}} H^0\left({}^\mu H^0(\phi_f \mathbb{C}^\bullet_X[k])\right)_{\mathbf{0}} = \dim_{\mathbb{C}} H^0\left(\phi_f \mathbb{C}^\bullet_X[k]\right)_{\mathbf{0}},$$

where the last equality is a result of the fact that $\mathbf{0}$ is an isolated point in the support of $\phi_f \mathbb{C}^\bullet_X[k]$. Therefore,

$$\mu_{\mathbf{0}}(f; {}^k\mathbf{P}^\bullet) = \dim_{\mathbb{C}} H^0\left(\phi_f \mathbb{C}^\bullet_X[k]\right)_{\mathbf{0}} = \dim_{\mathbb{C}} H^k\left(\phi_f \mathbb{C}^\bullet_X\right)_{\mathbf{0}} = \dim \widetilde{H}^k(F_{f,\mathbf{0}};\ \mathbb{C}).$$

That we also have the equality

$$\mu_{\mathbf{0}}(f; {}^k\mathbf{P}^\bullet) = (-1)^{\dim X}\left(\operatorname{im} d\tilde{f} \cdot \operatorname{Ch}({}^k\mathbf{P}^\bullet)\right)_{(\mathbf{0}, d_{\mathbf{0}}\tilde{f})}$$

is precisely the content of Theorem 3.10, interpreted as in the last paragraph of Remark 3.11.

The remaining equalities in ii) follow from the description of $\mathrm{Ch}({}^k\mathbf{P}^\bullet)$ given in Proposition 4.3 and Remark 4.4. □

Remark 4.11. The formulas from 4.10 provide a topological/algebraic method for "calculating" the Betti numbers of the Milnor fibre for isolated critical points on arbitrary spaces. It should not be surprising that the data that one needs is not just the algebraic data – coming from the polar curves and intersection numbers – but also includes topological data about the underlying space: one has to know the Betti numbers of the complex links of strata.

Example 4.12. The most trivial, non-trivial case where one can apply 4.10 is the case where X is an irreducible local, complete intersection with an isolated singularity (that is, X is an irreducible i.c.i.s). Let us assume that $\mathbf{0} \in X$ is the only singular point of X and that f has an isolated $\mathbb{C}$-critical point at $\mathbf{0}$. Let d denote the dimension of X.

Let us write $\mathbb{L}_{X,\mathbf{0}}$ for the complex link of X at $\mathbf{0}$. By [**Lê1**], $\mathbb{L}_{X,\mathbf{0}}$ has the homotopy-type of a finite bouquet of $(d-1)$-spheres. Applying 4.10.ii, we see, then, that the reduced cohomology of $F_{f,\mathbf{0}}$ is concentrated in degree $(d-1)$, and the $(d-1)$-th Betti number of $F_{f,\mathbf{0}}$ is equal to

$$\tilde{b}_{d-1}(\mathbb{L}_{X,\mathbf{0}})\left(\operatorname{im} d\tilde{f}\ \cdot\ T^*_{\mathbf{0}}\mathcal{U}\right)_{(\mathbf{0},d_{\mathbf{0}}\tilde{f})} + \left(\operatorname{im} d\tilde{f}\ \cdot\ \overline{T^*_{X_{\mathrm{reg}}}\mathcal{U}}\right)_{(\mathbf{0},d_{\mathbf{0}}\tilde{f})} =$$

$$\tilde{b}_{d-1}(\mathbb{L}_{X,\mathbf{0}}) + \left(\Gamma^1_{f,L}(X_{\mathrm{reg}})\cdot V(f)\right)_{\mathbf{0}} - \left(\Gamma^1_{f,L}(X_{\mathrm{reg}})\cdot V(L)\right)_{\mathbf{0}},$$

for generic linear L.

Now, the polar curve and the intersection numbers are quite calculable in practice; see Remark 1.8 and Example 1.9 of [**Ma1**]. However, there remains the question of how one can compute $\tilde{b}_{d-1}(\mathbb{L}_{X,\mathbf{0}})$. Corollary 5.6 and Example 5.4 of [**Ma1**] provide an inductive method for computing the **Euler characteristic** of $\mathbb{L}_{X,\mathbf{0}}$ (the induction is on the codimension of X in $\mathcal{U}$) and, since we know that $\mathbb{L}_{X,\mathbf{0}}$ has the homotopy-type of a bouquet of spheres, knowing the Euler characteristic is equivalent to knowing $\tilde{b}_{d-1}(\mathbb{L}_{X,\mathbf{0}})$.

The obstruction to using 4.10 to calculate Betti numbers in the general case is that, if X is not an l.c.i., then a formula for the Euler characteristic of the link of a stratum does not tell us the Betti numbers of the link.

Example 4.13. In this example, X will be a hypersurface with a non-isolated singularity. Use (x, y, z) as coordinates for $\mathcal{U} := \mathbb{C}^3$, and let $X := V(xy)$. Let $\tilde{f} := x^\alpha + y^\beta + z^\gamma$, where $\alpha, \beta, \gamma \geqslant 2$.

The strata are $S_0 := V(x, y)$, $S_1 := V(x) - V(y)$, and $S_2 := V(y) - V(x)$, with corresponding links $\mathbb{L}_0 =$ two points, $\mathbb{L}_1 = \emptyset$, and $\mathbb{L}_2 = \emptyset$. As $\mathrm{Ch}({}^k\mathbf{P}^\bullet) = (-1)^d \sum_\alpha \tilde{b}_{k-d_\alpha}(\mathbb{L}_\alpha)\left[\overline{T^*_{S_\alpha}\mathcal{U}}\right]$, we see that $\mathrm{Ch}({}^k\mathbf{P}^\bullet) = 0$ unless $k = 1$, and

$$\mathrm{Ch}({}^1\mathbf{P}^\bullet) = \left[\overline{T^*_{S_0}\mathcal{U}}\right] + \left[\overline{T^*_{S_1}\mathcal{U}}\right] + \left[\overline{T^*_{S_2}\mathcal{U}}\right] = [V(x, y, w_2)] + [V(x, w_1, w_2)] + [V(y, w_0, w_2)].$$

Now, we have that

$$\operatorname{im} d\tilde{f} = V(w_0 - \alpha x^{\alpha-1}, w_1 - \beta y^{\beta-1}, w_2 - \gamma z^{\gamma-1})$$

and $\operatorname{im} d\tilde{f} \cap |\operatorname{Ch}({}^1\mathbf{P}^\bullet)| = (\mathbf{0}, \mathbf{0})$. Therefore, $\dim_{\mathbf{0}} \Sigma_{\mathbb{C}} f = 0$, the only non-zero reduced Betti number of the Milnor fibre of f at $\mathbf{0}$ is

$$\tilde{b}_1(F_{f,\mathbf{0}}) = \left(\operatorname{im} d\tilde{f} \cdot \operatorname{Ch}({}^1\mathbf{P}^\bullet)\right)_{(\mathbf{0},\mathbf{0})} =$$

$$(c-1) + (b-1)(c-1) + (a-1)(c-1) = (c-1)(a+b-1).$$

One can actually verify this computation. The Milnor fibre $F_{f,\mathbf{0}}$ is easily seen to be the union of the Milnor fibre, F_1, of $y^\beta + z^\gamma$ restricted to $V(x)$ and the Milnor fibre, F_2, of $x^\alpha + z^\gamma$ restricted to $V(y)$; these two fibres intersect in c distinct points. The classical calculation of the Milnor numbers tells us that F_1 is homotopy-equivalent to a bouquet of $(\beta-1)(\gamma-1)$ 1-spheres, while F_2 is homotopy-equivalent to a bouquet of $(\alpha-1)(\gamma-1)$ 1-spheres. Applying the Mayer-Vietoris exact sequence, we recover the equality above.

Example 4.14. In this example, X will be the simplest non-l.c.i. Use (u, x, y, z) as coordinates for $\mathcal{U} := \mathbb{C}^4$, and let $X := V(u, x) \cup V(y, z)$. Let $\tilde{f} := u^\alpha + x^\beta + y^\gamma + z^\delta$, where $\alpha, \beta, \gamma, \delta \geqslant 2$.

The strata are $S_0 := \{\mathbf{0}\}$, $S_1 := V(u, x) - \{\mathbf{0}\}$, and $S_2 := V(y, z) - \{\mathbf{0}\}$, with corresponding links $\mathbb{L}_0 =$ two complex disks (sets of complex dimension one), $\mathbb{L}_1 = \emptyset$, and $\mathbb{L}_2 = \emptyset$.

We see that $\operatorname{Ch}({}^k\mathbf{P}^\bullet) = 0$ unless $k = 0$ or 1, and

$$\operatorname{Ch}({}^1\mathbf{P}^\bullet) = \left[\overline{T^*_{S_1}\mathcal{U}}\right] + \left[\overline{T^*_{S_2}\mathcal{U}}\right] = [V(u, x, w_2, w_3)] + [V(y, z, w_0, w_1)],$$

while

$$\operatorname{Ch}({}^0\mathbf{P}^\bullet) = \left[\overline{T^*_{S_0}\mathcal{U}}\right] = [V(u, x, y, z)].$$

Now, we find

$$\operatorname{im} d\tilde{f} = V(w_0 - \alpha u^{\alpha-1}, w_1 - \beta x^{\beta-1}, w_2 - \gamma y^{\gamma-1}, w_3 - \delta z^{\delta-1}),$$

$\operatorname{im} d\tilde{f} \cap |\operatorname{Ch}({}^1\mathbf{P}^\bullet)| = \{\mathbf{0}\}$, and $\operatorname{im} d\tilde{f} \cap |\operatorname{Ch}({}^0\mathbf{P}^\bullet)| = \{\mathbf{0}\}$.

Therefore, $\dim_{\mathbf{0}} \Sigma_{\mathbb{C}} f = 0$, the only non-zero reduced Betti numbers of the Milnor fibre of f at $\mathbf{0}$ are $\tilde{b}_1$ and $\tilde{b}_0$, and

$$\tilde{b}_1(F_{f,\mathbf{0}}) = \left(\operatorname{im} d\tilde{f} \cdot \operatorname{Ch}({}^1\mathbf{P}^\bullet)\right)_{(\mathbf{0},\mathbf{0})} = (\gamma-1)(\delta-1) + (\alpha-1)(\beta-1)$$

and

$$\tilde{b}_0(F_{f,\mathbf{0}}) = \left(\operatorname{im} d\tilde{f} \cdot \operatorname{Ch}({}^0\mathbf{P}^\bullet)\right)_{(\mathbf{0},\mathbf{0})} = 1.$$

Again, one can actually verify this computation. The Milnor fibre $F_{f,\mathbf{0}}$ is easily seen to be the disjoint union of the Milnor fibre, F_1, of $y^\gamma + z^\delta$ restricted to $V(u, x)$ and the Milnor fibre, F_2, of $u^\alpha + x^\beta$ restricted to $V(y, z)$. The classical calculation of the Milnor numbers tells us that F_1 is homotopy-equivalent to a bouquet of $(\gamma-1)(\delta-1)$ 1-spheres, while F_2 is homotopy-equivalent to a bouquet of $(\alpha-1)(\beta-1)$ 1-spheres. Thus, we recover the equalities above.

Chapter 5. THOM'S a_f CONDITION.

We continue with the notation from Chapter 3.

In this section, we explain the fundamental relationship between Thom's a_f condition and the vanishing cycles of f.

Definition 5.1. Let M and N be analytic submanifolds of X such that f has constant rank on N. Then, the pair (M, N) *satisfies Thom's a_f condition at a point* $\mathbf{x} \in N$ if and only if we have the containment $\left(\overline{T^*_{f_{|M}}\mathcal{U}}\right)_{\mathbf{x}} \subseteq \left(T^*_{f_{|N}}\mathcal{U}\right)_{\mathbf{x}}$ of fibres over $\mathbf{x}$.

In particular, if f is, in fact, constant on N, then the pair (M, N) satisfies Thom's a_f condition at a point $\mathbf{x} \in N$ if and only if we have the containment $\left(\overline{T^*_{f_{|M}}\mathcal{U}}\right)_{\mathbf{x}} \subseteq \left(T^*_N\mathcal{U}\right)_{\mathbf{x}}$ of fibres over $\mathbf{x}$.

We have been slightly more general in the above definition than is sometimes the case; we have not required that the rank of f be constant on M. Thus, if X is an analytic space, we may write that (X_{reg}, N) satisfies the a_f condition, instead of writing the much more cumbersome $(X_{\text{reg}} - \Sigma(f_{|X_{\text{reg}}}), N)$ satisfies the a_f condition. If f is not constant on any irreducible component of X, it is easy to see that these statements are equivalent:

Let $\overset{\circ}{X} := X_{\text{reg}} - \Sigma(f_{|X_{\text{reg}}})$, which is dense in X_{reg} (as f is not constant on any irreducible components of X). We claim that $\overline{T^*_{f_{|\overset{\circ}{X}}}\mathcal{U}} = \overline{T^*_{f_{|X_{\text{reg}}}}\mathcal{U}}$; clearly, this is equivalent to showing that $T^*_{f_{|X_{\text{reg}}}}\mathcal{U} \subseteq \overline{T^*_{f_{|\overset{\circ}{X}}}\mathcal{U}}$. This is simple, for if $\mathbf{x} \in \Sigma(f_{|X_{\text{reg}}})$, then $(\mathbf{x}, \eta) \in T^*_{f_{|X_{\text{reg}}}}\mathcal{U}$ if and only if $(\mathbf{x}, \eta) \in T^*_{X_{\text{reg}}}\mathcal{U}$, and $T^*_{X_{\text{reg}}}\mathcal{U} \subseteq \overline{T^*_{\overset{\circ}{X}}\mathcal{U}} \subseteq \overline{T^*_{f_{|\overset{\circ}{X}}}\mathcal{U}}$.

The link between Theorem 3.10 and the a_f condition is provided by the following theorem, which describes the fibre in the relative conormal in terms of the exceptional divisor in the blow-up of $\operatorname{im} d\tilde{f}$. Originally, we needed to assume Whitney's condition a) as an extra hypothesis; however, T. Gaffney showed us how to remove this assumption by using a re-parameterization trick.

Theorem 5.2. *Let $\pi : \mathcal{U} \times \mathbb{C}^{n+1} \times \mathbb{P}^n \to \mathcal{U} \times \mathbb{P}^n$ denote the projection.*

Suppose that f is not constant on any irreducible component of X. Let E denote the exceptional divisor in $\operatorname{Bl}_{\operatorname{im} d\tilde{f}} \overline{T^*_{X_{\text{reg}}}\mathcal{U}} \subseteq \mathcal{U} \times \mathbb{C}^{n+1} \times \mathbb{P}^n$.

Then, for all $\mathbf{x} \in X$, there is an inclusion of fibres over $\mathbf{x}$ given by

$$(\pi(E))_{\mathbf{x}} \subseteq \left(\mathbb{P}(\overline{T^*_{f_{|X_{\text{reg}}}}\mathcal{U}})\right)_{\mathbf{x}}.$$

Moreover, if $\mathbf{x} \in \Sigma_{\text{Nash}} f$, then this inclusion is actually an equality.

Proof. By 3.12, it does not matter what extension of f we use.

That $\left(\pi(E)\right)_{\mathbf{x}} \subseteq \left(\mathbb{P}(\overline{T^*_{f_{|X_{\text{reg}}}}\mathcal{U}})\right)_{\mathbf{x}}$ is easy. Suppose that $(\mathbf{x}, [\eta]) \in \pi(E)$, i.e., $(\mathbf{x}, d_{\mathbf{x}}\tilde{f}, [\eta]) \in E$. Then, there exists a sequence $(\mathbf{x}_i, \omega_i) \in \overline{T^*_{X_{\text{reg}}}\mathcal{U}} - \operatorname{im} d\tilde{f}$ such that $(\mathbf{x}_i, \omega_i, [\omega_i - d_{\mathbf{x}_i}\tilde{f}]) \to (\mathbf{x}, d_{\mathbf{x}}\tilde{f}, [\eta])$. Hence, there exist scalars a_i such that $a_i(\omega_i - d_{\mathbf{x}_i}\tilde{f}) \to \eta$, and these $a_i(\omega_i - d_{\mathbf{x}_i}\tilde{f})$ are relative conormal covectors whose projective class approaches that of η. Thus, $\left(\pi(E)\right)_{\mathbf{x}} \subseteq \left(\mathbb{P}(\overline{T^*_{f_{|X_{\text{reg}}}}\mathcal{U}})\right)_{\mathbf{x}}$.

We must now show that $\left(\mathbb{P}(\overline{T^*_{f_{|X_{\text{reg}}}}\mathcal{U}})\right)_{\mathbf{x}} \subseteq \left(\pi(E)\right)_{\mathbf{x}}$, provided that $\mathbf{x} \in \Sigma_{\text{Nash}} f$.

Let $\overset{\circ}{X} := X_{\text{reg}} - \Sigma(f_{|X_{\text{reg}}})$. Suppose that $(\mathbf{x}, [\eta]) \in \mathbb{P}(\overline{T^*_{f_{|\overset{\circ}{X}}}\mathcal{U}})$. Then, there exists a complex analytic path $\alpha(t) = (\mathbf{x}(t), \eta_t) \in \overline{T^*_{f_{|\overset{\circ}{X}}}\mathcal{U}}$ such that $\alpha(0) = (\mathbf{x}, \eta)$ and $\alpha(t) \in T^*_{f_{|\overset{\circ}{X}}}\mathcal{U}$ for $t \neq 0$. As f has no critical points on $\overset{\circ}{X}$, each η_t can be written uniquely as $\eta_t = \omega_t + \lambda(\mathbf{x}(t)) d_{\mathbf{x}(t)}\tilde{f}$, where $\omega_t(T_{\mathbf{x}(t)}\overset{\circ}{X}) = 0$ and $\lambda(\mathbf{x}(t))$ is a scalar. By evaluating each side on $\mathbf{x}'(t)$, we find that $\lambda(\mathbf{x}(t)) = \frac{\eta_t(\mathbf{x}'(t))}{\frac{d}{dt} f(\mathbf{x}(t))}$.

Thus, as $\lambda(\mathbf{x}(t))$ is a quotient of two analytic functions, there are only two possibilities for what happens to $\lambda(\mathbf{x}(t))$ as $t \to 0$.

Case 1: $|\lambda(\mathbf{x}(t))| \to \infty$ as $t \to 0$.

In this case, since $\eta_t \to \eta$, it follows that $\dfrac{\eta_t}{\lambda(\mathbf{x}(t))} \to 0$ and, hence, $-\dfrac{\omega_t}{\lambda(\mathbf{x}(t))} \to d_{\mathbf{x}}\tilde{f}$. Therefore,

$$\left(\mathbf{x}(t), -\frac{\omega_t}{\lambda(\mathbf{x}(t))}, \left[-\frac{\omega_t}{\lambda(\mathbf{x}(t))} - d_{\mathbf{x}(t)}\tilde{f}\right]\right) = \left(\mathbf{x}(t), -\frac{\omega_t}{\lambda(\mathbf{x}(t))}, [\eta_t(\mathbf{x}(t))]\right) \to (\mathbf{x}, d_{\mathbf{x}}\tilde{f}, [\eta]),$$

and so $(\mathbf{x}, [\eta]) \in \pi(E)$.

Case 2: $\lambda(\mathbf{x}(t)) \to \lambda_0$ as $t \to 0$.

In this case, ω_t must possess a limit as $t \to 0$. For t small and unequal to zero, let proj_t denote the complex orthogonal projection from the fibre $\left(T^*_{f_{|\overset{\circ}{X}}}\mathcal{U}\right)_{\mathbf{x}(t)}$ to the fibre $\left(T^*_{\overset{\circ}{X}}\mathcal{U}\right)_{\mathbf{x}(t)}$. Let $\gamma_t := \operatorname{proj}_t(\eta_t) = \omega_t + \lambda(\mathbf{x}(t)) \operatorname{proj}_t(d_{\mathbf{x}(t)}\tilde{f})$. Since $\mathbf{x} \in \Sigma_{\text{Nash}} f$, we have that $\operatorname{proj}_t(d_{\mathbf{x}(t)}\tilde{f}) \to d_{\mathbf{x}}\tilde{f}$ and, thus, $\gamma_t \to \eta$.

As η is not zero (since it represents a projective class), we may define the (real, non-negative) scalar

$$a_t := \sqrt{\frac{||\operatorname{proj}_t(d_{\mathbf{x}(t)}\tilde{f}) - d_{\mathbf{x}(t)}\tilde{f}||}{||\gamma_t||}}.$$

One now verifies easily that

$$(\mathbf{x}(t),\ a_t\gamma_t + \mathrm{proj}_t(d_{\mathbf{x}(t)}\tilde{f}),\ [a_t\gamma_t + \mathrm{proj}_t(d_{\mathbf{x}(t)}\tilde{f}) - d_{\mathbf{x}(t)}\tilde{f}]) \longrightarrow (\mathbf{x}, d_{\mathbf{x}}\tilde{f}, [\eta]),$$

and, hence, that $(\mathbf{x}, [\eta]) \in \pi(E)$. □

Remark 5.3. In a number of results throughout the remainder of Part III, the reader will find the hypotheses that $\mathbf{x} \in \Sigma_{\mathrm{Nash}} f$ or that $\mathbf{x} \in \Sigma_{\mathrm{alg}} f$. While Theorem 5.2 explains why the hypothesis $\mathbf{x} \in \Sigma_{\mathrm{Nash}} f$ is important, it may not be so clear why the hypothesis $\mathbf{x} \in \Sigma_{\mathrm{alg}} f$ is of interest.

If Y is an analytic subset of X, then one shows easily that $Y \cap \Sigma_{\mathrm{alg}} f \subseteq \Sigma_{\mathrm{alg}}(f_{|_Y})$. The Nash critical locus does not possess such an inheritance property. Thus, the easiest hypothesis to make in order to guarantee that a point, $\mathbf{x}$, is in the Nash critical locus of any analytic subset containing $\mathbf{x}$ is the hypothesis that $\mathbf{x} \in \Sigma_{\mathrm{alg}} f$, for then if $\mathbf{x} \in Y$, we conclude that $\mathbf{x} \in \Sigma_{\mathrm{alg}}(f_{|_Y}) \subseteq \Sigma_{\mathrm{Nash}}(f_{|_Y})$.

A further remark is that the fibre $\big(\pi(E)\big)_{\mathbf{x}}$ being non-empty is trivially seen to be equivalent to $\mathbf{x} \in \Sigma_{\mathrm{cnr}} f$. As the fibre $\Big(\mathbb{P}\big(\overline{T^*_{f_{|X_{\mathrm{reg}}}}\mathcal{U}}\big)\Big)_{\mathbf{x}}$ is always non-empty, the equality $\big(\pi(E)\big)_{\mathbf{x}} = \Big(\mathbb{P}\big(\overline{T^*_{f_{|X_{\mathrm{reg}}}}\mathcal{U}}\big)\Big)_{\mathbf{x}}$ implies that $\mathbf{x} \in \Sigma_{\mathrm{cnr}} f$. This is slightly short of being a converse to the statement in the theorem, unless we are in a situation where we know that $\Sigma_{\mathrm{cnr}} f = \Sigma_{\mathrm{Nash}} f$.

We come now to the result which tells one how the topological information provided by the sheaf of vanishing cycles controls the a_f condition.

Corollary 5.4. *Let N be a submanifold of X such that $N \subseteq V(f)$, and let $\mathbf{x} \in N$*

Let $\mathrm{Ch}(\mathbf{F}^\bullet) = \sum_\alpha m_\alpha \big[\overline{T^*_{M_\alpha}\mathcal{U}}\big]$, where $\{M_\alpha\}$ is a collection of connected analytic submanifolds of X such that either $m_\alpha \geqslant 0$ for all α, or $m_\alpha \leqslant 0$ for all α. Let $\mathrm{Ch}(\phi_f\mathbf{F}^\bullet) = \sum_\beta k_\beta \big[\overline{T^*_{R_\beta}\mathcal{U}}\big]$, where $\{R_\beta\}$ is a collection of connected analytic submanifolds. Finally, suppose that, for all β, there is an inclusion of fibres over $\mathbf{x}$ given by $\Big(\overline{T^*_{R_\beta}\mathcal{U}}\Big)_{\mathbf{x}} \subseteq \big(T^*_N\mathcal{U}\big)_{\mathbf{x}}$.

Then, the pair $\Big(\big(\overline{M_\alpha}\big)_{\mathrm{reg}}, N\Big)$ satisfies Thom's a_f condition at $\mathbf{x}$ for every M_α for which $f_{|M_\alpha} \not\equiv 0$, $m_\alpha \neq 0$, and $\mathbf{x} \in \Sigma_{\mathrm{Nash}}(f_{|\overline{M_\alpha}})$.

Proof. Let $\{S_\gamma\}$ be a Whitney stratification for X such that each $\overline{M_\alpha}$ is a union of strata and such that $\Sigma(f_{|S_\gamma}) = \emptyset$ unless $S_\gamma \subseteq V(f)$. Hence, for each α, there exists a unique S_γ such that $\overline{M_\alpha} = \overline{S_\gamma}$; denote this stratum by S_α. It follows at once that $\mathrm{Ch}(\mathbf{F}^\bullet) = \sum_\alpha m_\alpha \big[\overline{T^*_{S_\alpha}\mathcal{U}}\big]$.

From Theorem 3.10, $E = \sum_\alpha m_\alpha E_\alpha \cong \mathbb{P}\big(\operatorname{Ch}(\phi_f \mathbf{F}^\bullet)\big)$. Thus, since all non-zero m_α have the same sign, if m_α is not zero, then E_α appears with a non-zero coefficient in $\mathbb{P}\big(\operatorname{Ch}(\phi_f \mathbf{F}^\bullet)\big)$.

The result now follows immediately by applying Theorem 5.2 to each $\overline{M_\alpha}$ in place of X. $\square$

Theorem 5.2 also allows us to prove an interesting relationship between the characteristic varieties of the vanishing and nearby cycles – provided that the complex of sheaves under consideration is perverse.

Corollary 5.5. *Let $\mathbf{P}^\bullet$ be a perverse sheaf on X. If $\mathbf{x} \in \Sigma_{\mathrm{alg}} f$ and $(\mathbf{x}, \eta) \in |\operatorname{Ch}(\psi_f \mathbf{P}^\bullet)|$, then $(\mathbf{x}, \eta) \in |\operatorname{Ch}(\phi_f \mathbf{P}^\bullet)|$.*

Proof. Let $\mathcal{S} := \{S_\alpha\}$ be a Whitney stratification with connected strata such that $\mathbf{P}^\bullet$ is constructible with respect to $\mathcal{S}$ and such that $V(f)$ is a union of strata. For the remainder of the proof, we will work in a neighborhood of $V(f)$ – a neighborhood in which, if $S_\alpha \not\subseteq V(f)$, then $\Sigma(f_{|_{S_\alpha}}) = \emptyset$.

Let $\operatorname{Ch}(\mathbf{P}^\bullet) = \sum m_\alpha \left[\overline{T^*_{S_\alpha}\mathcal{U}}\right]$. As $\mathbf{P}^\bullet$ is perverse, all non-zero m_α have the same sign. Thus, 3.10 tells us – using the notation from 3.10 – that

$$(\dagger) \qquad |\mathbb{P}(\operatorname{Ch}(\phi_f \mathbf{P}^\bullet))| = \bigcup_{m_\alpha \neq 0} \pi(E_\alpha),$$

where E_α denotes the exceptional divisor in the blow-up of $\overline{T^*_{S_\alpha}\mathcal{U}}$ along $\operatorname{im} d\tilde{f}$ (in a neighborhood of $V(f)$). In addition, 3.3 tells us that

$$|\operatorname{Ch}(\psi_f \mathbf{P}^\bullet)| = \big(V(f) \times \mathbb{C}^{n+1}\big) \cap \bigcup_{\substack{m_\alpha \neq 0 \\ S_\alpha \not\subseteq V(f)}} \overline{T^*_{f_{|_{S_\alpha}}}\mathcal{U}}.$$

Assume $(\mathbf{x}, \eta) \in |\operatorname{Ch}(\psi_f \mathbf{P}^\bullet)|$. Then, there exists $S_\alpha \not\subseteq V(f)$ such that $m_\alpha \neq 0$ and $(\mathbf{x}, \eta) \in \overline{T^*_{f_{|_{S_\alpha}}}\mathcal{U}}$. Clearly, then, $(\mathbf{x}, \eta) \in \overline{T^*_{f_{|_{(\overline{S}_\alpha)_{\mathrm{reg}}}}}\mathcal{U}}$. Now, if $\mathbf{x} \in \Sigma_{\mathrm{alg}} f$ and $\eta \neq 0$, then $\mathbf{x} \in \Sigma_{\mathrm{alg}}(f_{|_{\overline{S}_\alpha}})$ and so Theorem 5.2 implies that $(\mathbf{x}, [\eta]) \in \pi(E_\alpha)$, where $[\eta]$ denotes the projective class of η and E_α denotes the exceptional divisor of the blow-up of $\overline{T^*_{(\overline{S}_\alpha)_{\mathrm{reg}}}\mathcal{U}} = \overline{T^*_{S_\alpha}\mathcal{U}}$ along $\operatorname{im} d\tilde{f}$. Thus, by $(\dagger)$, $(\mathbf{x}, \eta) \in |\operatorname{Ch}(\phi_f \mathbf{P}^\bullet)|$.

We are left with the trivial case of when $(\mathbf{x}, 0) \in |\operatorname{Ch}(\psi_f \mathbf{P}^\bullet)|$. Note that, if $(\mathbf{x}, 0) \in |\operatorname{Ch}(\psi_f \mathbf{P}^\bullet)|$, then there must exist some non-zero η such that $(\mathbf{x}, \eta) \in |\operatorname{Ch}(\psi_f \mathbf{P}^\bullet)|$. For, otherwise, the stratum (in some Whitney stratification) of $\operatorname{supp} \psi_f \mathbf{P}^\bullet$ containing $\mathbf{x}$ must be all of $\mathcal{U}$. However, $\psi_f \mathbf{P}^\bullet$ is supported on $V(f)$, and so f would have to be zero on all of $\mathcal{U}$; but, this implies that $|\operatorname{Ch}(\psi_f \mathbf{P}^\bullet)| = \emptyset$. Now, if we have some non-zero η such that $(\mathbf{x}, \eta) \in |\operatorname{Ch}(\psi_f \mathbf{P}^\bullet)|$, then by the above argument, $(\mathbf{x}, \eta) \in |\operatorname{Ch}(\phi_f \mathbf{P}^\bullet)|$ and, thus, certainly $(\mathbf{x}, 0) \in |\operatorname{Ch}(\phi_f \mathbf{P}^\bullet)|$. $\square$

The following result helps to illuminate the connection between the Lê-Iomdine (-Vogel) cycles and Thom's a_f condition (see II.6 and IV.2). The result tells us that adding a large power of a second function, g, to f reduces the critical locus, but expands the fibre of the relative conormal. For a generic choice of g, we can obtain effective lower bounds on the power to which g must be raised (see II.4.3.iii and IV.2.1.ii); however, the g below is completely general.

Corollary 5.6 (Thom reduction). *Suppose that* $\mathbf{x} \in \Sigma_{\text{Nash}} f$. *Assume that* $f(\mathbf{x}) = 0$, *and that we have a second function* $g : X \to \mathbb{C}$ *such that* $g(\mathbf{x}) = 0$. *Suppose that* $[\eta] \in \left(\mathbb{P}(\overline{T^*_{f_{|X_{\text{reg}}}}\mathcal{U}})\right)_{\mathbf{x}}$,

Then, for all $j \gg 2$, $[\eta] \in \left(\mathbb{P}(\overline{T^*_{(f+g^j)_{|X_{\text{reg}}}}\mathcal{U}})\right)_{\mathbf{x}}$, *and there exists a neighborhood* $\mathcal{W}$ *of* $\mathbf{x}$ *in* X *such that, in* $\mathcal{W}$, $\Sigma_{\text{reg}}(f + g^j) \subseteq V(g) \cap \Sigma_{\text{reg}} f$.

Proof. Let $\tilde{g}$ denote a local extension of g to $\mathcal{U}$. Let E denote the exceptional divisor in $\text{Bl}_{\text{im}\, d\tilde{f}} \overline{T^*_{X_{\text{reg}}}\mathcal{U}} \subseteq \mathcal{U} \times \mathbb{C}^{n+1} \times \mathbb{P}^n$, and let E_j denote the exceptional divisor in $\text{Bl}_{\text{im}\, d(\tilde{f}+\tilde{g}^j)} \overline{T^*_{X_{\text{reg}}}\mathcal{U}} \subseteq \mathcal{U} \times \mathbb{C}^{n+1} \times \mathbb{P}^n$. Let $\pi : \mathcal{U} \times \mathbb{C}^{n+1} \times \mathbb{P}^n \to \mathcal{U} \times \mathbb{P}^n$ denote the projection.

It is trivial to show that if $[\eta] \in \left(\pi(E)\right)_{\mathbf{x}}$, then, for all $j \gg 2$, $[\eta] \in \left(\pi(E_j)\right)_{\mathbf{x}}$. For $[\eta] \in \left(\pi(E)\right)_{\mathbf{x}}$ if and only if $(\mathbf{x}, d_{\mathbf{x}}\tilde{f}, [\eta]) \in E$, which means that there is an analytic path $\alpha(t) = (\mathbf{p}(t), \omega(t))$ in $\overline{T^*_{X_{\text{reg}}}\mathcal{U}}$ such that $\alpha(0) = (\mathbf{x}, d_{\mathbf{x}}\tilde{f})$, $\alpha(t) \in T^*_{X_{\text{reg}}}\mathcal{U} - \text{im}\, d\tilde{f}$ for $t \neq 0$, and $[\omega(t) - d_{\mathbf{p}(t)}\tilde{f}] \to [\eta]$. Clearly, since $g(\mathbf{x}) = 0$, we may now choose j large enough so that $[\omega(t) - d_{\mathbf{p}(t)}\tilde{f} - j\tilde{g}^{j-1}(\mathbf{p}(t))d_{\mathbf{p}(t)}\tilde{g}] \to [\eta]$. Moreover, if $\alpha(t) \in \text{im}\, d(\tilde{f} + \tilde{g}^j)$ for two different j's, then $\tilde{g}(\mathbf{p}(t)) \equiv 0$ and we are finished with the proof of the first statement; otherwise, $\alpha(t) \notin \text{im}\, d(\tilde{f} + \tilde{g}^j)$ for large j, and we are once again finished.

Therefore, if $[\eta] \in \left(\pi(E)\right)_{\mathbf{x}}$, then, for all $j \gg 2$, $[\eta] \in \left(\pi(E_j)\right)_{\mathbf{x}}$. Now, one shows easily that $\mathbf{x} \in \Sigma_{\text{Nash}} f$ implies that $\mathbf{x} \in \Sigma_{\text{Nash}}(f + g^j)$ for all $j \geqslant 2$. One now applies Theorem 5.2 twice to conclude the first part of the corollary.

It is somewhat lengthier to prove that there exists a neighborhood $\mathcal{W}$ of $\mathbf{x}$ in X such that, in $\mathcal{W}$, $\Sigma_{\text{reg}}(f + g^j) \subseteq V(g) \cap \Sigma_{\text{reg}} f$, but the idea is simple: we prove it first when X is smooth at $\mathbf{x}$ (using an inequality of Łojasiewicz), and then we resolve the singularity in the general case.

So, assume that X is smooth at $\mathbf{x}$. Perform an analytic change of coordinates to place ourselves in an open subset of affine space. By an inequality of Łojasiewicz ([**Loj**], p. 238), there exists a neighborhood $\mathcal{W}$ of x and a real θ, with $0 < \theta < 1$, such that, for $\mathbf{p} \in \mathcal{W}$, $|f(\mathbf{p})|^{\theta} \leqslant |\text{grad}\, f(\mathbf{p})|$. We will show how to pick j large depending on the size of θ.

Suppose that $\Sigma(f + g^j) \not\subseteq V(g) \cap \Sigma f$; we wish to derive a contradiction. Then, there would exist an analytic path $\alpha(t) \in X$ such that $\alpha(0) = \mathbf{x}$ and, for $t \neq 0$,

$$\alpha(t) \in \Sigma(f + g^j) - V(g) \cap \Sigma f.$$

By Remark 1.7, we know that, near $\mathbf{x}$, $\Sigma(f + g^j) \subseteq V(f + g^j)$. Thus, along $\alpha(t)$, $\text{grad}\, f = -jg^{j-1}\,\text{grad}\, g$ and $f = -g^j$. Hence, along $\alpha(t)$, $|g|^{j\theta} \leqslant j|g|^{j-1}|\,\text{grad}\, g|$ and so, as $g(\alpha(t)) \not\equiv 0$, we conclude that $|g|^{j\theta-j+1} \leqslant j|\,\text{grad}\, g|$. As $g(\alpha(0)) = 0$, we would have a contradiction if

$j\theta - j + 1 < 0$. Therefore, if $j > 1/(1-\theta)$, we obtain the desired conclusion. Actually, in the smooth case, we have shown the stronger result that there exists a single neighborhood $\mathcal{W}$ which can be used for all large j.

Now, allow X to be singular at $\mathbf{x}$. Let $\widetilde{X} \xrightarrow{\pi} X$ be a local analytic resolution of the singularities of X, i.e., a proper map from the smooth space $\widetilde{X}$ such that π is an isomorphism over X_{reg}. As π is proper, $\pi^{-1}(\mathbf{x})$ is compact. Applying the smooth case to $f \circ \pi$ and $g \circ \pi$ at each point of $\pi^{-1}(\mathbf{x})$, and using compactness, we conclude that there is a neighborhood, $\widetilde{\mathcal{W}}$, of $\pi^{-1}(\mathbf{x})$ such that, in $\widetilde{\mathcal{W}}$, for all $j \gg 2$, $\Sigma((f+g^j)\circ\pi) \subseteq V(g\circ\pi)$; fix a j this large.

As in the smooth case, suppose that $\Sigma(f+g^j) \not\subseteq V(g) \cap \Sigma f$; we wish to derive a contradiction. Then, there would exist an analytic path $\alpha(t) \in X$ such that $\alpha(0) = \mathbf{x}$ and, for $t \neq 0$, $\alpha(t) \in \Sigma_{\text{reg}}(f+g^j) - V(g) \cap \Sigma_{\text{reg}} f$. Let Γ denote the image of α. The proper transform, $\widetilde{\Gamma}$, of Γ is a curve which intersects $\pi^{-1}(\mathbf{x})$ in a unique point; the existence of such a curve contradicts the choice of j and $\widetilde{\mathcal{W}}$. $\square$

Chapter 6. CONTINUOUS FAMILIES OF CONSTRUCTIBLE COMPLEXES.

We wish to prove statements of the form: the constancy of certain data in a family implies that some nice geometric facts hold. As the reader should have gathered from the last section, it is very advantageous to use complexes of sheaves for cohomology coefficients; in particular, being able to use perverse coefficients is very desirable. The question arises: what should a family of complexes mean?

Let X be a d-dimensional analytic space, let $t : X \to \mathbb{C}$ be an analytic function, and let $\mathbf{F}^\bullet$ be a bounded, constructible complex of $\mathbb{C}$-vector spaces. We could say that $\mathbf{F}^\bullet$ and t form a "nice" family of complexes, since, for all $a \in \mathbb{C}$, we can consider the complex $\mathbf{F}^\bullet_{|_{t^{-1}(a)}}$ on the space $X_{|_{t^{-1}(a)}}$. However, this does yield a satisfactory theory, because there may be absolutely no relation between $\mathbf{F}^\bullet_{|_{t^{-1}(0)}}$ and $\mathbf{F}^\bullet_{|_{t^{-1}(a)}}$ for a close to 0. What we need is a notion of *continuous* families of complexes – we want $\mathbf{F}^\bullet_{|_{t^{-1}(0)}}$ to equal the "limit" of $\mathbf{F}^\bullet_{|_{t^{-1}(a)}}$ as a approaches 0. Fortunately, such a notion already exists; it just is not normally thought of as continuity.

Definition 6.1. Let X, t, and $\mathbf{F}^\bullet$ be as above. We define the *limit of* $\mathbf{F}^\bullet_a := \mathbf{F}^\bullet_{|_{t^{-1}(a)}}[-1]$ *as* a *approaches* b, $\lim\limits_{a \to b} \mathbf{F}^\bullet_a$, to be the nearby cycles $\psi_{t-b}\mathbf{F}^\bullet[-1]$.

We say that the family $\mathbf{F}^\bullet_a$ is *continuous at the value* b if the comparison map from $\mathbf{F}^\bullet_b$ to $\psi_{t-b}\mathbf{F}^\bullet[-1]$ is an isomorphism, i.e., if the vanishing cycles $\phi_{t-b}\mathbf{F}^\bullet[-1] = 0$. We say that the family $\mathbf{F}^\bullet_a$ is *continuous* if it is continuous for all values b.

We say that the family $\mathbf{F}^\bullet_a$ is *continuous at the point* $\mathbf{x} \in X$ if there is an open neighborhood $\mathcal{W}$ of $\mathbf{x}$ such that the family defined by restricting $\mathbf{F}^\bullet$ to $\mathcal{W}$ is continuous at the value $t(\mathbf{x})$.

If $\mathbf{P}^\bullet$ is a perverse sheaf on X and $\mathbf{P}^\bullet_a := \mathbf{P}^\bullet_{|_{t^{-1}(a)}}[-1]$ is a continuous family of complexes, then we say that $\mathbf{P}^\bullet_a$ is a continuous family of perverse sheaves.

Remark 6.2. The reason for the shifts by -1 in the families is so that if $\mathbf{P}^\bullet$ is perverse, and $\mathbf{P}^\bullet_a$ is a continuous family, then each $\mathbf{P}^\bullet_a$ is, in fact, a perverse sheaf (since $\mathbf{P}^\bullet_a \cong \psi_{t-a}\mathbf{P}^\bullet[-1]$).

It is not difficult to show that: if the family $\mathbf{F}^\bullet_a$ is continuous at the value b, and, for all $a \neq b$, each $\mathbf{F}^\bullet_a$ is perverse, then, near the value b, the family $\mathbf{F}^\bullet_a$ is a continuous family of perverse sheaves.

For the remainder of this section, we will be using the following additional notation. Let $\tilde{t}$ be an analytic function on $\mathcal{U}$, and let t denote its restriction to X. Let $\mathbf{P}^\bullet$ be a perverse sheaf on X. Consider the families of spaces, functions, and sheaves given by $X_a := X \cap V(t-a)$, $f_a := f_{|X_a}$, and $\mathbf{P}^\bullet_a := \mathbf{P}^\bullet_{|X_a}[-1]$ (normally, if we are not looking at a specific value for t, we write X_t, f_t, and $\mathbf{P}^\bullet_t$ for these families). Note that, if we have as an hypothesis that $\mathbf{P}^\bullet_t$ is continuous, then the family $\mathbf{P}^\bullet_t$ is actually a family of **perverse** sheaves.

We will now prove three fundamental lemmas; all of them have trivial proofs, but they are nonetheless extremely useful.

The first lemma uses Theorem 4.2 to characterize continuity at a point for families of perverse sheaves.

Lemma 6.3. *Let* $\mathbf{x} \in X$. *The following are equivalent:*

i) *The family* $\mathbf{P}^\bullet_t$ *is continuous at* $\mathbf{x}$;

ii) $\mathbf{x} \notin \overline{\Sigma_{\mathbf{P}^\bullet} t}$;

iii) $(\mathbf{x}, d_{\mathbf{x}}\tilde{t}) \notin |\operatorname{Ch}(\mathbf{P}^\bullet)|$ *for some local extension,* $\tilde{t}$, *of* t *to* $\mathcal{U}$ *in a neighborhood of* $\mathbf{x}$; *and*

iv) $(\mathbf{x}, d_{\mathbf{x}}\tilde{t}) \notin |\operatorname{Ch}(\mathbf{P}^\bullet)|$ *for every local extension,* $\tilde{t}$, *of* t *to* $\mathcal{U}$ *in a neighborhood of* $\mathbf{x}$.

Proof. The equivalence of i) and ii) follows from their definitions, together with Remark 1.7. The equivalence between ii), iii), and iv) follows immediately from Theorem 4.2. □

The next lemma is a necessary step in several proofs.

Lemma 6.4. *Suppose that the family* $\mathbf{P}^\bullet_t$ *is continuous at* $t = b$, *and that the characteristic cycle of* $\mathbf{P}^\bullet$ *is given by* $\sum_\alpha m_\alpha \left[\overline{T^*_{S_\alpha}\mathcal{U}}\right]$. *Then,* $S_\alpha \not\subseteq V(t-b)$ *if* $m_\alpha \neq 0$.

Proof. This follows immediately from 6.3. □

The last of our three lemmas is the **stability of continuity** result.

Lemma 6.5 (Stability of Continuity). *Suppose that the family* $\mathbf{P}^\bullet_t$ *is continuous at* $\mathbf{x} \in X$. *Then,* $\mathbf{P}^\bullet_t$ *is continuous at all points near* $\mathbf{x}$. *In addition, if* $\overset{\circ}{\mathbb{D}}$ *is an open disk around the origin in* $\mathbb{C}$, $h : \overset{\circ}{\mathbb{D}} \times X \to \mathbb{C}$ *is an analytic function,* $h_c(\mathbf{z}) := h(c, \mathbf{z})$, *and* $h_0 = t$, *then the family* $\mathbf{P}^\bullet_{h_c}$ *is continuous at* $\mathbf{x}$ *for all* c *sufficiently close to* 0.

Proof. Let $\tilde{t}$ be an extension of t to a neighborhood of $\mathbf{x}$ in $\mathcal{U}$, and let $\Pi_1 : T^*\mathcal{U} \to \mathcal{U}$ be the cotangent bundle. As $T^*\mathcal{U}$ is isomorphic to $\mathcal{U} \times \mathbb{C}^{n+1}$, there is a second projection $\Pi_2 : T^*\mathcal{U} \to \mathbb{C}^{n+1}$.

Now, $\Pi_1^{-1}(\mathbf{x}) \cap |\operatorname{Ch}(\mathbf{P}^\bullet)|$ and $\Pi_2^{-1}(d_{\mathbf{x}}\tilde{t}) \cap |\operatorname{Ch}(\mathbf{P}^\bullet)|$ are closed sets. Therefore, the lemma follows immediately from 6.3. □

The following lemma allows us to use intersection-theoretic arguments for families of generalized isolated critical points.

Lemma 6.6. *Suppose that the family* $\mathbf{P}^\bullet_t$ *is continuous at* $\mathbf{x} \in X$. *Let* $b := t(\mathbf{x})$. *Let* $\{S_\alpha\}$ *be a Whitney stratification of* X *with connected strata with respect to which* $\mathbf{P}^\bullet$ *is constructible. Suppose that* $\operatorname{Ch}(\mathbf{P}^\bullet)$ *is given by* $\sum_\alpha m_\alpha \left[\overline{T^*_{S_\alpha}\mathcal{U}}\right]$. *If* $\dim_{\mathbf{x}} \Sigma_{\mathbf{P}^\bullet_b} f_b \leqslant 0$, *then there exists an open neighborhood* $\mathcal{W}$ *of* $\mathbf{x}$ *in* $\mathcal{U}$ *such that:*

i) $\operatorname{im} d\tilde{f}$ *properly intersects* $\sum_\alpha m_\alpha \left[\overline{T^*_{t_{|S_\alpha}}\mathcal{U}}\right]$ *in* $\mathcal{W}$;

ii) for all $\mathbf{y} \in X \cap \mathcal{W}$, $V(t - t(\mathbf{y}))$ *properly intersects*

$$\operatorname{im} d\tilde{f} \cdot \sum_\alpha m_\alpha \left[\overline{T^*_{t_{|S_\alpha}}\mathcal{U}}\right]$$

at $(\mathbf{y}, d_{\mathbf{y}}\tilde{f})$ *in (at most) an isolated point; and*

iii) for all $\mathbf{y} \in X \cap \mathcal{W}$, *if* $a := t(\mathbf{y})$, *then* $\dim_{\mathbf{y}} \Sigma_{\mathbf{P}^\bullet_a} f_a \leqslant 0$ *and*

$$\mu_{\mathbf{y}}(f_a; \mathbf{P}^\bullet_a) = (-1)^d \left[\left(\operatorname{im} d\tilde{f} \cdot \sum_\alpha m_\alpha \left[\overline{T^*_{t_{|S_\alpha}}\mathcal{U}}\right]\right) \cdot V(t-a)\right]_{(\mathbf{y}, d_{\mathbf{y}}\tilde{f})}.$$

Proof. First, note that we may assume that $X = \operatorname{supp} \mathbf{P}^\bullet$; for, otherwise, we would immediately replace X by $\operatorname{supp} \mathbf{P}^\bullet$. We may refine our stratification so that $V(t-b)$ is a union of strata; for by Lemma 6.4, if $S_\alpha \subseteq V(t-b)$, then $m_\alpha = 0$. This also explains why we may index over all strata in the formulas. Finally, 6.4 implies that $V(t-b)$ does not contain an entire irreducible component of X; thus, $\dim X_0 = d - 1$.

We use $\tilde{f}$ as a common extension of f_t to $\mathcal{U}$, for all t. Proposition 4.10 tells us that $\mu_{\mathbf{x}}(f_b; \mathbf{P}^\bullet_b) = (-1)^{d-1}\left(\operatorname{im} d\tilde{f} \cdot \operatorname{Ch}(\mathbf{P}^\bullet_b)\right)_{(\mathbf{b}, d_{\mathbf{b}}\tilde{f})}$. Then, continuity, implies that $\operatorname{Ch}(\mathbf{P}^\bullet_b) = \operatorname{Ch}(\psi_{t-b}[-1]\mathbf{P}^\bullet)$, and

$$(*) \quad \operatorname{Ch}(\psi_{t-b}[-1]\mathbf{P}^\bullet) = -\operatorname{Ch}(\psi_{t-b}\mathbf{P}^\bullet) = -\left(V(t-b) \times \mathbb{C}^{n+1}\right) \cdot \sum_{S_\alpha \not\subseteq V(t-b)} m_\alpha \left[\overline{T^*_{t_{|S_\alpha}}\mathcal{U}}\right],$$

by Theorem 3.3.

Therefore,

$$(\dagger) \quad \mu_{\mathbf{x}}(f_b; \mathbf{P}^\bullet_b) = (-1)^d \left(\operatorname{im} d\tilde{f} \cdot \left(V(t-b) \times \mathbb{C}^{n+1}\right) \cdot \sum_\alpha m_\alpha \left[\overline{T^*_{t_{|S_\alpha}}\mathcal{U}}\right]\right)_{(\mathbf{x}, d_{\mathbf{x}}\tilde{f})} =$$

$$(-1)^d \left(\left(\operatorname{im} d\tilde{f} \cdot \sum_\alpha m_\alpha \left[\overline{T^*_{t_{|S_\alpha}}\mathcal{U}}\right]\right) \cdot \left(V(t-b) \times \mathbb{C}^{n+1}\right)\right)_{(\mathbf{x}, d_{\mathbf{x}}\tilde{f})}.$$

Thus,

$$C := (-1)^d\Big(\operatorname{im} d\tilde{f}\ \cdot\ \sum_\alpha m_\alpha\Big[\overline{T^*_{t_{|S_\alpha}}\mathcal{U}}\Big]\Big)$$

is a non-negative cycle such that $(\mathbf{x}, d_\mathbf{x}\tilde{f})$ is an isolated point in (or, is not in) $C \cdot V(t-b)$. Statements i) and ii) of the lemma follow immediately.

Now, Lemma 6.5 tells us that the family $\mathbf{P}^\bullet_t$ is continuous at all points near $\mathbf{x}$; therefore, if $\mathbf{y}$ is close to $\mathbf{x}$ and $a := t(\mathbf{y})$, then, by repeating the argument for $(*)$, we find that

$$\operatorname{Ch}(\mathbf{P}^\bullet_a) = -\operatorname{Ch}(\psi_{t-a}\mathbf{P}^\bullet) = -\big(V(t-a)\times\mathbb{C}^{n+1}\big)\ \cdot\ \sum_\alpha m_\alpha\Big[\overline{T^*_{t_{|S_\alpha}}\mathcal{U}}\Big]$$

and we know that the intersection of this cycle with $\operatorname{im} d\tilde{f}$ is (at most) zero-dimensional at $(\mathbf{y}, d_\mathbf{y}\tilde{f})$ (since $C\cap V(t-b)$ is (at most) zero-dimensional at $\mathbf{x}$). By considering $\tilde{f}$ an extension of f_a and applying Theorem 4.2, we conclude that $\dim_\mathbf{y}\Sigma_{\mathbf{P}^\bullet_a} f_a \leqslant 0$.

Finally, now that we know that $\mathbf{P}^\bullet_t$ is continuous at $\mathbf{y}$ and that $\dim_\mathbf{y}\Sigma_{\mathbf{P}^\bullet_a} f_a \leqslant 0$, we may argue as we did at $\mathbf{x}$ to conclude that (†) holds with $\mathbf{x}$ replaced by $\mathbf{y}$ and b replaced by a. This proves iii). □

We can now prove an **additivity/upper-semicontinuity** result. We prove this result for a more general type of family of perverse sheaves; instead of parametrizing by the values of a function, we parametrize implicitly. We will need this more general perspective in Theorem 6.10.

Theorem 6.7. *Suppose that the family $\mathbf{P}^\bullet_t$ is continuous at $\mathbf{x}\in X$. Let $b := t(\mathbf{x})$, and suppose that $\dim_\mathbf{x}\Sigma_{\mathbf{P}^\bullet_b} f_b \leqslant 0$.*

Let $\overset{\circ}{\mathbb{D}}$ be an open disk around the origin in $\mathbb{C}$, let $h : \overset{\circ}{\mathbb{D}}\times X\to\mathbb{C}$ be an analytic function, for all $c\in\overset{\circ}{\mathbb{D}}$, let $h_c(\mathbf{z}) := h(c,\mathbf{z})$, let ${}_c\mathbf{P}^\bullet := \mathbf{P}^\bullet_{|V(h_c-b)}[-1]$ and ${}_cf := f_{|V(h_c-b)}$. Suppose that $h_0 = t$.

Then, there exists an open neighborhood $\mathcal{W}$ of $\mathbf{x}$ in $\mathcal{U}$ such that, for all small c, for all $\mathbf{y}\in V(h_c-b)\cap\mathcal{W}$, $\dim_\mathbf{y}\Sigma_{{}_c\mathbf{P}^\bullet}\,{}_cf\leqslant 0$.

Moreover, for fixed c close to 0, there are a finite number of points $\mathbf{y}\in V(h_c-b)\cap\mathcal{W}$ such that $\mu_\mathbf{y}({}_cf;{}_c\mathbf{P}^\bullet)\neq 0$ and

$$\mu_\mathbf{x}({}_bf;{}_b\mathbf{P}^\bullet) = \sum_{\mathbf{y}\in V(h_c-b)\cap\mathcal{W}}\mu_\mathbf{y}({}_cf;{}_c\mathbf{P}^\bullet).$$

In particular, for all small c, for all $\mathbf{y}\in V(h_c-b)\cap\mathcal{W}$, $\mu_\mathbf{y}({}_cf;{}_c\mathbf{P}^\bullet)\leqslant\mu_\mathbf{x}({}_bf;{}_b\mathbf{P}^\bullet)$.

Proof. We continue to let $\mathbf{P}^\bullet_c = \mathbf{P}^\bullet_{|V(t-c)}[-1]$ and $f_c = f_{|V(t-c)}$. Note that, if we let $h(w,\mathbf{z}) := t(\mathbf{z}) - w$, then the statement of the theorem would reduce to a statement about the ordinary families $\mathbf{P}^\bullet_c$ and f_c. Moreover, this statement about the families $\mathbf{P}^\bullet_c$ and f_c

follows immediately from Lemma 6.6. We wish to see that this apparently weak form of the theorem actually implies the stronger form.

Shrinking $\overset{\circ}{\mathbb{D}}$ and $\mathcal{U}$ if necessary, let $\tilde{h} : \overset{\circ}{\mathbb{D}} \times \mathcal{U} \to \mathbb{C}$ denote a local extension of h to $\overset{\circ}{\mathbb{D}} \times \mathcal{U}$. We use w as our coordinate on $\overset{\circ}{\mathbb{D}}$. Note that replacing $h(w, \mathbf{z})$ by $h(w^2, \mathbf{z})$ does not change the statement of the theorem. Therefore, we can, and will, assume that $d_{(0,\mathbf{x})}\tilde{h}$ vanishes on $\mathbb{C} \times \{\mathbf{0}\}$.

Let $\tilde{p} : \overset{\circ}{\mathbb{D}} \times \mathcal{U} \to \mathcal{U}$ denote the projection, and let $p := \tilde{p}_{|_{\overset{\circ}{\mathbb{D}} \times X}}$. Let $\mathbf{Q}^\bullet := p^*\mathbf{P}^\bullet[1]$; as $\mathbf{P}^\bullet$ is perverse, so is $\mathbf{Q}^\bullet$. Let $Y := (\overset{\circ}{\mathbb{D}} \times X) \cap V(h - b)$, and let $\widehat{w} : Y \to \overset{\circ}{\mathbb{D}}$ denote the projection. Let $\mathbf{R}^\bullet := \mathbf{Q}^\bullet_{|_Y}[-1]$. Let $\hat{f} : Y \to \mathbb{C}$ be given by $\hat{f}(w, \mathbf{z}) := f(\mathbf{z})$. As we already know that the theorem is true for ordinary families of functions, we wish to apply it to the family of functions $\hat{f}_{\hat{w}}$ and the family of sheaves $\mathbf{R}^\bullet_{\hat{w}}$; this would clearly prove the desired result.

Thus, we need to prove two things: that $\mathbf{R}^\bullet$ is perverse near $(0, \mathbf{x})$, and that the family $\mathbf{R}^\bullet_{\hat{w}}$ is continuous at $(0, \mathbf{x})$.

Let $\{S_\alpha\}$ be a Whitney stratification, with connected strata, of X with respect to which $\mathbf{P}^\bullet$ is constructible. Refining the stratification if necessary, assume that $V(t - b)$ is a union of strata. Let $\operatorname{Ch}(\mathbf{P}^\bullet) = \sum m_\alpha \left[\overline{T^*_{S_\alpha}\mathcal{U}}\right]$. Clearly, $\mathbf{Q}^\bullet$ is constructible with respect to the Whitney stratification $\{\overset{\circ}{\mathbb{D}} \times S_\alpha\}$, and the characteristic cycle of $\mathbf{Q}^\bullet$ in $T^*(\overset{\circ}{\mathbb{D}} \times \mathcal{U})$ is given by $\operatorname{Ch}(\mathbf{Q}^\bullet) = -\sum m_\alpha \left[\overline{T^*_{\overset{\circ}{\mathbb{D}} \times S_\alpha}(\overset{\circ}{\mathbb{D}} \times \mathcal{U})}\right]$.

Note that, for all $(\mathbf{z}, \eta) \in T^*\mathcal{U}$, $(\mathbf{z}, \eta) \in \overline{T^*_{S_\alpha}\mathcal{U}}$ if and only if

$$(0, \mathbf{z}, \eta \circ d_{(0,\mathbf{z})}p) \in \overline{T^*_{\overset{\circ}{\mathbb{D}} \times S_\alpha}(\overset{\circ}{\mathbb{D}} \times \mathcal{U})}.$$

As we are assuming that $d_{(0,\mathbf{x})}\tilde{h}$ vanishes on $\mathbb{C} \times \{\mathbf{0}\}$ and that $h_0 = t$, we know that $d_{(0,\mathbf{x})}\tilde{h} = d_\mathbf{x}\tilde{t} \circ d_{(0,\mathbf{z})}\tilde{p}$. Thus, $(\mathbf{x}, d_\mathbf{x}\tilde{t}) \in \overline{T^*_{S_\alpha}\mathcal{U}}$ if and only if

$$(0, \mathbf{x}, d_{(0,\mathbf{x})}\tilde{h}) \in \overline{T^*_{\overset{\circ}{\mathbb{D}} \times S_\alpha}(\overset{\circ}{\mathbb{D}} \times \mathcal{U})}.$$

Therefore, $(\mathbf{x}, d_\mathbf{x}\tilde{t}) \in |\operatorname{Ch}(\mathbf{P}^\bullet)|$ if and only if $(0, \mathbf{x}, d_{(0,\mathbf{x})}\tilde{h}) \in |\operatorname{Ch}(\mathbf{Q}^\bullet)|$. As we are assuming that the family $\mathbf{P}^\bullet_t$ is continuous at $\mathbf{x}$, we may apply Lemma 6.3 to conclude that $(\mathbf{x}, d_\mathbf{x}\tilde{t}) \notin |\operatorname{Ch}(\mathbf{P}^\bullet)|$ and, hence, $(0, \mathbf{x}, d_{(0,\mathbf{x})}\tilde{h}) \notin |\operatorname{Ch}(\mathbf{Q}^\bullet)|$. It follows that, for all $(w, \mathbf{z})$ near $(0, \mathbf{x})$, $(w, \mathbf{z}, d_{(w,\mathbf{z})}\tilde{h}) \notin |\operatorname{Ch}(\mathbf{Q}^\bullet)|$ and that the family $\mathbf{Q}^\bullet_h$ is continuous at $(0, \mathbf{x})$; that is, there exists an open neighborhood, $\Omega \times \mathcal{W}$, of $(0, \mathbf{x})$ in $\overset{\circ}{\mathbb{D}} \times \mathcal{U}$, in which $\phi_{h-b}[-1]\mathbf{Q}^\bullet = 0$ and such that, if $(w, \mathbf{z}) \in \Omega \times \mathcal{W}$ and $m_\alpha \neq 0$, then

$$(w, \mathbf{z}, d_{(w,\mathbf{z})}\tilde{h}) \notin \overline{T^*_{\overset{\circ}{\mathbb{D}} \times S_\alpha}(\overset{\circ}{\mathbb{D}} \times \mathcal{U})}.$$

For the remainder of the proof, we assume that $\overset{\circ}{\mathbb{D}}$ and $\mathcal{U}$ have been rechosen to be small enough to use for Ω and $\mathcal{W}$.

As $\phi_{h-b}[-1]\mathbf{Q}^\bullet = 0$, $\mathbf{R}^\bullet \cong \psi_{h-b}[-1]\mathbf{Q}^\bullet$ is a perverse sheaf on Y. It remains for us to show that the family $\mathbf{R}^\bullet_{\hat{w}}$ is continuous at $(0, \mathbf{x})$.

Of course, we appeal to Lemma 6.3 again – we need to show that $(0, \mathbf{x}, d_{(0,\mathbf{x})}w) \notin |\operatorname{Ch}(\mathbf{R}^\bullet)|$. Now, $|\operatorname{Ch}(\mathbf{R}^\bullet)| = |\operatorname{Ch}(\psi_{h-b}[-1]\mathbf{Q}^\bullet)|$, and we wish to use Theorem 3.3 to describe this characteristic variety. If $(w, \mathbf{z}) \in \Omega \times \mathcal{W}$ and $m_\alpha \neq 0$, then $(w, \mathbf{z}, d_{(w,\mathbf{z})}\tilde{h}) \notin \overline{T^*_{\overset{\circ}{\mathbb{D}} \times S_\alpha}(\overset{\circ}{\mathbb{D}} \times \mathcal{U})}$; thus, if $m_\alpha \neq 0$, then h has no critical points when restricted to $\overset{\circ}{\mathbb{D}} \times S_\alpha$, and, using the notation of 3.2 and 3.3,

$$T^*_{h-b,\mathbf{Q}^\bullet}(\overset{\circ}{\mathbb{D}} \times \mathcal{U}) = \sum_\alpha m_\alpha \left[\overline{T^*_{h_{|_{\overset{\circ}{\mathbb{D}} \times S_\alpha}}}(\overset{\circ}{\mathbb{D}} \times \mathcal{U})}\right].$$

Now, using Theorem 3.3, we find that

$$|\operatorname{Ch}(\mathbf{R}^\bullet)| = \big(V(h-b) \times \mathbb{C}^{n+2}\big) \cap \bigcup_{m_\alpha \neq 0} \overline{T^*_{h_{|_{\overset{\circ}{\mathbb{D}} \times S_\alpha}}}(\overset{\circ}{\mathbb{D}} \times \mathcal{U})}.$$

We will be finished if we can show that, if $m_\alpha \neq 0$, then $(0, \mathbf{x}, d_{(0,\mathbf{x})}w) \notin \overline{T^*_{h_{|_{\overset{\circ}{\mathbb{D}} \times S_\alpha}}}(\overset{\circ}{\mathbb{D}} \times \mathcal{U})}$.

Fix an S_α for which $m_\alpha \neq 0$. Suppose that

$$(0, \mathbf{x}, \eta) \in \overline{T^*_{h_{|_{\overset{\circ}{\mathbb{D}} \times S_\alpha}}}(\overset{\circ}{\mathbb{D}} \times \mathcal{U})}.$$

Then, there exists a sequence $(w_i, \mathbf{z}_i, \eta_i) \in T^*_{h_{|_{\overset{\circ}{\mathbb{D}} \times S_\alpha}}}(\overset{\circ}{\mathbb{D}} \times \mathcal{U})$ such that $(w_i, \mathbf{z}_i, \eta_i) \to (0, \mathbf{x}, \eta)$. Thus, $\eta_i\big((\mathbb{C} \times T_{\mathbf{z}_i}S_\alpha) \cap \ker d_{(w_i,\mathbf{z}_i)}\tilde{h}\big) = 0$. By taking a subsequence, if necessary, we may assume that $T_{\mathbf{z}_i}S_\alpha$ converges to some $\mathcal{T}$ in the appropriate Grassmanian. Now, we know that $\ker d_{(w_i,\mathbf{z}_i)}\tilde{h} \to \ker d_{(0,\mathbf{x})}\tilde{h} = \mathbb{C} \times \ker d_{\mathbf{x}}\tilde{t}$. As $(\mathbf{x}, d_{\mathbf{x}}\tilde{t}) \notin \overline{T^*_{S_\alpha}\mathcal{U}}$, $\mathbb{C} \times \ker d_{\mathbf{x}}\tilde{t}$ transversely intersects $\mathbb{C} \times \mathcal{T}$. Therefore, $(\mathbb{C} \times T_{\mathbf{z}_i}S_\alpha) \cap \ker d_{(w_i,\mathbf{z}_i)}\tilde{h} \to (\mathbb{C} \times \mathcal{T}) \cap (\mathbb{C} \times \ker d_{\mathbf{x}}\tilde{t})$, and so $\mathbb{C} \times \{\mathbf{0}\} \subseteq \ker \eta$. However, $\ker d_{(0,\mathbf{x})}w = \{0\} \times \mathbb{C}^{n+1}$, and we are finished. □

We would like to translate Theorem 6.7 into a statement about Milnor fibres and the constant sheaf. First, though, it will be convenient to prove a lemma.

Lemma 6.8. *Let $\mathbf{x} \in X$, and let $b := t(\mathbf{x})$. Suppose that $\dim_{\mathbf{x}}\big(V(t-b) \cap \overline{\Sigma_{\mathbb{C}}t}\big) \leqslant 0$. Fix an integer k. If $\widetilde{H}^k(F_{t,\mathbf{x}};\ \mathbb{C}) = 0$, then the family ${}^k\mathbf{P}^\bullet_t$ is continuous at $\mathbf{x}$. In addition, if $\widetilde{H}^k(F_{t,\mathbf{x}};\ \mathbb{C}) = 0$ and $\widetilde{H}^{k-1}(F_{t,\mathbf{x}};\ \mathbb{C}) = 0$, then ${}^k\mathbf{P}^\bullet_b \cong {}^\mu H^0(\mathbb{C}^\bullet_{X_b}[k])$ near $\mathbf{x}$.*

Proof. By Remark 1.7, the assumption that $\dim_{\mathbf{x}}\big(V(t-b) \cap \overline{\Sigma_{\mathbb{C}}t}\big) \leqslant 0$ is equivalent to $\dim_{\mathbf{x}}\overline{\Sigma_{\mathbb{C}}t} \leqslant 0$ and, by Theorem 4.6, this is equivalent to $\dim_{\mathbf{x}}\overline{\Sigma_{{}^j\mathbf{P}^\bullet}t} \leqslant 0$ for all j. Thus,

$\operatorname{supp}\phi_{t-b}[-1]^k\mathbf{P}^\bullet \subseteq \{\mathbf{x}\}$ near $\mathbf{x}$. We claim that the added assumption that $\widetilde{H}^k(F_{t,\mathbf{x}};\ \mathbb{C}) = 0$ implies that, in fact, $\phi_{t-b}[-1]^k\mathbf{P}^\bullet = 0$ near $\mathbf{x}$.

For, near $\mathbf{x}$, $\operatorname{supp}\phi_{t-b}[-1]\mathbb{C}^\bullet_X[k+1] \subseteq \{\mathbf{x}\}$, and so

$$\phi_{t-b}[-1]^k\mathbf{P}^\bullet = \phi_{t-b}[-1]^{\mu}H^0(\mathbb{C}^\bullet_X[k+1]) \cong$$

$$^{\mu}H^0(\phi_{t-b}[-1]\mathbb{C}^\bullet_X[k+1]) \cong \mathbf{H}^0(\phi_{t-b}[-1]\mathbb{C}^\bullet_X[k+1]).$$

Near $\mathbf{x}$, $\phi_{t-b}[-1]\mathbb{C}^\bullet_X[k+1]$ is supported at, at most, the point $\mathbf{x}$ and, hence, $\phi_{t-b}[-1]^k\mathbf{P}^\bullet = 0$ provided that $\mathbf{H}^0(\phi_{t-b}[-1]\mathbb{C}^\bullet_X[k+1])_{\mathbf{x}} = 0$, i.e., provided that $\widetilde{H}^k(F_{t,\mathbf{x}};\ \mathbb{C}) = 0$. This proves the first claim in the lemma.

Now, if the family ${}^k\mathbf{P}^\bullet_t$ is continuous at $\mathbf{x}$, then, near $\mathbf{x}$,

$${}^k\mathbf{P}^\bullet_b = {}^k\mathbf{P}^\bullet_{|V(t-b)}[-1] \cong \psi_{t-b}[-1]^{\mu}H^0(\mathbb{C}^\bullet_X[k+1]) \cong {}^{\mu}H^0(\psi_{t-b}[-1]\mathbb{C}^\bullet_X[k+1]),$$

and we claim that, if $\widetilde{H}^k(F_{t,\mathbf{x}};\ \mathbb{C}) = 0$ and $\widetilde{H}^{k-1}(F_{t,\mathbf{x}};\ \mathbb{C}) = 0$, then there is an isomorphism (in the derived category) ${}^{\mu}H^0(\psi_{t-b}[-1]\mathbb{C}^\bullet_X[k+1]) \cong {}^{\mu}H^0(\mathbb{C}^\bullet_{X_b}[k])$.

To see this, consider the canonical distinguished triangle

$$\mathbb{C}^\bullet_{X_b}[k] \ \to\ \psi_{t-b}[-1]\mathbb{C}^\bullet_X[k+1] \ \to\ \phi_{t-b}[-1]\mathbb{C}^\bullet_X[k+1] \ \xrightarrow{[1]}\ \mathbb{C}^\bullet_{X_b}[k].$$

A portion of the long exact sequence (in the category of perverse sheaves) resulting from applying perverse cohomology is given by

$${}^{p}H^{-1}(\phi_{t-b}[-1]\mathbb{C}^\bullet_X[k+1]) \to {}^{p}H^0(\mathbb{C}^\bullet_{X_b}[k]) \to$$

$${}^{\mu}H^0(\psi_{t-b}[-1]\mathbb{C}^\bullet_X[k+1]) \to {}^{\mu}H^0(\phi_{t-b}[-1]\mathbb{C}^\bullet_X[k+1]).$$

We would be finished if we knew that the terms on both ends of the above were zero. However, since $\phi_{t-b}[-1]\mathbb{C}^\bullet_X[k+1]$ has no support other than $\mathbf{x}$ (near $\mathbf{x}$), we proceed as we did above to show that ${}^{p}H^{-1}(\phi_{t-b}[-1]\mathbb{C}^\bullet_X[k+1])$ and ${}^{p}H^0(\phi_{t-b}[-1]\mathbb{C}^\bullet_X[k+1])$ are zero precisely when $\widetilde{H}^{k-1}(F_{t,\mathbf{x}};\ \mathbb{C})$ and $\widetilde{H}^k(F_{t,\mathbf{x}};\ \mathbb{C})$ are zero. □

Theorem 6.9. *Let $\mathbf{x} \in X$ and $b := t(\mathbf{x})$. Suppose that $\mathbf{x} \notin \overline{\Sigma_{\mathbb{C}} t}$, and that $\dim_{\mathbf{x}} \Sigma_{\mathbb{C}}(f_b) \leqslant 0$.*

Then, there exists a neighborhood, $\mathcal{W}$, of $\mathbf{x}$ in X such that, for all a near b, there are a finite number of points $\mathbf{y} \in \mathcal{W} \cap V(t-a)$ for which $\widetilde{H}^*(F_{f_a,\mathbf{y}};\ \mathbb{C}) \neq 0$; moreover, for all integers, k, $\tilde{b}_{k-1}(F_{f_a,\mathbf{y}}) = \mu_{\mathbf{y}}(f_a; {}^k\mathbf{P}^\bullet_a)$, and

$$\tilde{b}_{k-1}(F_{f_b,\mathbf{x}}) = \sum_{\mathbf{y}\in\mathcal{W}\cap V(t-a)} \tilde{b}_{k-1}(F_{f_a,\mathbf{y}}),$$

where $\widetilde{H}^*()$ and $\tilde{b}_*()$ are as in Remark 4.4.

Proof. Let $v := f_b(\mathbf{x})$. Fix an integer k.

By the lemma, the family ${}^k\mathbf{P}^\bullet_t$ is continuous at $\mathbf{x}$ and ${}^k\mathbf{P}^\bullet_b \cong {}^\mu H^0(\mathbb{C}^\bullet_{X_b}[k])$ near $\mathbf{x}$. Thus,

$$\phi_{f_b-v}[-1]{}^k\mathbf{P}^\bullet_b \cong \phi_{f_b-v}[-1]{}^\mu H^0(\mathbb{C}^\bullet_{X_b}[k]) \cong {}^\mu H^0(\phi_{f_b-v}[-1]\mathbb{C}^\bullet_{X_b}[k]).$$

We are assuming that $\dim_{\mathbf{x}} \Sigma_{\mathbb{C}}(f_b) \leqslant 0$; this is equivalent to: $\operatorname{supp}\phi_{f_b-v}[-1]\mathbb{C}^\bullet_{X_b}[k] \subseteq \{\mathbf{x}\}$ near $\mathbf{x}$, it follows from the above line and Theorem 4.6 that $\dim_{\mathbf{x}} \Sigma_{{}^k\mathbf{P}^\bullet_b} f_b \leqslant 0$ and that

$$(\ddagger) \qquad \mu_{\mathbf{x}}(f_b; {}^k\mathbf{P}^\bullet_b) = \dim H^0(\phi_{f_b-v}[-1]\mathbb{C}^\bullet_{X_b}[k])_{\mathbf{x}} = \tilde{b}_{k-1}(F_{f_b,\mathbf{x}}).$$

Applying Theorem 6.7, we find that there exists an open neighborhood $\mathcal{W}'$ of $\mathbf{x}$ in $\mathcal{U}$ such that, for all $\mathbf{y} \in \mathcal{W}'$, if $a := t(\mathbf{y})$, then $(*)$ $\dim_{\mathbf{y}} \Sigma_{{}^k\mathbf{P}^\bullet_a} f_a \leqslant 0$, and, for fixed a close to b, there are a finite number of points $\mathbf{y} \in \mathcal{W}' \cap V(t-a)$ such that $\mu_{\mathbf{y}}(f_a; {}^k\mathbf{P}^\bullet_a) \neq 0$ and

$$(\dagger) \qquad \mu_{\mathbf{x}}(f_b; {}^k\mathbf{P}^\bullet_b) = \sum_{\mathbf{y} \in \mathcal{W}' \cap V(t-a)} \mu_{\mathbf{y}}(f_a; {}^k\mathbf{P}^\bullet_a).$$

Now, using the above argument for all k with $0 \leqslant k \leqslant d-1$ and intersecting the resulting $\mathcal{W}'$-neighborhoods, we obtain an open neighborhood $\mathcal{W}$ of $\mathbf{x}$ such that $(*)$ and $(\dagger)$ hold for all such k. We claim that, if a is close to b, then $\mathcal{W} \cap \overline{\Sigma_{\mathbb{C}} f_a}$ consists of isolated points, i.e., the points $\mathbf{y} \in \mathcal{W} \cap V(t-a)$ for which $\widetilde{H}^*(F_{f_a,\mathbf{y}};\ \mathbb{C}) \neq 0$ are isolated.

If $a = b$, then there is nothing to show. So, assume that $a \neq b$, and assume that we are working in $\mathcal{W}$ throughout. By Remark 1.7, t satisfies the hypotheses of Lemma 6.8 at $t = a$; hence, for all k, not only is ${}^k\mathbf{P}^\bullet_t$ continuous at $t = a$, but we also know that ${}^k\mathbf{P}^\bullet_a \cong {}^\mu H^0(\mathbb{C}^\bullet_{X_a}[k])$. By Theorem 4.6, $\overline{\Sigma_{\mathbb{C}} f_a} = \bigcup \overline{\Sigma_{{}^k\mathbf{P}^\bullet_a} f_a}$, where the union is over k where $0 \leqslant k \leqslant \dim X_a$. As $\dim X_a \leqslant d-1$, the claim follows from $(*)$ and the definition of $\mathcal{W}$.

Now that we know that ${}^k\mathbf{P}^\bullet_t$ is continuous at $t = a$ and that $\mathcal{W} \cap \overline{\Sigma_{\mathbb{C}} f_a}$ consists of isolated points, we may use the argument that produced $(\ddagger)$ to conclude that $\mu_{\mathbf{y}}(f_a; {}^k\mathbf{P}^\bullet_a) = \tilde{b}_{k-1}(F_{f_a,\mathbf{y}})$. The theorem follows from this, $(\ddagger)$, $(*)$, and $(\dagger)$. □

We want to prove a result which generalizes that of Lê and Saito [**L-S**]. We need to make the assumption that the Milnor number is constant along a curve that is embedded in X. Hence, it will be convenient to use a *local section* of $t : X \to \mathbb{C}$ at a point $\mathbf{x} \in X$; that is, an analytic function $\mathbf{r}$ from an open neighborhood, $\mathcal{V}$, of $t(\mathbf{x})$ in $\mathbb{C}$ into X such that $\mathbf{r}(t(\mathbf{x})) = \mathbf{x}$ and $t \circ \mathbf{r}$ equals the inclusion morphism of $\mathcal{V}$ into $\mathbb{C}$. Note that existence of such a local section implies that $\mathbf{x} \notin \Sigma_{\mathrm{alg}} t$; in particular, $V(\tilde{t} - \tilde{t}(\mathbf{x}))$ is smooth at $\mathbf{x}$.

Theorem 6.10. *Suppose that the family $\mathbf{P}^\bullet_t$ is continuous at $\mathbf{x} \in X$. Let $b := t(\mathbf{x})$, and let $v := f_b(\mathbf{x})$. Let $\mathbf{r} : \mathcal{V} \to X$ be a local section of t at $\mathbf{x}$, and let $C := \operatorname{im}\mathbf{r}$. Assume that $C \subseteq V(f-v)$, that $\dim_{\mathbf{x}} \Sigma_{\mathbf{P}^\bullet_b} f_b \leqslant 0$, and that, for all a close to b, the Milnor number $\mu_{\mathbf{r}(a)}(f_a; \mathbf{P}^\bullet_a)$ is non-zero and is independent of a; denote this common value by μ.*

Then, C is smooth at $\mathbf{x}$, $V(\tilde{t} - b)$ transversely intersects C in $\mathcal{U}$ at $\mathbf{x}$, and there exists a neighborhood, $\mathcal{W}$, of $\mathbf{x}$ in X such that $\mathcal{W} \cap \overline{\Sigma_{\mathbf{P}^\bullet} f} \subseteq C$ and $(\phi_{f-v}[-1]\mathbf{P}^\bullet)_{|_{\mathcal{W}\cap C}} \cong (\mathbb{C}^\mu_{\mathcal{W}\cap C}[1])^\bullet$.

In particular, if we let $\hat{t}$ denote the restriction of t to $V(f-v)$, then the family $\left(\phi_{f-v}[-1]\mathbf{P}^\bullet\right)_{\hat{t}}$ is continuous at $\mathbf{x}$.

If, in addition to the other hypotheses, we assume that $\mathbf{x} \in \Sigma_{\mathrm{alg}} f$, then the two families $\left(\psi_{f-v}[-1]\mathbf{P}^\bullet\right)_{\hat{t}}$ and $\left(\mathbf{P}^\bullet_{|V(f-v)}[-1]\right)_{\hat{t}}$ are continuous at $\mathbf{x}$. (Though $\mathbf{P}^\bullet_{|V(f-v)}[-1]$ need not be perverse.)

Proof. Let us first prove that the last statement of the theorem follows easily from the first portion of the theorem. So, assume that $\phi_{\hat{t}-b}[-1]\phi_{f-v}[-1]\mathbf{P}^\bullet = 0$ near $\mathbf{x}$. Therefore, working near $\mathbf{x}$, we have that $\phi_{\hat{t}-b}[-1]\left(\mathbf{P}^\bullet_{|V(f-v)}[-1]\right) \cong \phi_{\hat{t}-b}[-1]\psi_{f-v}[-1]\mathbf{P}^\bullet$, and we need to show that this is the zero-sheaf. By Lemma 6.3, what we need to show is that $(\mathbf{x}, d_{\mathbf{x}}\tilde{t}) \notin |\operatorname{Ch}(\psi_{f-v}[-1]\mathbf{P}^\bullet)| = |\operatorname{Ch}(\psi_{f-v}\mathbf{P}^\bullet)|$. As we are assuming that $\mathbf{x} \in \Sigma_{\mathrm{alg}} f$, we may apply Corollary 5.5 to find that it suffices to show that $(\mathbf{x}, d_{\mathbf{x}}\tilde{t}) \notin |\operatorname{Ch}(\phi_{f-v}\mathbf{P}^\bullet)| = |\operatorname{Ch}(\phi_{f-v}[-1]\mathbf{P}^\bullet)|$. By 6.3, this is equivalent to $\phi_{\hat{t}-b}[-1]\phi_{f-v}[-1]\mathbf{P}^\bullet = 0$ near $\mathbf{x}$, which we already know to be true. This proves the last statement of the theorem.

Before proceeding with the remainder of the proof, we wish to make some simplifying assumptions. As $\mathbf{x} \notin \Sigma_{\mathrm{alg}} t$, we may certainly perform an analytic change of coordinates in $\mathcal{U}$ to reduce ourselves to the case where t is simply the restriction to X of a linear form $\tilde{t}$. Moreover, it is notational convenient to assume, without loss of generality, that $\mathbf{x} = \mathbf{0}$ and that b and v are both zero.

Let $\{S_\alpha\}$ be a Whitney stratification of X with connected strata with respect to which $\mathbf{P}^\bullet$ is constructible and such that $V(t)$ and $V(f)$ are each unions of strata. Suppose that $\operatorname{Ch}(\mathbf{P}^\bullet)$ is given by $\sum_\alpha m_\alpha \left[\overline{T^*_{S_\alpha}\mathcal{U}}\right]$.

Let $\widetilde{C} := \{(\mathbf{r}(a), d_{\mathbf{r}(a)}\tilde{f}) \mid a \in \mathcal{V}\}$; the projection, ρ, onto the first component induces an isomorphism from $\widetilde{C}$ to C. By Lemma 6.6, the assumption that the Milnor number, $\mu_{\mathbf{r}(a)}(f_a; {}^k\mathbf{P}^\bullet_a)$, is independent of a is equivalent to:

(†) there exists an open neighborhood $\widetilde{\mathcal{W}}$ of $(\mathbf{0}, d_{\mathbf{0}}\tilde{f})$ in $T^*\mathcal{U}$ in which $\widetilde{C}$ equals

$$\operatorname{im} d\tilde{f} \cap \bigcup_{m_\alpha \neq 0} \overline{T^*_{t_{|S_\alpha}}\mathcal{U}}$$

and $\widetilde{C}$ is a smooth curve at $(\mathbf{0}, d_{\mathbf{0}}\tilde{f})$ such that $(\mathbf{0}, d_{\mathbf{0}}\tilde{f}) \notin \Sigma(t \circ \rho_{|\widetilde{C}})$.

It follows immediately that C is smooth at $\mathbf{0}$ and $\mathbf{0} \notin \Sigma(t_{|C})$. We need to show that (†) implies that $\mathcal{W} \cap \overline{\Sigma_{\mathbf{P}^\bullet} f} \subseteq C$ and $\left(\phi_f[-1]\mathbf{P}^\bullet\right)_{|\mathcal{W}\cap C} \cong \left(\mathbb{C}^\mu_{\mathcal{W}\cap C}[1]\right)^\bullet$, where $\mathcal{W} := \rho(\widetilde{\mathcal{W}})$.

As $\overline{T^*_{S_\alpha}\mathcal{U}} \subseteq \overline{T^*_{t_{|S_\alpha}}\mathcal{U}}$, we have that $|\operatorname{Ch}(\mathbf{P}^\bullet)| \subseteq \bigcup_{m_\alpha \neq 0} \overline{T^*_{t_{|S_\alpha}}\mathcal{U}}$ and, thus, $\operatorname{im} d\tilde{f} \cap |\operatorname{Ch}(\mathbf{P}^\bullet)| \subseteq \widetilde{C}$ inside $\widetilde{\mathcal{W}}$. It follows from Theorem 4.2 that $\mathcal{W} \cap \overline{\Sigma_{\mathbf{P}^\bullet} f} \subseteq C$.

It remains for us to show that $\left(\phi_f[-1]\mathbf{P}^\bullet\right)_{|\mathcal{W}\cap C} \cong \left(\mathbb{C}^\mu_{\mathcal{W}\cap C}[1]\right)^\bullet$. As $\phi_f[-1]\mathbf{P}^\bullet$ is perverse and we have just shown that the support of $\phi_f[-1]\mathbf{P}^\bullet$, near $\mathbf{0}$, is a smooth curve, it follows from the work of MacPherson and Vilonen in [**M-V**] that what we need to show is that, for a generic linear form L, $\mathbf{Q}^\bullet := \phi_L[-1]\phi_f[-1]\mathbf{P}^\bullet = 0$ near $\mathbf{0}$. By definition of the characteristic cycle (and since $\mathbf{0}$ is an isolated point in the support of $\mathbf{Q}^\bullet$), this is the same as showing that the coefficient of $T^*_{\{\mathbf{0}\}}\mathcal{U}$ in $\mathrm{Ch}(\phi_f[-1]\mathbf{P}^\bullet)$ equals zero. To show this, we will appeal to Theorem 3.4 and use the notation from there.

We need to show that $m_\mathbf{0}(\phi_f[-1]\mathbf{P}^\bullet) = 0$. By 3.4, if suffices to show that $m_\mathbf{0}(\mathbf{P}^\bullet) = 0$ and $\Gamma^1_{f,L}(S_\alpha) = \emptyset$ near $\mathbf{0}$, for all S_α which are not contained in $V(f)$ and for which $m_\alpha \neq 0$ (where L still denotes a generic linear form). As $\mathbf{P}^\bullet_t$ is continuous at $\mathbf{0}$, Lemma 6.3 tells us that $m_\mathbf{0}(\mathbf{P}^\bullet) = 0$. Now, near $\mathbf{0}$, if $\mathbf{y} \in \Gamma^1_{f,L}(S_\alpha) - \{\mathbf{0}\}$, then $(\mathbf{y}, d_\mathbf{y}\tilde{f}) \in T^*_{L_{|S_\alpha}}\mathcal{U}$. If we knew that, near $(\mathbf{0}, d_\mathbf{0}\tilde{f})$, $\widetilde{C}$ equals $\operatorname{im} d\tilde{f} \cap \bigcup_{m_\alpha \neq 0} \overline{T^*_{L_{|S_\alpha}}\mathcal{U}}$, then we would be finished – for C is contained in $V(f)$ while S_α is not; hence, $\Gamma^1_{f,L}(S_\alpha)$ would have to be empty near $\mathbf{0}$.

Looking back at (†), we see that what we still need to show is that if $\widetilde{C}$ equals

$$\operatorname{im} d\tilde{f} \cap \bigcup_{m_\alpha \neq 0} \overline{T^*_{t_{|S_\alpha}}\mathcal{U}}$$

near $(\mathbf{0}, d_\mathbf{0}\tilde{f})$, then the same statement holds with t replaced by a generic linear form L. We accomplish this by perturbing t until it is generic, and by then showing that this perturbed t satisfies the hypotheses of the theorem.

As C is smooth and transversely intersected by $V(\tilde{t})$ at $\mathbf{0}$, by performing an analytic change of coordinates, we may assume that $\tilde{t} = z_0$, that C is the z_0-axis, and that $r(a) = (a, \mathbf{0})$. Since the set of linear forms for which 3.4 holds is generic, there exists an open disk, $\overset{\circ}{\mathbb{D}}$, around the origin in $\mathbb{C}$ and an analytic family $\tilde{h} : (\overset{\circ}{\mathbb{D}} \times \mathcal{U}, \overset{\circ}{\mathbb{D}} \times \{\mathbf{0}\}) \to (\mathbb{C}, 0)$ such that $\tilde{h}_0(\mathbf{z}) := \tilde{h}(0, \mathbf{z}) = \tilde{t}(\mathbf{z})$ and such that, for all small non-zero c, $\tilde{h}_c(\mathbf{z}) := \tilde{h}(c, \mathbf{z})$ is a linear form for which Theorem 3.4 holds. Let $h := \tilde{h}_{|_{\overset{\circ}{\mathbb{D}} \times X}}$.

As the family $\mathbf{P}^\bullet_t$ is continuous at $\mathbf{0}$, Lemma 6.5 tells us that $\mathbf{P}^\bullet_{h_c}$ is continuous at $\mathbf{0}$ for all small c. As we are now considering these two different families with the same underlying sheaf, the expression $\mathbf{P}^\bullet_a$ for a fixed value of a is ambiguous, and we need to adopt some new notation. We continue to let $\mathbf{P}^\bullet_a := \mathbf{P}^\bullet_{|V(t-a)}[-1]$ and $f_a := f_{|V(t-a)}$, and let ${}_c\mathbf{P}^\bullet_a := \mathbf{P}^\bullet_{|V(h_c-a)}[-1]$ and ${}_cf_a := f_{|V(h_c-a)}$.

Since $V(\tilde{h}_0) = V(z_0)$ transversely intersects C at $\mathbf{0}$ in $\mathcal{U}$, for all small c, $V(h_c)$ transversely intersects C at $\mathbf{0}$ in $\mathcal{U}$. Hence, for all small c, there exists a local section $\mathbf{r}_c(a)$ for h_c at $\mathbf{0}$ such that $\operatorname{im} \mathbf{r}_c \subseteq C$.

We claim that, for all small c:

i) $\dim_\mathbf{0} \Sigma_{{}_c\mathbf{P}^\bullet_0}({}_cf_0) \leqslant 0$ and $\mu_\mathbf{0}({}_cf_0;\ {}_c\mathbf{P}^\bullet_0) \leqslant \mu_\mathbf{0}({}_0f_0;\ {}_0\mathbf{P}^\bullet_0) = \mu_\mathbf{0}(f_0;\ \mathbf{P}^\bullet_0)$;

ii) for all small a, $\dim_{\mathbf{r}_c(a)} \Sigma_{{}_c\mathbf{P}^\bullet_a}({}_cf_a) \leqslant 0$ and $\mu_{\mathbf{r}_c(a)}({}_cf_a;\ {}_c\mathbf{P}^\bullet_a) \leqslant \mu_\mathbf{0}({}_0f_0;\ {}_0\mathbf{P}^\bullet_0)$; and

iii) for all small $a \neq 0$, $\mu_{\mathbf{r}_c(a)}({}_cf_a;\ {}_c\mathbf{P}^\bullet_a) = \mu_{\mathbf{r}_c(a)}(f_{z_0(\mathbf{r}_c(a))};\ \mathbf{P}^\bullet_{z_0(\mathbf{r}_c(a))})$.

Note that proving i), ii), and iii) would complete the proof of the theorem, for they imply that the hypotheses of the theorem hold with t replaced by h_c for all small c. To be precise, we would know that $\mathbf{P}^\bullet_{h_c}$ is continuous at $\mathbf{0}$, $\dim_\mathbf{0} \Sigma_{{}_c\mathbf{P}^\bullet_0}({}_cf_0) \leqslant 0$, and, for all small a, $\mu_{\mathbf{r}_c(a)}({}_cf_a;\ {}_c\mathbf{P}^\bullet_a) = \mu_\mathbf{0}({}_cf_0;\ {}_c\mathbf{P}^\bullet_0)$; this last equality follows from i), ii), and iii), since, for all small $a \neq 0$, we would have

$$\mu = \mu_{\mathbf{r}_c(a)}(f_{z_0(\mathbf{r}_c(a))};\ \mathbf{P}^\bullet_{z_0(\mathbf{r}_c(a))}) = \mu_{\mathbf{r}_c(a)}({}_cf_a;\ {}_c\mathbf{P}^\bullet_a) \leqslant \mu_{\mathbf{0}}({}_cf_0;\ {}_c\mathbf{P}^\bullet_0) \leqslant \mu_{\mathbf{0}}(f_0;\ \mathbf{P}^\bullet_0) = \mu.$$

However, i), ii) and iii) are easy to prove. i) and ii) follow immediately from Theorem 6.7, and iii) follows simply from the fact that, for all small $a \neq 0$, $V(z_0 - z_0(\mathbf{r}_c(a)))$ and $V(\tilde{h}_c - \tilde{h}_c(\mathbf{r}_c(a)))$ are smooth and transversely intersect all strata of any analytic stratification of X in a neighborhood of $(0, \mathbf{0})$. This concludes the proof. □

Corollary 6.11. *Suppose that the family $\mathbf{P}^\bullet_t$ is continuous at $\mathbf{x} \in X$. Let $b := t(\mathbf{x})$, and let $v := f_b(\mathbf{x})$. Let $\mathbf{r} : \mathcal{V} \to X$ be a local section of t at $\mathbf{x}$, and let $C := \operatorname{im}\mathbf{r}$. Assume that $C \subseteq V(f - v)$, that $\dim_{\mathbf{x}} \Sigma_{\mathbf{P}^\bullet_b} f_b \leqslant 0$, and that, for all a close to b, the Milnor number $\mu_{\mathbf{r}(a)}(f_a; \mathbf{P}^\bullet_a)$ is non-zero and is independent of a. Let $\operatorname{Ch}(\mathbf{P}^\bullet) = \sum_\alpha m_\alpha \left[\overline{T^*_{S_\alpha}\mathcal{U}}\right]$, where $\{S_\alpha\}$ is a collection of connected analytic submanifolds of $\mathcal{U}$.*

Then, C is smooth at $\mathbf{x}$, and there exists a neighborhood, $\mathcal{W}$, of $\mathbf{x}$ in X such that, for all S_α for which $S_\alpha \not\subseteq V(f - v)$ and $m_\alpha \neq 0$:

$\mathcal{W} \cap \Sigma(f_{|_{(\overline{S_\alpha})_{\mathrm{reg}}}}) \subseteq C$ and, if $\mathbf{x} \in \Sigma_{\mathrm{Nash}}(f_{|_{\overline{S_\alpha}}})$, then the pair $\left((\overline{S_\alpha})_{\mathrm{reg}},\ C\right)$ satisfies Thom's a_f condition at $\mathbf{x}$.

Proof. One applies Theorem 6.10. The fact that $\mathcal{W} \cap \Sigma(f_{|_{(\overline{S_\alpha})_{\mathrm{reg}}}}) \subseteq C$, for all S_α for which $m_\alpha \neq 0$ follows from Theorem 4.2, since $\mathcal{W} \cap \overline{\Sigma_{\mathbf{P}^\bullet} f} \subseteq C$. The remainder of the corollary follows by applying Corollary 5.4, where one uses C for the submanifold N. □

Just as we used perverse cohomology to translate Theorem 6.7 into a statement about the constant sheaf in Theorem 6.9, we can use perverse cohomology to translate Corollary 6.11. We will use the notation and results from Proposition 4.3 and Remark 4.4.

Corollary 6.12. *Let $b := t(\mathbf{x})$, and let $v := f_b(\mathbf{x})$. Suppose that $\mathbf{x} \notin \overline{\Sigma_{\mathbb{C}} t}$. Suppose, further, that, $\dim_{\mathbf{x}} \Sigma_{\mathbb{C}}(f_b) \leqslant 0$.*

Let $\mathbf{r} : \mathcal{V} \to X$ be a local section of t at $\mathbf{x}$, and let $C := \operatorname{im}\mathbf{r}$. Assume that $C \subseteq V(f - v)$.

Let S_α be a visible stratum of X of dimension d_α, not contained in $V(f - v)$, and let j be an integer such that $\tilde{b}_{j-1}(\mathbb{L}_\alpha) \neq 0$. Let $Y := \overline{S_\alpha}$ and let $k := d_\alpha + j - 1$. In particular, Y could be any irreducible component of X, j could be zero, and k would be $(\dim Y) - 1$.

Suppose that the reduced Betti number $\tilde{b}_{k-1}(F_{f_a, \mathbf{r}(a)})$ is independent of a for all small a, and that either

a) $\mathbf{x} \in \Sigma_{\mathrm{Nash}}(f_{|_Y})$; or that

b) $\mathbf{x} \notin \Sigma_{\mathrm{cnr}}(f_{|_Y})$, C is smooth at $\mathbf{x}$, and (Y_{reg}, C) satisfies Whitney's condition a) at $\mathbf{x}$.

Then, C is smooth at $\mathbf{x}$, and the pair $(Y_{\mathrm{reg}},\ C)$ satisfies the a_f condition at $\mathbf{x}$.

Moreover, in case a), $\tilde{b}_{k-1}(F_{f_a,\mathbf{r}(a)}) \neq 0$, C is transversely intersected by $V(\tilde{t}-b)$ at $\mathbf{x}$, and $\Sigma(f_{|_{Y_{\mathrm{reg}}}}) \subseteq C$ near $\mathbf{x}$.

In addition, if $\mathbf{x} \in \Sigma_{\mathrm{alg}} f$ and, for all small a and for all i, $\tilde{b}_i(F_{f_a,\mathbf{r}(a)})$ is independent of a, then $\mathbf{x} \notin \overline{\Sigma_{\mathbb{C}}(t_{|_{V(f-v)}})}$.

Proof. We will dispose of case b) first. Suppose that $\mathbf{x} \notin \Sigma_{\mathrm{cnr}}(f_{|_Y})$, C is smooth at $\mathbf{x}$, and (Y_{reg}, C) satisfies Whitney's condition a) at $\mathbf{x}$. Let $\overset{\circ}{Y} := Y_{\mathrm{reg}} - \Sigma(f_{|_{Y_{\mathrm{reg}}}})$.

Suppose that we have an analytic path $(\mathbf{x}(t), \eta_t) \in \overline{T^*_{f_{|\overset{\circ}{Y}}}\mathcal{U}}$, where $(\mathbf{x}(0), \eta_0) = (\mathbf{x}, \eta)$ and, for $t \neq 0$, $(\mathbf{x}(t), \eta_t) \in T^*_{f_{|\overset{\circ}{Y}}}\mathcal{U}$. We wish to show that $(\mathbf{x}, \eta) \in T^*_C\mathcal{U}$.

For $t \neq 0$, $\mathbf{x}(t) \in \overset{\circ}{Y}$, and thus η_t can be written uniquely as $\eta_t = \omega_t + \lambda_t d_{\mathbf{x}_t}\tilde{f}$, where $\omega_t \in T^*_{\overset{\circ}{Y}}\mathcal{U}$ and $\lambda_t \in \mathbb{C}$. As we saw in Theorem 5.2, this implies that either $|\lambda_t| \to \infty$ or that $\lambda_t \to \lambda_0$, for some $\lambda_0 \in \mathbb{C}$. If $|\lambda_t| \to \infty$, then $\frac{\eta_t}{\lambda_t} \to 0$ and, therefore, $-\frac{\omega_t}{\lambda_t} \to d_{\mathbf{x}}\tilde{f}$; however, this implies that $\mathbf{x} \in \Sigma_{\mathrm{cnr}}(f_{|_Y})$, contrary to our assumption. Thus, we must have that $\lambda_t \to \lambda_0$.

It follows at once that ω_t converges to some ω_0. By Whitney's condition a), $(\mathbf{x}, \omega_0) \in T^*_C\mathcal{U}$. As $C \subseteq V(f-v)$, $(\mathbf{x}, d_{\mathbf{x}}\tilde{f}) \in T^*_C\mathcal{U}$. Hence, $(\mathbf{x}, \eta) \in T^*_C\mathcal{U}$ and we have finished with case b).

We must now prove the results in case a). The main step is to prove that $\tilde{b}_{k-1}(F_{f_b,\mathbf{x}}) \neq 0$.

We may refine our stratification, if necessary, so that $V(t-b)$ is a union of strata. By the first part of Theorem 6.9, $\tilde{b}_{k-1}(F_{f_b,\mathbf{x}}) = \mu_{\mathbf{x}}(f_b; {}^k\mathbf{P}^\bullet_b)$. Hence, by Lemma 6.6.iii, $\tilde{b}_{k-1}(F_{f_b,\mathbf{x}})$ would be unequal to zero if we knew, for some S_β for which $m_\beta({}^k\mathbf{P}^\bullet) \neq 0$, that $(\mathbf{x}, d_{\mathbf{x}}\tilde{f}) \in \overline{T^*_{t_{|S_\beta}}\mathcal{U}}$. However, our fixed S_α is such a stratum, for $b_{k+1-d_\alpha}(\mathbb{N}_\alpha, \mathbb{L}_\alpha) \neq 0$ and, since $\mathbf{x} \in \Sigma_{\mathrm{Nash}}(f_{|_Y})$, $\mathbf{x} \in \Sigma_{\mathrm{cnr}}(f_{|_Y})$ and so $(\mathbf{x}, d_{\mathbf{x}}\tilde{f}) \in \overline{T^*_{S_\alpha}\mathcal{U}} \subseteq \overline{T^*_{t_{|S_\alpha}}\mathcal{U}}$.

Now, applying the first part of 6.9 again, we have that $\mu_{\mathbf{r}(a)}(f_a;\ {}^k\mathbf{P}^\bullet_a) = \tilde{b}_{k-1}(F_{f_a,\mathbf{r}(a)})$ for all small a. The conclusions in case a) follow from Corollary 6.11.

We must still demonstrate the last statement of corollary.

Suppose that if $\tilde{b}_i(F_{f_a,\mathbf{r}(a)})$ is independent of a for all small a and for all i. Let $\hat{t}$ denote the restriction of t to $V(f-v)$. We will work in a small neighborhood of $\mathbf{x}$. Applying the last two sentences of Theorem 6.10, we find that $\phi_{\hat{t}-b}[-1]\phi_{f-v}[-1]{}^i\mathbf{P}^\bullet = 0$ and $\phi_{\hat{t}-b}[-1]\psi_{f-v}[-1]{}^i\mathbf{P}^\bullet = 0$ for all i. Commuting nearby and vanishing cycles with perverse cohomology, we find that

$$ {}^\mu H^0\big(\phi_{\hat{t}-b}[-1]\phi_{f-v}[-1]\mathbb{C}^\bullet_X[i+1]\big) = 0 \qquad \text{and} \qquad {}^\mu H^0\big(\phi_{\hat{t}-b}[-1]\psi_{f-v}[-1]\mathbb{C}^\bullet_X[i+1]\big) = 0, $$

for all i. Therefore, $\phi_{\hat{t}-b}[-1]\phi_{f-v}[-1]\mathbb{C}^\bullet_X = 0$ and $\phi_{\hat{t}-b}[-1]\psi_{f-v}[-1]\mathbb{C}^\bullet_X = 0$. It follows from the existence of the distinguished triangle (relating nearby cycles, vanishing cycles, and restriction to the hypersurface) that $\phi_{\hat{t}-b}[-1]\mathbb{C}^\bullet_{V(f-v)}[-1] = 0$. This proves the last statement of the corollary. $\square$

Remark 6.13. If X is a connected l.c.i., then each $\mathbb{L}_\alpha$ has (possibly) non-zero cohomology concentrated in middle degree. Hence, for each visible S_α, $\tilde{b}_{j-1}(\mathbb{L}_\alpha) \neq 0$ only when $j = \operatorname{codim}_X S_\alpha$; this corresponds to $k = d-1$. Therefore, the degree $d-2$ reduced Betti number of $F_{f_a, \mathbf{r}(a)}$ controls the a_f condition between **all** visible strata and C.

Corollary 6.14. *Let W be an analytic subset of an open subset of $\mathbb{C}^n$. Let Z be a d-dimensional irreducible component of W. Let $X := \overset{\circ}{\mathbb{D}} \times W$ be the product of an open disk about the origin with W, and let $Y := \overset{\circ}{\mathbb{D}} \times Z$. Let $f : (X, \overset{\circ}{\mathbb{D}} \times \{\mathbf{0}\}) \to (\mathbb{C}, 0)$ be an analytic function such that $f_{|_Y} \not\equiv 0$, and let $f_t(\mathbf{z}) := f(t, \mathbf{z})$.*

Suppose that $\mathbf{0}$ is an isolated point of $\Sigma_{\mathbb{C}}(f_0)$ and that, for all small a, the reduced Betti number $\tilde{b}_{d-1}(F_{f_a,(a,\mathbf{0})})$ is independent of a.

If either a) $\mathbf{0} \in \Sigma_{\mathrm{Nash}}(f_{|_Y})$ or b) $\mathbf{0} \notin \Sigma_{\mathrm{cnr}}(f_{|_Y})$, then the pair $(Y_{\mathrm{reg}}, \overset{\circ}{\mathbb{D}} \times \{\mathbf{0}\})$ satisfies Thom's a_f condition at $\mathbf{0}$.

Moreover, in case a), $\tilde{b}_{d-1}(F_{f_a,(a,\mathbf{0})}) \neq 0$ and, near $\mathbf{0}$, $\Sigma(f_{|_{Y_{\mathrm{reg}}}}) \subseteq \overset{\circ}{\mathbb{D}} \times \{\mathbf{0}\}$.

Remark 6.15. A question naturally arises: how effective is the criterion appearing in Corollary 6.14 that $\tilde{b}_{d-1}(F_{f_a,(a,\mathbf{0})})$ is independent of a?

By Proposition 4.10, if $\{R_\beta\}$ is a Whitney stratification of W, then (using the notation from 4.10)

$$\tilde{b}_{d-1}(F_{f_a,(a,\mathbf{0})}) =$$

$$\tilde{b}_{d-1}(\mathbb{L}_{\{\mathbf{0}\}}) + \sum_{\substack{R_\beta \text{ visible} \\ \dim R_\beta \geqslant 1}} \tilde{b}_{d-1-d_\beta}(\mathbb{L}_\beta) \left(\left(\Gamma^1_{f_a,L}(R_\beta) \cdot V(f_a)\right)_{\mathbf{0}} - \left(\Gamma^1_{f_a,L}(R_\beta) \cdot V(L)\right)_{\mathbf{0}} \right),$$

where $\mathbb{L}_{\{\mathbf{0}\}}$ denotes the complex link of the origin. As the Betti numbers do not vary with a, $\tilde{b}_{d-1}(F_{f_a,(a,\mathbf{0})})$ will be independent of a provided that

$$\left(\Gamma^1_{f_a,L}(R_\beta) \cdot V(f_a)\right)_{\mathbf{0}} - \left(\Gamma^1_{f_a,L}(R_\beta) \cdot V(L)\right)_{\mathbf{0}}$$

is independent of a for all visible strata, R_β, of dimension at least one.

This condition is certainly very manageable to check if the dimension of the singular set of X at the origin is zero or one.

The final statement of Corollary 6.12 has as its conclusion that the constant sheaf on $X \cap V(f - v)$, parametrized by the restriction of t, is continuous at $\mathbf{x}$; this is useful for inductive arguments, since the hypothesis on the ambient space in Corollary 6.12 is that the constant sheaf, parametrized by t, should be continuous at $\mathbf{x}$. For instance, we can prove the following corollary.

Corollary 6.16. *Suppose that $f^1, \dots, f^k$ are analytic functions from $\mathcal{U}$ into $\mathbb{C}$ which define a sequence of local complete intersections at the origin, i.e., are such that, for all i with $1 \leqslant i \leqslant k$, the space $X^{n+1-i} := V(f^1, \dots, f^i)$ is a local complete intersection of dimension $n+1-i$ at the origin. If, for all i, X_t^{n+1-i} has an isolated singularity at the origin and the restrictions $f_t^{i+1} : X_t^{n+1-i} \to \mathbb{C}$ are such that $\dim_{\mathbf{0}} \Sigma_{\text{can}} f_t^{i+1} \leqslant 0$ and have Milnor numbers (in the sense of* [**Loo**]*) which are independent of t, then $\Sigma\big(f_{|_{X_{\text{reg}}^{n+1-(k-1)}}}\big) \subseteq \mathbb{C} \times \{\mathbf{0}\}$ and the pair $\big(X_{\text{reg}}^{n+1-(k-1)}, \mathbb{C} \times \{\mathbf{0}\}\big)$ satisfies the a_{f^k} condition at the origin.*

Proof. Recall that $\mathbb{C}_X^\bullet[d]$ is a perverse sheaf if X is a local complete intersection. The "ordinary" Milnor number of f_t^{i+1} at the origin is equal to $\mu_{\mathbf{0}}(f_t^{i+1}; \mathbb{C}^\bullet_{X_t^{n+1-i}}[n-i])$. Hence, using Proposition 4.10.ii, this Milnor number is equal to the degree $n-i-1$ (the "middle" degree) reduced Betti number of the Milnor fibre of f_t^{i+1} at the origin – the only possible non-zero reduced Betti number. Now, use Corollary 6.12 and induct; the inductive requirement on the Milnor fibre of z_0 follows from the last statement of the corollary. □

Remark 6.17. In [**G-K**], Gaffney and Kleiman deal with families of local complete intersections as above. In this setting, they obtain the result of Corollary 6.16 using multiplicities of modules.

Part IV. NON-ISOLATED CRITICAL POINTS OF FUNCTIONS ON SINGULAR SPACES

Chapter 0. INTRODUCTION

In Part II, we generalized many results from the study of isolated critical points to the case of non-isolated critical points. We accomplished this by developing the Lê cycles and Lê numbers of a non-isolated critical point; the Lê numbers are a generalization of the Milnor number of an isolated critical point. However, throughout Part II, the domains of our analytic functions were required to be open subsets of affine space.

In Part III, we investigated what an "isolated critical point" of a function on an arbitrarily singular space should mean, and we developed a theory of Milnor numbers.

In Part IV, we wish to use our construction of the Lê cycles and numbers as a guide in order to decide how to generalize our work in Part III to the non-isolated case. We will produce Lê-Vogel cycles and numbers, and use them to generalize many previous results.

Chapter 1. LÊ-VOGEL CYCLES

We will adopt some notation that we will use throughout Part IV; much of this notation was used in Part III.

We let $\mathcal{U}$ be an open subset of $\mathbb{C}^{n+1}$, and let X be a (not necessarily purely) d-dimensional analytic subset of $\mathcal{U}$. Let $\tilde{f} : \mathcal{U} \to \mathbb{C}$ be an analytic function, and let $f := \tilde{f}_{|X}$. Let $\{S_\alpha\}$ denote a Whitney stratification, with connected strata, of X. We let $d_\alpha := \dim S_\alpha$.

As in Part III, Chapter 3, we let $\hat{d}_v := 1 + \dim V(f - v)$, and will usually denote $\hat{d}_0$ by simply $\hat{d}$. If we work locally, or assume that X is pure-dimensional, and require f not to vanish on a component of X, then $\hat{d}$ will have attain its "expected" value of d.

We use $z_0, \dots, z_n$ as coordinates on $\mathcal{U}$. We let $\eta : T^*\mathcal{U} \to \mathcal{U}$ denote the cotangent bundle, and we identify the cotangent space $T^*\mathcal{U}$ with $\mathcal{U} \times \mathbb{C}^{n+1}$ by using $dz_0, \dots, dz_n$ as a basis. We use $w_0, \dots, w_n$ as coordinates for the cotangent vectors, i.e., a cotangent vector is $w_0 dz_0 + \cdots + w_n dz_n$. If $\hat{\mathbf{z}}$ is a linear change of coordinates applied to $\mathbf{z}$, then we let $\widehat{\mathbf{w}}$ denote cotangent coordinates with respect to the new basis $d\hat{z}_0, \dots, d\hat{z}_n$, i.e., if A is in $\mathrm{Gl}_{n+1}(\mathbb{C})$ and $\hat{\mathbf{z}} := A\mathbf{z}$, then $\mathbf{w} = A^t \widehat{\mathbf{w}}$.

We shall be blowing-up subspaces of $T^*\mathcal{U} \cong \mathcal{U} \times \mathbb{C}^{n+1}$ along $(n+1)$-tuples. This blow-up will lie in $\mathcal{U} \times \mathbb{C}^{n+1} \times \mathbb{P}^n$; we let $\pi : \mathcal{U} \times \mathbb{C}^{n+1} \times \mathbb{P}^n \to \mathcal{U} \times \mathbb{P}^n$, $\tau : \mathcal{U} \times \mathbb{C}^{n+1} \times \mathbb{P}^n \to \mathcal{U} \times \mathbb{C}^{n+1}$, and $\nu : \mathcal{U} \times \mathbb{P}^n \to \mathcal{U}$ denote the projections. Note that $\eta \circ \tau = \nu \circ \pi$.

We remind the reader that we slightly modified the definition of the reduced Betti number $\tilde{b}_j()$ in III.4.4, so that the empty set has a non-zero reduced Betti number precisely in degree -1.

Recall, from Part III, that we defined ${}^k\mathbf{P}^\bullet := {}^\mu H^0(\mathbb{C}^\bullet_X[k+1])$. In Part IV, a different shift will be of more use to us. Thus, we define ${}^k\mathbf{Q}^\bullet := {}^{d-k-1}\mathbf{P}^\bullet = {}^\mu H^0(\mathbb{C}^\bullet_X[\dim X - k])$.

We wish to produce Lê-Vogel (LêVo) cycles in much the same way that we produce the Lê cycles: by taking the Vogel cycles of the Jacobian tuple. We immediately run into the problem of what ideal we should use. Theorem III.3.10 provides us with a clue: the vanishing cycles of the constant sheaf along f are integrally related to blowing-up $\operatorname{im} d\tilde{f}$ in $\overline{T^*_{S_\alpha}\mathcal{U}}$ for various strata S_α. Hence, we make the following definition.

Definition 1.1. The *conormal Jacobian tuple of $\tilde{f}$ with respect to* $\mathbf{z}$ (and the corresponding choice of $\mathbf{w}$) is given by

$$J^*_{\mathbf{z}}(\tilde{f}) := \left(w_0 - \frac{\partial \tilde{f}}{\partial z_0}, \dots, w_n - \frac{\partial \tilde{f}}{\partial z_n} \right) \in (\mathcal{O}_{T^*\mathcal{U}})^{n+1}.$$

Thus, $\operatorname{im} d\tilde{f}$ is the zero-locus of $J^*_{\mathbf{z}}(\tilde{f})$.

We shall normally be blowing-up $\overline{T^*_{S_\alpha}\mathcal{U}}$ along the restriction of $J^*_{\mathbf{z}}(\tilde{f})$ to $(\mathcal{O}_{\overline{T^*_{S_\alpha}\mathcal{U}}})^{n+1}$; we will follow the standard practice of simply writing $\mathrm{Bl}_{\operatorname{im} d\tilde{f}} \overline{T^*_{S_\alpha}\mathcal{U}}$.

In Part I, we defined the gap cycles and Vogel cycles with respect to given cycle M; we developed the theory in this generality precisely so that we could now make the appropriate choice(s) for M. Our choice of M is guided by our work in Part III; in particular, we use III.4.3.

Definition 1.2. Let ${}^kM$ be the cycle in $T^*\mathcal{U}$ given by

$$ {}^kM := (-1)^d \operatorname{Ch}\left({}^k\mathbf{Q}^\bullet\right) = \sum_\alpha \tilde{b}_{d-k-1-d_\alpha}(\mathbb{L}_\alpha)\left[\overline{T^*_{S_\alpha}\mathcal{U}}\right]. $$

Note that ${}^kM$ will be zero unless $0 \leqslant k \leqslant d$. Note also that if X is purely d-dimensional, then the final expression for ${}^kM$ above can be written more simply as

$$ \sum_\alpha \tilde{b}_{(\dim \mathbb{L}_\alpha)-k}(\mathbb{L}_\alpha)\left[\overline{T^*_{S_\alpha}\mathcal{U}}\right], $$

where we mean that $\dim \mathbb{L}_\alpha = -1$ if $\mathbb{L}_\alpha = \emptyset$.

We define ${}^km_\alpha := \tilde{b}_{d-k-1-d_\alpha}(\mathbb{L}_\alpha)$, and so ${}^kM = \sum_\alpha {}^km_\alpha\left[\overline{T^*_{S_\alpha}\mathcal{U}}\right]$.

It is also convenient to define ${}^kX$, the image in X of ${}^kM$, i.e.,

$$ {}^kX := |\eta_*({}^kM)| = \bigcup_{\tilde{b}_{d-k-1-d_\alpha}(\mathbb{L}_\alpha)\neq 0} \overline{S_\alpha}. $$

We can now define polar and Lê-Vogel cycles in the cotangent space by using the theory of gap and Vogel cycles developed in Part I. We can then push-forward these "conormal" Lê-Vogel cycles to arrive at the Lê-Vogel cycles in X. We will get one set of Lê-Vogel cycles for each ${}^kM$; note that each ${}^kM > 0$ and that all the components of ${}^kM$ have dimension n+1.

Definition 1.3. The *i-th k-shifted conormal polar cycle of $\tilde{f}$ with respect to* $\mathbf{z}$ *in* $T^*\mathcal{U}$, ${}^k\check{\Gamma}^i_{\tilde{f},\mathbf{z}}$, is defined to be $\widehat{\Pi}^i_{J^*_{\mathbf{z}}(\tilde{f})}({}^kM)$, the i-th inductive gap cycle of $J^*_{\mathbf{z}}(\tilde{f})$ with respect to ${}^kM$.

The *i-th k-shifted conormal Lê-Vogel (LêVo) cycle of $\tilde{f}$ with respect to* $\mathbf{z}$ *in* $T^*\mathcal{U}$, ${}^k\check{\Lambda}^i_{\tilde{f},\mathbf{z}}$, is defined to be $\Delta^i_{J^*_{\mathbf{z}}(\tilde{f})}({}^kM)$, the i-th Vogel cycle of $J^*_{\mathbf{z}}(\tilde{f})$ with respect to ${}^kM$, provided that these Vogel cycles exist (see I.2.14).

If the k-shifted conormal LêVo cycles of $\tilde{f}$ with respect to $\mathbf{z}$ exist, then each $|{}^k\check{\Lambda}^i_{\tilde{f},\mathbf{z}}|$ is contained in $\operatorname{im} d\tilde{f}$. Therefore, the proper push-forward η_* induces an isomorphism between ${}^k\check{\Lambda}^i_{\tilde{f},\mathbf{z}}$ and its image in $\mathcal{U}$; we define the *i-th k-shifted Lê-Vogel (LêVo) cycle of $\tilde{f}$ with respect to* $\mathbf{z}$, ${}^k\Lambda^i_{\tilde{f},\mathbf{z}}$, by ${}^k\Lambda^i_{\tilde{f},\mathbf{z}} := \eta_*\left({}^k\check{\Lambda}^i_{\tilde{f},\mathbf{z}}\right)$. Note that there is no claim at this point that these cycles are independent of the extension $\tilde{f}$.

Note, also, that to say "the k-shifted conormal LêVo cycles of $\tilde{f}$ with respect to $\mathbf{z}$ exist" is equivalent to saying "the k-shifted LêVo cycles of $\tilde{f}$ with respect to $\mathbf{z}$ exist"; hence, we will usually say the latter.

Proposition 1.4. *Suppose that the k-shifted Lê-Vogel cycles of $\tilde{f}$ with respect to $\mathbf{z}$ exist. Then,*

0) ${}^k\breve{\Gamma}^i_{\tilde{f},\mathbf{z}}$, ${}^k\breve{\Lambda}^i_{\tilde{f},\mathbf{z}}$, *and* ${}^k\Lambda^i_{\tilde{f},\mathbf{z}}$ *are non-negative and purely i-dimensional;*

i) $\displaystyle\bigcup_i |{}^k\breve{\Lambda}^i_{\tilde{f},\mathbf{z}}| \;=\; |{}^kM| \cap \operatorname{im} d\tilde{f}$;

ii) $\displaystyle\bigcup_i |{}^k\Lambda^i_{\tilde{f},\mathbf{z}}| \;=\; \overline{\Sigma_{{}^k\mathbf{Q}^\bullet} f}$; *and*

iii) $\displaystyle\bigcup_{i,k} |{}^k\Lambda^i_{\tilde{f},\mathbf{z}}| \;=\; \overline{\Sigma_{\mathbb{C}} f}$.

Proof. As ${}^kM$ is non-negative, all of the cycles defined in 1.3 are non-negative. I.2.2.i implies that ${}^k\breve{\Gamma}^i_{\tilde{f},\mathbf{z}}$ is purely i-dimensional, and I.2.15 implies that ${}^k\breve{\Lambda}^i_{\tilde{f},\mathbf{z}}$ and ${}^k\Lambda^i_{\tilde{f},\mathbf{z}}$ are purely i-dimensional.

i) follows from I.2.4 (and I.2.15). By applying η to each side of i), and using that each ${}^k\breve{\Lambda}^i_{\tilde{f},\mathbf{z}}$ is purely i-dimensional, we obtain that

$$\bigcup_{i,k} |{}^k\Lambda^i_{\tilde{f},\mathbf{z}}| \;=\; \{\mathbf{x} \in X \mid (\mathbf{x}, d_{\mathbf{x}}\tilde{f}) \in |{}^kM| = |\operatorname{Ch}({}^k\mathbf{Q}^\bullet)|\}.$$

ii) now follows immediately from III.4.2. Finally, iii) follows from ii) by Theorem III.4.6. □

Example 1.5. Consider the case where $\dim_{\mathbf{x}} \overline{\Sigma_{\mathbb{C}} f} \leqslant 0$. By III.4.6, this is equivalent to requiring that, for all k, $\dim_{(\mathbf{x}, d_{\mathbf{x}}\tilde{f})} \big(|{}^kM| \cap \operatorname{im} d\tilde{f}\big) \leqslant 0$.

Now, fix k. If $(\mathbf{x}, d_{\mathbf{x}}\tilde{f}) \notin |{}^kM|$, then, for all i, ${}^k\breve{\Lambda}^i_{\tilde{f},\mathbf{z}}$ and ${}^k\Lambda^i_{\tilde{f},\mathbf{z}}$ are defined and are zero near $(\mathbf{x}, d_{\mathbf{x}}\tilde{f})$ and $\mathbf{x}$, respectively. Suppose, then, that $(\mathbf{x}, d_{\mathbf{x}}\tilde{f})$ is an isolated point of $|{}^kM| \cap \operatorname{im} d\tilde{f}$.

Then, as we saw in I.2.8 and I.2.16, near $(\mathbf{x}, d_{\mathbf{x}}\tilde{f})$, the conormal LêVo cycles, ${}^k\breve{\Lambda}^i_{\tilde{f},\mathbf{z}}$, are defined for all i, ${}^k\breve{\Lambda}^i_{\tilde{f},\mathbf{z}} = 0$ for $i \geqslant 1$, and, at $(\mathbf{x}, d_{\mathbf{x}}\tilde{f})$,

$${}^k\breve{\Lambda}^0_{\tilde{f},\mathbf{z}} = {}^kM \cdot V\Big(w_n - \frac{\partial \tilde{f}}{\partial z_n}\Big) \cdot \ldots \cdot V\Big(w_0 - \frac{\partial \tilde{f}}{\partial z_0}\Big) =$$

$${}^kM \cdot V\Big(w_0 - \frac{\partial \tilde{f}}{\partial z_0}, \ldots, w_n - \frac{\partial \tilde{f}}{\partial z_n}\Big) \;=\; {}^kM \cdot \operatorname{im} d\tilde{f} \;=\; \tilde{b}_{d-1-k}(F_{f,\mathbf{x}})\big[(\mathbf{x}, d_{\mathbf{x}}\tilde{f})\big],$$

where the second equality follows from the fact that $w_0 - \frac{\partial \tilde{f}}{\partial z_0}, \ \dots, \ w_n - \frac{\partial \tilde{f}}{\partial z_n}$ must determine a regular sequence in $\mathcal{O}_{\mathcal{U}}$ at points in $|{}^kM|$, and the last equality follows from III.4.10.

Thus, the LêVo cycles of $\tilde{f}$ are all defined, ${}^k\Lambda^i_{\tilde{f},\mathbf{z}} = 0$ for $i \geqslant 1$, and, at $\mathbf{x}$,

$$ {}^k\Lambda^0_{\tilde{f},\mathbf{z}} = \tilde{b}_{d-1-k}(F_{f,\mathbf{x}})\big[\mathbf{x}\big]. $$

Remark 1.6. In Remark 2.16, we discussed how one actually calculates Vogel cycles in practice; we wish to do this again in our present setting, in order to describe how one calculates the LêVo cycles.

By definition,

$$ {}^k\breve{\Lambda}^i_{\tilde{f},\mathbf{z}} = \Delta^i_{J^*_{\mathbf{z}}(\tilde{f})}({}^kM) = \sum_\alpha \tilde{b}_{d-k-1-d_\alpha}(\mathbb{L}_\alpha)\left[\Delta^i_{J^*_{\mathbf{z}}(\tilde{f})}(\overline{T^*_{S_\alpha}\mathcal{U}})\right]. $$

So, how does one calculate $\Delta^i_\alpha := \Delta^i_{J^*_{\mathbf{z}}(\tilde{f})}(\overline{T^*_{S_\alpha}\mathcal{U}})$? One begins with

$$ \widehat{\Pi}^{n+1}_\alpha \ := \ \widehat{\Pi}^{n+1}_{J^*_{\mathbf{z}}(\tilde{f})}(\overline{T^*_{S_\alpha}\mathcal{U}}) \ = \ \overline{T^*_{S_\alpha}\mathcal{U}} \neg V\left(w_n - \frac{\partial \tilde{f}}{\partial z_n}\right); $$

thus, $\widehat{\Pi}^{n+1}_{J^*_{\mathbf{z}}(\tilde{f})}(\overline{T^*_{S_\alpha}\mathcal{U}})$ is either 0 or $\overline{T^*_{S_\alpha}\mathcal{U}}$. Next, one calculates the intersection $\widehat{\Pi}^{n+1}_\alpha \cdot V\left(w_n - \frac{\partial \tilde{f}}{\partial z_n}\right)$. This intersection cycle has components contained in

$$ W := V\left(w_n - \frac{\partial \tilde{f}}{\partial z_n}, \dots, w_n - \frac{\partial \tilde{f}}{\partial z_n}\right) $$

and components which are not contained in W. By I.2.12, the sum of the components which are not contained in W is precisely $\widehat{\Pi}^n_\alpha$ and the sum of the components which are contained in W is $\Delta^n_\alpha := \Delta^n_{J^*_{\mathbf{z}}(\tilde{f})}(\overline{T^*_{S_\alpha}\mathcal{U}})$. Having calculated $\widehat{\Pi}^{n+1}_\alpha \cdot V\left(w_n - \frac{\partial \tilde{f}}{\partial z_n}\right) = \widehat{\Pi}^n_\alpha + \Delta^n_\alpha$, we use our newly found $\widehat{\Pi}^n_\alpha$ in the next step: the calculation of $\widehat{\Pi}^n_\alpha \cdot V\left(w_{n-1} - \frac{\partial \tilde{f}}{\partial z_{n-1}}\right) = \widehat{\Pi}^{n-1}_\alpha + \Delta^{n-1}_\alpha$. One proceeds downward inductively.

As we pointed out in I.2.15, if one is working near a point of $\overline{T^*_{S_\alpha}\mathcal{U}} \cap \operatorname{im} d\tilde{f}$, the slightly subtle point here is that – to know that this method of calculation is valid– one has only to verify that each Δ^i_α is purely i-dimensional **as** one performs the calculations.

We demonstrate such a calculation in Example 1.14, after we have discussed when the LêVo cycles are independent of the extension of f.

If X is a pure-dimensional l.c.i. (e.g., a connected l.c.i.), then, by [**Lê9**], the complex links of strata of X have the homotopy-types of bouquets of spheres of middle dimension. Consequently, for such a space, the only ${}^kM$ which can be non-zero is ${}^0M$ and, therefore, it is only ${}^0\breve{\Gamma}^i_{\tilde{f},\mathbf{z}}$, ${}^0\breve{\Lambda}^i_{\tilde{f},\mathbf{z}}$, and ${}^0\Lambda^i_{\tilde{f},\mathbf{z}}$ which are of interest.

Note that there is still no claim that the LêVo cycles of $\tilde{f}$ are independent of the extension of f. However, we will now use the Segre-Vogel Relation of I.2.22 to find a manageable criterion guaranteeing the existence of the LêVo cycles, to relate the LêVo cycles to our work in Part III, and to see that, under reasonable hypotheses, the LêVo cycles are independent of the extension $\tilde{f}$. In fact, we shall prove an analog of Theorem II.1.26, and so we need to define analogs of good stratifications and prepolar coordinates in our current, more general, setting.

Definition 1.7 Let $\mathcal{R} := \{R_\beta\}$ be a Whitney stratification of X with connected strata.

Then, $\mathcal{R}$ is a *Lê-Vogel stratification* for f provided that, for all visible $R_\beta \in \mathcal{R}$,

i) $\Sigma_{\text{cnr}}(f_{|_{\overline{R_\beta}}})$ is a union of strata; and

ii) for all $R_\gamma \subseteq \Sigma_{\text{cnr}}(f_{|_{\overline{R_\beta}}})$, the pair (R_β, R_γ) satisfies Thom's a_f condition.

Since, near a point $\mathbf{x}$, $\Sigma_{\text{cnr}}(f_{|_{\overline{R_\beta}}}) \subseteq f^{-1}f(\mathbf{x})$ (see III.1.7), the condition that the pair (R_β, R_γ) satisfies Thom's a_f condition in ii) is equivalent to: $\left(\overline{T^*_{f_{|R_\beta}}\mathcal{U}}\right)_{|R_\gamma} \subseteq T^*_{R_\gamma}\mathcal{U}$.

We call the strata comprising $\Sigma_{\text{cnr}}(f_{|_{\overline{R_\beta}}})$ the *good strata of $\mathcal{R}$ associated to R_β* (with respect to f). As $\overline{\Sigma_{\mathbb{C}} f} = \bigcup_{\text{visible } R_\beta} \Sigma_{\text{cnr}}(f_{|_{\overline{R_\beta}}})$, we refer to any stratum contained in $\overline{\Sigma_{\mathbb{C}} f}$ as a *good stratum* (of $\mathcal{R}$ with respect to f).

Let $\mathbf{x} \in X$, and let $\{R_\beta\}$ be a Lê-Vogel stratification for f in an open neighborhood of $\mathbf{x}$. The tuple $(z_0, \dots, z_k)$ is a *Lê-Vogel tuple for f at $\mathbf{x}$ with respect to* $\{R_\beta\}$ provided that, for all i with $0 \leqslant i \leqslant k$, if $R_\beta \subseteq \overline{\Sigma_{\mathbb{C}} f}$ and $\dim R_\beta \geqslant i+1$, then $V(z_0 - z_0(\mathbf{x}), \dots, z_i - z_i(\mathbf{x}))$ transversely intersects R_β near $\mathbf{x}$; as we saw in II.1.24, this is equivalent to there existing an open neighborhood Ω of $\mathbf{x}$ such that

$$\mathbb{P}(T^*_{R_\beta}\Omega) \cap \big(V(z_0 - z_0(\mathbf{x}), \dots, z_i - z_i(\mathbf{x})) \times \mathbb{P}^i \times \{\mathbf{0}\}\big) = \emptyset.$$

Naturally, we define a *Lê-Vogel tuple for f at* $\mathbf{x}$ to be a tuple $(z_0, \dots, z_k)$ such that there exists a Lê-Vogel stratification for f near $\mathbf{x}$ with respect to which $(z_0, \dots, z_k)$ is a Lê-Vogel tuple at $\mathbf{x}$.

Proposition 1.8. *Lê-Vogel stratifications always exist and, for all $\mathbf{x} \in X$, for a generic linear reorganization, $\hat{\mathbf{z}}$, of $\mathbf{z}$, $\hat{\mathbf{z}}$ is a Lê-Vogel tuple for f at $\mathbf{x}$.*

In particular, if $\dim_{\mathbf{x}} \left(\overline{\Sigma_{\mathbb{C}} f} \cap V(z_0 - z_0(\mathbf{x}))\right) \leqslant 0$, *then $\mathbf{z}$ is a Lê-Vogel tuple for f at $\mathbf{x}$.*

Proof. By **[Hi]**, Thom stratifications always exist (since f has codomain $\mathbb{C}$); if we now refine a Thom stratification, $\{R_\beta\}$, so that, for all visible R_β, $\Sigma_{\text{cnr}}(f_{|_{\overline{R_\beta}}})$ is a union of strata, then we will have a Lê-Vogel stratification. Now that we know that we can always produce a Lê-Vogel stratification, it is completely trivial, and standard, to conclude that a generic linear reorganization will be a Lê-Vogel tuple. We leave it as an exercise for the reader.

The last sentence follows from the fact that if $\dim_{\mathbf{x}}\left(\overline{\Sigma_{\mathbb{C}} f} \cap V(z_0 - z_0(\mathbf{x}))\right) \leqslant 0$, then $\dim_{\mathbf{x}} \overline{\Sigma_{\mathbb{C}} f} \leqslant 1$ and $V(z_0 - z_0(\mathbf{x}))$ does not contain a 1-dimensional component of $\overline{\Sigma_{\mathbb{C}} f}$ through $\mathbf{x}$. If C_i are the irreducible components of $\overline{\Sigma_{\mathbb{C}} f}$ of dimension 1 through $\mathbf{x}$, then, in a neighborhood of $\mathbf{x}$, we may refine any Lê-Vogel stratification so that $C_i - \{\mathbf{x}\}$ and $\{\mathbf{x}\}$ are the good strata. The conclusion follows. $\square$

Recalling the notation that we used in I.2.22, we let

$$\mathrm{Bl}_{\operatorname{im} d\tilde{f}}({}^kM) := (-1)^d\, \mathrm{Bl}_{\operatorname{im} d\tilde{f}}\left(\mathrm{Ch}\left({}^k\mathbf{Q}^\bullet\right)\right) := \sum_{\alpha} {}^k m_\alpha\, \mathrm{Bl}_{\operatorname{im} d\tilde{f}}\left(\overline{T^*_{S_\alpha}\mathcal{U}}\right) \subseteq \mathcal{U} \times \mathbb{C}^{n+1} \times \mathbb{P}^n$$

and

$$E_{\operatorname{im} d\tilde{f}}({}^kM) := (-1)^d E_{\operatorname{im} d\tilde{f}}\left(\mathrm{Ch}\left({}^k\mathbf{Q}^\bullet\right)\right) := \sum_{\alpha} {}^k m_\alpha E_{\operatorname{im} d\tilde{f}}\left(\overline{T^*_{S_\alpha}\mathcal{U}}\right),$$

where Bl denotes the blow-up and E denotes the corresponding exceptional divisor.

The following theorem is analogous to the first part of Theorem II.1.2. It tells us that $\mathbf{z}$ being a Lê-Vogel tuple implies the correct hypotheses hold for us to apply the Segre-Vogel Relation of Part I.

Theorem 1.9. *Let $\mathbf{x} \in X$, and suppose that $\mathbf{z}$ is a Lê-Vogel tuple for f at $\mathbf{x}$. Then, there exists an open neighborhood, Ω, of $\mathbf{x}$ such that, for all i, $|\pi_*(E_{\operatorname{im} d\tilde{f}}({}^kM))|$ properly intersects $\Omega \times (\mathbb{P}^i \times \{\mathbf{0}\})$.*

Proof. Our goal is to reduce the proof to the point where it precisely follows the proofs of Lemma II.1.25 and the first part of Theorem II.1.26.

As ${}^kM$ is independent of the Whitney stratification, we may assume that $\{S_\alpha\}$ is a Lê-Vogel stratification for f at $\mathbf{x}$, and that $\mathbf{z}$ is a Lê-Vogel tuple for f at $\mathbf{x}$ with respect to $\{S_\alpha\}$.

Fix an S_α such that $\overline{T^*_{S_\alpha}\mathcal{U}}$ appears in ${}^kM$; such an S_α is necessarily visible. Let $E_\alpha := E_{\operatorname{im} d\tilde{f}}(\overline{T^*_{S_\alpha}\mathcal{U}})$. We need to show that $\pi(E_\alpha)$ properly intersects $\Omega \times (\mathbb{P}^i \times \{\mathbf{0}\})$, for all i.

By III.5.2, $\pi(E_\alpha) \subseteq \mathbb{P}\left(\overline{T^*_{f_{|S_\alpha}}\mathcal{U}}\right)$. Since we are assuming that we have a Lê-Vogel stratification, if $S_\beta \subseteq \Sigma_{\mathrm{cnr}}(f_{|\overline{S_\alpha}})$, then $\left(\overline{T^*_{f_{|S_\alpha}}\mathcal{U}}\right)_{|S_\beta} \subseteq T^*_{S_\beta}\mathcal{U}$. Therefore, if $S_\beta \subseteq \Sigma_{\mathrm{cnr}}(f_{|\overline{S_\alpha}})$, then $\nu^{-1}(S_\beta) \cap \pi(E_\alpha) \subseteq \mathbb{P}(T^*_{S_\beta}\mathcal{U})$. As $\nu(\pi(E_\alpha)) = \Sigma_{\mathrm{cnr}}(f_{|\overline{S_\alpha}})$, it follows that

$$\pi(E_\alpha) \subseteq \bigcup_{S_\beta \subseteq \Sigma_{\mathrm{cnr}}(f_{|\overline{S_\alpha}})} \mathbb{P}(T^*_{S_\beta}\mathcal{U}).$$

The proof is now exactly the arguments of Lemma II.1.25 and the first part of Theorem II.1.26. □

Theorem 1.10. *The analytic set* $|E_{\operatorname{im} d\tilde{f}}({}^kM)|$ *properly intersects* $\mathcal{U}\times\mathbb{C}^{n+1}\times(\mathbb{P}^i\times\{\mathbf{0}\})$ *for all* i *if and only if* $|\pi_*(E_{\operatorname{im} d\tilde{f}}({}^kM))| = |\sum_v \mathbb{P}(\operatorname{Ch}(\phi_{f-v}({}^k\mathbf{Q}^\bullet)))|$ *properly intersects* $\mathcal{U}\times(\mathbb{P}^i\times\{\mathbf{0}\})$ *for all* i*, and whenever these equivalent conditions hold:*

i) *the* k*-shifted LêVo and conormal LêVo cycles of* $\tilde{f}$ *with respect to* $\mathbf{z}$ *exist;*

ii) *for all* i, ${}^k\check{\Lambda}^i_{\tilde{f},\mathbf{z}} = \tau_*\big(E_{\operatorname{im} d\tilde{f}}({}^kM)\ \cdot\ (\mathcal{U}\times\mathbb{C}^{n+1}\times(\mathbb{P}^i\times\{\mathbf{0}\}))\big)$;

iii) *for all* i,

$$
{}^k\Lambda^i_{\tilde{f},\mathbf{z}} = \eta_*\tau_*\big(E_{\operatorname{im} d\tilde{f}}({}^kM)\ \cdot\ (\mathcal{U}\times\mathbb{C}^{n+1}\times(\mathbb{P}^i\times\{\mathbf{0}\}))\big)
$$

$$
\nu_*\Big(\pi_*\big(E_{\operatorname{im} d\tilde{f}}({}^kM)\big)\ \cdot\ (\mathcal{U}\times(\mathbb{P}^i\times\{\mathbf{0}\}))\Big) =
$$

$$
\nu_*\Big(\sum_v(-1)^{\hat{d}_v}\mathbb{P}(\operatorname{Ch}(\phi_{f-v}({}^k\mathbf{Q}^\bullet)))\ \cdot\ (\mathcal{U}\times(\mathbb{P}^i\times\{\mathbf{0}\}))\Big);
$$

and there exists a neighborhood Ω *of* $|{}^kM|\cap\operatorname{im} d\tilde{f}$ *such that*

iv) *the* k*-shifted conormal polar cycles of* $\tilde{f}_{|\Omega}$ *with respect to* $\mathbf{z}$ *exist inside* Ω*;*

v) *for all* i, $|\operatorname{Bl}_{\operatorname{im} d\tilde{f}}({}^kM)|$ *properly intersects* $\Omega\times(\mathbb{P}^i\times\{\mathbf{0}\})$ *in* $\Omega\times\mathbb{P}^n$*;*

vi) *inside* Ω*, for all* i, ${}^k\check{\Gamma}^{i+1}_{\tilde{f},\mathbf{z}} = \tau_*\big(\operatorname{Bl}_{\operatorname{im} d\tilde{f}}({}^kM)\ \cdot\ (\mathcal{U}\times\mathbb{C}^{n+1}\times(\mathbb{P}^i\times\{\mathbf{0}\}))\big)$.

Note that iii) implies that the k*-th shifted LêVo cycles are independent of the extension* $\tilde{f}$.

Proof. That $\pi_*\big(E_{\operatorname{im} d\tilde{f}}({}^kM)\big) = \sum_v(-1)^{\hat{d}_v}\mathbb{P}(\operatorname{Ch}(\phi_{f-v}({}^k\mathbf{Q}^\bullet)))$ follows from Theorem III.3.10. The equivalence of the two intersection conditions is a trivial consequence of the fact that the points in $E_{\operatorname{im} d\tilde{f}}({}^kM)$ lie above the graph $\operatorname{im} d\tilde{f}$.

Now, i), ii), iv), v), and vi) follow immediately from the Segre-Vogel Relation (I.2.20). iii) follows by applying η_* to each side of ii), using that $\nu\circ\pi = \eta\circ\tau$, and using again that $\pi_*\big(E_{\operatorname{im} d\tilde{f}}({}^kM)\big) = \sum_v(-1)^{\hat{d}_v}\mathbb{P}(\operatorname{Ch}(\phi_{f-v}({}^k\mathbf{Q}^\bullet)))$. □

Remark 1.11. Statement iii) of 1.10 tells us that – under the hypotheses of the theorem – ${}^k\Lambda^i_{\tilde{f},\mathbf{z}}$ only depends on f and the choice of $w_0,\dots,w_i$. As $w_0,\dots,w_i$ are determined by $z_0, \dots, z_i$, one might be tempted to reference only $z_0, \dots, z_i$ in the notation for ${}^k\Lambda^i_{\tilde{f},\mathbf{z}}$, e.g., ${}^k\Lambda^0_{\tilde{f},z_0}$. Note, however, the hypotheses of the theorem put conditions on all the z's. This should not be surprising – the Vogel cycles are defined in terms of the **inductive** gap cycles, which are defined by downward induction. Thus, the higher-dimensional data needs to behave well before we can work with the lower-dimensional data.

Of course, we could use 1.10.iii to **define** ${}^k\Lambda^i_{\tilde{f},\mathbf{z}}$, and thereby avoid needing to impose conditions on **all** the coordinates and also avoid referring to $\tilde{f}$ at all. However, we prefer the algorithmic, Vogel cycle definition, because it is the most useful for calculation. Moreover, in general, we will not be interested in working in isolation with individual LêVo cycles, but, rather, will want to require that **all** of them are well-behaved.

On the other hand, it **is** desirable to have the LêVo cycles be independent of the extension of f. Therefore, we make the following definition.

Definition 1.12. If $|E_{\operatorname{im} d\tilde{f}}({}^kM)|$ properly intersects $\mathcal{U} \times \mathbb{C}^{n+1} \times (\mathbb{P}^i \times \{\mathbf{0}\})$ for all i, then we say that the *k-shifted Lê-Vogel (LêVo) cycles of f with respect to* $\mathbf{z}$ *exist*; we write ${}^k\Lambda^i_{f,\mathbf{z}}$ in place of ${}^k\Lambda^i_{\tilde{f},\mathbf{z}}$ and refer to it as the *i-th k-shifted Lê-Vogel (LêVo) cycle of f with respect to* $\mathbf{z}$ (that is, we eliminate the reference to the extension $\tilde{f}$).

If $\mathbf{x} \in X$ and the k-shifted Lê-Vogel cycles of f with respect to $\mathbf{z}$ exist in a neighborhood of $\mathbf{x}$, then we say that *i-th k-shifted Lê-Vogel (LêVo) number of f at* $\mathbf{x}$ *with respect to* $\mathbf{z}$ exists provided that $|{}^k\Lambda^i_{f,\mathbf{z}}|$ properly intersects $V(z_0 - x_0, \dots, z_{i-1} - x_{i-1})$ at $\mathbf{x}$, and then we define this Lê-Vogel number to be ${}^k\lambda^i_{f,\mathbf{z}}(\mathbf{x}) := \left({}^k\Lambda^i_{f,\mathbf{z}} \cdot V(z_0 - x_0, \dots, z_{i-1} - x_{i-1})\right)_{\mathbf{x}}$. When $i = 0$, we mean that ${}^k\lambda^0_{f,\mathbf{z}}(\mathbf{x}) = ({}^k\Lambda^0_{f,\mathbf{z}})_{\mathbf{x}}$.

Note that the LêVo cycles and numbers are only (possibly) non-zero for $0 \leqslant k \leqslant d$ and $0 \leqslant i \leqslant d$, and they exist near a point $\mathbf{x}$ provided that $\mathbf{z}$ is a Lê-Vogel tuple for f at $\mathbf{x}$.

Example 1.13. As in Example 1.5, consider the case where $\dim_{\mathbf{x}} \overline{\Sigma_{\mathcal{C}} f} \leqslant 0$. Thus, for all k, $\dim_{(\mathbf{x}, d_{\mathbf{x}}\tilde{f})}\left(|{}^kM| \cap \operatorname{im} d\tilde{f}\right) \leqslant 0$. It follows that, in a neighborhood of $\mathbf{x}$, $|\pi_*\left(E_{\operatorname{im} d\tilde{f}}({}^kM)\right)| = \mathbb{P}(T^*_{\{\mathbf{x}\}}\mathcal{U}) = \{\mathbf{x}\} \times \mathbb{P}^n$, which certainly properly intersects $\mathcal{U} \times (\mathbb{P}^i \times \{\mathbf{0}\})$ for all i.

Therefore, the equivalent hypotheses of Theorem 1.10 hold, and so the LêVo cycles of f (not $\tilde{f}$) exist, and they equal the LêVo cycles of $\tilde{f}$ as given in 1.5.

Example 1.14. We return to the underlying space of Example III.4.14, the simplest non-l.c.i., but use a function with non-isolated critical points.

Use (u, x, y, z) as coordinates for $\mathcal{U} := \mathbb{C}^4$, and let $X := V(u, x) \cup V(y, z)$. Let

$$\tilde{f} := (u^\alpha + x^\beta)^\tau + y^\gamma + z^\delta,$$

where $\alpha, \beta, \gamma, \delta, \tau \geqslant 2$.

Since $d = 2$, ${}^k\mathbf{Q}^\bullet := {}^{1-k}\mathbf{P}^\bullet$ and our calculation in III.4.14 tells us that $\mathrm{Ch}({}^k\mathbf{Q}^\bullet) = 0$ unless $k = 0$ or 1, and

$$\mathrm{Ch}({}^0\mathbf{Q}^\bullet) = [V(u, x, w_2, w_3)] + [V(y, z, w_0, w_1)]\,,$$

while

$$\mathrm{Ch}({}^1\mathbf{Q}^\bullet) = [V(u, x, y, z)]\,.$$

One easily shows that

$$\operatorname{im} d\tilde{f} = V(w_0 - \tau(u^\alpha + x^\beta)^{\tau-1}\alpha u^{\alpha-1},\ w_1 - \tau(u^\alpha + x^\beta)^{\tau-1}\beta x^{\beta-1},\ w_2 - \gamma y^{\gamma-1},\ w_3 - \delta z^{\delta-1}),$$

$$\operatorname{im} d\tilde{f}\ \cap\ V(u, x, w_2, w_3) = \{\mathbf{0}\},$$

$$\operatorname{im} d\tilde{f}\ \cap\ V(y, z, w_0, w_1) = V(u^\alpha + x^\beta, y, z, w_0, w_1, w_2, w_3),$$

$$\operatorname{im} d\tilde{f}\ \cap\ |\mathrm{Ch}({}^0\mathbf{Q}^\bullet)| = V(u^\alpha + x^\beta, y, z, w_0, w_1, w_2, w_3)$$

and

$$\operatorname{im} d\tilde{f}\ \cap\ |\mathrm{Ch}({}^1\mathbf{Q}^\bullet)| = \operatorname{im} d\tilde{f}\ \cap\ V(u, x, y, z) = \{\mathbf{0}\}.$$

Thus, $\overline{\Sigma_{\mathbb{C}} f}$ is the 1-dimensional set $V(u^\alpha + x^\beta, y, z)$, and we calculate in the manner discussed in Remark 1.6.

Let $C_0 := V(u, x, y, z)$, $C_1 := V(u, x, w_2, w_3)$, and $C_2 := V(y, z, w_0, w_1)$. We need to calculate $\widehat{\Pi}^i_{J^*_{\mathbf{z}}(\tilde{f})}$ and $\Delta^i_{J^*_{\mathbf{z}}(\tilde{f})}$ for each C_j; let us denote the corresponding inductive gap cycles and Vogel cycles by simply $\widehat{\Pi}^i_j$ and Δ^i_j.

As $\operatorname{im} d\tilde{f}$ intersects C_0 and C_1 in the isolated point $\mathbf{0}$, Δ^i_0 and Δ^i_1 are easy to calculate – they are both 0 unless $i = 0$ and, then,

$$\Delta^0_0 = (\operatorname{im} d\tilde{f}\ \cdot\ V(u, x, y, z))_{\mathbf{0}}[\mathbf{0}] = [\mathbf{0}],$$

and

$$\Delta^0_1 = (\operatorname{im} d\tilde{f}\ \cdot\ V(u, x, w_2, w_3))_{\mathbf{0}}[\mathbf{0}] = (\gamma - 1)(\delta - 1)[\mathbf{0}].$$

Now,

$$\widehat{\Pi}^4_2 = [V(y, z, w_0, w_1)],$$

$$\widehat{\Pi}^4_2\ \cdot\ V(w_3 - \delta z^{\delta-1}) = [V(y, z, w_0, w_1, w_3)] = \widehat{\Pi}^3_2,$$

$$\widehat{\Pi}^3_2\ \cdot\ V(w_2 - \gamma y^{\gamma-1}) = [V(y, z, w_0, w_1, w_2, w_3)] = \widehat{\Pi}^2_2,$$

$$\widehat{\Pi}^2_2\ \cdot\ V(w_1 - \tau(u^\alpha + x^\beta)^{\tau-1}\beta x^{\beta-1}) =$$
$$(\beta - 1)[V(x, y, z, w_0, w_1, w_2, w_3)] + (\tau - 1)[V(u^\alpha + x^\beta, y, z, w_0, w_1, w_2, w_3)] =$$

$$\widehat{\Pi}^1_2 + \Delta^1_2,$$

$$\widehat{\Pi}^1_2 \cdot V(w_0 - \tau(u^\alpha + x^\beta)^{\tau-1}\alpha u^{\alpha-1}) =$$
$$(\beta-1)[\alpha(\tau-1)+(\alpha-1)][\mathbf{0}] = (\beta-1)(\alpha\tau-1)[\mathbf{0}] = \Delta^0_2.$$

Therefore, the only non-zero LêVo cycles are

$${}^0\Lambda^1_{f,\mathbf{z}} = (\tau-1)[V(u^\alpha + x^\beta, y, z)],$$

$${}^0\Lambda^0_{f,\mathbf{z}} = \big((\gamma-1)(\delta-1) + (\beta-1)(\alpha\tau-1)\big)[\mathbf{0}],$$

and

$${}^1\Lambda^0_{f,\mathbf{z}} = [\mathbf{0}].$$

The non-zero LêVo numbers at the origin are

$${}^0\lambda^1_{f,\mathbf{z}}(\mathbf{0}) = \beta(\tau-1),$$

$${}^0\lambda^0_{f,\mathbf{z}}(\mathbf{0}) = (\gamma-1)(\delta-1) + (\beta-1)(\alpha\tau-1),$$

and

$${}^1\lambda^0_{f,\mathbf{z}}(\mathbf{0}) = 1.$$

We have the following analog of Theorem II.7.2.

Theorem 1.15. *Let $\mathbf{x} \in X$, let Ω be an open neighborhood of $\mathbf{x}$, and let $\{R_\beta\}$ be a Lê-Vogel stratification for $f_{|\Omega}$. Suppose that $\mathbf{z}$ is a Lê-Vogel tuple for f with respect to $\{R_\beta\}$ at all points of Ω. Then, inside Ω, the k-shifted LêVo cycles of f, ${}^k\Lambda^i_{f,\mathbf{z}}$, exist and*

$$|{}^k\Lambda^i_{f,\mathbf{z}}| \subseteq \bigcup_{\substack{R_\beta \subseteq \overline{\Sigma_{\mathbb{C}} f} \\ \dim R_\beta \leqslant i}} R_\beta.$$

Proof. The existence of ${}^k\Lambda^i_{f,\mathbf{z}}$ follows from 1.7 and 1.8. As we saw in the proof of 1.7, if R_β is visible,

$$\pi\big(E_{\operatorname{im} d\tilde{f}}(\overline{T^*_{R_\beta}\mathcal{U}})\big) \subseteq \bigcup_{R_\gamma \subseteq \Sigma_{\mathrm{cnr}}(f_{|\overline{R_\beta}})} \mathbb{P}(\overline{T^*_{R_\gamma}\mathcal{U}}).$$

The assumption on the coordinates $\mathbf{z}$ is that, if R_β is visible, $R_\gamma \subseteq \Sigma_{\mathrm{cnr}}(f_{|\overline{R_\beta}})$, and the dimension of R_γ is at least $i+1$, then

$$\mathbb{P}(\overline{T^*_{R_\gamma}\mathcal{U}}) \cap \big(\Omega \times (\mathbb{P}^i \times \{\mathbf{0}\})\big) = \emptyset.$$

The result now follows at once from 1.8.iii. $\square$

Chapter 2. LÊ-IOMDINE FORMULAS AND THOM'S CONDITION

We developed the Lê-Iomdine-Vogel formulas in extreme generality in I.3.4 specifically so that we would be able to apply them to the LêVo cycles at this point. As we saw in Part II, Chapter 4, such formulas allow us to reduce questions concerning non-isolated singularities to questions about the isolated case. Hence, we will be able to use III.6.12 in order give conditions which imply that Thom's a_f condition holds.

We continue with our notation from the previous chapter.

For simplicity, we assume in the following two results that $\mathbf{x} \in V(z_0)$; clearly, this causes no loss of generality.

Theorem 2.1. *Let $\mathbf{x} \in V(z_0) \cap |\eta_*({}^kM)| = V(z_0) \cap {}^kX$. Let a be a non-zero complex number, and let $j \geqslant 1$ be an integer. Let $\hat{\mathbf{z}}$ denote the "rotated" coordinates $(z_1, \dots, z_n, z_0)$.*

Suppose that the k-shifted LêVo cycles of f, ${}^k\Lambda^i_{f,\mathbf{z}}$, exist in a neighborhood of $\mathbf{x}$ and that, for all $i \geqslant 1$, $V(z_0)$ properly intersects each ${}^k\Lambda^i_{f,\mathbf{z}}$ at $\mathbf{x}$.

Then, ${}^k\lambda^0_{f,\mathbf{z}}(\mathbf{x})$ and ${}^k\lambda^1_{f,\mathbf{z}}(\mathbf{x})$ exist and, if $j \geqslant 1 + {}^k\lambda^0_{f,\mathbf{z}}(\mathbf{x})$, then, in a neighborhood of $\mathbf{x}$,

i) $\overline{\Sigma_{{}^k\mathbf{Q}^\bullet}(f + az_0^{j+1})} = V(z_0) \cap \overline{\Sigma_{{}^k\mathbf{Q}^\bullet} f}$;

ii) $\dim_{\mathbf{x}} \overline{\Sigma_{{}^k\mathbf{Q}^\bullet}(f + az_0^{j+1})} = \left(\dim_{\mathbf{x}} \overline{\Sigma_{{}^k\mathbf{Q}^\bullet} f}\right) - 1$, *provided that* $\dim_{\mathbf{x}} \overline{\Sigma_{{}^k\mathbf{Q}^\bullet} f} \geqslant 1$;

iii) the k-shifted LêVo cycles of $f + az_0^{j+1}$ with respect to $\hat{\mathbf{z}}$ exist; and

iv) ${}^k\lambda^0_{f+az_0^{j+1}, \hat{\mathbf{z}}}(\mathbf{x}) = {}^k\lambda^0_{f,\mathbf{z}}(\mathbf{x}) + j({}^k\lambda^1_{f,\mathbf{z}}(\mathbf{x}))$ *and, for* $1 \leqslant i \leqslant n-1$,

$${}^k\Lambda^i_{f+az_0^{j+1}, \hat{\mathbf{z}}} = j\left({}^k\Lambda^{i+1}_{f,\mathbf{z}} \cdot V(z_0)\right).$$

Proof. This is simply a translation I.3.4 in our current situation. We have also used that $(\mathbf{p}, d_{\mathbf{p}}\tilde{f}) \in |{}^kM| \cap V\left(w_0 - \frac{\partial \tilde{f}}{\partial z_0}, \dots, w_n - \frac{\partial \tilde{f}}{\partial z_n}\right)$ if and only if $\mathbf{x} \in \overline{\Sigma_{{}^k\mathbf{Q}^\bullet} f}$. $\square$

We immediately conclude

Corollary 2.2 (Lê-Iomdine Formulas). *Let $\mathbf{x} \in V(z_0) \cap |\eta_*({}^kM)| = V(z_0) \cap {}^kX$. Let a be a non-zero complex number, and let $j \geqslant 1$ be an integer. Let $\hat{\mathbf{z}}$ denote the "rotated" coordinates $(z_1, \dots, z_n, z_0)$.*

Suppose that all of the k-shifted LêVo numbers at $\mathbf{x}$ of f, ${}^k\lambda^i_{f,\mathbf{z}}(\mathbf{x})$, exist and that $j \geqslant 1 + {}^k\lambda^0_{f,\mathbf{z}}(\mathbf{x})$.

Then, in a neighborhood of $\mathbf{x}$,

i) $\overline{\Sigma_{{}_k\mathbf{Q}^\bullet}(f+az_0^{j+1})} = V(z_0)\cap\overline{\Sigma_{{}_k\mathbf{Q}^\bullet}f}$;

ii) $\dim_{\mathbf{x}}\overline{\Sigma_{{}_k\mathbf{Q}^\bullet}(f+az_0^{j+1})} = \left(\dim_{\mathbf{x}}\overline{\Sigma_{{}_k\mathbf{Q}^\bullet}f}\right)-1$, *provided that* $\dim_{\mathbf{x}}\overline{\Sigma_{{}_k\mathbf{Q}^\bullet}f}\geqslant 1$;

iii) the k-shifted LêVo numbers of $f+az_0^{j+1}$ *with respect to* $\hat{\mathbf{z}}$ *exist; and*

iv) ${}^k\lambda^0_{f+az_0^{j+1},\hat{\mathbf{z}}}(\mathbf{x}) = {}^k\lambda^0_{f,\mathbf{z}}(\mathbf{x}) + j({}^k\lambda^1_{f,\mathbf{z}}(\mathbf{x}))$ *and, for* $1\leqslant i\leqslant n-1$,

$$ {}^k\lambda^i_{f+az_0^{j+1},\hat{\mathbf{z}}}(\mathbf{x}) = j\left({}^k\lambda^{i+1}_{f,\mathbf{z}}(\mathbf{x})\right). $$

Just as our generalized Lê-Saito Theorem of II.6.5 followed immediately by applying the Lê-Iomdine formulas to the actual result of Lê and Saito on families of isolated affine hypersurface singularities, so too does a "super" general Lê-Saito result follow by applying 2.2 above to Corollary III.6.12.

Since we wish to apply 2.2 to families, we first need to introduce some new notation. Let $\overset{\circ}{\mathbb{D}}$ be an open disc about the origin in $\mathbb{C}$, let $\Omega := \overset{\circ}{\mathbb{D}}\times\mathcal{U}$, let $\tilde{t}:\Omega\to\overset{\circ}{\mathbb{D}}$ denote the projection, let $\tilde{g}:\Omega\to\mathbb{C}$ be an analytic function, let $\mathcal{X}$ be a $(d+1)$-dimensional analytic subset of Ω, let t denote the restriction of $\tilde{t}$ to $\mathcal{X}$, and let g denote the restriction of $\tilde{g}$ to $\mathcal{X}$. We suppose that $\overset{\circ}{\mathbb{D}}\times\{\mathbf{0}\}\subseteq\mathcal{X}$ and that $g(\overset{\circ}{\mathbb{D}}\times\{\mathbf{0}\})=0$. For $a\in\overset{\circ}{\mathbb{D}}$, use $\mathbf{z}$ as coordinates on each $\tilde{t}^{-1}(a)\cong\mathcal{U}$, let $X_a := t^{-1}(a)$, and let $g_a : X_a\to\mathbb{C}$ be given by $g_a := g_{|X_a}$.

Theorem 2.3 (General Lê-Saito Theorem). *Suppose that* $\mathbf{0}\notin\overline{\Sigma_{\mathbb{C}}t}$.

Let S_α *be a visible stratum of* $\mathcal{X}$ *of dimension* d_α *such that* $g_{|S_\alpha}\neq 0$, *and let* j *be an integer such that* $\tilde{b}_{j-1}(\mathbb{L}_\alpha)\neq 0$. *Let* $Y:=\overline{S_\alpha}$ *and let* $k:=d-d_\alpha-j$. *In particular,* Y *could be any irreducible component of* $\mathcal{X}$, j *could be zero, and* k *would be* 0.

Suppose that, for all i, ${}^k\lambda^i_{g_a,\mathbf{z}}(\mathbf{0})$ *is independent of* a, *for all small* a, *and that either*

a) $\mathbf{0}\in\Sigma_{\text{Nash}}(g_{|Y})$; *or that*

b) $\mathbf{0}\notin\Sigma_{\text{cnr}}(g_{|Y})$, *and* $(Y_{\text{reg}},\overset{\circ}{\mathbb{D}}\times\{\mathbf{0}\})$ *satisfies Whitney's condition a) at* $\mathbf{0}$.

Then, the pair $(Y_{\text{reg}},\overset{\circ}{\mathbb{D}}\times\{\mathbf{0}\})$ *satisfies the* a_g *condition at* $\mathbf{0}$. *Moreover, in case a), there exists an* i *such that* ${}^k\lambda^i_{g_a,\mathbf{z}}(\mathbf{0})\neq 0$.

Proof. The argument in case b) is exactly that of III.6.12; so, assume that we are in case a).

Let $s := \dim_{\mathbf{0}} \Sigma_{\mathcal{C}}(g_0)$. Let $0 \ll j_0 \ll \cdots \ll j_s$, and let $h := g + z_0^{j_0} + \cdots + z_s^{j_s}$. Certainly, $h_{|S_\alpha} \not\equiv 0$, since $g_{|S_\alpha} \not\equiv 0$.

Consider the family $h_a := g_a + z_0^{j_0} + \cdots + z_s^{j_s}$. By 2.2.ii, $\dim_{\mathbf{0}} \Sigma_{\mathcal{C}}(h_0) = 0$ and so, by III.6.7, $\dim_{\mathbf{0}} \Sigma_{\mathcal{C}}(h_a) \leqslant 0$ for all small a. By 1.5 and 1.13, ${}^k\lambda^0_{h_a,\mathbf{z}}(\mathbf{0}) = \tilde{b}_{d_\alpha+j-2}(F_{h_a,\mathbf{0}})$. By an inductive application of the Lê-Iomdine formulas, ${}^k\lambda^0_{h_a,\mathbf{z}}(\mathbf{0})$ is a function of only $\{{}^k\lambda^i_{g_a,\mathbf{z}}(\mathbf{0})\}_i$ (and the fixed j's); thus, $\tilde{b}_{d_\alpha+j-2}(F_{h_a,\mathbf{0}})$ is independent of a for small a.

It is trivial to show that, since $\mathbf{0} \in \Sigma_{\text{Nash}}(g_{|Y})$, $\mathbf{0} \in \Sigma_{\text{Nash}}(h_{|Y})$. Therefore, we may apply III.6.2 to conclude that $(Y_{\text{reg}}, \mathring{\mathbb{D}} \times \{\mathbf{0}\})$ satisfies the a_h condition at $\mathbf{0}$ and $\tilde{b}_{d_\alpha+j-2}(F_{h_a,\mathbf{0}}) \neq 0$.

As ${}^k\lambda^0_{h_a,\mathbf{z}}(\mathbf{0}) = \tilde{b}_{d_\alpha+j-2}(F_{h_a,\mathbf{0}}) \neq 0$, the Lê-Iomdine formulas imply that there exists an i such that ${}^k\lambda^i_{g_a,\mathbf{z}}(\mathbf{0}) \neq 0$.

As in II.6.5, by inducting, we would be finished if we could show that:

$\mathbf{0} \in \Sigma_{\text{Nash}}(g_{|Y})$ and $[\eta] \in \mathbb{P}\big(\overline{T^*_{\tilde{g}_{|Y_{\text{reg}}}}\Omega}\big)_{\mathbf{0}}$ implies that, for all j sufficiently large,

$$[\eta] \in \mathbb{P}\big(\overline{T^*_{(\tilde{g}+z_0^j)_{|Y_{\text{reg}}}}\Omega}\big)_{\mathbf{0}}.$$

By III.5.2, what we need to show is that $[\eta] \in p(E_{\tilde{g}})_{\mathbf{0}}$ implies that, for all j sufficiently large, $[\eta] \in \mathbb{P}\big(\overline{T^*_{(\tilde{g}+z_0^j)_{|Y_{\text{reg}}}}\Omega}\big)_{\mathbf{0}}$, where $E_{\tilde{g}}$ denotes the exceptional divisors of $\text{Bl}_{\text{im}\, d\tilde{g}} \overline{T^*_{Y_{\text{reg}}}\Omega}$, and p denotes the projection $\Omega \times \mathbb{C}^{n+2} \times \mathbb{P}^{n+1} \to \Omega \times \mathbb{P}^{n+1}$.

Now, $[\eta] \in p(E_{\tilde{g}})_{\mathbf{0}}$ if and only if there exists an analytic path $\gamma(u) = (\mathbf{x}(u), \omega(u))$ in $\overline{T^*_{Y_{\text{reg}}}\Omega}$ such that $\gamma(0) = (\mathbf{0}, d_{\mathbf{0}}\tilde{g})$, $\gamma(u) \in \overline{T^*_{Y_{\text{reg}}}\Omega} - \text{im}\, d\tilde{g}$, and $[\omega(u) - d_{\mathbf{x}(u)}\tilde{g}] \to [\eta]$ as $u \to 0$. That $[\omega(u) - d_{\mathbf{x}(u)}\tilde{g}] \to [\eta]$ is equivalent to

$$\xi \cdot \frac{\omega(u) - d_{\mathbf{x}(u)}\tilde{g}}{|\omega(u) - d_{\mathbf{x}(u)}\tilde{g}|} \to \frac{\eta}{|\eta|},$$

for some root of unity ξ. One concludes easily that

$$\frac{|\eta|\xi}{|\omega(u) - d_{\mathbf{x}(u)}(\tilde{g} + z_0^j)|}\big(\omega(u) - d_{\mathbf{x}(u)}(\tilde{g} + z_0^j)\big) \;\to\; \eta,$$

for all large j. However, the terms on the left side of the above expression are clearly elements of $\big(\overline{T^*_{(\tilde{g}+z_0^j)_{|Y_{\text{reg}}}}\Omega}\big)_{\mathbf{x}(u)}$ whose projective class approaches that of η. $\square$

Chapter 3. LE-VOGEL CYCLES AND THE EULER CHARACTERISTIC

In this final chapter, we will relate the Lê-Vogel cycles to the Euler characteristic of the Milnor fibre, in a way that generalizes our result in II.10.3. In order to accomplish this, we must recall a definition and a result from [**Mas11**].

We continue with the notation from the previous two chapters.

Proposition/Definition 3.1. *Let* $\mathbf{p} \in X$ *and let* $\mathbf{F}^\bullet$ *be a bounded, constructible complex on* X*. Then, for a generic choice of the coordinates* $\mathbf{z}$*, there exists an open neighborhood,* $\mathcal{W}$*, of* $\mathbf{p}$ *and cycles* $\Lambda^i_{\mathbf{F}^\bullet,\mathbf{z}}$ *in* $\mathcal{W}$ *such that each* $\Lambda^i_{\mathbf{F}^\bullet,\mathbf{z}}$ *is purely* i*-dimensional,* $\Lambda^i_{\mathbf{F}^\bullet,\mathbf{z}}$ *properly intersects* $V(z_0 - z_0(\mathbf{x}), \dots, z_{i-1} - z_{i-1}(\mathbf{x}))$ *and, for all* $\mathbf{x} \in \mathcal{W}$,

$$\chi(\mathbf{F}^\bullet)_\mathbf{x} = (-1)^d \sum_i (-1)^i \left(\Lambda^i_{\mathbf{F}^\bullet,\mathbf{z}} \cdot V(z_0 - z_0(\mathbf{x}), \dots, z_{i-1} - z_{i-1}(\mathbf{x}))\right)_\mathbf{x},$$

(here, when $i = 0$*, we mean that the intersection number is simply* $\left(\Lambda^0_{\mathbf{F}^\bullet,\mathbf{z}}\right)_\mathbf{x}$*).*

Moreover, whenever such cycles exist, they are unique.

In the case where $\dim_\mathbf{p} \operatorname{supp} \mathbf{F}^\bullet = 1$*, such cycles exist if*

$$\dim_\mathbf{p} \left(V(z_0 - z_0(\mathbf{p})) \cap \operatorname{supp} \mathbf{F}^\bullet\right) \leqslant 0.$$

We call $\Lambda^i_{\mathbf{F}^\bullet,\mathbf{z}}$ the *i-dimensional characteristic polar cycle of* $\mathbf{F}^\bullet$ *with respect to* $\mathbf{z}$, and refer to $\lambda^i_{\mathbf{F}^\bullet,\mathbf{z}}(\mathbf{x}) := \left(\Lambda^i_{\mathbf{F}^\bullet,\mathbf{z}} \cdot V(z_0 - z_0(\mathbf{x}), \dots, z_{i-1} - z_{i-1}(\mathbf{x}))\right)_\mathbf{x}$ as the *i-th characteristic polar number of* $\mathbf{F}^\bullet$ *with respect to* $\mathbf{z}$ *at* $\mathbf{x}$.

Proof. The statement about the case where $\dim_\mathbf{p} \operatorname{supp} \mathbf{F}^\bullet \leqslant 1$ is trivial. The remaining statements are a combination of Propositions 2.4 and 3.1 from [**Mas11**]. □

Below, we refer to the absolute polar varieties of Lê and Teissier ([**L-T2**], [**Te4**], [**Te5**]); however, we need to explain our notation. We let $\Gamma^i_\mathbf{z}(\overline{S_\alpha})$ denote the i-dimensional polar variety of $\overline{S_\alpha}$ with respect to the flag

$$\{\mathbf{0}\} \subseteq V(z_0, z_1, \dots, z_{n-1}) \subseteq \dots V(z_0, z_1) \subseteq V(z_0) \subseteq \mathbb{C}^{n+1}.$$

Thus, as a set, $\Gamma^i_\mathbf{z}(\overline{S_\alpha}) = \overline{\operatorname{crit}\left((z_0, \dots z_i)_{|_{(\overline{S_\alpha})_{\mathrm{reg}}}}\right)}$. See also our treatment in Section 7 of [**Mas11**].

Theorem 3.2. *Let* $\mathbf{p} \in X$ *and let* $\mathbf{F}^\bullet$ *be a bounded complex on* X*, which is constructible with respect to* $\{S_\alpha\}$*. Suppose that* $\operatorname{Ch}(\mathbf{F}^\bullet) = \sum_\alpha m_\alpha \left[\overline{T^*_{S_\alpha} \mathcal{U}}\right]$.

Then, for a generic choice of the coordinates $\mathbf{z}$, *there exists an open neighborhood,* $\mathcal{W}$, *of* $\mathbf{p}$ *in which all of the* $\Lambda^i_{\mathbf{F}^\bullet,\mathbf{z}}$ *exist, such that, for all* i, $\mathbb{P}(\mathrm{Ch}(\mathbf{F}^\bullet))$ *and* $\mathcal{W}\times(\mathbb{P}^i\times\{\mathbf{0}\})$ *intersect properly in* $\mathbb{P}(T^*\mathcal{W})$, *and the restriction of* ν *to* $\mathcal{W}\times\mathbb{P}^n$ *yields the equalities*

$$\Lambda^i_{\mathbf{F}^\bullet,\mathbf{z}} = \sum_\alpha m_\alpha \Gamma^i_{\mathbf{z}}(\overline{S_\alpha}) = \nu_*\big(\mathbb{P}(\mathrm{Ch}(\mathbf{F}^\bullet)) \;\cdot\; (\mathcal{W}\times(\mathbb{P}^i\times\{\mathbf{0}\}))\big).$$

In the case where $Y := \operatorname{supp}\mathbf{F}^\bullet$ *is* 1*-dimensional at* $\mathbf{p}$, *the conclusions hold if* z_0 *is finite at* $\mathbf{p}$, *i.e., if* $\dim_{\mathbf{p}}(V(z_0 - z_0(\mathbf{p}))\cap Y)_{\mathbf{p}} = 0$.

Proof. Aside from the statement about the case where $\dim_{\mathbf{p}}\operatorname{supp}\mathbf{F}^\bullet = 1$, this follows immediately from Theorem 7.5 of [**Mas11**]. However, when $\dim_{\mathbf{p}}\operatorname{supp}\mathbf{F}^\bullet = 1$, the hypotheses of 7.5 of [**Mas11**] are strictly stronger than saying that z_0 is finite at $\mathbf{p}$, so we must provide a proof of this portion.

Assume that $Y := \operatorname{supp}\mathbf{F}^\bullet$ is 1-dimensional at $\mathbf{p}$ and that z_0 is finite at $\mathbf{p}$. Let C denote a component of Y through $\mathbf{p}$, and let $\mathbf{x}_C$ denote a point in $C - \{\mathbf{p}\}$. By shrinking our neighborhood, we may assume that $C-\{\mathbf{p}\}$ is smooth, and that $\chi(\mathbf{F}^\bullet)_{\mathbf{x}_C}$ is independent of the choice of $\mathbf{x}_C$. As z_0 is not constant along C, we may also assume that $(\mathbf{x}, d_{\mathbf{x}}z_0)\notin \overline{T^*_{C_{\mathrm{reg}}}\mathcal{U}}$ for $\mathbf{x}\neq\mathbf{p}$. It follows at once that $\mathbb{P}(\mathrm{Ch}(\mathbf{F}^\bullet))$ and $\mathcal{W}\times(\mathbb{P}^i\times\{\mathbf{0}\})$ intersect properly for all i. Also, since the hypotheses of 7.5 of [**Mas11**] are satisfied at all $\mathbf{x}\neq\mathbf{p}$, 7.5 of [**Mas11**] implies that

$$\nu_*\big(\mathbb{P}(\mathrm{Ch}(\mathbf{F}^\bullet)) \;\cdot\; (\mathcal{W}\times(\mathbb{P}^1\times\{\mathbf{0}\}))\big) = \sum_C \chi(\mathbf{F}^\bullet)_{\mathbf{x}_C}[C].$$

Moreover, by applying Theorem III.3.10 to the function z_0, we find that

$$\nu_*\big(\mathbb{P}(\mathrm{Ch}(\mathbf{F}^\bullet)) \;\cdot\; (\mathcal{W}\times(\mathbb{P}^0\times\{\mathbf{0}\}))\big) = \big(\mathrm{Ch}(\mathbf{F}^\bullet)\cdot\operatorname{im} dz_0\big)_{\mathbf{p}}[\mathbf{p}] = \chi(\phi_{z_0-z_0(\mathbf{p})}\mathbf{F}^\bullet)_{\mathbf{p}}[\mathbf{p}].$$

As

$$\chi(\phi_{z_0-z_0(\mathbf{p})}\mathbf{F}^\bullet)_{\mathbf{p}} = \chi(\psi_{z_0-z_0(\mathbf{p})}\mathbf{F}^\bullet)_{\mathbf{p}} - \chi(\mathbf{F}^\bullet)_{\mathbf{p}} = -\chi(\mathbf{F}^\bullet)_{\mathbf{p}} + \sum_C (C\cdot V(z_0 - z_0(\mathbf{p})))_{\mathbf{p}}\,\chi(\mathbf{F}^\bullet)_{\mathbf{x}_C},$$

the conclusion follows. □

We relate the LêVo cycles and numbers to the Euler characteristic by the following theorem.

Theorem 3.3. *Let* $\mathbf{p}\in X$ *and assume that* $f(\mathbf{p}) = 0$. *Fix a* k *and let* $\mathbf{F}^\bullet := \phi_f({}^k\mathbf{Q}^\bullet)$. *Then, for a generic choice of the coordinates* $\mathbf{z}$, *there exists an open neighborhood,* $\mathcal{W}$, *of* $\mathbf{p}$ *in which all of the* $\Lambda^i_{\mathbf{F}^\bullet,\mathbf{z}}$ *exist, all of the* k*-shifted LêVo numbers,* ${}^k\lambda^i_{f,\mathbf{z}}$, *exist, and such that* ${}^k\Lambda^i_{f,\mathbf{z}} = (-1)^d\Lambda^i_{\mathbf{F}^\bullet,\mathbf{z}}$ *and* ${}^k\lambda^i_{f,\mathbf{z}}(\mathbf{x}) = (-1)^d\lambda^i_{\mathbf{F}^\bullet,\mathbf{z}}(\mathbf{x})$ *for all* $\mathbf{x}\in\mathcal{W}$, *i.e.,*

$$\chi\big(\phi_f[-1]({}^k\mathbf{Q}^\bullet)\big)_{\mathbf{x}} = \sum_i (-1)^i\, {}^k\lambda^i_{f,\mathbf{z}}(\mathbf{x}).$$

In the case where $Y := \operatorname{supp} \mathbf{F}^\bullet$ *is* 1*-dimensional at* $\mathbf{p}$*, the conclusions hold if*

$$\dim_{\mathbf{p}} \left(V(z_0 - z_0(\mathbf{p})) \cap Y\right)_{\mathbf{p}} = 0.$$

Proof. Throughout the proof, we will work in an arbitrarily small neighborhood of $\mathbf{p}$.
From 1.10.iii, we have

$${}^k\Lambda^i_{f,\mathbf{z}} = (-1)^{\hat{d}}\, \nu_*\Big(\mathbb{P}(\operatorname{Ch}(\phi_f({}^k\mathbf{Q}^\bullet))) \;\cdot\; (\mathcal{W} \times (\mathbb{P}^i \times \{\mathbf{0}\}))\Big).$$

Hence, by 3.2, ${}^k\Lambda^i_{f,\mathbf{z}} = (-1)^{\hat{d}}\, \Lambda^i_{\mathbf{F}^\bullet,\mathbf{z}}$ and ${}^k\lambda^i_{f,\mathbf{z}}(\mathbf{x}) = (-1)^{\hat{d}}\, \lambda^i_{\mathbf{F}^\bullet,\mathbf{z}}(\mathbf{x})$.

By definition, $\chi(\mathbf{F}^\bullet)_{\mathbf{x}} = (-1)^{\hat{d}-1} \sum_i (-1)^i \lambda^i_{\mathbf{F}^\bullet,\mathbf{z}}(\mathbf{x})$. Therefore,

$$\chi(\mathbf{F}^\bullet)_{\mathbf{x}} = -\sum_i (-1)^i\, {}^k\lambda^i_{f,\mathbf{z}}(\mathbf{x}),$$

or, equivalently,

$$\chi(\mathbf{F}^\bullet[-1])_{\mathbf{x}} = \sum_i (-1)^i\, {}^k\lambda^i_{f,\mathbf{z}}(\mathbf{x}). \quad \square$$

Before we can connect Theorem 3.3 to the ordinary cohomology of the Milnor fibre with constant coeeficients, we need to prove a lemma. While we suspect that this lemma is well-known, we can find no reference.

Lemma 3.4. *Let* $\mathbf{F}^\bullet$ *be a bounded, constructible complex on* X*. Then, for all* $\mathbf{x} \in X$*,*

$$\chi(\mathbf{F}^\bullet)_{\mathbf{x}} = \sum_k (-1)^k \chi\left({}^\mu H^0(\mathbf{F}^\bullet[k])\right)_{\mathbf{x}}.$$

Proof. For convenience, assume that $\mathbf{x} = \mathbf{0}$. The proof is by induction on the dimension of X at $\mathbf{0}$.

If $\dim_{\mathbf{0}} X = 0$, then $H^i({}^\mu H^0(\mathbf{F}^\bullet[k]))_{\mathbf{0}} = 0$ unless $i = 0$, and then

$$H^0({}^\mu H^0(\mathbf{F}^\bullet[k]))_{\mathbf{0}} \cong H^0(\mathbf{F}^\bullet[k])_{\mathbf{0}} = H^k(\mathbf{F}^\bullet)_{\mathbf{0}}.$$

Thus, the lemma holds if $\dim_{\mathbf{0}} X = 0$.

Now, assume the lemma for spaces of dimension j, and suppose that $\dim_{\mathbf{0}} X = j+1$. Let L be a generic linear form. Consider the distinguished triangle

$$\left({}^{\mu}H^0(\mathbf{F}^\bullet[k])\right)_{|V(L)}[-1] \longrightarrow \psi_L[-1]{}^{\mu}H^0(\mathbf{F}^\bullet[k]) \longrightarrow \phi_L[-1]{}^{\mu}H^0(\mathbf{F}^\bullet[k]) \xrightarrow{[1]};$$

it yields the equality

$$\chi\big(\psi_L[-1]{}^{\mu}H^0(\mathbf{F}^\bullet[k])\big)_{\mathbf{0}} = \chi\Big(\big({}^{\mu}H^0(\mathbf{F}^\bullet[k])\big)_{|V(L)}[-1]\Big)_{\mathbf{0}} + \chi\big(\phi_L[-1]{}^{\mu}H^0(\mathbf{F}^\bullet[k])\big)_{\mathbf{0}}.$$

Thus,

$$\chi\big({}^{\mu}H^0(\psi_L[-1]\mathbf{F}^\bullet[k])\big)_{\mathbf{0}} = -\chi\big({}^{\mu}H^0(\mathbf{F}^\bullet[k])\big)_{\mathbf{0}} + \chi\big({}^{\mu}H^0(\phi_L[-1]\mathbf{F}^\bullet[k])\big)_{\mathbf{0}},$$

and so

$$\sum_k(-1)^k\chi\big({}^{\mu}H^0(\psi_L[-1]\mathbf{F}^\bullet[k])\big)_{\mathbf{0}} =$$

$$-\sum_k(-1)^k\chi\big({}^{\mu}H^0(\mathbf{F}^\bullet[k])\big)_{\mathbf{0}} + \sum_k(-1)^k\chi\big({}^{\mu}H^0(\phi_L[-1]\mathbf{F}^\bullet[k])\big)_{\mathbf{0}}.$$

Applying our inductive hypothesis twice, we obtain

$$\sum_k(-1)^k\chi\big({}^{\mu}H^0(\mathbf{F}^\bullet[k])\big)_{\mathbf{0}} = \chi\big(\phi_L[-1]\mathbf{F}^\bullet\big)_{\mathbf{0}} - \chi\big(\psi_L[-1]\mathbf{F}^\bullet\big)_{\mathbf{0}} =$$

$$-\chi\big(\phi_L\mathbf{F}^\bullet\big)_{\mathbf{0}} + \chi\big(\psi_L\mathbf{F}^\bullet\big)_{\mathbf{0}} = \chi(\mathbf{F}^\bullet)_{\mathbf{0}},$$

and we are finished. □

The relationship between the characteristic polar cycles of $\phi_f\mathbb{C}^\bullet_X$ and the k-shifted LêVo cycles is given in

Theorem 3.5. *Let $\mathbf{p} \in X$ and suppose that $f(\mathbf{p}) = 0$.*

Then, for a generic choice of the coordinates $\mathbf{z}$, for all k, there exists an open neighborhood, $\mathcal{W}$, of $\mathbf{p}$ in which all of the $\Lambda^i_{\phi_f\mathbb{C}^\bullet_X,\mathbf{z}}$ exist, all of the k-shifted LêVo numbers, ${}^k\lambda^i_{f,\mathbf{z}}$, exist, and such that

$$\Lambda^i_{\phi_f\mathbb{C}^\bullet_X,\mathbf{z}} = (-1)^{d-\hat{d}}\sum_k(-1)^k\, {}^k\Lambda^i_{f,\mathbf{z}}$$

and, for all $\mathbf{x} \in \mathcal{W} \cap V(f)$,

$$\widetilde{\chi}(F_{f,\mathbf{x}}) = \sum_k(-1)^k\sum_i(-1)^{d-i-1}\, {}^k\lambda^i_{f,\mathbf{z}}(\mathbf{x}),$$

where $F_{f,\mathbf{x}}$ denotes the Milnor fibre of f at $\mathbf{x}$.

In the case where $Y := \overline{\Sigma_{\mathbb{C}}f}$ has dimension 1 at $\mathbf{p}$, the conclusions hold for all k provided that $\dim_{\mathbf{p}}\big(V(z_0 - z_0(\mathbf{p})) \cap Y\big)_{\mathbf{p}} = 0$.

Proof. Using 3.3 and 3.4,

$$\widetilde{\chi}(F_{f,\mathbf{x}}) = \chi(\phi_f\mathbb{C}^\bullet_X)_\mathbf{x} = \sum_k(-1)^k\chi\big({}^\mu H^0(\phi_f\mathbb{C}^\bullet_X[k])\big)_\mathbf{x} =$$

$$\sum_k(-1)^k\chi\big(\phi_f[-1]^\mu H^0(\mathbb{C}^\bullet_X[k+1])\big)_\mathbf{x} =$$

$$\sum_k(-1)^k\chi\big(\phi_f[-1]\,{}^{d-k-1}\mathbf{Q}^\bullet\big)_\mathbf{x} = \sum_k(-1)^k\sum_i(-1)^i\,{}^{d-k-1}\lambda^i_{f,\mathbf{z}}(\mathbf{x}) =$$

$$\sum_k(-1)^k\sum_i(-1)^{d-i-1}\,{}^k\lambda^i_{f,\mathbf{z}}(\mathbf{x}).$$

That $\Lambda^i_{\phi_f\mathbb{C}^\bullet_X,\mathbf{z}} = (-1)^{d-\hat{d}}\sum_k(-1)^k\,{}^k\Lambda^i_{f,\mathbf{z}}$ follows immediately. □

Example 3.6. Recall Example 1.14 where $X = V(u,x)\cup V(y,z)\subseteq\mathbb{C}^4$, and

$$\tilde{f} := (u^\alpha + x^\beta)^\tau + y^\gamma + z^\delta,$$

where $\alpha,\beta,\gamma,\delta,\tau\geqslant 2$.

Applying Theorem 3.5, we obtain

$$\widetilde{\chi}(F_{f,\mathbf{0}}) = -{}^0\lambda^0_{f,\mathbf{z}}(\mathbf{0}) + {}^0\lambda^1_{f,\mathbf{z}}(\mathbf{0}) + {}^1\lambda^0_{f,\mathbf{z}}(\mathbf{0}) =$$

$$-(\gamma-1)(\delta-1) - (\beta-1)(\alpha\tau-1) + \beta(\tau-1) + 1 \;=\; -(\gamma-1)(\delta-1) + \tau(-\alpha\beta+\alpha+\beta).$$

We can verify this calculation. One easily sees that $F_{f,\mathbf{0}}$ is the disjoint union of F_1, the Milnor fibre of $(u^\alpha+x^\beta)^\tau$ restricted to $V(x,y)$, and F_2, the Milnor fibre of $y^\gamma+z^\delta$ restricted to $V(u,x)$. Thus, F_1 is homotopy-equivalent to the disjoint union of τ copies of a bouquet of $(\alpha-1)(\beta-1)$ circles, and F_2 is homotopy-equivalent to a bouquet of $(\gamma-1)(\delta-1)$ circles. Therefore,

$$\widetilde{\chi}(F_{f,\mathbf{0}}) = -\tau(\alpha-1)(\beta-1) - (\gamma-1)(\delta-1) + \tau + 1 - 1,$$

where the last -1 is due to the fact that we use the **reduced** cohomology. One sees, then, that the calculations agree.

Appendix A: Analytic Cycles and Intersections

We wish to consider schemes, cycles, and sets. Frequently, we will be in the algebraic setting and, hence, we may use algebraic schemes, cycles, and sets. However, as we wish to treat the more general analytic case, we should clarify what we mean by the terms scheme and cycle.

In the analytic setting, by *scheme*, we actually mean a (not necessarily reduced) complex analytic space, $(X, \mathcal{O}_X)$, in the sense of [**G-R1**] and [**G-R2**]. By the irreducible components of X, we mean simply the irreducible components of the underlying analytic set X. If we concentrate our attention on the germ of X at some point $\mathbf{p}$, then we may discuss embedded subvarieties and (non-embedded, or isolated) components of the germ of X at $\mathbf{p}$ – these correspond to non-minimal and minimal primes, respectively, in the set of associated primes of the Noetherian local ring $\mathcal{O}_{X,\mathbf{p}}$.

If X is a complex space and α is a coherent sheaf of ideals in $\mathcal{O}_X$, then we write $V(\alpha)$ for the possibly non-reduced analytic subspace defined by the vanishing of α.

By the intersection of a collection of closed subschemes, we mean the scheme defined by the sum of the underlying ideal sheaves. By the union of a finite collection of closed subschemes, we mean the scheme defined by the intersection (not the product) of the underlying ideal sheaves. We say that two subschemes, V and W, are equal up to embedded subvariety provided that, in each stalk, the isolated components of the defining ideals (those corresponding to minimal primes) are equal. Our main concern with this last notion is that it implies that the cycles [V] and [W] are equal (see below).

A.1 Given an analytic space X (with its reduced structure), an *analytic cycle* in X is a formal sum $\sum m_V[V]$, where the V's are irreducible analytic subsets of X, the m_V's are integers, and the collection $\{V\}$ is a locally finite collection of subsets of X. As a cycle is a locally finite sum, and as we will normally be concentrating on the germ of an analytic space at a point, usually we can safely assume that a cycle is actually a finite formal sum.

Throughout this book, whenever we write a cycle $\sum m_V[V]$, we shall assume that the V's are distinct and that none of the m_V's are zero. This is the same as saying that the presentation is minimal, in the sense that no further cancellations are possible.

We say that a cycle $\sum m_V[V]$ is *positive* if $m_V > 0$ for all V; a cycle is *non-negative* if it is the zero-cycle or is positive.

A.2 Given an analytic space, $(X, \mathcal{O}_X)$, we wish to define the (positive) cycle associated to $(X, \mathcal{O}_X)$. In the algebraic context, this is given by Fulton in [**Fu**, 1.5] as

$$[X] := \sum m_V[V],$$

where the V's run over all the irreducible components of X, and m_V equals the length of the ring $\mathcal{O}_{X,V}$, the local ring of X along V. In the analytic context, we wish to use the same definition, but we must be more careful in defining the m_V.

Define m_V as follows. Take a point $\mathbf{p}$ in V. The germ of V at $\mathbf{p}$ breaks up into irreducible germ components $(V_\mathbf{p})_i$. Take any one of the $(V_\mathbf{p})_i$ and let m_V equal the length of the ring $(\mathcal{O}_{X,\mathbf{p}})_{(V_\mathbf{p})_i}$ (that is, the local ring of X at $\mathbf{p}$ localized at the prime corresponding to $(V_\mathbf{p})_i$). This number is independent of the point $\mathbf{p}$ in V and the choice of $(V_\mathbf{p})_i$.

Note that any embedded subvarieties of a scheme do **not** contribute to the associated cycle.

One can easily show that, if $f, g \in \mathcal{O}_X$, then $[V(fg)] = [V(f)] + [V(g)]$; in particular, $[V(f^m)] = m[V(f)]$.

If Y is an analytic subset of X and C is a cycle in Y, then we may naturally consider C as a cycle in X.

We shall be dealing with analytic schemes, cycles, and analytic sets. For clarification of what structure we are considering, we shall at times enclose cycles in square brackets, [ ], and analytic sets in a pair of vertical lines,| |. Occasionally, when the notation becomes cumbersome, we shall simply state explicitly whether we are considering V as a scheme, a cycle, or a set.

We say that two cycles are equal at a point, $\mathbf{p}$, provided that the portions of each cycle which pass through $\mathbf{p}$ are equal. When we say that a space, X, is purely k-dimensional at a point, $\mathbf{p}$, we mean to allow for the vacuous case where X has no components through $\mathbf{p}$.

We wish to describe some aspects of intersection theory. Of course, [**Fu**] is the definitive reference for this subject. However, we deal only with cycles, **not** cycle classes, and we deal only with **proper** intersections inside complex **manifolds**; this makes much of the theory fairly trivial to describe.

A.3 If V and W are irreducible subschemes of a connected complex manifold, M, and Z is an irreducible component of $V \cap W$ such that $\operatorname{codim}_M Z = \operatorname{codim}_M V + \operatorname{codim}_M W$, then we say that V *and* W *intersect properly along* Z, or that Z *is a proper component of* $V \cap W$. Two irreducible subschemes V and W in a connected complex manifold, M, are said to *intersect properly* in M provided that they intersect properly along each component of $V \cap W$; when this is the case, the *intersection product*, $([V] \cdot [W]; M)$, of $[V]$ and $[W]$ in M is characterized axiomatically by four properties listed below: openness, transversality, projection, and continuity (see [**Fu**], Example 11.4.4).

A.4 If α is a coherent sheaf of ideals in $\mathcal{O}_M$ and $f \in \mathcal{O}_M$ is such that $V(f)$ contains no embedded subvarieties or irreducible components of $V(\alpha)$, then $V(\alpha)$ and $V(f)$ intersect properly in M and $[V(\alpha)] \cdot [V(f)] = [V(\alpha + \langle f \rangle)]$ (see [**Fu**], 7.1.b). This statement immediately implies one which, a priori, seems stronger: if α and f are as before and $V(f)$ contains no irreducible component of $V(\alpha)$ and contains no embedded subvariety **which is of codimension one inside some irreducible component of** $V(\alpha)$, then $V(\alpha)$ and $V(f)$ intersect properly in M and $[V(\alpha)] \cdot [V(f)] = [V(\alpha + \langle f \rangle)]$

More generally, if $W := V(\alpha)$ is a subscheme of M and $f_1, \ldots, f_k \in \mathcal{O}_M$ determine regular sequences in the stalks $\mathcal{O}_{W,\mathbf{p}}$ and $\mathcal{O}_{M,\mathbf{p}}$ at all points $\mathbf{p} \in W \cap V(f_1, \ldots, f_k)$, then

$$[V(\alpha)] \cdot [V(f_1, \ldots, f_k)] = [V(\alpha + \langle f_1, \ldots, f_k \rangle)].$$

The two paragraphs above allow one to define the intersection with a hypersurface (or, more generally, a Cartier divisor) without having to refer to an ambient manifold. Suppose that $V(\alpha)$ is a subscheme of an analytic space X, that X is contained in an analytic manifold M, and that $f \in \mathcal{O}_X$. Then, locally, $\mathcal{O}_X \cong \mathcal{O}_M/\gamma$ for some coherent sheaf of ideals $\gamma \subseteq \mathcal{O}_M$. Let $\tilde{\alpha} \subseteq \mathcal{O}_M$ be a coherent sheaf of ideals such that $\gamma \subseteq \tilde{\alpha}$ and such that $\tilde{\alpha}/\gamma$ corresponds to α, i.e., $\tilde{\alpha}$ is such that $V(\alpha) = V(\tilde{\alpha})$. Let $\tilde{f}$ be an extension of f to M. If $V(f)$ contains no embedded subvarieties or isolated components of $V(\alpha)$, then $V(\tilde{f})$ contains no embedded subvarieties or isolated components of $V(\tilde{\alpha})$ and so, by the previous paragraph, $\big([V(\tilde{\alpha})]\cdot[V(\tilde{f})];\ M\big) = [V(\tilde{\alpha} + < \tilde{f} >)]$, which defines the same cycle in X as does $[V(\alpha + < f >)]$. Therefore, we may unambiguously define $([V(\alpha)] \cdot [V(f)];\ X)$ by setting it equal to $[V(\alpha + < f >)]$.

A.5 Two cycles $\sum m_i[V_i]$ and $\sum n_j[W_j]$ are said to intersect properly if V_i and W_j intersect properly for all i and j; when this is the case, the intersection product is extended bilinearly by defining

$$\sum m_i[V_i] \cdot \sum n_j[W_j] = \sum m_i n_j \left([V_i] \cdot [W_j]\right).$$

Occasionally it is useful to include the ambient manifold in the notation; in these cases we write $(C_1 \cdot C_2;\ M)$ for the proper intersection of cycles C_1 and C_2 in M.

If two cycles C_1 and C_2 intersect properly and $C_1 \cdot C_2 = \sum n_k[Z_k]$, where the Z_k are irreducible, then the *intersection number* of C_1 and C_2 at Z_k, $(C_1 \cdot C_2)_{Z_k}$, is defined to be n_k; that is, the number of times Z_k occurs in the intersection, counted with multiplicity. Note that, when C_1 and C_2 have complementary codimensions, all the Z_k are merely points.

If V is irreducible at $\mathbf{p}$, then the multiplicity of V at $\mathbf{p}$, $\mathrm{mult}_{\mathbf{p}}V$, is the minimum value of $([V] \cdot [W])_{\mathbf{p}}$, where W ranges over all analytic subsets which are irreducible at $\mathbf{p}$ and which have $\mathbf{p}$ as a component of the proper intersection of V and W; in fact, when working in affine space, W may be chosen to be a generic affine linear subspace through $\mathbf{p}$ of dimension complementary to that of V.

Suppose that $\mathbf{p}$ is an isolated point in the proper intersection of V and W, where V and W are irreducible. Then, $([V] \cdot [W])_{\mathbf{p}} \geqslant \big(\mathrm{mult}_{\mathbf{p}}V\big)\big(\mathrm{mult}_{\mathbf{p}}W\big)$ with equality holding if and only if the projectivized tangent cones $\mathbb{P}(T_{\mathbf{p}}V)$ and $\mathbb{P}(T_{\mathbf{p}}W)$ are disjoint.

A.6 It is fundamental that $(C_1 \cdot C_2)_{Z_k}$ can be calculated locally; that is, if $\mathcal{U}$ is an open subset of M such that $Z_k \cap \mathcal{U} \neq \emptyset$, then

(openness) $$(C_1 \cap \mathcal{U} \ \cdot\ C_2 \cap \mathcal{U};\ \mathcal{U})_{Z_k \cap \mathcal{U}} = (C_1 \cdot C_2;\ M)_{Z_k}$$

(see [**Fu**], 11.4.4).

A.7 If V and W are two irreducible subvarieties of M and P is an irreducible component of $V \cap W$, we say that V and W are *generically transverse along P in M* provided that V and W are reduced and, at generic points of P, V and W are smooth and intersect transversely in M; naturally, we say that V and W are *generically transverse* in M provided they are generically transverse along every component of the intersection. Another fundamental property of intersection numbers is the **transversality characterization**:

if V and W are irreducible subschemes of M which intersect properly along an irreducible component P, then $(V \cdot W)_P = 1$ if and only if V and W are generically transverse along P in M ([**Fu**], 8.2.c and 11.4.4).

A.8 If C_1, C_2, and C_3 are positive cycles such that C_1 and C_2 intersect properly, and C_3 properly intersects $C_1 \cdot C_2$, then C_2 and C_3 intersect properly, C_1 properly intersects $C_2 \cdot C_3$, and

(**associativity**) $$(C_1 \cdot C_2) \cdot C_3 = C_1 \cdot (C_2 \cdot C_3).$$

We wish to introduce a slight generalization of proper intersections of cycles. If V and W are irreducible subschemes of a connected complex manifold, M, and Z is an irreducible component of $V \cap W$ along which V and W intersect properly, then, for every open neighborhood $\mathcal{U} \subseteq M$ such that $V \cap W \cap \mathcal{U} = Z \cap \mathcal{U}$, the value of $(V \cap \mathcal{U} \ \cdot \ W \cap \mathcal{U};\ \mathcal{U})_{Z \cap \mathcal{U}}$ is independent of $\mathcal{U}$; we define $(V \cdot W)_Z$ to be this common value. If Z is a proper component of $V \cap W$, then, by [**Fu**], 8.2.a, $(V \cdot W)_Z \leqslant [V \cap W]_Z$.

We define $[V] \cdot_p [W] := \sum_Z (V \cdot W)_Z$, where the sum is over all Z along which V and W intersect properly. We extend bilinearly

$$\sum m_i[V_i] \cdot_p \sum n_j[W_j] = \sum m_i n_j \left([V_i] \cdot_p [W_j]\right).$$

We refer to this as *the proper intersection* of the two cycles.

One easily verifies that, if C_1, C_2, and C_3 are positive cycles, then

$$(C_1 \cdot_p C_2) \cdot_p C_3 = C_1 \cdot_p (C_2 \cdot_p C_3).$$

A.9 Given a point $\mathbf{p} \in M$, a curve $W = V(\alpha)$ in M which is reduced and irreducible at $\mathbf{p}$, and a hypersurface $V(f) \subseteq M$ which intersects W properly at $\mathbf{p}$, there is a very useful way to calculate the intersection number $([W] \cdot [V(f)])_{\mathbf{p}}$. One takes a local parameterization $\phi(t)$ of W which takes 0 to $\mathbf{p}$, and then $([W] \cdot [V(f)])_{\mathbf{p}} = \operatorname{mult}_t f(\phi(t))$, the degree of the lowest non-zero term. This is easy to see, for composition with ϕ induces an isomorphism

$$\frac{\mathcal{O}_{M,\mathbf{p}}}{\alpha + < f >} \xrightarrow{\circ\phi} \frac{\mathbb{C}\{t\}}{f(\phi(t))}.$$

Of course, if c is small and unequal to zero, $\operatorname{mult}_t f(\phi(t))$ is precisely the number of roots of $f(\phi(t)) - c$ which occur near zero.

More generally, given a point $\mathbf{p} \in M$, a curve $W = V(\alpha)$ in M (which need not be reduced or irreducible at $\mathbf{p}$), and a hypersurface $V(f) \subseteq M$ which intersects W properly at $\mathbf{p}$, consider the map given by multiplication by f

$$\frac{\mathcal{O}_{M,\mathbf{p}}}{\alpha} \xrightarrow{\cdot f} \frac{\mathcal{O}_{M,\mathbf{p}}}{\alpha};$$

the intersection number $([W] \cdot [V(f)])_{\mathbf{p}} = \dim_{\mathbb{C}}(\operatorname{coker}(\cdot f)) - \dim_{\mathbb{C}}(\ker(\cdot f))$.

A.10 Combining this with transversality, we obtain the following *dynamic intersection property*:

$$(V(\alpha)\cdot V(f))_{\mathbf{p}} = \sum_{\mathbf{q}\in \overset{\circ}{B}_\epsilon \cap V(\alpha)\cap V(f-c)} \big(V(\alpha)\cdot V(f-c)\big)_{\mathbf{q}},$$

where $\epsilon > 0$ is sufficiently small, $\overset{\circ}{B}_\epsilon$ is an open ball of radius ϵ centered at $\mathbf{p}$, and $|c| \ll \epsilon$.

This formula may seem ridiculously complex, since all the $\big(V(\alpha)\cdot V(f-c)\big)_{\mathbf{q}}$ equal 1; however, it is the form which generalizes nicely: if C is a purely one-dimensional cycle and $V(f)$ properly intersects $|C|$ at $\mathbf{p}$, then

$$(C\cdot V(f))_{\mathbf{p}} = \sum_{\mathbf{q}\in \overset{\circ}{B}_\epsilon \cap |C|\cap V(f-c)} \big(C\cdot V(f-c)\big)_{\mathbf{q}}.$$

This is a special case of *conservation of number*, which we shall discuss more generally below.

A.11 The *projection formula* ([**Fu**], 11.4.4.iii) allows us to calculate intersections inside normal slices. Let $C = \sum n_i[V_i]$ be a cycle in M and let N be a closed submanifold of M such that N generically transversely intersects each V_i in M (this is equivalent to: for each component Z of $V_i \cap N$, $(V_i \cdot N)_Z = 1$).

We may consider $(C\cdot N;\ M)$ as a cycle in N; denote this cycle by $\widetilde{C}$. Let B be a cycle in N; we may also consider B as a cycle in M. Then, the projection formula states:

(projection formula) $\qquad (\widetilde{C} \cdot_p B;\ N) = (C \cdot_p B;\ M).$

The projection formula lets us reduce the problem of calculating intersection numbers to the case where the intersection consists of isolated points. To see this, suppose that C_1 and C_2 are two cycles in M which intersect properly and let $C_1 \cdot C_2 = \sum n_i[V_i]$. To calculate n_{i_0}, first let $\mathbf{p}$ be a smooth point of V_{i_0} which is not contained in any other V_i of dimension less than or equal to that of V_{i_0}. Now, take a *normal slice*, N, to V_{i_0} at a smooth point, $\mathbf{p}$, of V_{i_0}; that is, in an open neighborhood, $\mathcal{U}$, of $\mathbf{p}$ in M, N is a closed submanifold of $\mathcal{U}$ of complementary codimension to V_{i_0} such that N transversely intersects V_{i_0} inside $\mathcal{U}$ in the single point $\mathbf{p}$ and such that N is generically transverse to all other V_i and to all components of C_1 and C_2 in $\mathcal{U}$. By locality $n_{i_0} = (C_1\cap\mathcal{U}\ \cdot\ C_2\cap\mathcal{U};\ \mathcal{U})_{V_{i_0}\cap\mathcal{U}}$. As $V_{i_0}\cap\mathcal{U}$ is the only component of $(C_1\cap\mathcal{U}\ \cdot\ C_2\cap\mathcal{U};\ \mathcal{U})$ whose intersection with N gives $\{\mathbf{p}\}$, the transversality characterization yields that $n_{i_0} = (C_1\cap\mathcal{U}\ \cdot\ C_2\cap\mathcal{U}\ \cdot N;\ \mathcal{U})_{\mathbf{p}}$. But, we wish to calculate this intersection inside of the normal slice N – this is what we get from the projection formula.

Replace the M, N, C, and B in the projection formula as stated above by letting $M = \mathcal{U}$, $N = N$, $C = C_1\cap\mathcal{U}$, and $B = (C_2\cap\mathcal{U})\cdot N$ (consider B as a cycle in N). Then, the formula yields that

$$\big((C_1\cap\mathcal{U})\cdot N)\ \cdot\ ((C_2\cap\mathcal{U})\cdot N);\ N\big) = (C_1\cap\mathcal{U}\ \cdot\ C_2\cap\mathcal{U}\ \cdot N;\ \mathcal{U}).$$

Thus, we see that taking normal slices reduces calculating proper intersections of cycles to the case where the dimension of the intersection is zero.

A.12 There is one last property of intersections of cycles that we need – continuity ([**Fu**], 11.4.4.iii). This property is what makes intersections dynamic; one can move the intersections in a family.

Let M be a analytic manifold and let $\overset{\circ}{\mathbb{D}}$ be an open disc centered at the origin in $\mathbb{C}$. Then, the projection $M \times \overset{\circ}{\mathbb{D}} \xrightarrow{\pi} \overset{\circ}{\mathbb{D}}$ determines a one-parameter family of spaces, and any subscheme $W \subseteq M \times \overset{\circ}{\mathbb{D}}$ determines a one-parameter family of schemes $W_t := W \cap (M \times \{t\})$. Hence, any cycle $C := \sum n_i[V_i]$ in $M \times \overset{\circ}{\mathbb{D}}$ determines a family of cycles $C_t := \sum n_i[(V_i)_t]$ in $M \cong M \times \{t\}$.

If a cycle C in $M \times \overset{\circ}{\mathbb{D}}$ has a component contained in $M \times \{t\}$ for some t, then that component does not "propagate" through the family; we wish to eliminate such "bad" components. For any cycle $C = \sum n_i[V_i]$ in $M \times \overset{\circ}{\mathbb{D}}$ and any analytic set $W \subseteq M \times \overset{\circ}{\mathbb{D}}$, let

$$C\neg W := \sum_{V_i \not\subseteq W} n_i[V_i],$$

and let

$$C_t^* := \Big((C\neg(M \times \{t\})) \cdot (M \times \{t\});\ M \times \overset{\circ}{\mathbb{D}}\Big) = \sum_{V_i \not\subseteq M \times \{t\}} n_i\,[(V_i)_t]\,.$$

Continuity of intersections states that, if C is a cycle in $M \times \overset{\circ}{\mathbb{D}}$ with no component contained in $M \times \{0\}$, and E is a cycle in M such that C_0 properly intersects E in M, then there exists a (possibly) smaller disk centered at the origin $\overset{\circ}{\mathbb{D}}{}' \subseteq \overset{\circ}{\mathbb{D}}$ such that C properly intersects $E \times \overset{\circ}{\mathbb{D}}{}'$ in $M \times \overset{\circ}{\mathbb{D}}{}'$ and, for all $t \in \overset{\circ}{\mathbb{D}}{}'$,

$$\textbf{(continuity)} \qquad \left((E \times \overset{\circ}{\mathbb{D}}{}') \cdot C;\ M \times \overset{\circ}{\mathbb{D}}{}'\right)_t^* = (E \cdot C_t;\ M).$$

A.13 We saw earlier how the projection formula allows us to reduce the calculation of intersection multiplicities to the case where the intersection is zero-dimensional. We wish to see now how continuity allows us to deform in a family in order to calculate zero-dimensional intersection multiplicities.

We will prove a dynamic formula for intersection multiplicities; this formula is known as **conservation of number**. Let E be a k-dimensional cycle in M and let $\mathbf{f} := (f_1, \dots, f_k) \in (\mathcal{O}_M)^k$ be such that E and $V(\mathbf{f})$ intersect properly in the single point $\mathbf{p}$. This implies that $V(\mathbf{f})$ is purely k-codimensional inside M at $\mathbf{p}$. In what follows, we assume that we are always working in an arbitrarily small neighborhood of $\mathbf{p}$.

Let $g_1(\mathbf{z}, t), \dots, g_k(\mathbf{z}, t) \in \mathcal{O}_{M \times \overset{\circ}{\mathbb{D}}}$ be such that $g_i(\mathbf{z}, 0) = f_i(\mathbf{z})$ for all i. Let C be the cycle in $M \times \overset{\circ}{\mathbb{D}}$ given by $\big[V(g_1(\mathbf{z}, t), \dots, g_k(\mathbf{z}, t))\big]$. Note that $C_0 = \big[V(\mathbf{f})\big]$ (in M) and that C has no components contained in $M \times \{0\}$, for otherwise $V(\mathbf{f})$ would have a component of dimension at least $(\dim M) + 1 - k$.

Applying continuity at $t = 0$, we find that

$$(E \cdot V(\mathbf{f});\ M) = \left((E \times \overset{\circ}{\mathbb{D}}{}') \cdot C;\ M \times \overset{\circ}{\mathbb{D}}{}'\right)_0^*$$

for a smaller disc $\overset{\circ}{\mathbb{D}}{}'$. Note that $(E \times \overset{\circ}{\mathbb{D}}{}') \cdot C$ is a purely 1-dimensional cycle, say $\sum_j m_j[W_j]$. Applying continuity at general t, we find that, for all $t \in \overset{\circ}{\mathbb{D}}{}'$,

$$(E \cdot C_t;\ M) = \sum_{W_j \not\subseteq M \times \{0\}} m_j \left[(W_j)_t\right].$$

Hence, by openness and transversality, we find that $m_j = (E \cdot C_t;\ M)_{\mathbf{q}}$ for sufficiently small $t \neq 0$ and $\mathbf{q} \in (W_j)_t$. Now, since $(E \cdot V(\mathbf{f});\ M) = \sum_{W_j \not\subseteq M \times \{0\}} m_j \left[(W_j)_0\right]$, we may apply our earlier special case of dynamic intersections between curves and hypersurfaces to conclude the general **conservation of number** formula

$$(E \cdot V(\mathbf{f}))_{\mathbf{p}} = \sum_{\mathbf{q} \in \overset{\circ}{B}_\epsilon \cap |E \cap C_t|} (E \cdot C_t)_{\mathbf{q}},$$

where $\epsilon > 0$ is sufficiently small, $\overset{\circ}{B}_\epsilon$ is an open ball of radius ϵ centered at $\mathbf{p}$, $|t| \ll \epsilon$, and C_t equals $\left[V(g_1(\mathbf{z}, t), \dots, g_k(\mathbf{z}, t))\right]$.

A.14 Finally, we need to define the *proper push-forward* of cycles (see [**Fu**, 1.4]). Let $f : X \to Y$ be a proper morphism of analytic spaces. Then, for each irreducible subvariety $V \subseteq X$, $W := f(V)$ is an irreducible subvariety of Y. There is an induced embedding of rational function fields $R(W) \hookrightarrow R(V)$, which is a finite field extension if V and W have the same dimension. Define the *degree of V over W* by

$$\deg(V/W) := \begin{cases} [R(V) : R(W)] & \text{if } \dim W = \dim V \\ 0 & \text{if } \dim W \neq \dim V, \end{cases}$$

where $[R(V) : R(W)]$ denotes the degree of the field extension, which equals the number of points in $V \cap f^{-1}(\mathbf{p})$ for a generic choice of $\mathbf{p} \in W$.

Define $f_*(V)$ by

$$f_*(V) = \deg(V/W)[W].$$

This extends linearly to a homomorphism which is called the proper push-forward of cycles:

$$f_*\left(\sum m_V[V]\right) = \sum m_V f_*(V).$$

We will need the following special case of the more general *push-forward formula* (see [**Fu**], 2.3.c). Let $\pi : M \to N$ be a proper map between analytic manifolds. Let $f \in \mathcal{O}_N$. Let C be a cycle in M which intersects $V(f \circ \pi)$ properly in M. Then, $V(f)$ properly intersects $\pi_*(C)$ in N and

(push-forward formula) $\qquad \pi_*\big(V(f \circ \pi) \cdot C\big) = V(f) \cdot \pi_*(C).$

We need one other formula involving the push-forward and graphs of morphisms. Suppose that we have an analytic map $f : M \to N$ between analytic manifolds. Then, for any irreducible subvariety $V \subseteq M$, the graph of $f_{|_V}$, $\operatorname{Gr}(f_{|_V})$, is isomorphic to V. Thus, one would expect that intersecting with V in M could be identified with intersecting with $\operatorname{Gr}(f_{|_V})$ in $M \times N$; this is, in fact the case.

Suppose that $A := \sum m_i[V_i]$ and B are properly intersecting cycles in M. Let $\operatorname{Gr}(A) := \sum m_i[\operatorname{Gr}(f_{|_{V_i}})]$ in $M \times N$, and let $\operatorname{pr} : M \times N \to M$ denote the projection. Then, $\operatorname{Gr}(A)$ properly intersects $B \times N$ in $M \times N$, and we have the *graph formula*:

(graph formula) $$(A \cdot B;\ M) = \operatorname{pr}_*(\operatorname{Gr}(A) \cdot (B \times N);\ M \times N).$$

This is easy to see: $\operatorname{Gr}(A) = (A \times N) \cdot \operatorname{Gr}(f)$, since $\operatorname{Gr}(f)$ determines a regular sequence in each $\mathcal{O}_{V_i \times N}$ (see A.4). Hence,

$$\operatorname{Gr}(A) \cdot (B \times N) = (A \times N) \cdot (B \times N) \cdot \operatorname{Gr}(f) = ((A \cdot B) \times N) \cdot \operatorname{Gr}(f),$$

where the last equality follows from normal slicing (see A.11). The graph formula follows easily by using A.4 again, together with the definition of the proper push-forward.

APPENDIX B:

THE DERIVED CATEGORY AND VANISHING CYCLES

This appendix contains a number of basic results on the derived category, perverse sheaves, and vanishing cycles. Primary sources for most of these results are [**BBD**], [**Br**], [**De**], [**G-M3**], [**K-S2**], [**Mac2**], [**M-V**], and [**Ve**].

This appendix is organized as follows:

§1. **Constructible Complexes** – This section contains general results on bounded, constructible complexes of sheaves and the derived category.

§2. **Perverse Sheaves** – This section contains the definition and basic results on perverse sheaves. Here, we also give the axiomatic characterization of the intersection cohomology complex. Finally in this section, we also give some results on the category of perverse sheaves. This categorical information is augmented by that in section 5.

§3. **Nearby and Vanishing Cycles** – In this section, we define and examine the complexes of sheaves of nearby and vanishing cycles of an analytic function. These complexes contain hypercohomological information on the Milnor fibre of the function under consideration.

§4. **Some Quick Applications** – In this section, we give three easy examples of results on Milnor fibres which follow from the machinery described in the previous three sections.

§5. **Truncation and Perverse Cohomology** – This section contains an informal discussion on t-structures. This enables us to describe truncation functors and the perverse cohomology of a complex. It also sheds some light on our earlier discussion of the categorical structure of perverse sheaves.

§1. Constructible Complexes

Much of this section is lifted directly from Goresky and MacPherson's paper "Intersection Homology II" [**G-M3**].

In this appendix, we are primarily interested in sheaves on complex analytic spaces, and we make an effort to state most results in this context. However, as one frequently wishes to do such things as intersect with a closed ball, one really needs to consider at least the real semi-analytic case (that is, spaces locally defined by finitely many real analytic inequalities). In fact, one can treat the subanalytic case. Generally, when we leave the analytic category we shall do so without comment, assuming the natural generalizations of any needed results. However, the precise statements in the subanalytic case can be found in [**G-M2**], [**G-M3**], and [**K-S2**].

Let R be a regular Noetherian ring with finite Krull dimension (e.g., $\mathbb{Z}, \mathbb{Q}$, or $\mathbb{C}$). A complex $(\mathbf{A}^\bullet, d^\bullet)$ (usually denoted simply by $\mathbf{A}^\bullet$ if the differentials are clear or arbitrary)

$$\cdots \to \mathbf{A}^{-1} \xrightarrow{d^{-1}} \mathbf{A}^0 \xrightarrow{d^0} \mathbf{A}^1 \xrightarrow{d^1} \mathbf{A}^2 \xrightarrow{d^2} \cdots$$

of sheaves of R-modules on a complex analytic space, X, is *bounded* if $\mathbf{A}^p = 0$ for $|p|$ large.

The cohomology sheaves $\mathbf{H}^p(\mathbf{A}^\bullet)$ arise by taking the (sheaf-theoretic) cohomology of the complex. The stalk of $\mathbf{H}^p(\mathbf{A}^\bullet)$ at a point x is written $\mathbf{H}^p(\mathbf{A}^\bullet)_x$ and is isomorphic to what one gets by first taking stalks and then taking cohomology, i.e., $H^p(\mathbf{A}^\bullet_x)$.

The complex $\mathbf{A}^\bullet$ is *constructible* with respect to a complex analytic stratification, $\mathcal{S} = \{S_\alpha\}$, of X provided that, for all α and i, the cohomology sheaves $\mathbf{H}^i(\mathbf{A}^\bullet_{|S_\alpha})$ are locally constant and have finitely-generated stalks; we write $\mathbf{A}^\bullet \in \mathbf{D}_{\mathcal{S}}(X)$. If $\mathbf{A}^\bullet \in \mathbf{D}_{\mathcal{S}}(X)$ and $\mathbf{A}^\bullet$ is bounded, we write $\mathbf{A}^\bullet \in \mathbf{D}^b_{\mathcal{S}}(X)$.

If $\mathbf{A}^\bullet \in \mathbf{D}^b_{\mathcal{S}}(X)$ for some stratification (and, hence, for any refinement of $\mathcal{S}$) we say that $\mathbf{A}^\bullet$ is a bounded, constructible complex and write $\mathbf{A}^\bullet \in \mathbf{D}^b_c(X)$. (Note, however, that $\mathbf{D}^b_c(X)$ actually denotes the *derived* category and, while the objects of this category are, in fact, the bounded, constructible complexes, the morphisms are not merely maps between complexes. We shall return to this.)

When it is important to indicate the base ring in the notation, we write $\mathbf{D}_{\mathcal{S}}(R_X)$, $\mathbf{D}^b_{\mathcal{S}}(R_X)$, and $\mathbf{D}^b_c(R_X)$.

A single sheaf $\mathbf{A}$ on X is considered a complex, $\mathbf{A}^\bullet$, on X by letting $\mathbf{A}^0 = \mathbf{A}$ and $\mathbf{A}^i = 0$ for $i \neq 0$; thus, $\mathbf{R}^\bullet_X$ denotes the constant sheaf on X.

The shifted complex $\mathbf{A}^\bullet[n]$ is defined by letting $(\mathbf{A}^\bullet[n])^k = \mathbf{A}^{n+k}$ and defining the differential $d^k_{[n]} = (-1)^n d^{k+n}$.

A map of complexes is a graded collection of sheaf maps $\phi^\bullet : \mathbf{A}^\bullet \to \mathbf{B}^\bullet$ which commute with the differentials. The shifted sheaf map $\phi^\bullet_{[n]} : \mathbf{A}^\bullet[n] \to \mathbf{B}^\bullet[n]$ is defined by $\phi^k_{[n]} := \phi^{k+n}$ (note the lack of a $(-1)^n$). A map of complexes is a *quasi-isomorphism* provided that the induced maps

$$\mathbf{H}^p(\phi^\bullet) : \mathbf{H}^p(\mathbf{A}^\bullet) \to \mathbf{H}^p(\mathbf{B}^\bullet)$$

are isomorphisms for all p. We use the term "quasi-isomorphic" to mean the equivalence relation generated by "existence of a quasi-isomorphism"; this is sometimes refered to as "generalized" quasi-isomorphic.

If $\phi^\bullet : \mathbf{A}^\bullet \to \mathbf{I}^\bullet$ is a quasi-isomorphism and each $\mathbf{I}^p$ is injective, then $\mathbf{I}^\bullet$ is called an *injective resolution* of $\mathbf{A}^\bullet$. Injective resolutions always exist (in our setting), and are unique up to chain homotopy. However, it is sometimes important to associate one particular resolution to a complex, so it is important that there is a *canonical injective resolution* which can be associated to any complex (we shall not describe the canonical resolution here).

If $\mathbf{A}^\bullet$ is a complex on X, then the *hypercohomology module*, $\mathbb{H}^p(X; \mathbf{A}^\bullet)$, is defined to be the p-th cohomology of the global section functor applied to the canonical injective resolution of $\mathbf{A}^\bullet$.

Note that if $\mathbf{A}$ is a single sheaf on X and we form $\mathbf{A}^\bullet$, then $\mathbb{H}^p(X; \mathbf{A}^\bullet) = H^p(X; \mathbf{A})$ is ordinary sheaf cohomology. In particular, $\mathbb{H}^p(X; \mathbf{R}^\bullet_X) = H^p(X; R)$.

Note also that if $\mathbf{A}^\bullet$ and $\mathbf{B}^\bullet$ are quasi-isomorphic, then $\mathbb{H}^*(X;\mathbf{A}^\bullet) \cong \mathbb{H}^*(X;\mathbf{B}^\bullet)$.

If Y is a subspace of X and $\mathbf{A}^\bullet \in \mathbf{D}_c^b(X)$, then one usually writes $\mathbb{H}^*(Y;\mathbf{A}^\bullet)$ in place of $\mathbb{H}^*(Y;\mathbf{A}^\bullet{}_{|Y})$.

The usual Mayer-Vietoris sequence is valid for hypercohomology; that is, if U and V form an open cover of X and $\mathbf{A}^\bullet \in \mathbf{D}_c^b(X)$, then there is an exact sequence

$$\cdots \to \mathbb{H}^i(X;\mathbf{A}^\bullet) \to \mathbb{H}^i(U;\mathbf{A}^\bullet) \oplus \mathbb{H}^i(V;\mathbf{A}^\bullet) \to \mathbb{H}^i(U\cap V;\mathbf{A}^\bullet) \to \mathbb{H}^{i+1}(X;\mathbf{A}^\bullet) \to \dots.$$

Of course, hypercohomology is not a homotopy invariant. However, it is true that: if $\mathcal{S}$ is a real analytic Whitney stratification of X, $\mathbf{A}^\bullet \in \mathbf{D}_{\mathcal{S}}^b(X)$, and $r : X \to [0,1)$ is a proper real analytic map such that, for all $S \in \mathcal{S}$, $r_{|S}$ has no critical values in $(0,1)$, then the inclusion $r^{-1}(0) \hookrightarrow X$ induces an isomorphism

$$\mathbb{H}^i(X;\mathbf{A}^\bullet) \cong \mathbb{H}^i(r^{-1}(0);\mathbf{A}^\bullet).$$

If R is a principal ideal domain, we may talk about the rank of a finitely-generated R-module. In this case, if $\mathbf{A}^\bullet \in \mathbf{D}_c^b(X)$, then the Euler characteristic, χ, of the stalk cohomology is defined as the alternating sum of the ranks of the cohomology modules, i.e., $\chi(\mathbf{A}^\bullet)_x = \sum(-1)^i \operatorname{rank} \mathbf{H}^i(\mathbf{A}^\bullet)_x$. If the hypercohomology modules are finitely-generated – for instance, if $\mathbf{A}^\bullet \in \mathbf{D}_c^b(X)$ and X is compact – then the Euler characteristic $\chi\big(\mathbb{H}^*(X;\mathbf{A}^\bullet)\big)$ is defined analogously.

If $\mathbf{H}^i(\mathbf{A}^\bullet) = 0$ for all but, possibly, one value of i - say, $i = p$, then $\mathbf{A}^\bullet$ is quasi-isomorphic to the complex that has $\mathbf{H}^p(\mathbf{A}^\bullet)$ in degree p and zero elsewhere. We reserve the term *local system* for a locally constant single sheaf or a complex which is concentrated in degree zero and is locally constant. If M is the stalk of a local system $\mathcal{L}$ on a path-connected space X, then $\mathcal{L}$ is determined up to isomorphism by a monodromy representation $\pi_1(X, \mathbf{x}) \to Aut(M)$, where x is a fixed point in X.

For any $\mathbf{A}^\bullet \in \mathbf{D}_c^b(X)$, there is an E_2 cohomological spectral sequence:

$$E_2^{p,q} = H^p(X;\mathbf{H}^q(\mathbf{A}^\bullet)) \Rightarrow \mathbb{H}^{p+q}(X;\mathbf{A}^\bullet).$$

If $\epsilon > 0$, $x \in X$, and (X,x) is locally embedded in some $\mathbb{C}^n$, then we let $\overset{\circ}{B}_\epsilon(x)$ denote the open ball of radius ϵ, centered at x, in $\mathbb{C}^n$; in this situation, if $\mathbf{A}^\bullet \in \mathbf{D}_c^b(X)$, then for all $\epsilon > 0$ small, the restriction map $\mathbb{H}^q(\overset{\circ}{B}_\epsilon(x);\mathbf{A}^\bullet) \to \mathbf{H}^q(\mathbf{A}^\bullet)_\mathbf{x}$ is an isomorphism. If, in addition, R is a principal ideal domain, then the Euler characteristic $\chi\big(\mathbb{H}^*(\overset{\circ}{B}_\epsilon(x) - x;\mathbf{A}^\bullet)\big)$ is defined and

$$\chi\big(\mathbb{H}^*(\overset{\circ}{B}_\epsilon(x) - x;\mathbf{A}^\bullet)\big) = \chi\big(\mathbb{H}^*(S_{\epsilon'}(x);\mathbf{A}^\bullet)\big) = 0,$$

where $0 < \epsilon' < \epsilon$ and $S_{\epsilon'}(x)$ denotes the sphere of radius ϵ' centered at x.

We now wish to say a little about the morphisms in the derived category $\mathbf{D}_c^b(X)$. The derived category is obtained by formally inverting the quasi-isomorphisms so that they become isomorphisms in $\mathbf{D}_c^b(X)$. Thus, $\mathbf{A}^\bullet$ and $\mathbf{B}^\bullet$ are isomorphic in $\mathbf{D}_c^b(X)$ provided that there exists a complex $\mathbf{C}^\bullet$ and quasi-isomorphisms $\mathbf{A}^\bullet \leftarrow \mathbf{C}^\bullet \rightarrow \mathbf{B}^\bullet$; $\mathbf{A}^\bullet$ and $\mathbf{B}^\bullet$ are then said to be *incarnations* of the same isomorphism class in $\mathbf{D}_c^b(X)$.

More generally, a morphism in $\mathbf{D}_c^b(X)$ from $\mathbf{A}^\bullet$ to $\mathbf{B}^\bullet$ is an equivalence class of diagrams of maps of complexes $\mathbf{A}^\bullet \leftarrow \mathbf{C}^\bullet \rightarrow \mathbf{B}^\bullet$ where $\mathbf{A}^\bullet \leftarrow \mathbf{C}^\bullet$ is a quasi-isomorphism. Two such diagrams,

$$\mathbf{A}^\bullet \xleftarrow{f_1} \mathbf{C}_1^\bullet \xrightarrow{g_1} \mathbf{B}^\bullet, \qquad \mathbf{A}^\bullet \xleftarrow{f_2} \mathbf{C}_2^\bullet \xrightarrow{g_2} \mathbf{B}^\bullet$$

are equivalent provided that there exists a third such diagram $\mathbf{A}^\bullet \xleftarrow{f} \mathbf{C}^\bullet \xrightarrow{g} \mathbf{B}^\bullet$ and a diagram

$$\begin{array}{ccccc} & & \mathbf{C}_1^\bullet & & \\ & f_1 \swarrow & \uparrow & \searrow g_1 & \\ \mathbf{A}^\bullet & \xleftarrow{\ f\ } & \mathbf{C}^\bullet & \xrightarrow{\ g\ } & \mathbf{B}^\bullet \\ & f_2 \nwarrow & \downarrow & \nearrow g_2 & \\ & & \mathbf{C}_2^\bullet & & \end{array}$$

which commutes up to (chain) homotopy.

Composition of morphisms in $\mathbf{D}_c^b(X)$ is not difficult to describe. If we have two representatives of morphisms, from $\mathbf{A}^\bullet$ to $\mathbf{B}^\bullet$ and from $\mathbf{B}^\bullet$ to $\mathbf{D}^\bullet$, respectively,

$$\mathbf{A}^\bullet \xleftarrow{f_1} \mathbf{C}_1^\bullet \xrightarrow{g_1} \mathbf{B}^\bullet, \qquad \mathbf{B}^\bullet \xleftarrow{f_2} \mathbf{C}_2^\bullet \xrightarrow{g_2} \mathbf{D}^\bullet$$

then we consider the pull-back $\mathbf{C}_1^\bullet \times_{\mathbf{B}^\bullet} \mathbf{C}_2^\bullet$ (in the category of chain complexes) and the projections π_1 and π_2 to $\mathbf{C}_1^\bullet$ and $\mathbf{C}_2^\bullet$, respectively. As f_2 is a quasi-isomorphism, so is π_1, and the composed morphism from $\mathbf{A}^\bullet$ to $\mathbf{D}^\bullet$ is represented by $\mathbf{A}^\bullet \xleftarrow{f_1\circ\pi_1} \mathbf{C}_1^\bullet \times_{\mathbf{B}^\bullet} \mathbf{C}_2^\bullet \xrightarrow{g_2\circ\pi_2} \mathbf{D}^\bullet$.

If we restrict ourselves to considering only injective complexes, by associating to any complex its canonical injective resolution, then morphisms in the derived category become easy to describe – they are chain-homotopy classes of maps between the injective complexes.

The moral is: in $\mathbf{D}_c^b(X)$, we essentially only care about complexes up to quasi-isomorphism. Note, however, that the objects of $\mathbf{D}_c^b(X)$ are **not** equivalence classes – this is one reason why it is important that to each complex we can associate a **canonical** injective resolution. It allows us to talk about certain functors in $\mathbf{D}_c^b(X)$ being **naturally** isomorphic. When we write $\mathbf{A}^\bullet \cong \mathbf{B}^\bullet$, we mean in $\mathbf{D}_c^b(X)$. As we shall discuss later, $\mathbf{D}_c^b(X)$ is an additive category, but is not Abelian.

Warning: While it is true that morphisms of complexes which induce isomorphisms on cohomology sheaves become isomorphisms in the derived category, there are morphisms of complexes which induce the zero map on cohomology sheaves but are not zero in the derived category. The easiest example of such a morphism is given by the following.

Let X be a space consisting of two complex lines L_1 and L_2 which intersect in a single point $\mathbf{p}$. For $i = 1, 2$, let $\widetilde{\mathbb{C}}_{L_i}$ denote the $\mathbb{C}$-constant sheaf on L_i extended by zero to all of X. There is a canonical map, α, from the sheaf $\mathbb{C}_X$ to the direct sum of sheaves $\widetilde{\mathbb{C}}_{L_1} \oplus \widetilde{\mathbb{C}}_{L_2}$, which on $L_1 - \mathbf{p}$ is $\mathrm{id} \oplus 0$, on $L_2 - \mathbf{p}$ is $0 \oplus \mathrm{id}$, and is the diagonal map on the stalk at $\mathbf{p}$. Consider the complex, $\mathbf{A}^\bullet$, which has $\mathbb{C}_X$ in degree 0, $\widetilde{\mathbb{C}}_{L_1} \oplus \widetilde{\mathbb{C}}_{L_2}$ in degree 1, zeroes elsewhere, and the coboundary map from degree 0 to degree 1 is α. This complex has cohomology only in degree 1. Nonetheless, the morphism of complexes from $\mathbf{A}^\bullet$ to $\mathbb{C}^\bullet_X$ which is the identity in degree 0 and is zero elsewhere determines a non-zero morphism in the derived category.

We now wish to describe *derived functors*; for this, we will need the derived category of an arbitrary Abelian category $\mathcal{C}$.

Let $\mathcal{C}$ be an Abelian category. Then, the derived category of bounded complexes in $\mathcal{C}$ is the category whose objects consist of bounded differential complexes of objects of $\mathcal{C}$, and where the morphisms are obtained exactly as in the case of $\mathbf{D}^b_c(X)$ – namely, by inverting the quasi-isomorphisms as we did above. Naturally, we denote this derived category by $\mathbf{D}^b(\mathcal{C})$.

We need some more general notions before we come back to complexes of sheaves. If $\mathcal{C}$ is an Abelian category, then we let $\mathbf{K}^b(\mathcal{C})$ denote the category whose objects are again bounded differential complexes of objects of $\mathcal{C}$, but where the morphisms are chain-homotopy classes of maps of differential complexes. A *triangle* in $\mathbf{K}^b(\mathcal{C})$ is a sequence of morphisms $\mathbf{A}^\bullet \to \mathbf{B}^\bullet \to \mathbf{C}^\bullet \to \mathbf{A}^\bullet[1]$, which is usually written in the more "triangular" form

$$\begin{array}{ccc} \mathbf{A}^\bullet & \longrightarrow & \mathbf{B}^\bullet \\ {\scriptstyle [1]}\nwarrow & & \swarrow \\ & \mathbf{C}^\bullet & \end{array}$$

A triangle in $\mathbf{K}^b(\mathcal{C})$ is called *distinguished* if it is isomorphic in $\mathbf{K}^b(\mathcal{C})$ to a diagram of maps of complexes

$$\begin{array}{ccc} \widetilde{\mathbf{A}}^\bullet & \xrightarrow{\phi} & \widetilde{\mathbf{B}}^\bullet \\ {\scriptstyle [1]}\nwarrow & & \swarrow \\ & \mathbf{M}^\bullet & \end{array}$$

where $\mathbf{M}^\bullet$ is the algebraic mapping cone of ϕ and $\widetilde{\mathbf{B}}^\bullet \to \mathbf{M}^\bullet \to \widetilde{\mathbf{A}}^\bullet[1]$ are the canonical maps. (Recall that the algebraic mapping cone is defined by

$$\mathbf{M}^k := \widetilde{\mathbf{A}}^{k+1} \oplus \widetilde{\mathbf{B}}^k \longrightarrow \widetilde{\mathbf{A}}^{k+2} \oplus \widetilde{\mathbf{B}}^{k+1} =: \mathbf{M}^{k+1}$$

$$(a, b) \longmapsto (-\partial a, \phi a + \delta b)$$

where ∂ and δ are the differentials of $\widetilde{\mathbf{A}}^\bullet$ and $\widetilde{\mathbf{B}}^\bullet$ respectively.) Note that if $\phi = 0$, then we have an equality $\mathbf{M}^\bullet = \mathbf{A}^\bullet[1] \oplus \mathbf{B}^\bullet$ (recall that the shifted complex $\mathbf{A}^\bullet[1]$ has as its differential the negated, shifted differential of $\mathbf{A}^\bullet$).

Now we can define derived functors. Let $\mathcal{C}$ denote the Abelian category of sheaves of R-modules on an analytic space X, and let $\mathcal{C}'$ be another Abelian category. Suppose that F is an additive, covariant functor from $\mathbf{K}^b(\mathcal{C})$ to $\mathbf{K}^b(\mathcal{C}')$ such that $F \circ [1] = [1] \circ F$ and such that F takes distinguished triangles to distinguished triangles (such an F is called a *functor of triangulated categories*). Suppose also that, for all complexes of injective sheaves $\mathbf{I}^\bullet \in \mathbf{K}^b(\mathcal{C})$ which are quasi-isomorphic to 0, $F(\mathbf{I}^\bullet)$ is also quasi-isomorphic to 0.

Then, F induces a morphism RF – the *right derived functor of* F – from $\mathbf{D}^b(X)$ to $\mathbf{D}^b(\mathcal{C}')$; for any $\mathbf{A}^\bullet \in \mathbf{D}^b(X)$, let $\mathbf{A}^\bullet \to \mathbf{I}^\bullet$ denote the canonical injective resolution of $\mathbf{A}^\bullet$, and define $RF(\mathbf{A}^\bullet) := F(\mathbf{I}^\bullet)$. The action of RF on the morphisms is the obvious associated one.

A morphism $F : \mathbf{K}^b(\mathcal{C}) \to \mathbf{K}^b(\mathcal{C}')$ as described above is frequently obtained by starting with a left-exact functor $T : \mathcal{C} \to \mathcal{C}'$ and then extending T in a term-wise fashion to be a functor from $\mathbf{K}^b(\mathcal{C})$ to $\mathbf{K}^b(\mathcal{C}')$. In this case, we naturally write RT for the derived functor.

This is the process which is applied to:

$\Gamma(X; \cdot)$ (global sections);

$\Gamma_c(X; \cdot)$ (global sections with compact support);

f_* (direct image);

$f_!$ (direct image with proper supports); and

f^* (pull-back or inverse image),

where $f : X \to Y$ is a continuous map (actually, in these notes, we would need an analytically constructible map; e.g., an analytic map).

If the functor T is an exact functor from sheaves to sheaves, then $RT(\mathbf{A}^\bullet) \cong T(\mathbf{A}^\bullet)$; in this case, we normally suppress the R. Hence, if $f : X \to Y$, $\mathbf{A}^\bullet \in \mathbf{D}^b(X)$, and $\mathbf{B}^\bullet \in \mathbf{D}^b(Y)$, we write:

$f^*\mathbf{B}^\bullet$;

$f_!\mathbf{A}^\bullet$, if f is the inclusion of a subspace and, hence, $f_!$ is extension by zero;

$f_*\mathbf{A}^\bullet$, if f is the inclusion of a closed subspace.

Note that hypercohomology is just the cohomology of the derived global section functor, i.e., $\mathbb{H}^*(X; \cdot) = H^* \circ R\Gamma(X; \cdot)$. The cohomology of the derived functor of global sections with compact support is the *compactly supported hypercohomology* and is denoted $\mathbb{H}^*_c(X; \mathbf{A}^\bullet)$.

If $f : X \to Y$ is the inclusion of a subset and $\mathbf{B}^\bullet \in \mathbf{D}^b(Y)$, then the *restriction of* $\mathbf{B}^\bullet$ to X is defined to be $f^*(\mathbf{B}^\bullet)$, and is usually denoted by $\mathbf{B}^\bullet{}_{|X}$.

If $f : X \to Y$ is continuous and $\mathbf{A}^\bullet \in \mathbf{D}^b_c(X)$, there is a canonical map

$$Rf_!\mathbf{A}^\bullet \to Rf_*\mathbf{A}^\bullet.$$

For $f : X \to Y$ continuous, there are canonical isomorphisms

$$R\Gamma(X; \mathbf{A}^\bullet) \cong R\Gamma(Y; Rf_*\mathbf{A}^\bullet) \text{ and } R\Gamma_c(X; \mathbf{A}^\bullet) \cong R\Gamma_c(Y; Rf_!\mathbf{A}^\bullet)$$

which lead to canonical isomorphisms

$$\mathbb{H}^*(X; \mathbf{A}^\bullet) \cong \mathbb{H}^*(Y; Rf_*\mathbf{A}^\bullet) \text{ and } \mathbb{H}^*_c(X; \mathbf{A}^\bullet) \cong \mathbb{H}^*_c(Y; Rf_!\mathbf{A}^\bullet)$$

for all $\mathbf{A}^\bullet$ in $\mathbf{D}_c^b(X)$.

If $f : X \to Y$ is continuous, $\mathbf{A}^\bullet \in \mathbf{D}_c^b(X)$, and $\mathbf{B}^\bullet \in \mathbf{D}_c^b(Y)$, there are natural maps induced by restriction of sections

$$\mathbf{B}^\bullet \to Rf_*f^*\mathbf{B}^\bullet \quad \text{and} \quad f^*Rf_*\mathbf{A}^\bullet \to \mathbf{A}^\bullet.$$

If $\{S_\alpha\}$ is a stratification of X, $\mathbf{A}^\bullet \in \mathbf{D}^b_{\{S_\alpha \times \mathbb{C}^k\}}(X \times \mathbb{C}^k)$, and $\pi : X \times \mathbb{C}^k \to X$ is the projection, then restriction of sections induces a quasi-isomorphism $\pi^* R\pi_* \mathbf{A}^\bullet \to \mathbf{A}^\bullet$.

It follows easily that if $j : X \hookrightarrow X \times \mathbb{C}^k$ is the zero section, then $\pi^* j^* \mathbf{A}^\bullet \cong \mathbf{A}^\bullet$. This says exactly what one expects: the complex $\mathbf{A}^\bullet$ has a product structure in the $\mathbb{C}^k$ directions.

An important consequence of this is the following: let $\mathcal{S} = \{S_\alpha\}$ be a Whitney stratification of X and let $\mathbf{A}^\bullet \in \mathbf{D}^b_{\mathcal{S}}(X)$. Let $x \in S_\alpha \subseteq X$. As S_α is a Whitney stratum, X has a product structure along S_α near x. By the above, $\mathbf{A}^\bullet$ itself also has a product structure along S_α. Hence, by taking a normal slice, many problems concerning the complex $\mathbf{A}^\bullet$ can be reduced to considering a zero-dimensional stratum.

Let $\mathbf{A}^\bullet, \mathbf{B}^\bullet \in \mathbf{D}_c^b(X)$. Define $\mathbf{A}^\bullet \otimes \mathbf{B}^\bullet$ to be the single complex which is associated to the double complex $\mathbf{A}^p \otimes \mathbf{B}^q$. The left derived functor $\mathbf{A}^\bullet \overset{L}{\otimes} *$ is defined by

$$\mathbf{A}^\bullet \overset{L}{\otimes} \mathbf{B}^\bullet = \mathbf{A}^\bullet \otimes \mathbf{J}^\bullet,$$

where $\mathbf{J}^\bullet$ is a flat resolution of $\mathbf{B}^\bullet$, i.e., the stalks of $\mathbf{J}^\bullet$ are flat R-modules and there exists a quasi-isomorphism $\mathbf{J}^\bullet \to \mathbf{B}^\bullet$.

For all $\mathbf{A}^\bullet, \mathbf{B}^\bullet \in \mathbf{D}_c^b(X)$, there is an isomorphism $\mathbf{A}^\bullet \overset{L}{\otimes} \mathbf{B}^\bullet \cong \mathbf{B}^\bullet \overset{L}{\otimes} \mathbf{A}^\bullet$.

For any map $f : X \to Y$ and any $\mathbf{A}^\bullet, \mathbf{B}^\bullet \in \mathbf{D}_c^b(Y)$,

$$f^*(\mathbf{A}^\bullet \overset{L}{\otimes} \mathbf{B}^\bullet) \cong f^*\mathbf{A}^\bullet \overset{L}{\otimes} f^*\mathbf{B}^\bullet.$$

Fix a complex $\mathbf{B}^\bullet$ on X. There are two covariant functors which we wish to consider: the functor $\mathbf{Hom}^\bullet(\mathbf{B}^\bullet, *)$ from the category of complexes of sheaves to complexes of sheaves and the functor $Hom^\bullet(\mathbf{B}^\bullet, *)$ from the category of complexes of sheaves to the category of complexes of R-modules. These functors are given by

$$(\mathbf{Hom}^\bullet(\mathbf{B}^\bullet, \mathbf{A}^\bullet))^n = \prod_{p \in \mathbb{Z}} \mathbf{Hom}(\mathbf{B}^p, \mathbf{A}^{n+p})$$

and

$$(Hom^\bullet(\mathbf{B}^\bullet, \mathbf{A}^\bullet))^n = \prod_{p \in \mathbb{Z}} Hom(\mathbf{B}^p, \mathbf{A}^{n+p})$$

with differential given by

$$[\partial^n f]^p = \partial^{n+p} f^p + (-1)^{n+1} f^{p+1}\partial^p$$

(there is an indexing error in [**Iv**, 12.4]). The associated derived functors are $R\mathbf{Hom}^\bullet(\mathbf{B}^\bullet, *)$ and $RHom^\bullet(\mathbf{B}^\bullet, *)$, respectively.

If $\mathbf{P}^\bullet \to \mathbf{B}^\bullet$ is a projective resolution of $\mathbf{B}^\bullet$, then, in $\mathbf{D}^b_c(X)$, $R\mathbf{Hom}^\bullet(\mathbf{B}^\bullet, \mathbf{A}^\bullet)$ is isomorphic to $\mathbf{Hom}^\bullet(\mathbf{P}^\bullet, \mathbf{A}^\bullet)$. For all k, $R\mathbf{Hom}^\bullet(\mathbf{B}^\bullet, \mathbf{A}^\bullet[k]) = R\mathbf{Hom}^\bullet(\mathbf{B}^\bullet, \mathbf{A}^\bullet)[k]$.

The functor $RHom^\bullet(\mathbf{B}^\bullet, *)$ is naturally isomorphic to the derived global sections functor applied to $R\mathbf{Hom}^\bullet(\mathbf{B}^\bullet, *)$, i.e., for any $\mathbf{A}^\bullet \in \mathbf{D}^b_c(X)$,

$$RHom^\bullet(\mathbf{B}^\bullet, \mathbf{A}^\bullet) \cong R\Gamma\left(X; R\mathbf{Hom}^\bullet(\mathbf{B}^\bullet, \mathbf{A}^\bullet)\right).$$

$H^0(RHom^\bullet(\mathbf{B}^\bullet, \mathbf{A}^\bullet))$ is naturally isomorphic as an R-module to the derived category homomorphisms from $\mathbf{B}^\bullet$ to $\mathbf{A}^\bullet$, i.e.,

$$H^0(RHom^\bullet(\mathbf{B}^\bullet, \mathbf{A}^\bullet)) \cong Hom_{\mathbf{D}^b_c(X)}(\mathbf{B}^\bullet, \mathbf{A}^\bullet).$$

If $\mathbf{B}^\bullet$ and $\mathbf{A}^\bullet$ are complexes of sheaves on X which have locally constant cohomology sheaves, then, for all $x \in X$, $R\mathbf{Hom}^\bullet(\mathbf{B}^\bullet, \mathbf{A}^\bullet)_x$ is naturally isomorphic to $RHom^\bullet(\mathbf{B}^\bullet_x, \mathbf{A}^\bullet_x)$.

For all $\mathbf{A}^\bullet, \mathbf{B}^\bullet, \mathbf{C}^\bullet \in \mathbf{D}^b_c(X)$, there is a natural isomorphism

$$R\mathbf{Hom}^\bullet(\mathbf{A}^\bullet \overset{L}{\otimes} \mathbf{B}^\bullet, \mathbf{C}^\bullet) \cong R\mathbf{Hom}^\bullet(\mathbf{A}^\bullet, R\mathbf{Hom}^\bullet(\mathbf{B}^\bullet, \mathbf{C}^\bullet)).$$

Moreover, if $\mathbf{C}^\bullet$ has locally constant cohomology sheaves, then there is an isomorphism

$$R\mathbf{Hom}^\bullet(\mathbf{A}^\bullet, \mathbf{B}^\bullet \overset{L}{\otimes} \mathbf{C}^\bullet) \cong R\mathbf{Hom}^\bullet(\mathbf{A}^\bullet, \mathbf{B}^\bullet) \overset{L}{\otimes} \mathbf{C}^\bullet.$$

For all j, we define $\mathbf{Ext}^j(\mathbf{B}^\bullet, \mathbf{A}^\bullet) := \mathbf{H}^j(R\mathbf{Hom}^\bullet(\mathbf{B}^\bullet, \mathbf{A}^\bullet))$ and define

$$Ext^j(\mathbf{B}^\bullet, \mathbf{A}^\bullet) := H^j(RHom^\bullet(\mathbf{B}^\bullet, \mathbf{A}^\bullet)).$$

It is immediate that we have isomorphisms of R-modules

$$Ext^j(\mathbf{B}^\bullet, \mathbf{A}^\bullet) = H^0(RHom^\bullet(\mathbf{B}^\bullet, \mathbf{A}^\bullet[j])) \cong Hom_{\mathbf{D}^b_c(X)}(\mathbf{B}^\bullet, \mathbf{A}^\bullet[j]).$$

If $X = point$, $\mathbf{B}^\bullet \in \mathbf{D}^b_c(X)$, and the base ring is a PID, then $\mathbf{B}^\bullet \cong \bigoplus_k \mathbf{H}^k(\mathbf{B}^\bullet)[-k]$ in $\mathbf{D}^b_c(X)$; if we also have $\mathbf{A}^\bullet \in \mathbf{D}^b_c(X)$, then

$$\mathbf{H}^i(\mathbf{A}^\bullet \overset{L}{\otimes} \mathbf{B}^\bullet) \cong \Big(\bigoplus_{p+q=i} \mathbf{H}^p(\mathbf{A}^\bullet) \otimes \mathbf{H}^q(\mathbf{B}^\bullet)\Big) \oplus \Big(\bigoplus_{r+s=i+1} \mathrm{Tor}(\mathbf{H}^r(\mathbf{A}^\bullet), \mathbf{H}^s(\mathbf{B}^\bullet))\Big).$$

If, in addition, the cohomology modules of $\mathbf{A}^\bullet$ are projective (hence, free), then

$$Hom_{\mathbf{D}^b_c(X)}(\mathbf{A}^\bullet, \mathbf{B}^\bullet) \cong \bigoplus_k Hom\left(\mathbf{H}^k(\mathbf{A}^\bullet), \mathbf{H}^k(\mathbf{B}^\bullet)\right).$$

If we have a map $f : X \to Y$, then the functors f^* and Rf_* are adjoints of each other in the derived category. In fact, for all $\mathbf{A}^\bullet$ on X and $\mathbf{B}^\bullet$ on Y, there is a canonical isomorphism in $\mathbf{D}_c^b(Y)$

$$R\mathbf{Hom}^\bullet(\mathbf{B}^\bullet, Rf_*\mathbf{A}^\bullet) \cong Rf_*R\mathbf{Hom}^\bullet(f^*\mathbf{B}^\bullet, \mathbf{A}^\bullet)$$

and so

$$Hom_{\mathbf{D}_c^b(Y)}(\mathbf{B}^\bullet, Rf_*\mathbf{A}^\bullet) \cong H^0\left(RHom^\bullet(\mathbf{B}^\bullet, Rf_*\mathbf{A}^\bullet)\right) \cong \mathbb{H}^0\left(Y; R\mathbf{Hom}^\bullet(\mathbf{B}^\bullet, Rf_*\mathbf{A}^\bullet)\right)$$

$$\cong \mathbb{H}^0\left(X; R\mathbf{Hom}^\bullet(f^*\mathbf{B}^\bullet, \mathbf{A}^\bullet)\right) \cong H^0\left(RHom^\bullet(f^*\mathbf{B}^\bullet, \mathbf{A}^\bullet)\right) \cong Hom_{\mathbf{D}_c^b(X)}(f^*\mathbf{B}^\bullet, \mathbf{A}^\bullet).$$

We wish now to describe an analogous adjoint for $Rf_!$

Let $\mathbf{I}^\bullet$ be a complex of injective sheaves on Y. Then, $f^!(\mathbf{I}^\bullet)$ is defined to be the sheaf associated to the presheaf given by

$$\Gamma(U; f^!\mathbf{I}^\bullet) = Hom^\bullet(f_!\mathbf{K}_U^\bullet, \mathbf{I}^\bullet),$$

for any open $U \subseteq X$, where $\mathbf{K}_U^\bullet$ denotes the canonical injective resolution of the constant sheaf $\mathbf{R}_U^\bullet$. For any $\mathbf{A}^\bullet \in \mathbf{D}_c^b(X)$, define $f^!\mathbf{A}^\bullet$ to be $f^!\mathbf{I}^\bullet$, where $\mathbf{I}^\bullet$ is the canonical injective resolution of $\mathbf{A}^\bullet$.

Now that we have this definition, we may state:

(Verdier Duality) If $f : X \to Y$, $\mathbf{A}^\bullet \in \mathbf{D}_c^b(X)$, and $\mathbf{B}^\bullet \in \mathbf{D}_c^b(Y)$, then there is a canonical isomorphism in $\mathbf{D}_c^b(Y)$:

$$Rf_*R\mathbf{Hom}^\bullet(\mathbf{A}^\bullet, f^!\mathbf{B}^\bullet) \cong R\mathbf{Hom}^\bullet(Rf_!\mathbf{A}^\bullet, \mathbf{B}^\bullet)$$

and so

$$Hom_{\mathbf{D}_c^b(X)}(\mathbf{A}^\bullet, f^!\mathbf{B}^\bullet) \cong Hom_{\mathbf{D}_c^b(Y)}(Rf_!\mathbf{A}^\bullet, \mathbf{B}^\bullet).$$

If $\mathbf{B}^\bullet$ and $\mathbf{C}^\bullet$ are in $\mathbf{D}_c^b(Y)$, then we have an isomorphism

$$f^!R\mathbf{Hom}^\bullet(\mathbf{B}^\bullet, \mathbf{C}^\bullet) \cong R\mathbf{Hom}^\bullet(f^*\mathbf{B}^\bullet, f^!\mathbf{C}^\bullet).$$

Let $f : X \to$ *point*. Then, the *dualizing complex*, $\mathbb{D}_X^\bullet$, is $f^!$ applied to the constant sheaf, i.e., $\mathbb{D}_X^\bullet = f^!\mathbf{R}_{pt}^\bullet$. For any complex $\mathbf{A}^\bullet \in \mathbf{D}_c^b(X)$, the *Verdier dual* (or, simply, *the dual*) of $\mathbf{A}^\bullet$ is $R\mathbf{Hom}^\bullet(\mathbf{A}^\bullet, \mathbb{D}_X^\bullet)$ and is denoted by $\mathcal{D}_X\mathbf{A}^\bullet$ (or just $\mathcal{D}\mathbf{A}^\bullet$). There is a canonical isomorphism between $\mathbb{D}_X^\bullet$ and the dual of the constant sheaf on X, i.e., $\mathbb{D}_X^\bullet \cong \mathcal{D}\mathbf{R}_X^\bullet$.

Let $\mathbf{A}^\bullet \in \mathbf{D}_c^b(X)$. The dual of $\mathbf{A}^\bullet$, $\mathcal{D}\mathbf{A}^\bullet$, is well-defined up to quasi-isomorphism by:

for any open $U \subseteq X$, there is a natural split exact sequence:

$$0 \to Ext(\mathbb{H}_c^{q+1}(U;\mathbf{A}^\bullet), R) \to \mathbb{H}^{-q}(U;\mathcal{D}\mathbf{A}^\bullet) \to Hom(\mathbb{H}_c^q(U;\mathbf{A}^\bullet), R) \to 0.$$

In particular, if R is a field, then $\mathbb{H}^{-q}(U;\mathcal{D}\mathbf{A}^\bullet) \cong \mathbb{H}_c^q(U;\mathbf{A}^\bullet)$, and so

$$\mathbf{H}^q(\mathcal{D}\mathbf{A}^\bullet)_x \cong \mathbb{H}^q(\mathring{B}_\epsilon(x);\mathcal{D}\mathbf{A}^\bullet) \cong \mathbb{H}_c^{-q}(\mathring{B}_\epsilon(x);\mathbf{A}^\bullet).$$

If, in addition, X is compact, $\mathbb{H}^{-q}(X;\mathcal{D}\mathbf{A}^\bullet) \cong \mathbb{H}^q(X;\mathbf{A}^\bullet)$.

Dualizing is a local operation, i.e., if $i : U \hookrightarrow X$ is the inclusion of an open subset and $\mathbf{A}^\bullet \in \mathbf{D}_c^b(X)$, then $i^*\mathcal{D}\mathbf{A}^\bullet \cong \mathcal{D}i^*\mathbf{A}^\bullet$.

If $\mathcal{L}$ is a local system on a connected real m-manifold, N, then $(\mathcal{D}\mathcal{L}^\bullet)[-m]$ is quasi-isomorphic to a local system; if, in addition, N is smooth and oriented, and $\mathcal{L}$ is actually locally free with stalks R^a and monodromy representation $\eta : \pi_1(N,\mathbf{p}) \to Aut(R^a)$, then $\mathcal{D}\mathcal{L}^\bullet[-m]$ is quasi-isomorphic to a local system with stalks R^a and monodromy

$${}^t\eta : \pi_1(N,\mathbf{p}) \to Aut(R^a),$$

where ${}^t\eta(\alpha) =$ transpose of $\eta(\alpha)$.

If $\mathbf{A}^\bullet \in \mathbf{D}_c^b(X)$, then $\mathcal{D}(\mathbf{A}^\bullet[n]) = (\mathcal{D}\mathbf{A}^\bullet)[-n]$.

If $\pi : X \times \mathbb{C}^n \to X$ is projection, then $\mathcal{D}(\pi^*\mathbf{A}^\bullet)[-n] \cong \pi^*(\mathcal{D}\mathbf{A}^\bullet)[n]$.

The dualizing complex, $\mathbb{D}_X$, is quasi-isomorphic to the complex of sheaves of singular chains on X which is associated to the complex of presheaves, $\mathbf{C}^\bullet$, given by $\Gamma(U;\mathbf{C}^{-p}) := C_p(X, X-U;R)$.

The cohomology sheaves of $\mathbb{D}_X^\bullet$ are non-zero in negative degrees only, with stalks

$$\mathbf{H}^{-p}(\mathbb{D}_X^\bullet)_x = H_p(X, X-x;R).$$

If X is a smooth, oriented, real m-manifold, then $\mathbb{D}_X^\bullet[-m]$ is quasi-isomorphic to $\mathbf{R}_X^\bullet$; hence, for any $\mathbf{A}^\bullet \in \mathbf{D}_c^b(X)$,

$$\mathcal{D}\mathbf{A}^\bullet \cong R\mathbf{Hom}^\bullet(\mathbf{A}^\bullet, \mathbf{R}_X^\bullet[m]) = \big(R\mathbf{Hom}^\bullet(\mathbf{A}^\bullet, \mathbf{R}_X^\bullet)\big)[m].$$

$\mathbb{D}_{V\times W}^\bullet$ is naturally isomorphic to $\pi_1^*\mathbb{D}_V^\bullet \overset{L}{\otimes} \pi_2^*\mathbb{D}_W^\bullet$, where π_1 and π_2 are the projections onto V and W, respectively.

$\mathbb{H}^*(X;\mathbb{D}_X^\bullet) \cong$ homology with closed supports $=$ Borel-Moore homology.

If X is a real, smooth, oriented m-manifold and $R = \mathbb{R}$, then $\mathbb{D}_X^\bullet[-m]$ is naturally isomorphic to the complex of real differential forms on X.

$\mathbb{D}^\bullet_X$ is constructible with respect to any Whitney stratification of X. It follows that if $\mathcal{S}$ is a Whitney stratification of X, then $\mathbf{A}^\bullet \in \mathbf{D}^b_{\mathcal{S}}(X)$ if and only if $\mathcal{D}\mathbf{A}^\bullet \in \mathbf{D}^b_{\mathcal{S}}(X)$.

The functor $\mathcal{D}$ from $\mathbf{D}^b_c(X)$ to $\mathbf{D}^b_c(X)$ is contravariant, and $\mathcal{D}\mathcal{D}$ is naturally isomorphic to the identity. For all $\mathbf{A}^\bullet, \mathbf{B}^\bullet \in \mathbf{D}^b_c(X)$, we have isomorphisms

$$R\mathbf{Hom}^\bullet(\mathbf{A}^\bullet, \mathbf{B}^\bullet) \cong R\mathbf{Hom}^\bullet(\mathcal{D}\mathbf{B}^\bullet, \mathcal{D}\mathbf{A}^\bullet) \cong \mathcal{D}\left(\mathcal{D}\mathbf{B}^\bullet \overset{L}{\otimes} \mathbf{A}^\bullet\right).$$

If $f : X \to Y$ is continuous, then we have natural isomorphisms

$$Rf_! \cong \mathcal{D}Rf_*\mathcal{D} \quad \text{and} \quad f^! \cong \mathcal{D}f^*\mathcal{D}.$$

If $Y \subseteq X$ and $f : X - Y \hookrightarrow X$ is the inclusion, we define

$$\mathbb{H}^k(X, Y; \mathbf{A}^\bullet) := \mathbb{H}^k(X; f_! f^! \mathbf{A}^\bullet).$$

Excision has the following form: if $Y \subseteq U \subseteq X$, where U is open in X and Y is closed in X, then

$$\mathbb{H}^k(X, X - Y; \mathbf{A}^\bullet) \cong \mathbb{H}^k(U, U - Y; \mathbf{A}^\bullet).$$

If $x \in X$ and $\mathbf{A}^\bullet \in \mathbf{D}^b_c(X)$, then for all $\epsilon > 0$ sufficiently small,

$$\mathbb{H}^q_c(\overset{\circ}{B}_\epsilon(x); \mathbf{A}^\bullet) \cong \mathbb{H}^q(\overset{\circ}{B}_\epsilon(x), \overset{\circ}{B}_\epsilon(x) - x; \mathbf{A}^\bullet) \cong \mathbb{H}^q(X, X - x; \mathbf{A}^\bullet)$$

and so, if R is a field,

$$\mathbf{H}^{-q}(\mathcal{D}\mathbf{A}^\bullet)_x \cong \mathbb{H}^q(\overset{\circ}{B}_\epsilon(x), \overset{\circ}{B}_\epsilon(x) - x; \mathbf{A}^\bullet) \cong \mathbb{H}^q(X, X - x; \mathbf{A}^\bullet).$$

If $f : X \to Y$ and $g : Y \to Z$, then there are natural isomorphisms

$$R(g \circ f)_* \cong Rg_* \circ Rf_* \qquad R(g \circ f)_! \cong Rg_! \circ Rf_!$$

and

$$(g \circ f)^* \cong f^* \circ g^* \qquad (g \circ f)^! \cong f^! \circ g^!.$$

Suppose that $f : Y \hookrightarrow X$ is inclusion of a subset. Then, if Y is open, $f^! = f^*$. If Y is closed, then $Rf_! = f_! = f_* = Rf_*$.

If $f : Y \hookrightarrow X$ is the inclusion of one complex manifold into another and $\mathbf{B}^\bullet \in \mathbf{D}_c^b(X)$ has locally constant cohomology on X, then $f^!\mathbf{B}^\bullet$ has locally constant cohomology on Y and

$$f^!\mathbf{B}^\bullet \cong f^*\mathbf{B}^\bullet[-2\ \mathrm{codim}_X Y].$$

(Here, we mean the complex codimension. There is an error here in [**G-M3**]; they have the negation of the correct shift.)

If $\pi : X \times \mathbb{C}^n \to X$ is projection and $\mathbf{A}^\bullet \in \mathbf{D}_c^b(X)$, then $(\pi^!\mathbf{A}^\bullet)[-2n] \cong \pi^*\mathbf{A}^\bullet$.

If $f : X \to Y$ is continuous, $\mathbf{A}^\bullet \in \mathbf{D}_c^b(X)$, and $\mathbf{B}^\bullet \in \mathbf{D}_c^b(Y)$, then dual to the canonical maps

$$\mathbf{B}^\bullet \to Rf_*f^*\mathbf{B}^\bullet \quad \text{and} \quad f^*Rf_*\mathbf{A}^\bullet \to \mathbf{A}^\bullet$$

are the canonical maps

$$Rf_!f^!\mathbf{B}^\bullet \to \mathbf{B}^\bullet \quad \text{and} \quad \mathbf{A}^\bullet \to f^!Rf_!\mathbf{A}^\bullet.$$

If

$$\begin{array}{ccc} Z & \xrightarrow{\hat{f}} & W \\ {\scriptstyle\hat{\pi}}\downarrow & & \downarrow{\scriptstyle\pi} \\ X & \xrightarrow{f} & S \end{array}$$

is a pull-back diagram (a.k.a. fibre square, Cartesian diagram), then, for all $\mathbf{A}^\bullet \in \mathbf{D}_c^b(X)$, $R\hat{f}_!\hat{\pi}^*\mathbf{A}^\bullet \cong \pi^*Rf_!\mathbf{A}^\bullet$ (there is an error in [**G-M3**]; they have lower $*$'s, not lower !'s, but see below for when these agree) and, dually, $R\hat{f}_*\hat{\pi}^!\mathbf{A}^\bullet \cong \pi^!Rf_*\mathbf{A}^\bullet$. In particular, if f is proper (and, hence, $\hat{f}$ is proper) or π is the inclusion of an open subset (and, hence, so is $\hat{\pi}$, up to homeomorphism), then $R\hat{f}_*\hat{\pi}^*\mathbf{A}^\bullet \cong \pi^*Rf_*\mathbf{A}^\bullet$; this is also true if $W = S \times \mathbb{C}^n$ and $\pi : W \to S$ is projection (and, hence, up to homeomorphism, $\hat{\pi}$ is projection from $X \times \mathbb{C}^n$ to X).

If we have $\mathbf{A}^\bullet \in \mathbf{D}_c^b(X)$ and $\mathbf{B}^\bullet \in \mathbf{D}_c^b(W)$, then we let $\mathbf{A}^\bullet \overset{L}{\boxtimes}_S \mathbf{B}^\bullet := \hat{\pi}^*\mathbf{A}^\bullet \overset{L}{\otimes} \hat{f}^*\mathbf{B}^\bullet$, assuming that the maps $\hat{\pi}$ and $\hat{f}$ are clear. If S is a point, so that $Z \cong X \times W$, then we omit the S in the notation and write simply $\mathbf{A}^\bullet \overset{L}{\boxtimes} \mathbf{B}^\bullet$.

There is a Künneth formula, which we now state in its most general form, in terms of maps over a base space S. Suppose that we have two maps $f_1 : X_1 \to Y_1$ and $f_2 : X_2 \to Y_2$ over S, i.e., there are commutative diagrams

$$\begin{array}{ccc} X_1 & \xrightarrow{f_1} & Y_1 \\ {\scriptstyle r_1}\searrow & & \swarrow{\scriptstyle t_1} \\ & \mathrm{S} & \end{array} \qquad \text{and} \qquad \begin{array}{ccc} X_2 & \xrightarrow{f_2} & Y_2 \\ {\scriptstyle r_2}\searrow & & \swarrow{\scriptstyle t_2} \\ & \mathrm{S} & \end{array} .$$

Then, there is an induced map $f = f_1 \times_S f_2 : X_1 \times_S X_2 \to Y_1 \times_S Y_2$. If $\mathbf{A}^\bullet \in \mathbf{D}_c^b(X_1)$ and $\mathbf{B}^\bullet \in \mathbf{D}_c^b(X_2)$, there is the **Künneth isomorphism**

$$Rf_!(\mathbf{A}^\bullet \overset{L}{\boxtimes}_S \mathbf{B}^\bullet) \cong Rf_{1!}\mathbf{A}^\bullet \overset{L}{\boxtimes}_S Rf_{2!}\mathbf{B}^\bullet.$$

Using the above notation, if S is a point and $\mathbf{F}^\bullet \in \mathbf{D}_c^b(Y_1)$ and $\mathbf{G}^\bullet \in \mathbf{D}_c^b(Y_2)$, then there is a natural isomorphism (the **adjoint Künneth isomorphism**)

$$f^!(\mathbf{F}^\bullet \overset{L}{\boxtimes} \mathbf{G}^\bullet) \cong f_1{}^!\mathbf{F}^\bullet \overset{L}{\boxtimes} f_2{}^!\mathbf{G}^\bullet.$$

If we let q_1 and q_2 denote the projections from $Y_1 \times Y_2$ onto Y_1 and Y_2, respectively, then the adjoint Künneth formula can be proved by using the following natural isomorphism twice

$$\mathcal{D}\mathbf{F}^\bullet \overset{L}{\boxtimes} \mathbf{G}^\bullet \cong R\mathbf{Hom}^\bullet(q_1^*\mathbf{F}^\bullet, q_2^!\mathbf{G}^\bullet).$$

Let Z be a locally closed subset of an analytic space X. There are two derived functors, associated to Z, that we wish to describe: the derived functors of restricting-extending to Z, and of taking the sections supported on Z. Let i denote the inclusion of Z into X.

If $\mathbf{A}$ is a (single) sheaf on X, then the restriction-extension of $\mathbf{A}$ to Z, $(\mathbf{A})_Z$, is given by $i_!i^*(\mathbf{A})$. Thus, up to isomorphism, $(\mathbf{A})_Z$ is characterized by $((\mathbf{A})_Z)_{|_Z} \cong \mathbf{A}_{|_Z}$ and $((\mathbf{A})_Z)_{|_{X-Z}} = 0$. This functor is exact, and so we also denote the derived functor by $()_Z$.

Now, we want to define the sheaf of sections of $\mathbf{A}$ supported by Z, $\Gamma_Z(\mathbf{A})$. If $\mathcal{U}$ is an open subset of X which contains Z, then we define

$$\Gamma_Z(\mathcal{U}; \mathbf{A}) := \ker\{\Gamma(\mathcal{U}; \mathbf{A}) \to \Gamma(\mathcal{U} - Z; \mathbf{A})\}.$$

Up to isomorphism, $\Gamma_Z(\mathcal{U}; \mathbf{A})$ is independent of the open set $\mathcal{U}$ (this uses that $\mathbf{A}$ is a sheaf, not just a presheaf). The sheaf $\Gamma_Z(\mathbf{A})$ is defined by, for all open $\mathcal{U} \subseteq X$, $\Gamma(\mathcal{U}; \Gamma_Z(\mathbf{A})) := \Gamma_{\mathcal{U}\cap Z}(\mathcal{U}; \mathbf{A})$.

One easily sees that $\operatorname{supp}\Gamma_Z(\mathbf{A}) \subseteq \overline{Z}$. It is also easy to see that, if Z is open, then $\Gamma_Z(\mathbf{A}) = i_*i^*(\mathbf{A})$.

The functor $\Gamma_Z()$ is left exact; of course, we denote the right derived functor by $R\Gamma_Z()$.

There is a canonical isomorphism $i^! \cong i^* \circ R\Gamma_Z$. It follows that, if Z is closed, then $R\Gamma_Z \cong i_!i^!$. In addition, if Z is open, then $R\Gamma_Z \cong Ri_*i^*$.

Avoiding Injective Resolutions:

To calculate right derived functors from the definition, one must use injective resolutions. However, this is inconvenient in many proofs if some functor involved in the proof does not take injective complexes to injective complexes. There are (at least) four "devices" which come to our aid, and enable one to prove many of the isomorphisms described earlier; these devices are **fine resolutions**, **flabby resolutions**, **c-soft resolutions**, and **injective subcategories with respect to a functor**.

If T is a left-exact functor on the category of sheaves on X, then the right derived functor RT is defined by applying T term-wise to the sheaves in a canonical injective resolution. The importance of saying that a certain subcategory of the category of sheaves on X is injective with respect to T is that one may take a resolution in which the individual sheaves are in the given subcategory, then apply T term-wise, and end up with a complex which is canonically isomorphic to that produced by RT.

Recall that a single sheaf $\mathbf{A}$ on X is:

fine, if partitions of unity of $\mathbf{A}$ subordinate to any given locally finite open cover of X exist;

flabby, if for every open subset $\mathcal{U} \subseteq X$, the restriction homomorphism $\Gamma(X; \mathbf{A}) \rightarrow \Gamma(\mathcal{U}; \mathbf{A})$ is a surjection;

c-soft, if for every compact subset $\mathcal{K} \subseteq X$, the restriction homomorphism $\Gamma(X; \mathbf{A}) \rightarrow \Gamma(\mathcal{K}; \mathbf{A})$ is a surjection;

Injective sheaves are flabby, and flabby sheaves are c-soft. In addition, fine sheaves are c-soft.

The subcategory of c-soft sheaves is injective with respect to the functors $f_!$, $\Gamma(X; *)$, and $\Gamma_c(X; *)$. The subcategory of flabby sheaves is injective with respect to the functor f_*.

If $\mathbf{A}^\bullet$ is a bounded complex of sheaves, then a bounded c-soft resolution of $\mathbf{A}^\bullet$ is given by $\mathbf{A}^\bullet \rightarrow \mathbf{A}^\bullet \otimes \mathbf{S}^\bullet$, where $\mathbf{S}^\bullet$ is a c-soft, bounded above, resolution of the base ring (which always exists in our context).

Triangles:

$\mathbf{D}_c^b(X)$ is an additive category, but is not an Abelian category. In place of short exact sequences, one has distinguished triangles, just as we did in $\mathbf{K}^b(\mathcal{C})$. A triangle of morphisms in $\mathbf{D}_c^b(X)$

$$\begin{array}{ccc} \mathbf{A}^\bullet & \longrightarrow & \mathbf{B}^\bullet \\ {\scriptstyle [1]} \nwarrow & & \swarrow \\ & \mathbf{C}^\bullet & \end{array}$$

(the [1] indicates a morphism shifted by one, i.e., a morphism $\mathbf{C}^\bullet \rightarrow \mathbf{A}^\bullet[1]$) is called *distinguished* if it is isomorphic in $\mathbf{D}_c^b(X)$ to a diagram of sheaf maps

$$\begin{array}{ccc} \tilde{\mathbf{A}}^\bullet & \xrightarrow{\phi} & \tilde{\mathbf{B}}^\bullet \\ {\scriptstyle [1]} \nwarrow & & \swarrow \\ & \mathbf{M}^\bullet & \end{array}$$

where $\mathbf{M}^\bullet$ is the algebraic mapping cone of ϕ and $\mathbf{B}^\bullet \rightarrow \mathbf{M}^\bullet \rightarrow \mathbf{A}^\bullet[1]$ are the canonical maps.

The "in-line" notation for a triangle is $\mathbf{A}^\bullet \rightarrow \mathbf{B}^\bullet \rightarrow \mathbf{C}^\bullet \rightarrow \mathbf{A}^\bullet[1]$ or $\mathbf{A}^\bullet \rightarrow \mathbf{B}^\bullet \rightarrow \mathbf{C}^\bullet \xrightarrow{[1]}$.

Any short exact sequence of complexes becomes a distinguished triangle in $\mathbf{D}_c^b(X)$. Any edge of a distinguished triangle determines the triangle up to (non-canonical) isomorphism in $\mathbf{D}^b_c(X)$; more specifically, we can "turn" the distinguished triangle:

$$\mathbf{A}^\bullet \xrightarrow{\alpha} \mathbf{B}^\bullet \xrightarrow{\beta} \mathbf{C}^\bullet \xrightarrow{\gamma} \mathbf{A}^\bullet[1]$$

is a distinguished triangle if and only if

$$\mathbf{B}^\bullet \xrightarrow{\beta} \mathbf{C}^\bullet \xrightarrow{\gamma} \mathbf{A}^\bullet[1] \xrightarrow{-\alpha[1]} \mathbf{B}^\bullet[1]$$

is a distinguished triangle.

Given two distinguished triangles and maps u and v which make the left-hand square of the following diagram commute

$$\begin{array}{cccc} \mathbf{A}^\bullet \to & \mathbf{B}^\bullet \to & \mathbf{C}^\bullet \to & \mathbf{A}^\bullet[1] \\ \downarrow u & \downarrow v & & \downarrow u[1] \\ \widetilde{\mathbf{A}^\bullet} \to & \widetilde{\mathbf{B}^\bullet} \to & \widetilde{\mathbf{C}^\bullet} \to & \widetilde{\mathbf{A}^\bullet}[1], \end{array}$$

there exists a (not necessarily unique) $w : \mathbf{C}^\bullet \to \widetilde{\mathbf{C}^\bullet}$ such that

$$\begin{array}{cccc} \mathbf{A}^\bullet \to & \mathbf{B}^\bullet \to & \mathbf{C}^\bullet \to & \mathbf{A}^\bullet[1] \\ \downarrow u & \downarrow v & \downarrow w & \downarrow u[1] \\ \widetilde{\mathbf{A}^\bullet} \to & \widetilde{\mathbf{B}^\bullet} \to & \widetilde{\mathbf{C}^\bullet} \to & \widetilde{\mathbf{A}^\bullet}[1] \end{array}$$

also commutes. We say that the original commutative square embeds in a morphism of distinguished triangles.

We will now give the *octahedral lemma*, which allows one to realize an isomorphism between mapping cones of two composed maps. Suppose that we have two distinguished triangles

$$\mathbf{A}^\bullet \xrightarrow{f} \mathbf{B}^\bullet \xrightarrow{g} \mathbf{C}^\bullet \xrightarrow{h} \mathbf{A}^\bullet[1]$$

and

$$\mathbf{B}^\bullet \xrightarrow{\beta} \mathbf{E}^\bullet \xrightarrow{\gamma} \mathbf{F}^\bullet \xrightarrow{\delta} \mathbf{B}^\bullet[1].$$

Then, there exists a complex $\mathbf{M}^\bullet$ and two distinguished triangles

$$\mathbf{A}^\bullet \xrightarrow{\beta\circ f} \mathbf{E}^\bullet \xrightarrow{\tau} \mathbf{M}^\bullet \xrightarrow{\omega} \mathbf{A}^\bullet[1]$$

and

$$\mathbf{C}^\bullet \xrightarrow{\sigma} \mathbf{M}^\bullet \xrightarrow{\nu} \mathbf{F}^\bullet \xrightarrow{g[1]\circ\delta} \mathbf{C}^\bullet[1].$$

such that the following diagram commutes

$$\begin{array}{ccccccc}
\mathbf{A}^\bullet & \xrightarrow{f} & \mathbf{B}^\bullet & \xrightarrow{g} & \mathbf{C}^\bullet & \xrightarrow{h} & \mathbf{A}^\bullet[1] \\
\mathrm{id}\downarrow & & \beta\downarrow & & \sigma\downarrow & & \mathrm{id}\downarrow \\
\mathbf{A}^\bullet & \xrightarrow{\beta\circ f} & \mathbf{E}^\bullet & \xrightarrow{\tau} & \mathbf{M}^\bullet & \xrightarrow{\omega} & \mathbf{A}^\bullet[1] \\
f\downarrow & & \mathrm{id}\downarrow & & \nu\downarrow & & f[1]\downarrow \\
\mathbf{B}^\bullet & \xrightarrow{\beta} & \mathbf{E}^\bullet & \xrightarrow{\gamma} & \mathbf{F}^\bullet & \xrightarrow{\delta} & \mathbf{B}^\bullet[1] \\
g\downarrow & & \tau\downarrow & & \mathrm{id}\downarrow & & g[1]\downarrow \\
\mathbf{C}^\bullet & \xrightarrow{\sigma} & \mathbf{M}^\bullet & \xrightarrow{\nu} & \mathbf{F}^\bullet & \xrightarrow{g[1]\circ\delta} & \mathbf{C}^\bullet[1].
\end{array}$$

It is somewhat difficult to draw this in its octahedral form (and worse to type it); moreover, it is no easier to read the relations from the octahedron. However, the interested reader can give it a try: the octahedron is formed by gluing together two pyramids along their square bases. One pyramid has $\mathbf{B}^\bullet$ at its top vertex, with $\mathbf{A}^\bullet$, $\mathbf{C}^\bullet$, $\mathbf{E}^\bullet$, and $\mathbf{F}^\bullet$ at the vertices of its base, and has the original two distinguished triangles as opposite faces. The other pyramid has $\mathbf{M}^\bullet$ at its top vertex, with $\mathbf{A}^\bullet$, $\mathbf{C}^\bullet$, $\mathbf{E}^\bullet$, and $\mathbf{F}^\bullet$ at the vertices of its base, and has the other two distinguished triangles (whose existence is asserted in the lemma) as opposite faces. The two pyramids are joined together by matching the vertices of the two bases, forming an octahedron in which the faces are alternately distinguished and commuting.

A distinguished triangle determines long exact sequences on cohomology and hypercohomology:

$$\cdots \to \mathbf{H}^p(\mathbf{A}^\bullet) \to \mathbf{H}^p(\mathbf{B}^\bullet) \to \mathbf{H}^p(\mathbf{C}^\bullet) \to \mathbf{H}^{p+1}(\mathbf{A}^\bullet) \to \cdots$$

$$\cdots \to \mathbb{H}^p(X;\mathbf{A}^\bullet) \to \mathbb{H}^p(X;\mathbf{B}^\bullet) \to \mathbb{H}^p(X;\mathbf{C}^\bullet) \to \mathbb{H}^{p+1}(X;\mathbf{A}^\bullet) \to \cdots .$$

If $f : X \to Y$ and $\mathbf{F}^\bullet \in \mathbf{D}^b_c(X)$, then the functors $Rf_*, Rf_!, f^*,\ f^!$, and $\mathbf{F}^\bullet \overset{L}{\otimes} *$ all take distinguished triangles to distinguished triangles (with all arrows in the same direction and the shift in the same place).

As for $R\mathbf{Hom}^\bullet$, if $\mathbf{F}^\bullet \in \mathbf{D}^b_c(X)$ and $\mathbf{A}^\bullet \to \mathbf{B}^\bullet \to \mathbf{C}^\bullet \to \mathbf{A}^\bullet[1]$ is a distinguished triangle in $\mathbf{D}^b_c(X)$, then we have distinguished triangles

$$\begin{array}{ccc}
R\mathbf{Hom}^\bullet(\mathbf{F}^\bullet,\mathbf{A}^\bullet) & \longrightarrow & R\mathbf{Hom}^\bullet(\mathbf{F}^\bullet,\mathbf{B}^\bullet) \\
{\scriptstyle[1]}\nwarrow & & \swarrow \\
& R\mathbf{Hom}^\bullet(\mathbf{F}^\bullet,\mathbf{C}^\bullet) &
\end{array}
\quad\text{and}\quad
\begin{array}{ccc}
R\mathbf{Hom}^\bullet(\mathbf{A}^\bullet,\mathbf{F}^\bullet) & \longleftarrow & R\mathbf{Hom}^\bullet(\mathbf{B}^\bullet,\mathbf{F}^\bullet) \\
{\scriptstyle[1]}\searrow & & \nearrow \\
& R\mathbf{Hom}^\bullet(\mathbf{C}^\bullet,\mathbf{F}^\bullet). &
\end{array}$$

By applying the right-hand triangle above to the special case where $\mathbf{F}^\bullet = \mathbb{D}^\bullet_X$, we find that the dualizing functor $\mathcal{D}$ also takes distinguished triangles to distinguished triangles, but with a reversal of arrows, i.e., if we have a distinguished triangle $\mathbf{A}^\bullet \to \mathbf{B}^\bullet \to \mathbf{C}^\bullet \to \mathbf{A}^\bullet[1]$ in $\mathbf{D}^b_c(X)$, then, by dualizing, we have distinguished triangles

$$\begin{array}{ccc} \mathcal{D}\mathbf{A}^\bullet & \longleftarrow & \mathcal{D}\mathbf{B}^\bullet \\ {\scriptstyle [1]}\searrow & & \nearrow \\ & \mathcal{D}\mathbf{C}^\bullet & \end{array} \qquad \text{or} \qquad \begin{array}{ccc} \mathcal{D}\mathbf{C}^\bullet & \longrightarrow & \mathcal{D}\mathbf{B}^\bullet \\ {\scriptstyle [1]}\nwarrow & & \swarrow \\ & \mathcal{D}\mathbf{A}^\bullet. & \end{array}$$

There are (at least) six distinguished triangles associated to the functors $()_Z$ and $R\Gamma_Z$. Let $\mathbf{F}^\bullet$ be in $\mathbf{D}_c^b(X)$, $\mathcal{U}_1$ and $\mathcal{U}_2$ be open subsets of X, Z_1 and Z_2 be closed subsets of X, Z be locally closed in X, and Z' be closed in Z. Then, we have the following distinguished triangles:

$$R\Gamma_{\mathcal{U}_1\cup\mathcal{U}_2}(\mathbf{F}^\bullet) \to R\Gamma_{\mathcal{U}_1}(\mathbf{F}^\bullet) \oplus R\Gamma_{\mathcal{U}_2}(\mathbf{F}^\bullet) \to R\Gamma_{\mathcal{U}_1\cap\mathcal{U}_2}(\mathbf{F}^\bullet) \xrightarrow{[1]}$$

$$R\Gamma_{Z_1\cap Z_2}(\mathbf{F}^\bullet) \to R\Gamma_{Z_1}(\mathbf{F}^\bullet) \oplus R\Gamma_{Z_2}(\mathbf{F}^\bullet) \to R\Gamma_{Z_1\cup Z_2}(\mathbf{F}^\bullet) \xrightarrow{[1]}$$

$$(\mathbf{F}^\bullet)_{\mathcal{U}_1\cap\mathcal{U}_2} \to (\mathbf{F}^\bullet)_{\mathcal{U}_1} \oplus (\mathbf{F}^\bullet)_{\mathcal{U}_2} \to (\mathbf{F}^\bullet)_{\mathcal{U}_1\cup\mathcal{U}_2} \xrightarrow{[1]}$$

$$(\mathbf{F}^\bullet)_{Z_1\cup Z_2} \to (\mathbf{F}^\bullet)_{Z_1} \oplus (\mathbf{F}^\bullet)_{Z_2} \to (\mathbf{F}^\bullet)_{Z_1\cap Z_2} \xrightarrow{[1]}$$

$$R\Gamma_{Z'}(\mathbf{F}^\bullet) \to R\Gamma_{Z}(\mathbf{F}^\bullet) \to R\Gamma_{Z-Z'}(\mathbf{F}^\bullet) \xrightarrow{[1]}$$

$$(\mathbf{F}^\bullet)_{Z-Z'} \to (\mathbf{F}^\bullet)_{Z} \to (\mathbf{F}^\bullet)_{Z'} \xrightarrow{[1]}.$$

If $j : Y \hookrightarrow X$ is the inclusion of a closed subspace and $i : U \hookrightarrow X$ the inclusion of the open complement, then for all $\mathbf{A}^\bullet \in \mathbf{D}_c^b(X)$, the last two triangles above give us distinguished triangles

$$\begin{array}{ccc} Ri_!i^!\mathbf{A}^\bullet & \longrightarrow & \mathbf{A}^\bullet \\ {\scriptstyle [1]}\nwarrow & & \swarrow \\ & Rj_*j^*\mathbf{A}^\bullet & \end{array} \qquad \text{and} \qquad \begin{array}{ccc} Rj_!j^!\mathbf{A}^\bullet & \longrightarrow & \mathbf{A}^\bullet \\ {\scriptstyle [1]}\nwarrow & & \swarrow \\ & Ri_*i^*\mathbf{A}^\bullet, & \end{array}$$

where the second triangle can be obtained from the first by dualizing. (Note that $Ri_! = i_!$, $Rj_* = j_* = j_! = Rj_!$, and $i^! = i^*$.) The associated long exact sequences on hypercohomology are those for the pairs $\mathbb{H}^*(X, Y; \mathbf{A}^\bullet)$ and $\mathbb{H}^*(X, U; \mathbf{A}^\bullet)$, respectively.

By applying these two triangles to $Ri_*i^*\mathbf{A}^\bullet$ and $Ri_!i^!\mathbf{A}^\bullet$, respectively, we obtain a natural isomorphism

$$Rj_!j^!Ri_!i^!\mathbf{A}^\bullet[1] \cong Rj_*j^*Ri_*i^*\mathbf{A}^\bullet.$$

As in our earlier discussion of the octahedral lemma, all of the morphisms of the last two paragraphs fit into the *fundamental octahedron of the pair* (X, Y). The four distinguished triangles making up the fundamental octahedron are the top pair

$$Ri_!i^!\mathbf{A}^\bullet \to \mathbf{A}^\bullet \to Rj_*j^*\mathbf{A}^\bullet \to Ri_!i^!\mathbf{A}^\bullet[1]$$

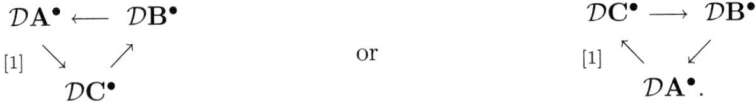

and

$$\mathbf{A}^\bullet \to Ri_*i^*\mathbf{A}^\bullet \to Rj_!j^!\mathbf{A}^\bullet[1] \to \mathbf{A}^\bullet[1]$$

and the bottom pair

$$Ri_!i^!\mathbf{A}^\bullet \to Ri_*i^*\mathbf{A}^\bullet \to \mathbf{M}^\bullet \to Ri_!i^!\mathbf{A}^\bullet[1]$$

and

$$Rj_*j^*\mathbf{A}^\bullet \to \mathbf{M}^\bullet \to Rj_!j^!\mathbf{A}^\bullet[1] \to Rj_*j^*\mathbf{A}^\bullet[1],$$

where $\mathbf{M}^\bullet \cong Rj_!j^!Ri_!i^!\mathbf{A}^\bullet[1] \cong Rj_*j^*Ri_*i^*\mathbf{A}^\bullet$.

§2. Perverse Sheaves

Suppose that $\mathbf{P}^\bullet \in \mathbf{D}_c^b(X)$. There are two non-equivalent definitions of what it means for $\mathbf{P}^\bullet$ to be *perverse*. The first one (which is actually **the** definition of perverse) is a purely local definition and, when the base ring is a field, is symmetric with respect to dualizing. This definition is definitely the more elegant of the two, but it gives cohomology groups only in negative dimensions; this seems non-intuitive from the topologist's point of view.

The second definition of perverse - which differs from the first only by a shift - has the advantage that the cohomology groups appear in non-negative dimensions only. Also, the constant sheaf on a local complete intersection is such a sheaf and, with this definition of perverse, the nearby and vanishing cycles (see §3) of a perverse sheaf are again perverse. Finally, if one wants intersection cohomology with its usual indexing (that is, the indexing that gives cohomology in non-negative dimensions) to be a perverse sheaf, then one must use this second definition of perverse.

Despite these advantages of this second definition of perverse, the fact that it does not localize well on non-pure-dimensional spaces complicates general statements in almost every case. Statements tend to be much cleaner using the first definition. Hence, below, we use the term *perverse sheaf* for this first definition, and use *positively perverse sheaf* for the second definition.

We shall give most statements in terms of perverse sheaves only; the reader may do the necessary shifts to obtain the positively perverse statements. The exceptions to this are those few statements which seem cleaner using positively perverse.

Definition: Let X be a complex analytic space, and for each $\mathbf{x} \in X$, let $j_{\mathbf{x}} : \mathbf{x} \hookrightarrow X$ denote the inclusion.

If $\mathbf{F}^\bullet \in \mathbf{D}_c^b(X)$, then the *support of* $\mathbf{H}^i(\mathbf{F}^\bullet)$ is the closure in X of

$$\{\mathbf{x} \in X |\ \mathbf{H}^i(\mathbf{F}^\bullet)_{\mathbf{x}} \neq 0\} = \{\mathbf{x} \in X |\ \mathbf{H}^i(j_{\mathbf{x}}^*\mathbf{F}^\bullet) \neq 0\};$$

we denote this by $\operatorname{supp}^{\mathrm{i}} \mathbf{F}^\bullet$.

The *i-th cosupport of* $\mathbf{F}^\bullet$ is the closure in X of

$$\{\mathbf{x} \in X |\ \mathbf{H}^i(j_{\mathbf{x}}^!\mathbf{F}^\bullet) \neq 0\} \quad = \quad \{\mathbf{x} \in X |\ \mathbb{H}^i(\overset{\circ}{B}_\epsilon(x), \overset{\circ}{B}_\epsilon(x) - x;\ \mathbf{F}^\bullet) \neq 0\};$$

we denote this by $\operatorname{cosupp}^{\mathrm{i}} \mathbf{F}^\bullet$.

If the base ring, R, is a field, then $\text{cosupp}^{\text{i}}\,\mathbf{F}^\bullet = \text{supp}^{\text{-i}}\,\mathcal{D}\mathbf{F}^\bullet$.

Definition: Let X be a complex analytic space (not necessarily pure dimensional). Then, $\mathbf{P}^\bullet \in \mathbf{D}_c^b(X)$ is *perverse* provided that for all i:

(support) $\dim(\text{supp}^{\text{-i}}\,\mathbf{P}^\bullet) \leqslant i$;

(cosupport) $\dim(\text{cosupp}^{\text{i}}\,\mathbf{P}^\bullet) \leqslant i$,

where we set the dimension of the empty set to be $-\infty$.

This definition is equivalent to: let $\{S_\alpha\}$ be any Whitney stratification of X with respect to which $\mathbf{P}^\bullet$ is constructible, and let $s_\alpha : S_\alpha \hookrightarrow X$ denote the inclusion. Then,

(support) $\mathbf{H}^k(s_\alpha^*\mathbf{P}^\bullet) = 0$ for $k > -\dim_{\mathbb{C}} S_\alpha$;

(cosupport) $\mathbf{H}^k(s_\alpha^!\mathbf{P}^\bullet) = 0$ for $k < -\dim_{\mathbb{C}} S_\alpha$.

(There is a missing minus sign in [**G-M2**, 6.A.5].)

If X is an n-dimensional space, then $\mathbf{P}^\bullet$ is *positively perverse* if and only if $\mathbf{P}^\bullet[n]$ is perverse.

From the definition, it is clear that being perverse is a local property.

If the base ring R is, in fact, a field, then the support and cosupport conditions can be written in the following form, which is symmetric with respect to dualizing:

(support) $\dim(\text{supp}^{\text{-i}}\,\mathbf{P}^\bullet) \leqslant i$;

(cosupport) $\dim(\text{supp}^{\text{-i}}\,\mathcal{D}\mathbf{P}^\bullet) \leqslant i$.

Suppose that $\mathbf{P}^\bullet$ is perverse on X, (X, x) is locally embedded in $\mathbb{C}^n$, S is a stratum of a Whitney stratification with respect to which $\mathbf{P}^\bullet$ is constructible, and $x \in S$. Let M be a normal slice of X at x; that is, let M be a smooth submanifold of $\mathbb{C}^n$ of dimension $n - \dim S$ which transversely intersects S at x. Then, for some open neighborhood U of x in X, $\mathbf{P}^\bullet_{|X\cap M\cap U}[-\dim S]$ is perverse on $X \cap M \cap U$.

Let $\mathbf{P}^\bullet \in \mathbf{D}_c^b(X)$; one can use this normal slicing proposition to prove:

if $\mathbf{P}^\bullet$ is perverse, then $\mathbf{H}^i(\mathbf{P}^\bullet) = 0$ for all $i < -\dim X$;

and so,

if $\mathbf{P}^\bullet$ is positively perverse, then $\mathbf{H}^i(\mathbf{P}^\bullet) = 0$ for all $i < 0$.

A converse to the normal slicing proposition is:

if $\pi : X \times \mathbb{C}^s \to X$ is projection and $\mathbf{P}^\bullet$ is positively perverse on X, then $\pi^*\mathbf{P}^\bullet$ is also positively perverse. Thus, if $\mathbf{P}^\bullet$ is perverse on X, then $\pi^*\mathbf{P}^\bullet[s]$ is perverse.

Suppose $\mathbf{P}^\bullet \in \mathbf{D}_c^b(X^n)$. Let $\Sigma = \text{supp } \mathbf{H}^*(\mathbf{P}^\bullet)$. Then, $\mathbf{P}^\bullet$ is perverse on X if and only if $\mathbf{P}^\bullet_{|_\Sigma}$ is perverse on Σ. Hence, $\mathbf{P}^\bullet$ is positively perverse on X if and only if $\mathbf{P}^\bullet_{|_\Sigma}[\text{codim}_X\Sigma]$ is positively perverse on Σ.

Another way of saying this is: if $j : \Sigma \to X$ is the inclusion of a closed subspace, then $\mathbf{Q}^\bullet$ is perverse on Σ if and only if $j_!\mathbf{Q}^\bullet$ is perverse on X; hence, $\mathbf{Q}^\bullet$ is positively perverse on Σ if and only if $j_!\mathbf{Q}^\bullet[-\text{codim}_X\Sigma]$ is positively perverse on X.

It follows that if $\mathbf{P}^\bullet$ is perverse, then $\mathbf{H}^i(\mathbf{P}^\bullet) = 0$ unless $-\dim \Sigma \leqslant i \leqslant 0$; and so, if $\mathbf{P}^\bullet$ is positively perverse, then $\mathbf{H}^i(\mathbf{P}^\bullet) = 0$ unless $\text{codim}_X\Sigma \leqslant i \leqslant n$. In particular, on an n-dimensional space, a positively perverse sheaf which is supported only at isolated points has cohomology only in dimension n (i.e., the *middle* dimension).

The constant sheaf $\mathbf{R}^\bullet_X$ is positively perverse provided that X is a pure-dimensional local complete intersection. More generally, if X is a pure-dimensional local complete intersection, and $\mathbf{M}$ is a locally free sheaf of R-modules, then $\mathbf{M}^\bullet$ is a positively perverse sheaf on X.

The other basic example of a perverse sheaf that we wish to give is that of intersection cohomology with local coefficients (with the perverse indexing, i.e., cohomology in degrees less than or equal to zero). Note that the definition below is shifted by $-\dim_{\mathbb{C}} X$ from the definition in [**G-M3**], and yields a perverse sheaf which has possibly non-zero cohomology only in degrees between $-\dim_{\mathbb{C}} X$ and -1, inclusive.

Let X be a n-dimensional complex analytic set, let $X^{(n)} = X_1 \cup \dots X_k$ be the union of the n-dimensional components of X, and let $\mathcal{L}$ be a local system on a smooth, open dense subset, $\overset{\circ}{X}$, of $X^{(n)}$. Then, in $\mathbf{D}_c^b(X)$, there is an object, $\mathbf{IC}^\bullet_X(\mathcal{L})$, called the *intersection cohomology with coefficients in* $\mathcal{L}$ which is uniquely determined up to quasi-isomorphism by:

0) $\quad \mathbf{IC}^\bullet_X(\mathcal{L})_{|_{X-X^{(n)}}} = 0;$

1) $\quad \mathbf{IC}^\bullet_X(\mathcal{L})_{|_{\overset{\circ}{X}}} = \mathcal{L}^\bullet[n];$

2) $\quad \mathbf{H}^i\left(\mathbf{IC}^\bullet_X(\mathcal{L})\right) = 0 \qquad \text{for } i < -n;$

3) $\quad \dim \text{supp}^{-i}\left(\mathbf{IC}^\bullet_X(\mathcal{L})\right) < i \qquad \text{for all } i < n;$

4) $\quad \dim \text{cosupp}^{i}\left(\mathbf{IC}^\bullet_X(\mathcal{L})\right) < i \qquad \text{for all } i < n.$

Note the **strict** inequalities in 3) and 4).

The uniqueness assertion implies that

$$\mathbf{IC}^\bullet_X(\mathcal{L}) \cong j^1_!\mathbf{IC}^\bullet_{X_1}(\mathcal{L}_{|_{X_1 \cap \overset{\circ}{X}}}) \oplus \dots \oplus j^k_!\mathbf{IC}^\bullet_{X_k}(\mathcal{L}_{|_{X_k \cap \overset{\circ}{X}}}),$$

where j^m denotes the inclusion of X_m into X.

In many sources, $\mathbf{IC}^\bullet_X(\mathcal{L})$ is only defined if X is pure-dimensional. We find it convenient to have the intersection cohomology complex defined in the general situation – though, condition 0) above says that our intersection cohomology complex is precisely the intersection cohomology complex on the pure-dimensional space $X^{(n)}$ extended by zero to all of X. See section 5 for more on $\mathbf{IC}^\bullet_X(\mathcal{L})$.

The uniqueness assertion which accompanied our axioms for intersection cohomology implies that $\mathbf{IC}^\bullet_X(\mathcal{L})$ is semi-simple. More precisely, suppose that X is a n-dimensional complex analytic set, let be the union of the n-dimensional components of X, and let $\mathcal{L}$ be a local system on a smooth, open dense subset, $\overset{\circ}{X}$, of $X^{(n)}$.

The Category of Perverse Sheaves (see, also, section 5)

The *category of perverse sheaves* on X, $Perv(X)$, is the full subcategory of $\mathbf{D}^b_c(X)$ whose objects are the perverse sheaves. Given a Whitney stratification, $\mathcal{S}$, of X, it is also useful to consider the category $Perv_{\mathcal{S}}(X) := Perv(X) \cap \mathbf{D}^b_{\mathcal{S}}(X)$ of perverse sheaves which are constructible with respect to $\mathcal{S}$.

$Perv(X)$ and $Perv_{\mathcal{S}}(X)$ are both Abelian categories in which the short exact sequences

$$0 \to \mathbf{A}^\bullet \to \mathbf{B}^\bullet \to \mathbf{C}^\bullet \to 0$$

are precisely the distinguished triangles

$$\begin{array}{ccc} \mathbf{A}^\bullet & \longrightarrow & \mathbf{B}^\bullet \\ {\scriptstyle [1]}\nwarrow & & \swarrow \\ & \mathbf{C}^\bullet & . \end{array}$$

If we have complexes $\mathbf{A}^\bullet, \mathbf{B}^\bullet$, and $\mathbf{C}^\bullet$ in $\mathbf{D}^b_c(X)$ (resp. $\mathbf{D}^b_{\mathcal{S}}(X)$), a distinguished triangle $\mathbf{A}^\bullet \to \mathbf{B}^\bullet \to \mathbf{C}^\bullet \to \mathbf{A}^\bullet[1]$, and $\mathbf{A}^\bullet$ and $\mathbf{C}^\bullet$ are perverse, then $\mathbf{B}^\bullet$ is also in $Perv(X)$ (resp. $Perv_{\mathcal{S}}(X)$).

If the Whitney stratification $\mathcal{S}$ has a finite number of strata, then $Perv_{\mathcal{S}}(X)$ is actually an Artinian category, which means that every perverse sheaf which is constructible with respect to $\mathcal{S}$ has a finite composition series in $Perv_{\mathcal{S}}(X)$ with uniquely determined simple subquotients. If X is compact, then $Perv(X)$ is also Artinian.

The simple objects in $Perv(X)$ (resp. $Perv_{\mathcal{S}}(X)$) are extensions by zero of intersection cohomology sheaves on irreducible analytic subvarieties (resp. connected components of strata) of X with coefficients in irreducible local systems. To be precise, let M be a connected analytic submanifold (resp. a connected component of a stratum) of X and let $\mathcal{L}_M$ be an irreducible local system on M; then, the pair $(\overline{M}, \mathcal{L}_M)$ is called an *irreducible enriched subvariety* of X (where $\overline{M}$ denotes the closure of M). Let $j : \overline{M} \hookrightarrow X$ denote the inclusion. Then, the simple objects of $Perv(X)$ (resp. $Perv_{\mathcal{S}}(X)$) are those of the form $j_!\mathbf{IC}^\bullet_{\overline{M}}(\mathcal{L}_M)$, where $(\overline{M}, \mathcal{L}_M)$ is an irreducible enriched subvariety (again, we are indexing intersection cohomology so that it is non-zero only in non-positive dimensions).

Finally, we wish to state the *decomposition theorem* of Beilinson, Bernstein, Deligne, and Gabber. For this statement, we must restrict ourselves to $R = \mathbb{Q}$. We give the statement as it appears in [**Mac2**], except that in [**Mac2**] intersection cohomology is defined as a **positively** perverse sheaf, and we must adjust by shifting. Note that, in [**Mac2**], the setting is algebraic; the analytic version appears in [**Sai**].

An algebraic map $f : X \to Y$ is called *projective* if it can be factored as an embedding $X \hookrightarrow Y \times \mathbb{P}^m$ (for some m) followed by projection $Y \times \mathbb{P}^m \to Y$.

The Decomposition Theorem [BBD, 6.2.5]: If $f : X \to Y$ is proper, then there exists a unique set of irreducible enriched subvarieties $\{(\overline{M}_\alpha, \mathcal{L}_\alpha)\}$ in Y and Laurent polynomials $\{\phi^\alpha = \dots + \phi^\alpha_{-2}t^{-2} + \phi^\alpha_{-1}t^{-1} + \phi^\alpha_0 + \phi^\alpha_1 t + \phi^\alpha_2 t^2 + \dots\}$ such that there is a quasi-isomorphism

$$Rf_*\mathbf{IC}^\bullet_X(\mathbb{Q}^\bullet_{\overset{\circ}{X}}) \cong \bigoplus_{\alpha, i} \mathbf{IC}^\bullet_{\overline{M}_\alpha}(\mathcal{L}_\alpha)[-i] \otimes \mathbb{Q}^{\phi^\alpha_i},$$

(here, $\mathbf{IC}^\bullet_{\overline{M}_\alpha}(\mathcal{L}_\alpha)$ actually equals $j_{\alpha!}\mathbf{IC}^\bullet_{\overline{M}_\alpha}(\mathcal{L}_\alpha)$, where $j_\alpha : \overline{M}_\alpha \hookrightarrow Y$ is the inclusion).

Moreover, if f is projective, then the coefficients of ϕ^α are palindromic around 0 (i.e., $\phi^\alpha(t^{-1}) = \phi^\alpha(t)$) and the even and odd terms are separately unimodal (i.e., if $i \leqslant 0$, then $\phi^\alpha_{i-2} \leqslant \phi^\alpha_i$).

Applying hypercohomology to each side, we obtain:

$$IH^k(X;\mathbb{Q}) = \bigoplus_{\alpha, i} (IH^{k-i}(\overline{M}_\alpha; \mathcal{L}_\alpha))^{\phi^\alpha_i}.$$

We now wish to describe the category of perverse sheaves on a one-dimensional space; this is a particularly nice case of the results obtained in [**M-V**]. Unfortunately, we will use the notions of vanishing cycles and nearby cycles, which are not covered until the next section. Nonetheless, it seems appropriate to place this material here.

We actually wish to consider perverse sheaves on the germ of a complex analytic space X at a point x. Hence, we assume that X is a one-dimensional complex analytic space with irreducible analytic components $X_1, \dots, X_d$ which all contain x, such that X_i is homeomorphic to a complex line and $X_i - \{x\}$ is smooth for all i. We wish to describe the category, $\mathcal{C}$, of perverse sheaves on X with complex coefficients which are constructible with respect to the stratification $\{X_1 - \{x\}, \dots, X_d - \{x\}, \{x\}\}$.

Since perverse sheaves are topological in nature, we may reduce ourselves to considering exactly the case where X consists of d complex lines through the origin in some $\mathbb{C}^N$. Let L denote a linear form on $\mathbb{C}^N$ such that $X \cap L^{-1}(0) = \{\mathbf{0}\}$.

Suppose now that $\mathbf{P}^\bullet$ is in $\mathcal{C}$, i.e., $\mathbf{P}^\bullet$ is perverse on X and constructible with respect to the stratification which has $\{\mathbf{0}\}$ as the only zero-dimensional stratum. Then $\mathbf{P}^\bullet_{|X-\{\mathbf{0}\}}$ consists of a collection of local systems, $\mathcal{L}_1, \dots \mathcal{L}_d$, in degree -1. These local systems are completely determined by monodromy isomorphisms $h_i : \mathbb{C}^{r_i} \to \mathbb{C}^{r_i}$ representing looping once around the origin in X_i. In terms of nearby cycles, the monodromy automorphism on $H^0(\psi_L\mathbf{P}^\bullet[-1])_\mathbf{0} \cong \bigoplus_i \mathbb{C}^{r_i}$ is given by $\bigoplus_i h_i$.

The vanishing cycles $\phi_L\mathbf{P}^\bullet[-1]$ are a perverse sheaf on a point, and so have possibly non-zero cohomology only in degree 0; say, $H^0(\phi_L\mathbf{P}^\bullet[-1])_\mathbf{0} \cong \mathbb{C}^\lambda$. We have the canonical map

$$r : H^0(\psi_L\mathbf{P}^\bullet[-1])_\mathbf{0} \to H^0(\phi_L\mathbf{P}^\bullet[-1])_\mathbf{0}$$

and the variation map

$$\mathrm{var} : H^0(\phi_L \mathbf{P}^\bullet[-1])_{\mathbf{0}} \to H^0(\psi_L \mathbf{P}^\bullet[-1])_{\mathbf{0}},$$

and $\mathrm{var} \circ r = \mathrm{id} - \bigoplus_i h_i$.

Thus, an object in $\mathcal{C}$ determines a vector space $W := H^0(\phi_L \mathbf{P}^\bullet[-1])_{\mathbf{0}}$, a vector space $V_i := \mathbb{C}^{r_i}$ for each irreducible component X_i, an automorphism h_i on V_i, and two linear maps $\alpha : \bigoplus_i V_i \to W$ and $\beta : W \to \bigoplus_i V_i$ such that $\beta \circ \alpha = \mathrm{id} - \bigoplus_i h_i$. This situation is nicely represented by a commutative triangle

$$\begin{array}{ccc} \oplus_i V_i & \xrightarrow{\mathrm{id} - \bigoplus_i h_i} & \oplus_i V_i \\ & \alpha \searrow \quad \nearrow \beta & \\ & W & \end{array} .$$

The category $\mathcal{C}$ is equivalent to the category of such triangles, where a morphism of triangles is defined in the obvious way: a morphism is determined by linear maps $\tau_i : V_i \to V_i'$ and $\eta : W \to W'$ such that

$$\begin{array}{ccccc} \oplus_i V_i & \xrightarrow{\alpha} & W & \xrightarrow{\beta} & \oplus_i V_i \\ \oplus_i \tau_i \downarrow & & \eta \downarrow & & \oplus_i \tau_i \downarrow \\ \oplus_i V_i' & \xrightarrow{\alpha'} & W' & \xrightarrow{\beta'} & \oplus_i V_i' \end{array}$$

commutes.

§3. Nearby and Vanishing Cycles

Historically, there has been some confusion surrounding the terminology *nearby (or neighboring) cycles* and *vanishing cycles*; now, however, the terminology seems to have stabilized. In the past, the term "vanishing cycles" was sometimes used to describe what are now called the nearby cycles (this is true, for instance, in [**A'C**], [**BBD**], and [**G-M1**].)

The two different indexing schemes for perverse sheaves also add to this confusion in statements such as "the nearby cycles of a perverse sheaf are perverse". Finally, a new piece of confusion has been added in [**K-S2**], where the sheaf of vanishing cycles is shifted by one from the usual definition (we will **not** use this new, shifted definition).

The point is: one should be very careful when reading works on nearby and vanishing cycles.

Let $\mathcal{S} = \{S_\alpha\}$ be a Whitney stratification of X and suppose $\mathbf{F}^\bullet \in \mathbf{D}^b_{\mathcal{S}}(X)$. Given an analytic map $f : X \to \mathbb{C}$, define a *(stratified) critical point* of f (with respect to $\mathcal{S}$) to be a point $x \in S_\alpha \subseteq X$ such that $f_{|S_\alpha}$ has a critical point at x; we denote the set of such critical points by $\Sigma_{\mathcal{S}} f$.

We wish to investigate how the cohomology of the level sets of f with coefficients in $\mathbf{F}^\bullet$ changes at a critical point (which we normally assume lies in $f^{-1}(0)$).

Consider the diagram

$$\begin{array}{ccc} E & \longrightarrow & \widetilde{\mathbb{C}^*} \\ {\scriptstyle \hat{\pi}}\downarrow & & \downarrow{\scriptstyle \pi} \\ X - f^{-1}(0) & \xrightarrow{\hat{f}} & \mathbb{C}^* \\ i\downarrow & & \\ f^{-1}(0) \underset{j}{\hookrightarrow} X & & \end{array}$$

where:

$j : f^{-1}(0) \hookrightarrow X$ is inclusion;

$i : X - f^{-1}(0) \hookrightarrow X$ is inclusion;

$\hat{f}$ = restriction of f;

$\widetilde{\mathbb{C}^*}$ = cyclic (universal) cover of $\mathbb{C}^*$;

and E denotes the pull-back.

The *nearby (or neighboring) cycles* of $\mathbf{F}^\bullet$ along f are defined to be

$$\psi_f\mathbf{F}^\bullet := j^*R(i\circ\hat{\pi})_*(i\circ\hat{\pi})^*\mathbf{F}^\bullet.$$

Note that this is a sheaf on $f^{-1}(0)$.

As $\psi_f\big(\mathbf{F}^\bullet[k]\big) = \big(\psi_f\mathbf{F}^\bullet\big)[k]$, we may write $\psi_f\mathbf{F}^\bullet[k]$ unambiguously. In fact, it is frequently useful to consider the functor where one first shifts the complex by k and then takes the nearby cycles; thus, we introduce the notation $\psi_f[k]$ to be the functor such that $\psi_f[k]\mathbf{F}^\bullet = \psi_f\mathbf{F}^\bullet[k]$ (and which has the corresponding action on morphisms). The functor ψ_f takes distinguished triangles to distinguished triangles.

If $\mathbf{P}^\bullet$ is a perverse sheaf on X, then $\psi_f[-1]\mathbf{P}^\bullet$ is perverse on $f^{-1}(0)$. (Actually, to conclude that $\psi_f[-1]\mathbf{P}^\bullet$ is perverse, we only need to assume that $\mathbf{P}^\bullet|_{X-f^{-1}(0)}$ is perverse.)

Because $\psi_f[-1]$ takes perverse sheaves to perverse sheaves, it is useful to include the shift by -1 in many statements about ψ_f. Consequently, we also want to shift $j^*\mathbf{F}^\bullet$ by -1 in many statements, and so we write $j^*[-1]$ for the functor which first shifts by -1 and then pulls-back by j.

As there is a canonical map $\mathbf{F}^\bullet \to Rg_*g^*\mathbf{F}^\bullet$ for any map $g : Z \to X$, there is a map

$$\mathbf{F}^\bullet \to R(i\circ\hat{\pi})_*(i\circ\hat{\pi})^*\mathbf{F}^\bullet$$

and, hence, a canonical map, called the *comparison map*:

$$j^*[-1]\mathbf{F}^\bullet \xrightarrow{c} j^*[-1]R(i\circ\hat{\pi})_*(i\circ\hat{\pi})^*\mathbf{F}^\bullet = \psi_f[-1]\mathbf{F}^\bullet.$$

For $x \in f^{-1}(0)$, the stalk cohomology of $\psi_f \mathbf{F}^\bullet$ at x is the cohomology of the Milnor fibre of f at x with coefficients in $\mathbf{F}^\bullet$, i.e., for all $\epsilon > 0$ small and all $\xi \in \mathbb{C}^*$ with $|\xi| << \epsilon$,

$$\mathbf{H}^i(\psi_f \mathbf{F}^\bullet)_x \cong \mathbb{H}^i(\overset{\circ}{B}_\epsilon(x) \cap X \cap f^{-1}(\xi); \mathbf{F}^\bullet),$$

where the open ball $\overset{\circ}{B}_\epsilon(x)$ is taken inside any local embedding of (X, x) in affine space. The sheaf $\psi_f \mathbf{F}^\bullet$ only depends on f and $\mathbf{F}^\bullet|_{X - f^{-1}(0)}$.

While the above definition of the nearby cycles treats all angular directions equally, it is perhaps more illuminating to fix an angle θ and describe the nearby cycles in terms of moving out slightly along the ray $e^{i\theta}[0, \infty)$. Consider the three inclusions $k_\theta : f^{-1}(e^{i\theta}(0, \infty)) \hookrightarrow f^{-1}(e^{i\theta}[0, \infty))$, $m_\theta : f^{-1}(0) \hookrightarrow f^{-1}(e^{i\theta}[0, \infty))$, and $l_\theta : f^{-1}(e^{i\theta}[0, \infty)) \hookrightarrow X$. Then, one can define the *nearby cycles at angle* θ to be $\psi_f^\theta \mathbf{F}^\bullet := m_\theta^* R k_{\theta *} k_\theta^* l_\theta^* \mathbf{F}^\bullet$.

For each θ, there is a canonical isomorphism $\psi_f \mathbf{F}^\bullet \cong \psi_f^\theta \mathbf{F}^\bullet$. By letting θ travel around a full circle, we obtain isomorphisms $\psi_f^\theta \mathbf{F}^\bullet \cong \psi_f^{\theta+2\pi} \mathbf{F}^\bullet$. These isomorphisms correspond to the *monodromy automorphism* $T_f : \psi_f[-1]\mathbf{F}^\bullet \to \psi_f[-1]\mathbf{F}^\bullet$, which comes from the deck transformation obtained in our definition of $\psi_f \mathbf{F}^\bullet$ (and, hence, $\psi_f[-1]\mathbf{F}^\bullet$) by traveling once around the origin in $\mathbb{C}$. Actually, T_f is a natural automorphism from the functor $\psi_f[-1]$ to itself; thus, strictly speaking, when we write $T_f : \psi_f[-1]\mathbf{F}^\bullet \to \psi_f[-1]\mathbf{F}^\bullet$, we should include $\mathbf{F}^\bullet$ in the notation for T_f – however, we shall normally omit the explicit reference to $\mathbf{F}^\bullet$ if the complex is clear.

There is a natural distinguished triangle

$$\begin{array}{ccc} j^*[-1]Ri_*i^*\mathbf{F}^\bullet & \longrightarrow & \psi_f[-1]\mathbf{F}^\bullet \\ {\scriptstyle [1]} \nwarrow & & \swarrow T_f - \mathrm{id} \\ & \psi_f[-1]\mathbf{F}^\bullet. & \end{array}$$

The associated long exact sequences on stalk cohomology are the Wang sequences.

The comparison map $j^*[-1]\mathbf{F}^\bullet \xrightarrow{c} \psi_f[-1]\mathbf{F}^\bullet$ is T_f-equivariant, i.e., $c = T_f \circ c$.

Since we have a map $c[1] : j^*\mathbf{F}^\bullet \to \psi_f \mathbf{F}^\bullet$, the third vertex of a distinguished triangle is defined up to quasi-isomorphism. We define the sheaf of *vanishing cycles*, $\phi_f \mathbf{F}^\bullet$, of $\mathbf{F}^\bullet$ along f to be this third vertex, i.e., there is a distinguished triangle

$$\begin{array}{ccc} j^*\mathbf{F}^\bullet & \longrightarrow & \psi_f \mathbf{F}^\bullet \\ {\scriptstyle [1]} \nwarrow & & \swarrow \\ & \phi_f \mathbf{F}^\bullet. & \end{array}$$

Letting $\phi_f[-1]$ denote the functor which first shifts by -1 and then applies ϕ_f, we can write the triangle above as

$$\begin{array}{ccc} j^*[-1]\mathbf{F}^\bullet & \xrightarrow{c} & \psi_f[-1]\mathbf{F}^\bullet \\ {\scriptstyle [1]}\nwarrow & & \swarrow \\ & \phi_f[-1]\mathbf{F}^\bullet. & \end{array}$$

Note that this is a triangle of sheaves on $f^{-1}(0)$. Note also that, by replacing $\mathbf{F}^\bullet$ with $i_!i^!\mathbf{F}^\bullet$, we conclude that there is a natural isomorphism $\psi_f[-1]\mathbf{F}^\bullet \cong \phi_f[-1](i_!i^!\mathbf{F}^\bullet)$. There is another natural isomorphism $\psi_f[-1]\mathbf{F}^\bullet \cong \phi_f[-1](Ri_*i^*\mathbf{F}^\bullet)$.

The functor ϕ_f takes distinguished triangles to distinguished triangles.

If $\mathbf{P}^\bullet$ is a perverse sheaf on X, then $\phi_f[-1]\mathbf{P}^\bullet$ is a perverse sheaf on $f^{-1}(0)$.

For $x \in f^{-1}(0)$, the stalk cohomology of $\phi_f\mathbf{F}^\bullet$ at x is the relative cohomology of the Milnor fibre of f at x with coefficients in $\mathbf{F}^\bullet$ and with a shift by one, i.e., for all $\epsilon > 0$ small and all $\xi \in \mathbb{C}^*$ with $|\xi| << \epsilon$,

$$\mathbf{H}^i(\phi_f\mathbf{F}^\bullet)_x \cong \mathbb{H}^{i+1}(\overset{\circ}{B}_\epsilon(x) \cap X, \overset{\circ}{B}_\epsilon(x) \cap X \cap f^{-1}(\xi); \mathbf{F}^\bullet).$$

As an example, if $X = \mathbb{C}^{n+1}$ and $\mathbf{F}^\bullet = \mathbb{C}^\bullet_X$, then for all $x \in f^{-1}(0)$, $H^i(\psi_f\mathbb{C}^\bullet_X)_x =$ i-th cohomology of the Milnor fibre of f at x (with $\mathbb{C}$ coefficients) $= H^i(F_{f,x};\mathbb{C})$, while $H^i(\phi_f\mathbb{C}^\bullet_X)_x =$ **reduced** i-th cohomology of the Milnor fibre of f at x $= \widetilde{H}^i(F_{f,x};\mathbb{C})$.

Just as we defined the nearby cycles at angle θ to be $\psi^\theta_f\mathbf{F}^\bullet := m^*_\theta Rk_{\theta*}k^*_\theta l^*_\theta\mathbf{F}^\bullet$, we can define the *vanishing cycles at angle* θ to be $\phi^\theta_f\mathbf{F}^\bullet := m^*_\theta m_{\theta!}m^!_\theta l^*_\theta\mathbf{F}^\bullet[1] = m^!_\theta l^*_\theta\mathbf{F}^\bullet[1]$. Then, $\phi^\theta_f\mathbf{F}^\bullet \cong \phi_f\mathbf{F}^\bullet$, and again there is a monodromy automorphism $\widetilde{T}_f : \phi_f[-1]\mathbf{F}^\bullet \to \phi_f[-1]\mathbf{F}^\bullet$. The monodromy $\widetilde{T}_f$ is actually a natural automorphism of the functor $\phi_f[-1]$.

If we let $Z_\theta := \{\mathbf{z} \in X \mid \operatorname{Re}\left(e^{i\theta}f(\mathbf{z})\right) \leqslant 0\}$, then there is a canonical isomorphism

$$\phi^\theta_f\mathbf{F}^\bullet \cong \left(R\Gamma_{Z_\theta}(\mathbf{F}^\bullet)\right)_{|_{f^{-1}(0)}}[1].$$

Thus, there is a monodromy automorphism on the distinguished triangle

$$j^*[-1]\mathbf{F}^\bullet \xrightarrow{c} \psi_f[-1]\mathbf{F}^\bullet \xrightarrow{r} \phi_f[-1]\mathbf{F}^\bullet \longrightarrow j^*\mathbf{F}^\bullet$$

given by $(\mathrm{id}, T_f, \widetilde{T}_f)$, i.e., a commutative diagram

$$\begin{array}{ccccccc} j^*[-1]\mathbf{F}^\bullet & \xrightarrow{c} & \psi_f[-1]\mathbf{F}^\bullet & \xrightarrow{r} & \phi_f[-1]\mathbf{F}^\bullet & \longrightarrow & j^*\mathbf{F}^\bullet \\ \mathrm{id}\downarrow & & T_f\downarrow & & \widetilde{T}_f\downarrow & & \mathrm{id}\downarrow \\ j^*[-1]\mathbf{F}^\bullet & \xrightarrow{c} & \psi_f[-1]\mathbf{F}^\bullet & \xrightarrow{r} & \phi_f[-1]\mathbf{F}^\bullet & \longrightarrow & j^*\mathbf{F}^\bullet. \end{array}$$

From this, it follows formally that there exists a *variation morphism*, $\mathrm{var} : \phi_f[-1]\mathbf{F}^\bullet \to \psi_f[-1]\mathbf{F}^\bullet$ such that $r \circ \mathrm{var} = \mathrm{id} - \widetilde{T}_f$ and $\mathrm{var} \circ r = \mathrm{id} - T_f$. (Note that, if we are not using field coefficients, then the variation morphism does not necessarily exist on the level of chain complexes – the derived category structure is necessary here.)

The monodromy isomorphisms T_f and $\widetilde{T}_f$ are natural automorphisms of the (shifted) nearby cycle and vanishing cycle functors, respectively, and the maps c, r, and var above are all natural maps.

The variation map can be described in a more concrete fashion. There is the canonical map from $\mathbf{F}^\bullet$ to $Ri_*i^*\mathbf{F}^\bullet$. Applying the shifted vanishing cycle functor, we obtain a natural map from $\phi_f[-1](\mathbf{F}^\bullet)$ to $\phi_f[-1](Ri_*i^*\mathbf{F}^\bullet)$, and as we mentioned above, there is a natural isomorphism $\phi_f[-1](Ri_*i^*\mathbf{F}^\bullet) \cong \psi_f[-1]\mathbf{F}^\bullet$. The variation map is the composition of these two natural maps. To make this more clear, we will describe the variation map on the stalk cohomology; this should also help clarify how one obtains the isomorphism $\phi_f[-1](Ri_*i^*\mathbf{F}^\bullet) \cong \psi_f[-1]\mathbf{F}^\bullet$.

We follow the construction in [**G-M1**]. Let x be a point in $f^{-1}(0)$, let N denote the intersection of X with a sufficiently small open ball around x (for some Reimannian metric), and let $\mathbb{D}_\eta$ be a complex disk of sufficiently small radius, η, centered at the origin so that, for all ξ with $0 < \xi \leqslant \eta$, $N \cap f^{-1}(\partial\mathbb{D}_\xi) \xrightarrow{f} \partial\mathbb{D}_\xi$ represents the Milnor fibration of f at x with coefficients in $\mathbf{F}^\bullet$. Let $W := N \cap f^{-1}(\{v \in \mathbb{D}_\eta \mid \operatorname{Re} v \geqslant 0\} - \{0\})$, let $Z := N \cap f^{-1}(\{v \in \mathbb{D}_\eta \mid \operatorname{Re} v \leqslant 0\} - \{0\})$, let $A := N \cap f^{-1}(\{v \in \mathbb{D}_\eta \mid \operatorname{Re} v = 0, \operatorname{Im} v > 0\})$, and let $B := N \cap f^{-1}(\{v \in \mathbb{D}_\eta \mid \operatorname{Re} v = 0, \operatorname{Im} v < 0\})$.

Then, we have isomorphisms:

$$H^i(\phi_f\mathbf{F}^\bullet)_x \cong \mathbb{H}^{i+1}(N \cap f^{-1}(\mathbb{D}_\eta), N \cap f^{-1}(\eta); \mathbf{F}^\bullet) \cong \mathbb{H}^{i+1}(N \cap f^{-1}(\mathbb{D}_\eta), W; \mathbf{F}^\bullet);$$

the map induced by inclusion of pairs:

$$\mathbb{H}^{i+1}(N \cap f^{-1}(\mathbb{D}_\eta), W; \mathbf{F}^\bullet) \to \mathbb{H}^{i+1}(N \cap f^{-1}(\mathbb{D}_\eta - \{0\}), W; \mathbf{F}^\bullet);$$

and isomorphisms:

$$\mathbb{H}^{i+1}(N \cap f^{-1}(\mathbb{D}_\eta - \{0\}), W; \mathbf{F}^\bullet) \cong \mathbb{H}^{i+1}(Z, A \cup B; \mathbf{F}^\bullet) \cong H^i(\psi_f\mathbf{F}^\bullet)_x,$$

where the first isomorphism is by excision, and the second is from the long exact sequence of the pair.

The map induced by the (shifted) variation on the stalk cohomology is the composition of the above maps.

Applying the shifted vanishing cycle functor to the distinguished triangle

$$\begin{array}{ccc} j_!j^!\mathbf{F}^\bullet & \longrightarrow & \mathbf{F}^\bullet \\ {\scriptstyle [1]}\nwarrow & & \swarrow \\ & Ri_*i^*\mathbf{F}^\bullet, & \end{array}$$

noting that $\phi_f[-1](j_!j^!\mathbf{F}^\bullet) \cong j^!\mathbf{F}^\bullet$, and using the natural isomorphism

$$\psi_f[-1]\mathbf{F}^\bullet \cong \phi_f[-1](Ri_*i^*\mathbf{F}^\bullet),$$

we obtain the distinguished triangle

$$\begin{array}{ccc} j^!\mathbf{F}^\bullet & \longrightarrow & \phi_f[-1]\mathbf{F}^\bullet \\ {\scriptstyle [1]}\nwarrow & & \swarrow {\scriptstyle \text{var}} \\ & \psi_f[-1]\mathbf{F}^\bullet. & \end{array}$$

Starting with the two distinguished triangles

$$\phi_f[-1]\mathbf{F}^\bullet \xrightarrow{\text{var}} \psi_f[-1]\mathbf{F}^\bullet \longrightarrow j^![1]\mathbf{F}^\bullet \longrightarrow \phi_f\mathbf{F}^\bullet$$

and

$$\psi_f[-1]\mathbf{F}^\bullet \xrightarrow{c} \phi_f[-1]\mathbf{F}^\bullet \longrightarrow j^*\mathbf{F}^\bullet \longrightarrow \psi_f\mathbf{F}^\bullet,$$

we may apply the octahedral lemma to conclude that there exists a complex $w_f\mathbf{F}^\bullet$ and two distinguished triangles

$$\phi_f[-1]\mathbf{F}^\bullet \xrightarrow{\text{id} - \widetilde{T}_f} \phi_f[-1]\mathbf{F}^\bullet \longrightarrow w_f\mathbf{F}^\bullet \longrightarrow \phi_f\mathbf{F}^\bullet$$

and

$$j^![1]\mathbf{F}^\bullet \longrightarrow w_f\mathbf{F}^\bullet \longrightarrow j^*\mathbf{F}^\bullet \xrightarrow{\tau} j^![2]\mathbf{F}^\bullet.$$

We refer to the morphism $\omega_f := \tau[-1]$ from $j^*[-1]\mathbf{F}^\bullet$ to $j^![1]\mathbf{F}^\bullet$ as the *Wang morphism of* f. The application of the octahedral lemma above tells us that the mapping cone of $\text{id} - \widetilde{T}_f$ is isomorphic to the mapping cone of ω_f. Note that, while $j^*[-1]\mathbf{F}^\bullet$ and $j^![1]\mathbf{F}^\bullet$ depend only on $f^{-1}(0)$ (and $\mathbf{F}^\bullet$), ω_f may change if f (or some factor of f) is raising to a power.

For any Whitney stratification, $\mathcal{S}$, with respect to which $\mathbf{F}^\bullet$ is constructible, the support of $\mathbf{H}^*(\phi_f\mathbf{F}^\bullet)$ is contained in the stratified critical locus of f, $\Sigma_{\mathcal{S}}f$. In addition, if $\mathcal{S}$ is a Whitney stratification with respect to which $\mathbf{F}^\bullet$ is constructible and such that $f^{-1}(0)$ is a union of strata, then – by [**BMM**] and [**P2**] – it follows that $\mathcal{S}$ also satisfies Thom's a_f condition; by Thom's second isotopy lemma, this implies that the entire situation locally trivializes over strata, and hence both $\psi_f\mathbf{F}^\bullet$ and $\phi_f\mathbf{F}^\bullet$ are constructible with respect to $\{S \in \mathcal{S} \mid S \subseteq f^{-1}(0)\}$.

Suppose we have $X \xrightarrow{\pi} Y \xrightarrow{f} \mathbb{C}$ where π is proper and $\hat{\pi} : \pi^{-1}f^{-1}(0) \to f^{-1}(0)$ is the restriction of π. Then, for all $\mathbf{A}^\bullet \in \mathbf{D}^b_c(X)$,

$$R\hat{\pi}_*(\psi_{f\circ\pi}\mathbf{A}^\bullet) \cong \psi_f(R\pi_*\mathbf{A}^\bullet) \quad \text{and} \quad R\hat{\pi}_*(\phi_{f\circ\pi}\mathbf{A}^\bullet) \cong \phi_f(R\pi_*\mathbf{A}^\bullet).$$

The Sebastiani-Thom Isomorphism

Let $f : X \to \mathbb{C}$ and $g : Y \to \mathbb{C}$ be complex analytic functions. Let π_1 and π_2 denote the projections of $X \times Y$ onto X and Y, respectively. Let $\mathbf{A}^\bullet$ and $\mathbf{B}^\bullet$ be bounded, constructible complexes of sheaves of R-modules on X and Y, respectively. In this situation, $\mathbf{A}^\bullet \overset{L}{\boxtimes} \mathbf{B}^\bullet := \pi_1^*\mathbf{A}^\bullet \overset{L}{\otimes} \pi_2^*\mathbf{B}^\bullet$. Let us adopt the similar notation $f \boxplus g := f \circ \pi_1 + g \circ \pi_2$.

Let p_1 and p_2 denote the projections of $V(f) \times V(g)$ onto $V(f)$ and $V(g)$, respectively, and let k denote the inclusion of $V(f) \times V(g)$ into $V(f \boxplus g)$.

Theorem (Sebastiani-Thom Isomorphism). *There is a natural isomorphism*

$$k^*\phi_{f\boxplus g}[-1]\big(\mathbf{A}^\bullet \overset{L}{\boxtimes} \mathbf{B}^\bullet\big) \;\cong\; \phi_f[-1]\mathbf{A}^\bullet \overset{L}{\boxtimes} \phi_g[-1]\mathbf{B}^\bullet,$$

and this isomorphism commutes with the corresponding monodromies.

Moreover, if we let $\mathbf{p} := (\mathbf{x}, \mathbf{y}) \in X \times Y$ be such that $f(\mathbf{x}) = 0$ and $g(\mathbf{y}) = 0$, then, in an open neighborhood of $\mathbf{p}$, the complex $\phi_{f\boxplus g}[-1]\big(\mathbf{A}^\bullet \overset{L}{\boxtimes} \mathbf{B}^\bullet\big)$ has support contained in $V(f) \times V(g)$, and, in any open set in which we have this containment, there are natural isomorphisms

$$\phi_{f\boxplus g}[-1]\big(\mathbf{A}^\bullet \overset{L}{\boxtimes} \mathbf{B}^\bullet\big) \;\cong\; k_!(\phi_f[-1]\mathbf{A}^\bullet \overset{L}{\boxtimes} \phi_g[-1]\mathbf{B}^\bullet) \;\cong\; k_*(\phi_f[-1]\mathbf{A}^\bullet \overset{L}{\boxtimes} \phi_g[-1]\mathbf{B}^\bullet).$$

If $\mathbf{Q}^\bullet$ is perverse on $X - f^{-1}(0)$ and $i : X - f^{-1}(0) \to X$ is the inclusion, then it is easy to see that $Ri_*\mathbf{Q}^\bullet$ satisfies the cosupport condition; moreover, by combining the fact that $\psi_f(Ri_*\mathbf{Q}^\bullet)[-1]$ is perverse on $f^{-1}(0)$ with the Wang sequences on stalk cohomology, one can prove that $Ri_*\mathbf{Q}^\bullet$ also satisfies the support condition - hence, $Ri_*\mathbf{Q}^\bullet$ is perverse. In an analogous fashion, one obtains that $Ri_!\mathbf{Q}^\bullet$ is perverse (if the base ring is a field, this can be obtained by dualizing).

If R is a field, then the operators $\psi_f[-1]$ and $\mathcal{D}$ commute, as do $\phi_f[-1]$ and $\mathcal{D}$; i.e.,

$$\mathcal{D}(\psi_f\mathbf{A}^\bullet[-1]) \;\cong\; \psi_f(\mathcal{D}\mathbf{A}^\bullet)[-1] \quad\text{and}\quad \mathcal{D}(\phi_f\mathbf{A}^\bullet[-1]) \;\cong\; \phi_f(\mathcal{D}\mathbf{A}^\bullet)[-1].$$

These isomorphisms in $\mathbf{D}_c^b(X)$ are non-canonical.

Let $\mathbf{A}^\bullet \in \mathbf{D}_c^b(X)$ and $f : X \to \mathbb{C}$. The monodromy automorphism

$$\psi_f[-1]\mathbf{A}^\bullet \xrightarrow{T_f} \psi_f[-1]\mathbf{A}^\bullet$$

induces a map on cohomology sheaves which is quasi-unipotent, i.e., letting T_f also denote the map on cohomology, this means that there exist integers k and j such that $(\mathrm{id} - T_f^k)^j = 0$.

Suppose that the base ring is a field; if m_x denotes the maximal ideal of X at x and $f \in m_x^2$, then the Lefschetz number of the map $\mathbf{H}^*(\psi_f[-1]\mathbf{A}^\bullet)_x \xrightarrow{T_f} \mathbf{H}^*(\psi_f[-1]\mathbf{A}^\bullet)_x$ equals 0, i.e.,

$$\sum_i (-1)^i \ \mathrm{Trace}\{\mathbf{H}^i(\psi_f[-1]\mathbf{A}^\bullet)_x \xrightarrow{T_f} \mathbf{H}^i(\psi_f[-1]\mathbf{A}^\bullet)_x\} \ = \ 0.$$

If $\psi_f[-1]\mathbf{A}^\bullet$ is a perverse sheaf, then we may use the Abelian structure of the category $Perv(X)$ to investigate the map $\psi_f[-1]\mathbf{A}^\bullet \xrightarrow{T_f} \psi_f[-1]\mathbf{A}^\bullet$. This morphism can be factored into $T_f = F \cdot (1+N)$, where F has finite order and N is nilpotent. It follows that there is a unique increasing filtration W^i on $\psi_f[-1]\mathbf{A}^\bullet$ such that N sends W^i to W^{i-2} and N^i takes $Gr^i\psi_f[-1]\mathbf{A}^\bullet$ isomorphically to $Gr^{-i}\psi_f[-1]\mathbf{A}^\bullet$, where Gr^i is the associated graded to the filtration $W^\bullet$. This is called the *nilpotent filtration* of $\psi_f[-1]\mathbf{A}^\bullet$. (The existence of such a filtration is just linear algebra; the interesting result is the following theorem, due to Gabber.)

Theorem: If we have $f : X \to \mathbb{C}$, $\mathcal{S}$ a Whitney stratification of X with a finite number of strata, and $\mathbf{A}^\bullet \in \mathbf{D}^b_{\mathcal{S}}(X)$ such that

$$\mathbf{A}^\bullet|_{X-f^{-1}(0)} \cong \mathbf{IC}^\bullet_{X-f^{-1}(0)}(\mathbb{C}^\bullet_{\overset{\circ}{X}}),$$

then the graded pieces of the nilpotent filtration of $\psi_f[-1]\mathbf{A}^\bullet$ are semi-simple in the category $Perv_{\mathcal{S}}(f^{-1}(0))$, i.e., they are direct sums of intersection cohomology sheaves of irreducible enriched subvarieties of $f^{-1}(0)$ (extended by zero).

In particular, if $X = \mathbb{C}^{n+1}$, then each $Gr^i\psi_f[-1]\mathbb{C}^\bullet_X[n+1]$ is semi-simple.

§4. Some Quick Applications

The applications of perverse sheaves are widespread and are frequently quite deep - particularly for those applications which rely on the decomposition theorem. For beautiful discussions of these applications, we highly recommend [**Mac1**] and [**Mac2**]. We shall not describe any of these applications here; rather we shall give some fairly easy results on general Milnor fibres. These results are "easy" now that we have all the machinery of the first three sections at our disposal. While the applications below could undoubtedly be proved without the general theory of perverse sheaves, with this theory in hand, the results and their proofs can be presented in a unified manner and, what is more, the proofs become mere exercises.

Consider the classical case of the Milnor fibre of a non-zero map $f : (\mathbb{C}^{n+1}, \mathbf{0}) \to (\mathbb{C}, 0)$. Let $X = \mathbb{C}^{n+1}$ and let $s = \dim \Sigma f$. Then, as X is a manifold, $\mathbb{C}^\bullet_X$ is a positively perverse sheaf and so $\phi_f\mathbb{C}^\bullet_X$ is positively perverse on $f^{-1}(0)$ with support only on Σf. It follows that the stalk cohomology of $\phi_f\mathbb{C}^\bullet_X$ is non-zero only for dimensions i with $n-s \leqslant i \leqslant n$; that is, we recover the well-known result that the reduced cohomology of the Milnor fibre is non-zero only in these dimensions.

A much more general case is just as easy to derive from the machinery that we have. Suppose that X is a purely $(n+1)$-dimensional local complete intersection with arbitrary singularities. Let $\mathcal{S}$ be a Whitney stratification of X. Let $\mathbf{p} \in X$ be such that $\dim_{\mathbf{p}} f^{-1}(0) = n$, and let $F_{f,\mathbf{p}}$ denote the Milnor fibre of f at $\mathbf{p}$. Then, as X is a local complete intersection, $\mathbb{C}^\bullet_X$ is a positively perverse sheaf and so $\phi_f \mathbb{C}^\bullet_X$ is positively perverse on $f^{-1}(0)$ with support only on $\Sigma_{\mathcal{S}} f$. It follows that the stalk cohomology of $\phi_f \mathbb{C}^\bullet_X$ is non-zero only for dimensions i with $n - \dim_{\mathbf{p}} \Sigma_{\mathcal{S}} f \leqslant i \leqslant n$. Hence, the reduced cohomology of $F_{f,\mathbf{p}}$ is non-zero only in these dimensions.

While this general statement could no doubt be proved by induction on hyperplane sections, the above proof via general techniques avoids the re-working of many technical lemmas on privileged neighborhoods and generic slices.

Another application relates to the homotopy-type of the complex link of a space at a point; for instance, for an s-dimensional local complete intersection, the complex link has the homotopy-type of a bouquet of spheres of real dimension $s-1$. In terms of vanishing cycles and perverse sheaves, we only obtain this result up to cohomology: let (X, x) be a germ of an analytic space embedded in some $\mathbb{C}^n$, and assume $s := \dim X = \dim_x X$. Suppose that we have a positively perverse sheaf, $\mathbf{P}^\bullet$, on X (e.g., the constant sheaf, if X is a local complete intersection). Let l be a generic linear form, and consider $\phi_{l-l(x)}\mathbf{P}^\bullet$; this is a perverse sheaf on an $s-1$ dimensional space and, as l is generic, it is supported at the single point x (because the hyperplane slice $l = l(x)$ can be chosen to transversely intersect all the strata of any stratification with respect to which $\mathbf{P}^\bullet$ is constructible - except, possibly, the point-stratum x itself). Hence, $H^*(\phi_{l-l(x)}\mathbf{P}^\bullet)_x$ is (possibly) non-zero only in dimension $s-1$. In the case of the constant sheaf on a local complete intersection, this gives the desired result.

For our final application, we wish to investigate functions with one-dimensional critical loci; we must first set up some notation.

Let $\mathcal{U}$ be an open neighborhood of the origin in $\mathbb{C}^{n+1}$ and suppose that $f : (\mathcal{U}, \mathbf{0}) \to (\mathbb{C}, 0)$ has a one-dimensional critical locus at the origin, i.e., $\dim_{\mathbf{0}} \Sigma f = 1$. The reduced cohomology of the Milnor fibre, $F_{f,\mathbf{0}}$, of f at the origin is possibly non-zero only in dimensions $n-1$ and n. We wish to show that the $n-1$-st cohomology group embeds inside another group which is fairly easy to describe; thus, we obtain a bound on the $n-1$-st Betti number of the Milnor fibre of f.

For each component ν of Σf, one may consider a generic hyperplane slice, H, at points $\mathbf{p} \in \nu - \mathbf{0}$ close to the origin; then, the restricted function, $f_{|_H}$, will have an isolated critical point at $\mathbf{p}$. By shrinking the neighborhood $\mathcal{U}$ if necessary, we may assume that the Milnor number of this isolated singularity of $f_{|_H}$ at $\mathbf{p}$ is independent of the point $\mathbf{p} \in \nu - \mathbf{0}$; denote this value by $\overset{\circ}{\mu}_\nu$. As $\nu - \mathbf{0}$ is homotopy-equivalent to a circle, there is a monodromy map from the Milnor fibre of $f_{|_H}$ at $\mathbf{p} \in \nu - \mathbf{0}$ to itself, which induces a map on the middle dimensional cohomology, i.e., a map $h_\nu : \mathbb{Z}^{\overset{\circ}{\mu}_\nu} \to \mathbb{Z}^{\overset{\circ}{\mu}_\nu}$. We wish to show that $H^{n-1}(F_{f,\mathbf{0}})$ (with integer coefficients) injects into $\oplus_\nu \ker(id - h_\nu)$.

Let j denote the inclusion of the origin into $X = V(f)$, let i denote the inclusion of $X - \mathbf{0}$ into X, and let $\mathbf{K}^\bullet$ denote $\phi_f(\mathbb{Z}^\bullet_{\mathcal{U}})$. As $\mathbb{Z}^\bullet_{\mathcal{U}}$ is positively perverse, $\phi_f(\mathbb{Z}^\bullet_{\mathcal{U}})$ is positively perverse with one-dimensional support (as we are assuming a one-dimensional critical locus). Also, we always have the distinguished triangle

$$\begin{array}{ccc} Rj_*j^!\mathbf{K}^\bullet & \longrightarrow & \mathbf{K}^\bullet \\ {\scriptstyle [1]}\nwarrow & & \swarrow \\ & Ri_*i^*\mathbf{K}^\bullet & \end{array}$$

We wish to examine the associated stalk cohomology exact sequence at the origin.

First, we have that $H^{n-1}((Rj_*j^!\mathbf{K}^\bullet)_\mathbf{0}) = H^{n-1}(j^!\mathbf{K}^\bullet)$ and so, by the cosupport condition for perverse sheaves, $H^{n-1}((Rj_*j^!\mathbf{K}^\bullet)_\mathbf{0}) = 0$.

Now, we need to look more closely at the sheaf $Ri_*i^*\mathbf{K}^\bullet$. $i^*\mathbf{K}^\bullet$ is the restriction of $\mathbf{K}^\bullet$ to $X - \mathbf{0}$; near the origin, this sheaf has cohomology only in degree $n-1$ with support on $\Sigma f - \mathbf{0}$. Moreover, the cohomology sheaf $\mathbf{H}^{n-1}(i^*\mathbf{K}^\bullet)$ is locally constant when restricted to $\Sigma f - \mathbf{0}$. It follows that $i^*\mathbf{K}^\bullet$ is naturally isomorphic in the derived category to the extension by zero of a local system of coefficients in dimension $n-1$ on $\Sigma - \mathbf{0}$.

To be more precise, let p denote the inclusion of the closed subset $\Sigma f - \mathbf{0}$ into $X - \mathbf{0}$. Then, there exists a locally constant (single) sheaf, $\mathcal{L}$, on $\Sigma f - \mathbf{0}$ such that when $\mathcal{L}$ is considered as a complex, $\mathcal{L}^\bullet$, we have that $p_!\mathcal{L}^\bullet[-(n-1)] \cong p_*\mathcal{L}^\bullet[-(n-1)]$ is naturally isomorphic to $i^*\mathbf{K}^\bullet$. For each component ν of Σf, the restriction of $\mathcal{L}$ to $\nu - \mathbf{0}$ is a local system with stalks $\mathbb{Z}^{\overset{\circ}{\mu}_\nu}$ which is completely determined by the monodromy map $h_\nu : \mathbb{Z}^{\overset{\circ}{\mu}_\nu} \rightarrow \mathbb{Z}^{\overset{\circ}{\mu}_\nu}$.

Therefore, inside a small open ball $\overset{\circ}{B}$,

$$H^0((Ri_*i^*\mathcal{L}^\bullet)_\mathbf{0}) \cong \oplus_\nu \mathbb{H}^0(\overset{\circ}{B} \cap (\nu - \mathbf{0}); \mathcal{L})$$

and these global sections are well-known to be given by $\ker(id - h_\nu)$. It follows that

$$H^{n-1}((Ri_*i^*\mathbf{K}^\bullet)_\mathbf{0}) \cong \oplus_\nu \ker(id - h_\nu).$$

Thus, when we consider the long exact sequence on stalk cohomology associated to our distinguished triangle, we find – starting in dimension $n-1$ – that it begins

$$0 \rightarrow H^{n-1}(F_{f,\mathbf{0}}) \rightarrow \oplus_\nu \ker(id - h_\nu) \rightarrow \dots .$$

The desired conclusion follows.

§5. **Truncation and Perverse Cohomology**

This section is taken entirely from [**BBD**], [**G-M3**], and [**K-S2**].

There are (at least) two forms of truncation associated to an object $\mathbf{F}^\bullet \in \mathbf{D}_c^b(X)$ – one form of truncation is related to the ordinary cohomology of the complex, while the other form leads to something called the *perverse cohomology* or *perverse projection.* These two types of truncation bear little resemblance to each other, except in the general framework of a t-structure on $\mathbf{D}_c^b(X)$.

Loosely speaking, a t-structure on $\mathbf{D}_c^b(X)$ consists of two full subcategories, denoted $\mathbf{D}^{\leqslant 0}(X)$ and $\mathbf{D}^{\geqslant 0}(X)$, such that for any $\mathbf{F}^\bullet \in \mathbf{D}_c^b(X)$, there exist $\mathbf{E}^\bullet \in \mathbf{D}^{\leqslant 0}(X)$, $\mathbf{G}^\bullet \in \mathbf{D}^{\geqslant 0}(X)$, and a distinguished triangle

$$\begin{array}{ccc} \mathbf{E}^\bullet & \longrightarrow & \mathbf{F}^\bullet \\ {\scriptstyle [1]}\nwarrow & & \swarrow \\ & \mathbf{G}^\bullet[-1] & \end{array} \quad ;$$

moreover, such $\mathbf{E}^\bullet$ and $\mathbf{G}^\bullet$ are required to be unique up to isomorphism in $\mathbf{D}_c^b(X)$.

Given a t-structure as above, and using the same notation, we write $\mathbf{E}^\bullet = \tau_{\leqslant 0}\mathbf{F}^\bullet$ (the *truncation of* $\mathbf{F}^\bullet$ *below* 0) and $\mathbf{G}^\bullet = \tau^{\geqslant 0}(\mathbf{F}^\bullet[1])$ (the *truncation of* $\mathbf{F}^\bullet[1]$ *above* 0); these are the basic truncation functors associated to the t-structure.

In addition, we write $\mathbf{D}^{\leqslant n}(X)$ for

$$\mathbf{D}^{\leqslant 0}(X)[-n] := \left\{\mathbf{F}^\bullet[-n] \mid \mathbf{F}^\bullet \in \mathbf{D}^{\leqslant 0}(X)\right\},$$

and we analogously write $\mathbf{D}^{\geqslant n}(X)$ for $\mathbf{D}^{\geqslant 0}(X)[-n]$.

Also, we define $\tau_{\leqslant n}\mathbf{F}^\bullet$ by

$$\tau_{\leqslant n}\mathbf{F}^\bullet = (\tau_{\leqslant 0}(\mathbf{F}^\bullet[n]))[-n] = ([-n]\circ\tau_{\leqslant 0}\circ[n])\mathbf{F}^\bullet,$$

and we analogously define $\tau^{\geqslant n}\mathbf{F}^\bullet$ as $([-n]\circ\tau^{\geqslant 0}\circ[n])\mathbf{F}^\bullet$.

Note that $\tau_{\leqslant n}\mathbf{F}^\bullet \in \mathbf{D}^{\leqslant n}(X)$, $\tau^{\geqslant n}\mathbf{F}^\bullet \in \mathbf{D}^{\geqslant n}(X)$ and, for all n, we have a distinguished triangle

$$\begin{array}{ccc} \tau_{\leqslant n}\mathbf{F}^\bullet & \longrightarrow & \mathbf{F}^\bullet \\ {\scriptstyle [1]}\nwarrow & & \swarrow \\ & \tau^{\geqslant n+1}\mathbf{F}^\bullet & \end{array} \quad .$$

Writing $\simeq$ to denote natural isomorphisms between functors: for all a and b,

$$\tau_{\leqslant b}\circ\tau^{\geqslant a} \simeq \tau^{\geqslant a}\circ\tau_{\leqslant b},$$

$$\tau_{\leqslant b}\circ\tau_{\leqslant a} \simeq \tau_{\leqslant a}\circ\tau_{\leqslant b},$$

and

$$\tau^{\geqslant b}\circ\tau^{\geqslant a} \simeq \tau^{\geqslant a}\circ\tau^{\geqslant b}.$$

If $a \geqslant b$, then

$$\tau_{\leqslant b}\circ\tau_{\leqslant a} \simeq \tau_{\leqslant b},$$

and

$$\tau^{\geqslant a}\circ\tau^{\geqslant b} \simeq \tau^{\geqslant a}.$$

Also, if $a > b$, then

$$\tau_{\leqslant b}\circ\tau^{\geqslant a} = \tau^{\geqslant a}\circ\tau_{\leqslant b} = 0.$$

The *heart* of the t-structure is defined to be the full subcategory $\mathcal{C} := \mathbf{D}^{\leqslant 0}(X)\cap\mathbf{D}^{\geqslant 0}(X)$; this is always an Abelian category. We wish to describe the kernels and cokernels in this category.

Let $\mathbf{E}^\bullet, \mathbf{F}^\bullet \in \mathcal{C}$ and let f be a morphism from $\mathbf{E}^\bullet$ to $\mathbf{F}^\bullet$. We can form a distinguished triangle in $\mathbf{D}_c^b(X)$

$$\begin{array}{ccc} \mathbf{E}^\bullet & \xrightarrow{f} & \mathbf{F}^\bullet \\ {\scriptstyle [1]}\nwarrow & & \swarrow \\ & \mathbf{G}^\bullet & \end{array} ,$$

where $\mathbf{G}^\bullet$ need not be in $\mathcal{C}$. Then, up to natural isomorphism,

$$\operatorname{coker} f = \tau^{\geqslant 0}\mathbf{G}^\bullet \text{ and } \ker f = \tau_{\leqslant 0}(\mathbf{G}^\bullet[-1]).$$

We define cohomology associated to a t-structure as follows. Define ${}^tH^0(\mathbf{F}^\bullet)$ to be $\tau^{\geqslant 0}\tau_{\leqslant 0}\mathbf{F}^\bullet$; this is naturally isomorphic to $\tau_{\leqslant 0}\tau^{\geqslant 0}\mathbf{F}^\bullet$. Now, define ${}^tH^n(\mathbf{F}^\bullet)$ to be

$${}^tH^0(\mathbf{F}^\bullet[n]) = \left(\tau^{\geqslant n}\tau_{\leqslant n}\mathbf{F}^\bullet\right)[n].$$

Note that this cohomology does not give back modules or even sheaves of modules, but rather gives back complexes which are objects in the heart of the t-structure.

If $\mathbf{F}^\bullet \in \mathbf{D}_c^b(X)$, then the following are equivalent:

1) $\mathbf{F}^\bullet \in \mathbf{D}^{\leqslant 0}(X)$ (resp. $\mathbf{D}^{\geqslant 0}(X)$);

2) the morphism $\tau_{\leqslant 0}\mathbf{F}^\bullet \to \mathbf{F}^\bullet$ is an isomorphism (resp. the morphism $\mathbf{F}^\bullet \to \tau^{\geqslant 0}\mathbf{F}^\bullet$ is an isomorphism);

3) $\tau^{\geqslant 1}\mathbf{F}^\bullet = 0$ (resp. $\tau_{\leqslant -1}\mathbf{F}^\bullet = 0$));

4) $\tau^{\geqslant i}\mathbf{F}^\bullet = 0$ for all $i \geqslant 1$ (resp. $\tau_{\leqslant i}\mathbf{F}^\bullet = 0$ for all $i \leqslant -1$);

5) there exists a such that $\mathbf{F}^\bullet \in \mathbf{D}^{\leqslant a}(X)$ and ${}^tH^i(\mathbf{F}^\bullet) = 0$ for all $i \geqslant 1$ (resp. there exists a such that $\mathbf{F}^\bullet \in \mathbf{D}^{\geqslant a}(X)$ and ${}^tH^i(\mathbf{F}^\bullet) = 0$ for all $i \leqslant -1$).

It follows that, if $\mathbf{F}^\bullet \in \mathbf{D}_c^b(X)$, then the following are equivalent:

1) $\mathbf{F}^\bullet \in \mathcal{C}$;

2) ${}^tH^0(\mathbf{F}^\bullet)$ is isomorphic to $\mathbf{F}^\bullet$;

3) there exist a and b such that $\mathbf{F}^\bullet \in \mathbf{D}^{\geqslant a}(X)$, $\mathbf{F}^\bullet \in \mathbf{D}^{\leqslant b}(X)$, and ${}^tH^n(\mathbf{F}^\bullet) = 0$ for all $n \neq 0$.

As the heart is an Abelian category, we may talk about exact sequences in $\mathcal{C}$. Any distinguished triangle in $\mathbf{D}_c^b(X)$ determines a long exact sequence of objects in the heart of the t-structure; if

$$\begin{array}{ccc} \mathbf{E}^\bullet & \longrightarrow & \mathbf{F}^\bullet \\ & {\scriptstyle [1]}\nwarrow \quad \swarrow & \\ & \mathbf{G}^\bullet & \end{array}$$

is a distinguished triangle in $\mathbf{D}_c^b(X)$, then the associated long exact sequence in $\mathcal{C}$ is

$$\cdots \to {}^tH^{-1}(\mathbf{G}^\bullet) \to {}^tH^0(\mathbf{E}^\bullet) \to {}^tH^0(\mathbf{F}^\bullet) \to {}^tH^0(\mathbf{G}^\bullet) \to {}^tH^1(\mathbf{E}^\bullet) \to \cdots.$$

We are finished now with our generalities on t-structures and wish to, at last, give our two primary examples.

The "ordinary" t-structure

The "ordinary" t-structure on $\mathbf{D}_c^b(X)$ is given by

$$\mathbf{D}^{\leqslant 0}(X) = \{\mathbf{F}^\bullet \in \mathbf{D}_c^b(X) \mid \mathbf{H}^i(\mathbf{F}^\bullet) = 0 \text{ for all } i > 0\}$$

and

$$\mathbf{D}^{\geqslant 0}(X) = \{\mathbf{F}^\bullet \in \mathbf{D}_c^b(X) \mid \mathbf{H}^i(\mathbf{F}^\bullet) = 0 \text{ for all } i < 0\}.$$

The associated truncation functors are the ordinary ones described in [**G-M3**]. If $\mathbf{F}^\bullet$ is in $D_c^b(X)$, then

$$(\tau_{\leqslant p}\mathbf{F}^\bullet)^n = \begin{cases} \mathbf{F}^n & \text{if } n < p \\ \mathbf{ker}\ d^p & \text{if } n = p \\ 0 & \text{if } n > p \end{cases}$$

and

$$\left(\tau^{\geqslant p}\mathbf{F}^\bullet\right)^n = \begin{cases} 0 & \text{if } n < p \\ \mathbf{coker}\ d^{p-1} & \text{if } n = p \\ \mathbf{F}^n & \text{if } n > p. \end{cases}$$

These truncated complexes are naturally quasi-isomorphic to the complexes

$$(\tilde{\tau}_{\leqslant p}\mathbf{F}^\bullet)^n = \begin{cases} \mathbf{F}^n & \text{if } n \leqslant p \\ \mathbf{Im}\ d^p & \text{if } n = p+1 \\ 0 & \text{if } n > p+1 \end{cases}$$

and

$$\left(\tilde{\tau}^{\geqslant p}\mathbf{F}^\bullet\right)^n = \begin{cases} 0 & \text{if } n < p-1 \\ \mathbf{Im}\ d^{p-1} & \text{if } n = p-1 \\ \mathbf{F}^n & \text{if } n \geqslant p. \end{cases}$$

If $\mathbf{A}^\bullet, \mathbf{B}^\bullet \in \mathbf{D}^b_c(X)$, then

1. $(\tau_{\leqslant p}\mathbf{A}^\bullet)_x = \tau_{\leqslant p}(\mathbf{A}^\bullet_x)$;

2. $$\mathbf{H}^k(\tau_{\leqslant p}\mathbf{A}^\bullet)_x = \begin{cases} \mathbf{H}^k(\mathbf{A}^\bullet)_x & \text{if } k \leqslant p \\ 0 & \text{for } k > p. \end{cases}$$

3. If $\phi : \mathbf{A}^\bullet \to \mathbf{B}^\bullet$ is a morphism of complexes of sheaves which induces isomorphisms on the associated cohomology sheaves
$$\phi^* : \mathbf{H}^n(\mathbf{A}^\bullet) \cong \mathbf{H}^n(\mathbf{B}^\bullet) \text{ for all } n \leqslant p,$$
then $\tau_{\leqslant p}\phi : \tau_{\leqslant p}\mathbf{A}^\bullet \to \tau_{\leqslant p}\mathbf{B}^\bullet$ is a quasi-isomorphism.

4. If $f : X \to Y$ is a continuous map and $\mathbf{C}^\bullet$ is a complex of sheaves on Y, then
$$\tau_{\leqslant p} f^*(\mathbf{C}^\bullet) \cong f^*\tau_{\leqslant p}(\mathbf{C}^\bullet).$$

5. If R is a field and $\mathbf{A}^\bullet$ is a complex of sheaves of R-modules on X with locally constant cohomology sheaves, then there are natural quasi-isomorphisms
$$\tau^{\geqslant -p}R\mathbf{Hom}^\bullet(\mathbf{A}^\bullet, \mathbf{R}^\bullet_X) \to \tau^{\geqslant -p}R\mathbf{Hom}^\bullet(\tau_{\leqslant p}\mathbf{A}^\bullet, \mathbf{R}^\bullet_X) \leftarrow R\mathbf{Hom}^\bullet(\tau_{\leqslant p}\mathbf{A}^\bullet, \mathbf{R}^\bullet_X).$$

The heart of this t-structure consists of those complexes which have non-zero cohomology sheaves only in degree 0; such complexes are quasi-isomorphic to complexes which are non-zero only in degree 0.

The t-structure cohomology of a complex $\mathbf{F}^\bullet$ is essentially the sheaf cohomology of $\mathbf{F}^\bullet$; ${}^tH^n(\mathbf{F}^\bullet)$ is quasi-isomorphic to a complex which has $\mathbf{H}^n(\mathbf{F}^\bullet)$ in degree 0 and is zero in all other degrees. With this identification, the t-structure long exact sequence associated to a distinguished triangle is merely the usual long exact sequence on sheaf cohomology.

We are now going to give the construction of the intersection cohomology complexes as it is presented in [**G-M3**]. Our indexing will look different from that of [**G-M3**] for several reasons.

First, we are dealing only with complex analytic spaces, X, and we are using only middle perversity; this accounts for some of the indexing differences. In addition, in this setting, the intersection cohomology complex defined in [**G-M3**] would have possibly non-zero cohomology only in degrees between $-2\dim_{\mathbb{C}} X$ and $-(\dim_{\mathbb{C}} X) - 1$, inclusive. The definition below is shifted by $-\dim_{\mathbb{C}} X$ from the [**G-M3**] definition, and yields a perverse sheaf which has possibly non-zero cohomology only in degrees between $-\dim_{\mathbb{C}} X$ and -1, inclusive.

Let X be a complex analytic n-dimensional space with a complex analytic Whitney stratification $\mathcal{S} = \{S_\alpha\}$. While we do not explicitly require that X is pure-dimensional, it will follow from the construction that components of X of dimension less than n will essentially be ignored.

For all k, let X^k denote the union of the strata of dimension less than or equal to k. By convention, we set $X^{-1} = \emptyset$. Hence, we have a filtration

$$\emptyset = X^{-1} \subseteq X^0 \subseteq X^1 \subseteq \cdots \subseteq X^{n-1} \subseteq X^n = X.$$

For all k, let $\mathcal{U}_k := X - X^{n-k}$, and let i_k denote the inclusion $\mathcal{U}_k \hookrightarrow \mathcal{U}_{k+1}$. Let $\mathcal{L}^\bullet_{\mathcal{U}_1}$ be a local system on the top-dimensional strata.

Then, the intersection cohomology complex on X with coefficients in $\mathcal{L}^\bullet_{\mathcal{U}_1}$, as described in section 2, is given by

$$\mathbf{IC}^\bullet_X(\mathcal{L}^\bullet_{\mathcal{U}_1}) := \tau_{\leqslant -1} Ri_{n*} \ldots \tau_{\leqslant 1-n} Ri_{2*} \tau_{\leqslant -n} Ri_{1*}(\mathcal{L}^\bullet_{\mathcal{U}_1}[n]).$$

Up to quasi-isomorphism, this complex is independent of the stratification. Note that the cohomology sheaves of $\mathbf{IC}^\bullet_X(\mathcal{L}^\bullet_{\mathcal{U}_1})$ are supported only in degrees k for which $-n \leqslant k \leqslant -1$ (unless X is 0-dimensional, and then $\mathbf{IC}^\bullet_X(\mathcal{L}^\bullet_{\mathcal{U}_1}) \cong \mathcal{L}^\bullet_{\mathcal{U}_1}$).

Also note that it follows from the construction that there is always a canonical map from the shifted constant sheaf $\mathbf{R}^\bullet_X[n]$ to $\mathbf{IC}^\bullet_X(\mathbf{R}^\bullet_{\mathcal{U}_1})$ which induces an isomorphism when restricted to $\mathcal{U}_1$.

To see this, consider the canonical morphism $\mathbf{R}^\bullet_{\mathcal{U}_{k+1}}[n] \to Ri_{k*}i_k^*\mathbf{R}^\bullet_{\mathcal{U}_{k+1}}[n]$ for each $k \geqslant 1$. As $i_k^*\mathbf{R}^\bullet_{\mathcal{U}_{k+1}}[n] \cong \mathbf{R}^\bullet_{\mathcal{U}_k}[n]$, we have a canonical map $\mathbf{R}^\bullet_{\mathcal{U}_{k+1}}[n] \to Ri_{k*}\mathbf{R}^\bullet_{\mathcal{U}_k}[n]$ and, hence, a canonical map between the truncations $\tau_{\leqslant k-n-1}(\mathbf{R}^\bullet_{\mathcal{U}_{k+1}}[n]) \to \tau_{\leqslant k-n-1} Ri_{k*}(\mathbf{R}^\bullet_{\mathcal{U}_k}[n])$. But,

$$\tau_{\leqslant k-n-1}(\mathbf{R}^\bullet_{\mathcal{U}_{k+1}}[n]) \cong \mathbf{R}^\bullet_{\mathcal{U}_{k+1}}[n]$$

and so we have a canonical map $\mathbf{R}^\bullet_{\mathcal{U}_{k+1}}[n] \to \tau_{\leqslant k-n-1} Ri_{k*}(\mathbf{R}^\bullet_{\mathcal{U}_k}[n])$. By piecing all of these maps together, one obtains the desired morphism.

The perverse t-structure

The *perverse* t-structure (with middle perversity μ) on $\mathbf{D}^b_c(X)$ is given by

$${}^{\mu}\mathbf{D}^{\leqslant 0}(X) = \{\mathbf{F}^\bullet \in \mathbf{D}^b_c(X) \mid \dim \operatorname{supp}^{\text{-j}} \mathbf{F}^\bullet \leqslant j \text{ for all } j\}$$

and

$${}^{\mu}\mathbf{D}^{\geqslant 0}(X) = \{\mathbf{F}^\bullet \in \mathbf{D}^b_c(X) \mid \dim \operatorname{cosupp}^{\text{j}} \mathbf{F}^\bullet \leqslant j \text{ for all } j\}.$$

Note that the heart of this t-structure is precisely $Perv(X)$. Thus, every distinguished triangle in $\mathbf{D}^b_c(X)$ determines a long exact sequence in the Abelian category $Perv(X)$.

We naturally call the t-structure cohomology associated to the perverse t-structure the *perverse cohomology* or *perverse projection* and denote it in degree n by ${}^{\mu}H^n(\mathbf{F}^\bullet)$.

Let d be an integer, and let $f : Y \to X$ be a morphism of complex spaces such that $\dim f^{-1}(x) \leqslant d$, for all $\mathbf{x} \in X$. Let $\dim Y/X := \dim Y - \dim X$. Then,

1) f^* sends ${}^\mu\mathbf{D}^{\leqslant 0}(X)$ to ${}^\mu\mathbf{D}^{\leqslant d}(Y)$, and sends ${}^\mu\mathbf{D}^{\geqslant 0}(X)$ to ${}^\mu\mathbf{D}^{\geqslant \dim Y/X}(Y)$;

2) $f^!$ sends ${}^\mu\mathbf{D}^{\geqslant 0}(X)$ to ${}^\mu\mathbf{D}^{\geqslant -d}(Y)$, and sends ${}^\mu\mathbf{D}^{\leqslant 0}(X)$ to ${}^\mu\mathbf{D}^{\leqslant -\dim Y/X}(Y)$;;

3) if $\mathbf{F}^\bullet \in {}^\mu\mathbf{D}^{\leqslant 0}(Y)$ and $Rf_!\mathbf{F}^\bullet \in \mathbf{D}_c^b(X)$, then $Rf_!\mathbf{F}^\bullet \in {}^\mu\mathbf{D}^{\leqslant d}(X)$;

4) if $\mathbf{F}^\bullet \in {}^\mu\mathbf{D}^{\geqslant 0}(Y)$ and $Rf_*\mathbf{F}^\bullet \in \mathbf{D}_c^b(X)$, then $Rf_*\mathbf{F}^\bullet \in {}^\mu\mathbf{D}^{\geqslant -d}(X)$.

Let $f : Y \to X$ be a morphism of complex spaces such that each point in X has an open neighborhood $\mathcal{U}$ such that $f^{-1}(\mathcal{U})$ is a Stein space (e.g., an affine map between algebraic varieties). Then,

1) if $\mathbf{F}^\bullet \in {}^\mu\mathbf{D}^{\leqslant 0}(Y)$ and $Rf_*\mathbf{F}^\bullet \in \mathbf{D}_c^b(X)$, then $Rf_*\mathbf{F}^\bullet \in {}^\mu\mathbf{D}^{\leqslant 0}(X)$;

2) if $\mathbf{F}^\bullet \in {}^\mu\mathbf{D}^{\geqslant 0}(Y)$ and $Rf_!\mathbf{F}^\bullet \in \mathbf{D}_c^b(X)$, then $Rf_!\mathbf{F}^\bullet \in {}^\mu\mathbf{D}^{\geqslant 0}(X)$.

If $f : X \to \mathbb{C}$ is an analytic map, then the functors $\psi_f[-1]$ and $\phi_f[-1]$ are *t-exact* with respect to the perverse t-structures; this means that if $\mathbf{E}^\bullet \in {}^\mu\mathbf{D}^{\leqslant 0}(X)$ and $\mathbf{F}^\bullet \in {}^\mu\mathbf{D}^{\geqslant 0}(X)$, then $\psi_f\mathbf{E}^\bullet[-1]$ and $\phi_f\mathbf{E}^\bullet[-1]$ are in ${}^\mu\mathbf{D}^{\leqslant 0}(f^{-1}(0))$, and $\psi_f\mathbf{F}^\bullet[-1]$ and $\phi_f\mathbf{F}^\bullet[-1]$ are in ${}^\mu\mathbf{D}^{\geqslant 0}(f^{-1}(0))$.

In particular, $\psi_f[-1]$ and $\phi_f[-1]$ take perverse sheaves to perverse sheaves and, for any $\mathbf{F}^\bullet \in \mathbf{D}_c^b(X)$,

$${}^\mu\!H^n(\psi_f\mathbf{F}^\bullet[-1]) \cong \psi_f{}^\mu\!H^n(\mathbf{F}^\bullet)[-1] \quad\text{and}\quad {}^\mu\!H^n(\phi_f\mathbf{F}^\bullet[-1]) \cong \phi_f{}^\mu\!H^n(\mathbf{F}^\bullet)[-1].$$

If the base ring is a field, then the functor ${}^\mu\!H^0$ also commutes with Verdier dualizing; that is, there is a natural isomorphism

$$\mathcal{D}\circ {}^\mu\!H^0 \cong {}^\mu\!H^0 \circ \mathcal{D}.$$

Let $\mathbf{F}^\bullet$ be a bounded complex of sheaves on X which is constructible with respect to a connected Whitney stratification $\{S_\alpha\}$ of X, and let $d_\alpha := \dim S_\alpha$. Then, ${}^\mu\!H^0(\mathbf{F}^\bullet)$ is also constructible with respect to $\mathcal{S}$, and $\left({}^\mu\!H^0(\mathbf{F}^\bullet)\right)_{|\mathbb{N}_\alpha}[-d_\alpha]$ is naturally isomorphic to ${}^\mu\!H^0(\mathbf{F}^\bullet_{|\mathbb{N}_\alpha}[-d_\alpha])$, where $\mathbb{N}_\alpha$ denotes a normal slice to S_α.

Let $S_{\max}$ be a maximal stratum contained in the support of $\mathbf{F}^\bullet$, and let $m = \dim S_{\max}$. Then, $\left({}^\mu\!H^0(\mathbf{F}^\bullet)\right)_{|S_{\max}}$ is isomorphic (in the derived category) to the complex which has $\left(\mathbf{H}^{-m}(\mathbf{F}^\bullet)\right)_{|S_{\max}}$ in degree $-m$ and zero in all other degrees.

In particular, $\operatorname{supp}\mathbf{F}^\bullet = \bigcup_i \operatorname{supp}{}^\mu\!H^i(\mathbf{F}^\bullet)$, and if $\mathbf{F}^\bullet$ is supported on an isolated point, $\mathbf{q}$, then $H^0({}^\mu\!H^0(\mathbf{F}^\bullet))_{\mathbf{q}} \cong H^0(\mathbf{F}^\bullet)_{\mathbf{q}}$. From this, and the fact that perverse cohomology commutes with nearby and vanishing cycles shifted by -1, one easily concludes that, at all points $\mathbf{x}\in X$,

$$\chi(\mathbf{F}^\bullet)_{\mathbf{x}} = \sum_k (-1)^k \chi\left({}^\mu\!H^k(\mathbf{F}^\bullet)\right)_{\mathbf{x}}.$$

Switching Coefficients

Suppose that the base ring R is a p.i.d. For each prime ideal $\mathfrak{p}$ of R, let $k_\mathfrak{p}$ denote the field of fractions of $R/\mathfrak{p}$, i.e., k_0 is the field of fractions of R, and for $\mathfrak{p} \neq 0$, $k_\mathfrak{p} = R/\mathfrak{p}$. There are the obvious functors $\delta_\mathfrak{p} : \mathbf{D}_c^b(R_X) \to \mathbf{D}_c^b((k_\mathfrak{p})_X)$, which sends $\mathbf{F}^\bullet$ to $\mathbf{F}^\bullet \overset{L}{\otimes} (k_\mathfrak{p})^\bullet_X$, and $\epsilon_\mathfrak{p} : \mathbf{D}_c^b((k_\mathfrak{p})_X) \to \mathbf{D}_c^b(R_X)$, which considers $k_\mathfrak{p}$-vector spaces as R-modules.

If $\mathbf{A}^\bullet$ is a complex of $k_\mathfrak{p}$-vector spaces, we may consider the perverse cohomology of $\mathbf{A}^\bullet$, ${}^\mu\!H^i_{k_\mathfrak{p}}(\mathbf{A}^\bullet)$, or the perverse cohomology of $\epsilon(\mathbf{A}^\bullet)$, which we denote by ${}^\mu\!H^i_R(\mathbf{A}^\bullet)$. If $\mathbf{A}^\bullet \in \mathbf{D}_c^b((k_\mathfrak{p})_X)$ and $S_{\max}$ is a maximal stratum contained in the support of $\mathbf{A}^\bullet$, then there is a canonical isomorphism

$$\epsilon\big(({}^\mu\!H^i_{k_\mathfrak{p}}(\mathbf{A}^\bullet))_{|S_\alpha}\big) \cong ({}^\mu\!H^i_R(\mathbf{A}^\bullet))_{|S_\alpha};$$

in particular, $\operatorname{supp} {}^\mu\!H^i_{k_\mathfrak{p}}(\mathbf{A}^\bullet) = \operatorname{supp} {}^\mu\!H^i_R(\mathbf{A}^\bullet)$.

If $\mathbf{F}^\bullet \in \mathbf{D}_c^b(R_X)$, $S_{\max}$ is a maximal stratum contained in the support of $\mathbf{F}^\bullet$, and $\mathbf{x} \in S_{\max}$, then for some prime ideal $\mathfrak{p} \subset R$ and for some integer i, $H^i(\mathbf{F}^\bullet)_\mathbf{x} \otimes k_\mathfrak{p} \neq 0$; it follows that $S_{\max}$ is also a maximal stratum in the support of $\mathbf{F}^\bullet \overset{L}{\otimes} (k_\mathfrak{p})^\bullet_X$. Thus,

$$\operatorname{supp} \mathbf{F}^\bullet = \bigcup_\mathfrak{p} \operatorname{supp}(\mathbf{F}^\bullet \overset{L}{\otimes} (k_\mathfrak{p})^\bullet_X)$$

and so

$$\operatorname{supp} \mathbf{F}^\bullet = \bigcup_{i,\mathfrak{p}} \operatorname{supp} {}^\mu\!H^i_{k_\mathfrak{p}}(\mathbf{F}^\bullet \overset{L}{\otimes} (k_\mathfrak{p})^\bullet_X),$$

where the boundedness and constructibility of $\mathbf{F}^\bullet$ imply that this union is locally finite.

APPENDIX C:

PRIVILEGED NEIGHBORHOODS AND LIFTING MILNOR FIBRATIONS

In this appendix, we prove a number of very technical results. These results tell us when we can use certain types of "nice" neighborhoods to define the Milnor fibre (at least, up to homotopy), and give conditions under which Milnor fibrations remain constant in a parameterized family. Lê numbers and cycles do not appear here, though we will use the relative polar curve.

Throughout, for convenience, we concentrate our attention at the origin. Let $\mathcal{U}$ be an open neighborhood of the origin in some $\mathbb{C}^{n+1}$ and let $h : (\mathcal{U}, \mathbf{0}) \to (\mathbb{C}, 0)$ be an analytic function.

In what sense the Milnor fibre and Milnor fibration of h are well-defined has been discussed in a number of places (see, for instance, [**Se-Th**]). If one is primarily interested in the ambient, local topology of the hypersurface $V(h)$ defined by h, then "the" Milnor fibre is only well-defined up to homotopy-type [**Lê6**]. Thus, we may make the weakest possible definition of the Milnor fibre of h at the origin as a homotopy-type:

Definition/Proposition C.1. A *system of Milnor neighborhoods* for h at the origin is a fundamental system of neighborhoods, $\{C_\alpha\}$, at the origin in $\mathcal{U}$ such that for all $C_\alpha \subseteq C_\beta$, there exists $\epsilon > 0$ such that for all complex ξ with $0 < |\xi| < \epsilon$, we have that the inclusion $C_\alpha \cap V(h-\xi) \hookrightarrow C_\beta \cup V(h-\xi)$ is a homotopy-equivalence. The *standard system of Milnor neighborhoods* for h at the origin is just the set of closed balls of sufficiently small radius centered at the origin (this system is independent of h except for how small the radii must be).

If $\{C_\alpha\}$ is a system of Milnor neighborhoods for h at the origin, then for each C_α there exists $\epsilon > 0$ such that the homotopy-type of $C_\alpha \cap V(h-\xi)$ is independent of the complex number ξ chosen as long as $0 < |\xi| < \epsilon$. Moreover, this homotopy-type is independent of the choice of the particular C_α and is, in fact, independent of the choice of the system of Milnor neighborhoods.

Proof. The proof is standard. Let $\{C_\alpha\}$ be a system of Milnor neighborhoods for h at the origin. We shall compare it with the standard system. Select any C_β. Now, pick C_α, B_η, and B_δ such that B_η and B_δ are in the standard system of Milnor neighborhoods for h at the origin and such that $B_\delta \subseteq C_\alpha \subseteq B_\eta \subseteq C_\beta$. We may certainly pick $\epsilon > 0$ such that, for all complex ξ with $0 < |\xi| < \epsilon$, the inclusion $C_\alpha \cap V(h-\xi) \hookrightarrow C_\beta \cap V(h-\xi)$ and the inclusion $B_\delta \cap V(h-\xi) \hookrightarrow B_\eta \cap V(h-\xi)$ are both homotopy-equivalences. It follows that the inclusion $C_\alpha \cap V(h-\xi) \hookrightarrow B_\eta \cap V(h-\xi)$ is a homotopy-equivalence for all small $\xi \neq 0$ and thus, as the homotopy-type of $B_\eta \cap V(h-\xi)$ is independent of ξ, so is that of $C_\alpha \cap V(h-\xi)$. The conclusion follows immediately. □

It is sometimes more convenient to prove that $C_\alpha \cap V(h-\xi) \hookrightarrow C_\beta \cap V(h-\xi)$ is a homotopy-equivalence whenever C_α is contained in the *interior* of C_β. It is easy to see by the proof above that this is enough to show that the system is a system of Milnor neighborhoods.

A system of Milnor neighborhoods allows one to discuss the Milnor fibre up to homotopy. However, one frequently wishes to use stratified, differential techniques to study the Milnor fibre and, hence, one would like for the Milnor fibre to have the structure of a smooth, compact manifold with (stratified) boundary and would also like to have some control over what happens on the boundary as one moves through a family of singularities. Furthermore, one would like to have a notion of the Milnor *fibration* – at least up to fibre-homotopy-type.

To gain this additional structure, we will use two types of (complex analytic) stratifications. One is the well-known Whitney stratification [**G-M2**], [**Mat**], [**Th**]. The second is a *good stratification*, as defined in 1.24. Note that any refinement of a good stratification which does not refine the smooth stratum is automatically a good stratification. This fact will be very useful when combined with the following proposition, which is Theorem 18.11 of [**W**] (or just a small portion of Theorem 1.7 of [**G-M2**]).

Proposition C.2. *Let X be an analytic subset of $\mathbb{C}^N$ and let Y be an analytic subset of X. Suppose that $\mathcal{D}$ and $\mathcal{F}$ are analytic stratifications for X and Y, respectively. Then, there exists an analytic stratification, $\mathcal{L}$, of X which is a common refinement of both $\mathcal{D}$ and $\mathcal{F}$, i.e. every stratum of $\mathcal{L}$ is contained in a stratum of $\mathcal{D}$, Y is a union of strata of $\mathcal{L}$, and every stratum of $\mathcal{L}$ which is contained in Y is contained in a stratum of $\mathcal{F}$.*

The $\mathcal{L}$ above is sometimes referred to as a refinement of $\mathcal{D}$ adapted to Y.

(The reader should note that when Goresky and MacPherson use the term "stratification", they mean that the Whitney conditions are satisfied. Hence, their Theorem 1.7 actually allows us to pick a common analytic, Whitney refinement.)

We now generalize the notion of a privileged polydisc as given in [**Lê3**]. This definition should be compared with [**L-T1**, 2.2.3].

Definition C.3. Let $\mathfrak{G}$ be a good stratification for h at the origin. A fundamental system of neighborhoods, $\{C_\alpha\}$, at the origin in $\mathcal{U}$ is a *system of privileged neighborhoods for h at $\mathbf{0}$ with respect to $\mathfrak{G}$* if and only if

i) $\{C_\alpha\}$ is a system of compact, Milnor neighborhoods for h at the origin;

and, for each C_α, there is an associated Whitney stratification, $\mathcal{S}_\alpha$, of C_α such that

ii) the interior of C_α, $\overset{\circ}{C}_\alpha$, in $\mathcal{U}$ is a stratum in $\mathcal{S}_\alpha$;

iii) C_α equals the closure of $\overset{\circ}{C}_\alpha$ in $\mathcal{U}$;

By ii) and iii) and the condition of the frontier, the boundary of C_α, ∂C_α, is a union of Whitney strata, and we make the final requirement:

iv) the boundary strata of each C_α transversely intersect all the strata of $\mathfrak{G}$.

A fundamental system of neighborhoods, $\mathcal{C} = \{C_\alpha\}$, is a *system of privileged neighborhoods* *for h at $\mathbf{0}$* if and only if there exists a good stratification, $\mathfrak{G}$, for h at $\mathbf{0}$ such that $\mathcal{C}$ is a system of privileged neighborhoods for h at $\mathbf{0}$ with respect to $\mathfrak{G}$.

A fundamental system of neighborhoods, $\mathcal{C} = \{C_\alpha\}$, satisfying i), ii), and iii) above is a *system of weakly privileged neighborhoods for h at $\mathbf{0}$* if and only if for each C_α, for all small $\xi \neq 0$, $V(h-\xi)$ transversely intersects the boundary strata of C_α. We shall see below that a system of privileged neighborhoods is automatically a system of weakly privileged neighborhoods.

A fundamental system of neighborhoods, $\mathcal{C} = \{C_\alpha\}$, is a *universal system of privileged neighborhoods for h at $\mathbf{0}$* if and only if for every good stratification, $\mathfrak{G}$, for h at $\mathbf{0}$, there exists an open neighborhood W of the origin such that $\{C_\alpha \in \mathcal{C} \mid C_\alpha \subseteq W\}$ is a system of privileged neighborhoods for h at $\mathbf{0}$ with respect to $\mathfrak{G}$.

One should note that the set of closed balls centered at the origin is a universal system of privileged neighborhoods for h, regardless of the function h – this is a very "universal" system, and this may seem like the more natural notion. This, however, seems to be too restrictive. Universal for h simply means that, locally, the fundamental system is privileged independent of the choice of good stratification for the particular function h.

Proposition C.4. *Suppose that $\mathcal{C} = \{C_\alpha\}$ is a system of privileged neighborhoods for h at $\mathbf{0}$. Then, $\mathcal{C} = \{C_\alpha\}$ is a system of weakly privileged neighborhoods for h at $\mathbf{0}$ and, hence, for all C_α, for all small $\delta > 0$, $C_\alpha \cap h^{-1}(\partial\mathbb{D}_\delta) \xrightarrow{h} \partial\mathbb{D}_\delta$ is a proper, stratified submersion and is thus a locally trivial fibration. The fibre-homotopy-type of this fibration is independent of the choice of the system of weakly privileged neighborhoods, $\mathcal{C}$, for h, the choice of C_α, and the choice of small $\delta > 0$.*

Proof. The proof is essentially that of Lê in [**Lê 4**]. Let $\mathfrak{G}$ be a good stratification for h at the origin with respect to which $\mathcal{C}$ is a system of privileged neighborhoods. Pick a C_α in $\mathcal{C}$. We shall actually show that there exists $\epsilon > 0$ such that $C_\alpha \cap h^{-1}(\mathring{\mathbb{D}}_\epsilon - 0) \xrightarrow{h} \mathring{\mathbb{D}}_\epsilon - 0$ is a proper, stratified submersion. It follows that, for all δ with $0 < \delta < \epsilon$, $C_\alpha \cap h^{-1}(\partial\mathbb{D}_\delta) \xrightarrow{h} \partial\mathbb{D}_\delta$ is a proper, stratified submersion and, hence, a locally trivial fibration with fibre-homotopy-type independent of the choice of δ. This certainly shows that $\mathcal{C}$ is a system of weakly privileged neighborhoods for h at $\mathbf{0}$.

Suppose to the contrary that no matter how small we choose $\epsilon > 0$ it is not the case that $C_\alpha \cap h^{-1}(\mathring{\mathbb{D}}_\epsilon - 0) \xrightarrow{h} \mathring{\mathbb{D}}_\epsilon - 0$ is a proper, stratified submersion. As each C_α is compact, clearly this map is always proper. So, by the local finiteness of the stratification, there must exist a single Whitney stratum, S, of C_α and a sequence of points $\mathbf{p}_i \in S$ such that the $\mathbf{p}_i$ converge to some point $\mathbf{p} \in V(h)$, $T_{\mathbf{p}_i}S$ converges to some T, $T_{\mathbf{p}_i}V(h - h(\mathbf{p}_i))$ converges to some $\mathcal{T}$, and $T_{\mathbf{p}_i}S \subseteq T_{\mathbf{p}_i}V(h - h(\mathbf{p}_i))$. Let G denote the good stratum of $\mathfrak{G}$ which contains $\mathbf{p}$ and let R denote the Whitney stratum of C_α which contains $\mathbf{p}$.

As $T_{\mathbf{p}_i}S \subseteq T_{\mathbf{p}_i}V(h-h(\mathbf{p}_i))$, we must have that $T \subseteq \mathcal{T}$. By the Thom condition, $T_{\mathbf{p}}G \subseteq \mathcal{T}$. By Whitney's condition a), $T_{\mathbf{p}}R \subseteq T$. Hence, $T_{\mathbf{p}}R$ and $T_{\mathbf{p}}G$ are both contained in $\mathcal{T}$ – a contradiction as R and G intersect transversely.

Thus, there exists $\epsilon > 0$ such that for all δ with $0 < \delta < \epsilon$, $C_\alpha \cap h^{-1}(\partial\mathbb{D}_\delta) \xrightarrow{h} \partial\mathbb{D}_\delta$ is a proper, stratified submersion and, hence, a locally trivial fibration with fibre-homotopy-type independent of the choice of δ. To see that the fibre-homotopy-type is independent of the choice of $\mathcal{C}$ and the choice of C_α, one may once again compare with the standard system of Milnor neighborhoods and then use the theorem of Dold [**Hu**, p.209], since we know that the inclusion of each fibre is a homotopy-equivalence by the proof of C.1. We leave the details to the reader. $\square$

Note that we have the implications: $\{C_\alpha\}$ is a universal system $\Rightarrow$ for all good stratifications $\mathfrak{G}$, $\{C_\alpha\}$ is a privileged system with respect to $\mathfrak{G} \Rightarrow \{C_\alpha\}$ is a privileged system $\Rightarrow$ $\{C_\alpha\}$ is a weakly privileged system $\Rightarrow \{C_\alpha\}$ is a Milnor system.

Definition C.5. If $\mathcal{C} = \{C_\alpha\}$ is a system of Milnor neighborhoods for h at $\mathbf{0}$, then a *Milnor pair* for h at $\mathbf{0}$ is a pair $(C_\alpha, \overset{\circ}{\mathbb{D}}_\delta)$ such that for all $\xi \in \overset{\circ}{\mathbb{D}}_\delta - \mathbf{0}$, $C_\alpha \cap V(h-\xi)$ has the homotopy-type of the Milnor fibre. If, in addition, $\mathcal{C}$ is a system of weakly privileged neighborhoods, then we also make the requirement that $C_\alpha \cap h^{-1}(\partial\mathbb{D}_\delta) \xrightarrow{h} \partial\mathbb{D}_\delta$ is a proper, stratified submersion.

We now wish to consider an analytic function $f : (\overset{\circ}{\mathbb{D}} \times \mathcal{U}, \overset{\circ}{\mathbb{D}} \times \mathbf{0}) \to (\mathbb{C}, 0)$ where $\overset{\circ}{\mathbb{D}}$ is an open complex disc centered at the origin and $\mathcal{U} \subseteq \mathbb{C}^{n+1}$. We use the coordinates $(t, z_0, \dots, z_n)$ for $\overset{\circ}{\mathbb{D}} \times \mathcal{U}$. We distinguish the t-coordinate because we will either be considering the particular hyperplane slice $V(t)$ or because we will be interested in the family $f_t(z_0, \dots, z_n) := f(t, z_0, \dots, z_n)$.

Proposition C.6. *Suppose that $V(t)$ is prepolar for f at the origin with respect to a good stratification $\mathfrak{G}$, and let $\{C_\alpha\}$ be a system of privileged neighborhoods with respect to the good stratification $\mathfrak{G} \cap V(t)$ for $f_{|_{V(t)}}$. Then, there exits an open neighborhood, W, of the origin in $V(t)$ such that, for all $C_\alpha \subseteq W$, there exists $\tau_\alpha > 0$ such that*

i) *there exists $\omega > 0$ such that*

$$\begin{array}{c} \overset{\circ}{\mathbb{D}}_{\tau_\alpha} \times \partial C_\alpha \ \cap \ \Psi^{-1}\big((\overset{\circ}{\mathbb{D}}_\omega - \mathbf{0}) \times \overset{\circ}{\mathbb{D}}_{\tau_\alpha}\big) \\ \big\downarrow \Psi := (f,t) \\ (\overset{\circ}{\mathbb{D}}_\omega - \mathbf{0}) \times \overset{\circ}{\mathbb{D}}_{\tau_\alpha} \end{array}$$

is a proper, stratified submersion;

ii) *for all δ with $0 < \delta < \tau_\alpha$, there exists $\xi > 0$ such that*

$$\begin{array}{c} \mathbb{D}_\delta \times C_\alpha \ \cap \ f^{-1}(\overset{\circ}{\mathbb{D}}_\xi - \mathbf{0}) \\ \big\downarrow f \\ \overset{\circ}{\mathbb{D}}_\xi - \mathbf{0} \end{array}$$

is a proper, stratified submersion, where the strata are the cross-product strata of $\overset{\circ}{\mathbb{D}}_\delta \times C_\alpha$ *together with those of* $\partial\mathbb{D}_\delta \times C_\alpha$*; and*

iii) $\{\mathbb{D}_\delta \times C_\alpha \mid 0 < \delta < \tau_\alpha\}$ *is a system of Milnor neighborhoods for* f *at the origin and hence, by ii), is in fact a system of weakly privileged neighborhoods.*

Proof. There exists an open neighborhood of the origin in $\overset{\circ}{\mathbb{D}} \times \mathcal{U}$ of the form $\overset{\circ}{\mathbb{D}}_\eta \times W$ such that $(\overset{\circ}{\mathbb{D}}_\eta \times W) \cap \Sigma f \subseteq V(f)$. As $V(t)$ is prepolar, we may assume that $\mathfrak{G}$ is defined inside $\overset{\circ}{\mathbb{D}}_\eta \times W$ and that $V(t)$ transversely intersects all strata of $\mathfrak{G}$, other than the origin, inside $\overset{\circ}{\mathbb{D}}_\eta \times W$. Finally, as $V(t)$ is prepolar, we may use Theorem 1.28 to conclude that $\gamma^1_{f,t}(\mathbf{0})$ exists and, hence, we may select $\overset{\circ}{\mathbb{D}}_\eta \times W$ so that $(0 \times W) \cap \Gamma^1_{f,t} \subseteq \{\mathbf{0}\}$. Let $C_\alpha \subseteq W$.

i) This follows the proof of Proposition 2.1 of [**Lê1**], applied to each stratum of ∂C_α. Suppose the contrary. Then, we would have a stratum S of C_α and a sequence of points $\mathbf{p}_i$ not in $V(f)$ but in $\mathbb{C} \times S$ such that $\mathbf{p}_i = (t_i, \mathbf{q}_i) \to \mathbf{p} = (0, \mathbf{q}) \in V(t) \cap V(f)$ and such that

$$(*) \qquad T_{\mathbf{p}_i} V(f - f(\mathbf{p}_i), t - t_i) + T_{\mathbf{p}_i}(\mathbb{C} \times S) \neq \mathbb{C}^{n+2}.$$

(That $T_{\mathbf{p}_i} V(f - f(\mathbf{p}_i), t - t_i)$ exists is not completely trivial – it follows from the assumptions made in the preceding paragraph.) Let G denote the good stratum of $V(f)$ containing $\mathbf{p}$. Note that G cannot be the point-stratum $\{\mathbf{0}\}$ as $\mathbf{p}$ is contained in $0 \times \partial C_\alpha$. Let R denote the stratum of ∂C_α containing $\mathbf{q}$.

By taking a subsequence if necessary, we may assume that $T_{\mathbf{p}_i} V(f - f(\mathbf{p}_i))$ converges to some T and that $T_{\mathbf{q}_i} S$ converges to some $\mathcal{T}$. By the Thom condition, $T_{\mathbf{p}} G \subseteq T$ and, by Whitney's condition a), $T_{\mathbf{q}} R \subseteq \mathcal{T}$. Furthermore, as $V(t)$ is prepolar, $V(t)$ transversely intersects G at $\mathbf{p}$.

Thus,

$$T_{\mathbf{p}_i}(\mathbb{C} \times S) \to \mathbb{C} \times \mathcal{T}$$

and

$$T_{\mathbf{p}_i} V(f - f(\mathbf{p}_i), t - t_i) \to T \cap T_{\mathbf{p}} V(t).$$

Also, we have that

$$T_{\mathbf{p}}(G \cap V(t)) = T_{\mathbf{p}} G \cap T_{\mathbf{p}} V(t) \subseteq T \cap T_{\mathbf{p}} V(t)$$

and we know that

$$T_{\mathbf{p}}(G \cap V(t)) + T_{\mathbf{p}}(0 \times S) = 0 \times \mathbb{C}^{n+1},$$

as $\{C_\alpha\}$ is a system of privileged neighborhoods with respect to $\mathfrak{G} \cap V(t)$. It follows at once that

$$T \cap T_{\mathbf{p}} V(t) + \mathbb{C} \times \mathcal{T} = \mathbb{C}^{n+2},$$

but this contradicts (*). This proves i).

ii) That f can be made a submersion on $\overset{\circ}{\mathbb{D}}_\delta \times \overset{\circ}{C}_\alpha$ follows from the fact that $\overset{\circ}{\mathbb{D}}_\delta \times \overset{\circ}{C}_\alpha \subseteq \overset{\circ}{\mathbb{D}}_\eta \times W$ and $(\overset{\circ}{\mathbb{D}}_\eta \times W) \cap \Sigma f \subseteq V(f)$.

That f can be made a stratified submersion on $\overset{\circ}{\mathbb{D}}_\delta \times \partial C_\alpha$ and on $\partial\mathbb{D}_\delta \times \partial C_\alpha$ is exactly the argument of i).

Thus, what remains to be shown is that f can be made a submersion on the stratum $\partial\mathbb{D}_\delta \times \overset{\circ}{C}_\alpha$. By Theorem 1.28 and Proposition 1.23, $\dim_{\mathbf{0}}(\Gamma^1_{f,t} \cap V(f)) \leqslant 0$ and thus we may assume that $\Gamma^1_{f,t} \cap V(f) \cap (\partial\mathbb{D}_\delta \times C_\alpha)$ is empty.

As $\Gamma^1_{f,t} \cap (\partial\mathbb{D}_\delta \times C_\alpha)$ is compact, $|f|$ obtains a minimum, $\xi > 0$, on $\Gamma^1_{f,t} \cap (\partial\mathbb{D}_\delta \times C_\alpha)$. Now, consider the critical points of f restricted to $\partial\mathbb{D}_\delta \times \overset{\circ}{C}_\alpha$ that occur in $f^{-1}(\overset{\circ}{\mathbb{D}}_\xi - 0)$. These points occur precisely on

$$\Gamma^1_{f,t} \cap (\partial\mathbb{D}_\delta \times \overset{\circ}{C}_\alpha) \cap f^{-1}(\overset{\circ}{\mathbb{D}}_\xi - 0)$$

which we know is empty. This proves ii).

iii) We first need two results.

a) for all ω_1, ω_2 with $0 < \omega_1 < \omega_2 < \tau_\alpha$, there exists $\xi > 0$ such that

$$\begin{array}{c} \mathbb{C} \times C_\alpha \ \cap \ \Phi^{-1}((\overset{\circ}{\mathbb{D}}_\xi - 0) \times [\omega_1^2, \omega_2^2]) \\ \Big\downarrow \Phi := (f, |t|^2) \\ (\overset{\circ}{\mathbb{D}}_\xi - 0) \times [\omega_1^2, \omega_2^2] \end{array}$$

is a proper, stratified submersion and thus, for all $\eta \in \overset{\circ}{\mathbb{D}}_\xi - 0$, the inclusion

$$(\mathbb{D}_{\omega_1} \times C_\alpha) \cap V(f - \eta) \hookrightarrow (\mathbb{D}_{\omega_2} \times C_\alpha) \cap V(f - \eta)$$

is a homotopy-equivalence; and

b) if $C_\alpha \subseteq \overset{\circ}{C}_\beta$, then there exist $\tau, \xi > 0$ such that, for all $\delta \in \overset{\circ}{\mathbb{D}}_\tau - 0$ and $\eta \in \overset{\circ}{\mathbb{D}}_\xi - 0$, the inclusion

$$(\mathbb{D}_\delta \times C_\alpha) \cap V(f - \eta) \hookrightarrow (\mathbb{D}_\delta \times C_\beta) \cap V(f - \eta)$$

is a homotopy-equivalence.

Assuming a) and b) for the moment, we proceed with the proof. Suppose that $\mathbb{D}_\sigma \times C_\alpha \subseteq \mathbb{D}_\rho \times \overset{\circ}{C}_\beta$.

By b), for all small, non-zero δ and η,

$$(\mathbb{D}_\delta \times C_\alpha) \cap V(f - \eta) \hookrightarrow (\mathbb{D}_\delta \times C_\beta) \cap V(f - \eta)$$

is a homotopy-equivalence. If we select δ so small that $\mathbb{D}_\delta$ is contained in both $\mathbb{D}_\sigma$ and $\mathbb{D}_\rho$, then we may apply a) twice to obtain that, for all small, non-zero η,

$$(\mathbb{D}_\delta \times C_\alpha) \cap V(f - \eta) \hookrightarrow (\mathbb{D}_\sigma \times C_\alpha) \cap V(f - \eta)$$

and

$$(\mathbb{D}_\delta \times C_\beta) \cap V(f-\eta) \hookrightarrow (\mathbb{D}_\sigma \times C_\beta) \cap V(f-\eta)$$

are homotopy-equivalences. The conclusion that

$$(\mathbb{D}_\delta \times C_\alpha) \cap V(f-\eta) \hookrightarrow (\mathbb{D}_\rho \times C_\beta) \cap V(f-\eta)$$

is a homotopy-equivalence now follows immediately by combining the three previous homotopy-equivalences.

We now prove a) and b).

Proof of a): That Φ is a stratified submersion on $\mathbb{C} \times \partial C_\alpha$ is once again exactly the proof of i). That Φ is a submersion on $\mathbb{C} \times \overset{\circ}{C}_\alpha$ is similar to our argument in ii): as $\dim_{\mathbf{0}}(\Gamma^1_{f,t} \cap V(f)) \leqslant 0$, we may assume that

$$\Gamma^1_{f,t} \cap V(f) \cap ((\mathbb{D}_{\omega_2} - \overset{\circ}{\mathbb{D}}_{\omega_1}) \times C_\alpha)$$

is empty. Therefore, by compactness, $|f|$ obtains a minimum, $\xi > 0$, on

$$\Gamma^1_{f,t} \cap ((\mathbb{D}_{\omega_2} - \overset{\circ}{\mathbb{D}}_{\omega_1}) \times C_\alpha).$$

Now, consider the critical points of Φ restricted to $\mathbb{C} \times \overset{\circ}{C}_\alpha$ that occur inside the set $\Phi^{-1}((\overset{\circ}{\mathbb{D}}_\xi - 0) \times [\omega_1^2, \omega_2^2])$. These points occur precisely in

$$\Gamma^1_{f,t} \cap f^{-1}(\overset{\circ}{\mathbb{D}}_\xi - 0) \cap ((\mathbb{D}_{\omega_2} - \overset{\circ}{\mathbb{D}}_{\omega_1}) \times C_\alpha)$$

which we know is empty. This proves a).

Proof of b): Let $C_\alpha \subseteq \overset{\circ}{C}_\beta$. Let τ be so small that inside $\mathbb{D}_\tau \times C_\beta$ all points of $\Gamma^1_{f,t}$ occur in $\mathbb{D}_\tau \times \overset{\circ}{C}_\alpha$. Further, choose $\tau < \min\{\tau_\alpha, \tau_\beta\}$ so that we may apply i) in both cases. Choose ξ so small that $(C_\alpha, \mathbb{D}_\xi)$ and $(C_\beta, \mathbb{D}_\xi)$ are Milnor pairs for $f_{|V(t)}$ and so small that we may apply i) to both $\mathbb{D}_\tau \times \partial C_\alpha$ and $\mathbb{D}_\tau \times \partial C_\beta$ over $(\overset{\circ}{\mathbb{D}}_\xi - 0) \times \overset{\circ}{\mathbb{D}}_\tau$. Fix some $\delta \in \overset{\circ}{\mathbb{D}}_\tau - 0$ and $\eta \in \overset{\circ}{\mathbb{D}}_\xi - 0$.

By i) or ii), $V(f-\eta)$ transversely intersects all the strata of $\mathbb{C} \times \partial C_\alpha$ and $\mathbb{C} \times \partial C_\beta$, so may Whitney stratify $(\mathbb{C} \times C_\beta) \cap V(f-\eta)$ by taking as strata the intersection of $V(f-\eta)$ with each of $\mathbb{C} \times \overset{\circ}{C}_\alpha$, $\mathbb{C} \times (\overset{\circ}{C}_\beta - C_\alpha)$, and the strata of $\mathbb{C} \times \partial C_\alpha$ and $\mathbb{C} \times \partial C_\beta$.

As $C_\alpha \cap V(f_{|V(t)} - \eta) \hookrightarrow C_\beta \cap V(f_{|V(t)} - \eta)$ is a homotopy-equivalence and $V(f_{|V(t)}) - \eta)$ transversely intersects ∂C_α and ∂C_β, for all small $\mathbb{D}_\mu$ we must have that

$$(\mathbb{D}_\mu \times C_\alpha) \cap V(f-\eta) \hookrightarrow (\mathbb{D}_\mu \times C_\beta) \cap V(f-\eta)$$

is also a homotopy-equivalence. We wish to pass from $\mathbb{D}_\mu$ to $\mathbb{D}_\delta$ by considering the function $|t|^2$ on the stratified space $(\mathbb{C} \times C_\beta) \cap V(f-\eta)$ (with the stratification given above).

By i), $|t|^2$ has no critical points on the strata of $(\mathbb{C}\times\partial C_\alpha)\cap V(f-\eta)$ and

$$(\mathbb{C}\times\partial C_\beta)\cap V(f-\eta)$$

when $|t|<\delta$. In addition, the critical points on the interior strata, $(\mathbb{C}\times\overset{\circ}{C}_\alpha)\cap V(f-\eta)$ and $(\mathbb{C}\times(\overset{\circ}{C}_\beta - C_\alpha))\cap V(f-\eta)$, occur on the polar curve and, hence, by our earlier requirement, these critical points all occur in $\mathbb{C}\times\overset{\circ}{C}_\alpha$. Therefore, using stratified Morse theory [**G-M2**] together with the homotopy-equivalence lemma 3.7 of [**Mi2**], we find that the inclusion

$$(\mathbb{D}_\delta\times C_\alpha)\cap V(f-\eta)\hookrightarrow(\mathbb{D}_\delta\times C_\beta)\cap V(f-\eta)$$

is a homotopy-equivalence. □

For a family of analytic functions $f_t:(\mathcal{U},\mathbf{0})\to(\mathbb{C},0)$, we are interested in how the Milnor fibre and fibration "jump" as we move from small non-zero t to $t=0$. Hence, we make the following definition.

Definition C.7. If we are considering the family $f_t:(\mathcal{U},\mathbf{0})\to(\mathbb{C},0)$, we refer to i) of C.6 by saying that the family satisfies the *conormal condition with respect to* $\{C_\alpha\}$.

The point of this condition is that it says that the Milnor fibration of f_0 lifts trivially in the family f_t on the boundary of the neighborhoods C_α.

Definition C.8. The *Thom set at the origin*, $\mathfrak{T}_f$, is the set of $(n+1)$-planes which occur as limits at the origin of the tangent spaces to level hypersurfaces of f, i.e. $T\in\mathfrak{T}_f$ if and only if there exists a sequence of points $\mathbf{p}_i$ in $\overset{\circ}{\mathbb{D}}\times\mathcal{U}-\Sigma f$ such that $\mathbf{p}_i\to\mathbf{0}$ and $T=\lim T_{\mathbf{p}_i}V(f-f(\mathbf{p}_i))$.

Equivalently, $\mathfrak{T}_f$ is the fibre over the origin in the Jacobian blow-up of f (see [**H-L**]). $\mathfrak{T}_f$ is thus a closed algebraic subset of the Grassmanian $G_{n+1}(\mathbb{C}^{n+2})=$ the projective space of $(n+1)$-planes in $\mathbb{C}^{n+2}$.

Proposition C.9. *Suppose that either $V(t)$ is a prepolar slice for f at $\mathbf{0}$ or that $V(t)=T_{\mathbf{0}}V(t)\notin\mathfrak{T}_f$. Then,*

i) $\dim_{\mathbf{0}}\Gamma^1_{f,t}\leqslant 1$, *and*

ii) the family f_t satisfies the conormal condition with respect to any universal system of privileged neighborhoods, $\mathcal{C}$, for f_0 at $\mathbf{0}$.

Moreover, whenever i) and ii) are satisfied, there is an inclusion of the Milnor fibre $F_{f_{t_0},\mathbf{0}}$ into the Milnor fibre $F_{f_0,\mathbf{0}}$ for all small non-zero t_0; the homotopy-type of this inclusion is independent of the choice of t_0 and the choice of the universal system of privileged neighborhoods, $\mathcal{C}$.

Proof. That there is such an inclusion whenever i) and ii) are satisfied is standard. One considers the map $\Psi := (f, t)$ and its restriction

$$(\mathbb{D}_\tau \times C) \cap \Psi^{-1}\left((\mathbb{D}_\xi - 0) \times \mathbb{D}_\tau\right)$$

$$\downarrow \Psi$$

$$(\mathbb{D}_\xi - 0) \times \mathbb{D}_\tau$$

for appropriately small choices of $C \in \mathcal{C}$, ξ, and τ. By the conormal condition, this is a stratified submersion on the boundary. As $\dim_{\mathbf{0}}\Gamma^1_{f,t} \leqslant 1$, the discriminant of Ψ, $\Psi(\Gamma^1_{f,t})$, is also at most one-dimensional. Thus, we may lift a path in the base which avoids the discriminant to get a diffeomorphism between the Milnor fibre of f_0 and $C \cap V(f_{t_0} - \eta)$ for all small t_0 and for all η with $0 < |\eta| \ll |t_0|$. And, though we do not know that $\mathcal{C}$ is a system of privileged neighborhoods for f_{t_0}, we may still take a small enough ball inside C to obtain the desired inclusion, which is clearly independent of the choice of t_0.

That the inclusion is independent of the choice of privileged neighborhoods follows similarly. Suppose that $\mathcal{C}'$ is second universal system of privileged neighborhoods for f_0. Let $C \in \mathcal{C}$ and let $C' \in \mathcal{C}'$ be such that $C' \subseteq \overset{\circ}{C}$, and such that C and C' are small enough to give the Milnor fibre, i.e. for all small non-zero ξ, the inclusion of $C' \cap V(f_0 - \xi)$ into $C \cap V(f_0 - \xi)$ is a homotopy-equivalence where both spaces are homotopy-equivalent to the Milnor fibre of f_0 at the origin. Then, as above, over a curve which avoids the discriminant, we have a proper, stratified submersion – where the strata are those of $\mathbb{D}_\tau \times \partial C$ together with those of $\mathbb{D}_\tau \times \partial C'$ plus the interior.

Hence, the homotopy-equivalence $C' \cap V(f_0 - \xi) \to C \cap V(f_0 - \xi)$ lifts to a homotopy-equivalence $C' \cap V(f_{t_0} - \xi) \to C \cap V(f_{t_0} - \xi)$. The independence statement now follows easily.

We must still show that if $V(t)$ is a prepolar slice for f at $\mathbf{0}$ or $V(t) = T_{\mathbf{0}}V(t) \notin \mathfrak{T}_f$, then i) and ii) hold.

If $V(t)$ is prepolar for f at $\mathbf{0}$, then i) follows from Theorem 1.28 and ii) follows from C.6.i. If $V(t) \notin \mathfrak{T}_f$, then clearly $\Gamma^1_{f,t}$ is empty near the origin. It remains for us to show that if $V(t) \notin \mathfrak{T}_f$, then the family f_t satisfies the conormal condition with respect to any universal system of privileged neighborhoods, $\mathcal{C}$, for f_0.

If $V(t) \notin \mathfrak{T}_f$, then $V(t)$ certainly transversely intersects the smooth part of $V(f)$ in a neighborhood of the origin. Hence, we may use Proposition C.2 to conclude that there exists a good stratification, $\mathfrak{G}$, for f at the origin such that the strata of $\mathfrak{G}$ which are contained in $V(t)$ form a good stratification for f_0 at the origin. The proof now proceeds like that of C.6.i.

Suppose to the contrary that, for arbitrarily small C_α in $\mathcal{C}$, there exists a stratum S of ∂C_α and a sequence of points $\mathbf{p}_i$ not in $V(f)$ but which are in $\mathbb{C} \times S$ such that $\mathbf{p}_i = (t_i, \mathbf{q}_i) \to \mathbf{p} := (0, \mathbf{q}) \in V(t) \cap V(f)$ and such that

$$(*) \qquad T_{\mathbf{p}_i}V(f - f(\mathbf{p}_i), t - t_i) + T_{\mathbf{p}_i}(\mathbb{C} \times S) \neq \mathbb{C}^{n+2}.$$

Let G denote the good stratum of $V(f)$ which contains $\mathbf{p}$. Note that G is contained in $V(t)$ by the nature of our good stratification and that G cannot be simply the stratum consisting of the origin since $\mathbf{p}$ is contained in $0 \times \partial C_\alpha$. Let R denote the stratum of ∂C_α containing $\mathbf{q}$.

By taking a subsequence if necessary, we may assume that $T_{\mathbf{p}_i}V(f - f(\mathbf{p}_i))$ converges to some T and that $T_{\mathbf{q}_i}S$ converges to some $\mathcal{T}$. By the Thom condition, $T_{\mathbf{p}}G \subseteq T$ and, by Whitney's condition a), $T_{\mathbf{q}}R \subseteq \mathcal{T}$. Furthermore, as $V(t) \notin \mathfrak{T}_f$, we may assume that $\mathbf{p}$ is close enough to the origin that $T \neq V(t)$.

Thus, $T_{\mathbf{p}_i}(\mathbb{C} \times S) \to \mathbb{C} \times \mathcal{T}$ and $T_{\mathbf{p}_i}V(f - f(\mathbf{p}_i), t - t - i) \to T \cap T_{\mathbf{p}_i}V(t)$. Also, we have that $T_{\mathbf{p}}G = T_{\mathbf{p}}G \cap T_{\mathbf{p}}V(t) \subseteq T \cap T_{\mathbf{p}}V(t)$, and we know that $T_{\mathbf{p}}G + T_{\mathbf{p}}(0 \times S) = 0 \times \mathbb{C}^{n+1}$, as $\{C_\alpha\}$ is a system of privileged neighborhoods with respect to $\mathfrak{G} \cap V(t)$. It follows at once that $T \cap T_{\mathbf{p}}V(t) \ + \ \mathbb{C} \times \mathcal{T} = \mathbb{C}^{n+2}$ – which contradicts $(*)$. □

If the polar curve, $\Gamma^1_{f,t}$, is empty, then the map Ψ which appears in the proof of Proposition C.9 is a stratified submersion over the entire base space and so, for all small $t_0 \neq 0$, we have a fibre-prerserving inclusion of the total space of the Milnor fibration of f_{t_0} into the total space of the Milnor fibration of f_0. Moreover, exactly as above, this inclusion is independent – up to homotopy – of all of the choices made. By the theorem of Dold (see [**Hu**, p. 209]), this inclusion is a fibre homotopy-equivalence if and only if the inclusion of each fibre is a homotopy-equivalence. Therefore, we make the following definitions.

Definition C.10 Whenever i) and ii) of C.9 hold, we say that the family, f_t, satisfies the *universal conormal condition.*

If f_t satisfies the universal conormal condition, we say that f_t has the *homotopy Milnor fibre lifting property* if and only if the inclusion of C.9 is a homotopy-equivalence.

If f_t satisfies the universal conormal condition, we say that f_t has the *homology Milnor fibre lifting property* if and only if the inclusion of C.9 induces isomorphisms on all integral homology groups.

The family, f_t, has the *homotopy Milnor fibration lifting property* if and only if f_t has the homotopy Milnor fibre lifting property and $\Gamma^1_{f,t} = \emptyset$ in a neighborhood of the origin. This definition makes sense in light of our above discussion concerning the result of Dold.

One may also discuss the Milnor fibre and Milnor fibration up to diffeomorphism if one is willing to restrict consideration to the standard universal system of Milnor neighborhoods, namely the set of closed balls centered at the origin. In this case, we may use the h-cobordism Theorem and the pseudo-isotopy result of Cerf [**Ce**] to translate the homotopy information into smooth information – provided that we are in a sufficiently high dimension and that the Milnor fibre and its boundary are sufficiently connected. More specifically, if $\mathcal{U}$ is an open neighborhood of the origin in $\mathbb{C}^{n+1}$, $f_t : (\mathcal{U}, \mathbf{0}) \to (\mathbb{C}, 0)$ has the homotopy Milnor fibration lifting property, $n \geqslant 3$, and the Milnor fibre and its boundary are simply-connected for each f_t for all small t, then the diffeomorphism-type of the Milnor fibrations is constant in the family near $t = 0$. This connectedness condition can be realized by requiring $n - \dim_{\mathbf{0}}\Sigma f_0 \geqslant 3$ (see [**K-M**] and [**Ra**]).

We wish to state the diffeomorphism results discussed above precisely. First, we give without proof Cerf's pseudo-isotopy result in the form that we shall need it.

Lemma C.11. *Let X be a smooth manifold with boundary $\partial X = X_0 \dot{\cup} X_1$ and let $\pi : X \to S^1$ be a smooth locally trivial fibration over a circle with fibre diffeomorphic to $M \times [0,1]$, where M is a closed, simply-connected, smooth manifold of dimension $\geqslant 5$.*

Then, the restriction of π to X_0 is a smooth locally trivial fibration with fibre diffeomorphic to M, and there exists a commutative diagram

$$\begin{array}{ccc} (X, X_0) & \xrightarrow[\text{diffeo.}]{\cong} & (X_0 \times [0,1], X_0 \times \{0\}) \\ \pi \searrow & & \swarrow \pi_{|X_0} \circ pr_1 \\ & S^1 & \end{array}$$

where the diffeomorphism is the identity on $X_0 = X_0 \times \{0\}$.

Proposition C.12. *Let $\mathcal{U}$ be an open neighborhood of the origin in $\mathbb{C}^{n+1}$. Suppose that the family $f_t : (\mathcal{U}, \mathbf{0}) \to (\mathbb{C}, 0)$ has the homotopy Milnor fibration lifting property and $n - \dim_{\mathbf{0}} \Sigma f_0 \geqslant 3$. Then, the diffeomorphism-type of the Milnor fibrations of f_t at the origin is independent of t for all small t.*

Proof. We shall use the notation from the proof of Proposition C.9. We fix the universal system of privileged neighborhoods to be the collection of closed balls centered at the origin.

As f_t has the homotopy Milnor fibration lifting property, the polar curve $\Gamma^1_{f,t}$ is empty and so the map Ψ in the proof of Proposition C.9 is a proper stratified submersion. Hence, for $0 < \xi, |t_0| \ll \epsilon$, the Milnor fibration of f_0 is diffeomorphic to $B_\epsilon \cap f_{t_0}^{-1}(\partial \mathbb{D}_\xi) \xrightarrow{f_{t_0}} \partial \mathbb{D}_\xi$. The problem, of course, is that B_ϵ may be too large a ball in which to define the Milnor fibration of f_{t_0}. Let F and E denote the fibre and the total space, respectively, of this previous fibration.

Let F' denote the Milnor fibre of f_{t_0} at the origin, where we again use closed balls for the Milnor neighborhoods. Let E' denote the total space of the Milnor fibration f_{t_0}.

As f_t has the homotopy Milnor fibration lifting property, the inclusion of E' into E induces an inclusion $F' \hookrightarrow F$ which is a homotopy-equivalence.

Since $n - \dim_{\mathbf{0}} \Sigma f_0 \geqslant 3$, F, F', ∂F, and $\partial F'$ are simply-connected (see [**Ra**]). Combining this with the fact that $F' \hookrightarrow F$ is a homotopy-equivalence, we may duplicate the argument of Lê and Ramanujam [**L-R**] to conclude that $\Delta T := E - \overset{\circ}{E'}$ is the total space of a differentiable fibration over $\partial \mathbb{D}_\xi$ with projection f_{t_0} and fibre $F - \overset{\circ}{F'}$ which is diffeomorphic to $\partial F \times [0,1]$ via the h-cobordism theorem.

Now, by Lemma C.11, $\Delta T \xrightarrow{f_{t_0}} \partial \mathbb{D}_\xi$ is diffeomorphic to

$$\partial E' \times [0,1] \xrightarrow{f_{t_0} \circ pr_1} \partial \mathbb{D}_\xi \times \{0\}$$

by a diffeomorphism which is the identity on $\partial E' = \partial E' \times \{0\}$. Combining this with a fibred collar of $\partial E'$ in E', we conclude that $E' \xrightarrow{f_{t_0}} \partial \mathbb{D}_\xi$ is diffeomorphic to $E \xrightarrow{f_{t_0}} \partial \mathbb{D}_\xi$, which we already know is diffeomorphic to the Milnor fibration of f_0 at $\mathbf{0}$. $\square$

We now wish to prove a fundamental result – namely, that if we have a family f_t in which the Milnor fibrations of a hyperplane slice are independent of t and the number of handles attached in passing from the Milnor fibre of the hyperplane slice to the entire Milnor fibre is constant, then the Milnor fibrations are constant in the family. Despite the fact that the dimension of the critical loci is allowed to be arbitrary, the argument is exactly that which we used in [**Mas3**] where the critical loci were all one-dimensional.

Theorem C.13. *Let $\mathcal{W}$ be an open neighborhood of the origin in $\mathbb{C}^{n+2}$ and let $g_t : (\mathcal{W}, \mathbf{0}) \to (\mathbb{C}, 0)$ be an analytic family. Let s denote $\dim_{\mathbf{0}} \Sigma g_0$. Assume that g_t satisfies the universal conormal condition and that L is a linear form such that $V(L)$ is prepolar for f_t at the origin for all small t and such that ${g_t}_{|_{V(L)}}$ satisfies the universal conormal condition. Suppose further that $\left(\Gamma^1_{g_t,L} \cdot V(g_t)\right)_{\mathbf{0}}$ is constant for all small t.*

Under the above assumptions, if ${g_t}_{|_{V(L)}}$ has the homology Milnor fibre lifting property, then g_t has the homology Milnor fibre lifting property.

Moreover, if $s \leqslant n-1$ and ${g_t}_{|_{V(L)}}$ has the homotopy Milnor fibre lifting property, then g_t has the homotopy Milnor fibre lifting property.

Proof. This is actually quite trivial. Let F_0 and F_{t_0} denote the Milnor fibre of g_0 and g_{t_0} for small non-zero t_0, respectively. The Milnor fibres of ${g_0}_{|_{V(L)}}$ and ${g_{t_0}}_{|_{V(L)}}$ are then $F_0 \cap V(L)$ and $F_{t_0} \cap V(L)$, respectively. Let γ denote the constant value of $\left(\Gamma^1_{g_t,L} \cdot V(g_t)\right)_{\mathbf{0}}$.

As g_t and ${g_t}_{|_{V(L)}}$ satisfy the universal conormal condition, we may repeat the argument of Proposition C.9 – lifting a path in the base which avoids the discriminants of both (g_t, t) and $({g_t}_{|_{V(L)}}, t)$ – to obtain compatible inclusions $F_{t_0} \hookrightarrow F_0$ and $F_{t_0} \cap V(L) \hookrightarrow F_0 \cap V(L)$.

Suppose that ${g_t}_{|_{V(L)}}$ has the homology Milnor fibre lifting property, i.e.,

$$F_{t_0} \cap V(L) \hookrightarrow F_0 \cap V(L)$$

induces isomorphisms on homology. We wish to show that $F_{t_0} \hookrightarrow F_0$ induces isomorphisms on homology. We will accomplish this by showing that $H_*(F_0, F_{t_0}) = 0$.

By considering the homology long exact sequence of the triple $(F_0, F_0 \cap V(L), F_{t_0} \cap V(L))$, we find that $H_i(F_0, F_{t_0} \cap V(L)) \cong H_i(F_0, F_0 \cap V(L))$ for all i. By Lê's attaching result (Theorem 0.9) or Theorem 3.1, $H_i(F_0, F_0 \cap V(L)) = 0$ unless $i = n+1$ and

$$H_{n+1}(F_0, F_0 \cap V(L)) \cong \mathbb{Z}^{\gamma}.$$

Now, we are going to consider the homology long exact sequence of the triple

$$(F_0, F_{t_0}, F_{t_0} \cap V(L)).$$

From the last paragraph, we know that $H_i(F_0, F_{t_0} \cap V(L)) = 0$ unless $i = n+1$. In addition, $H_i(F_{t_0}, F_{t_0} \cap V(L)) = 0$ unless $i = n+1$. Moreover,

$$H_{n+1}(F_0, F_0 \cap V(L)) \cong H_{n+1}(F_{t_0}, F_{t_0} \cap V(L)) \cong \mathbb{Z}^{\gamma}.$$

Thus, in the long exact sequence of the triple $(F_0, F_{t_0}, F_{t_0} \cap V(L))$, all terms are zero except in the portion

$$0 \to \mathbb{Z}^{\gamma} \to \mathbb{Z}^{\gamma} \to H_{n+1}(F_0, F_{t_0}) \to 0.$$

But, as in the proof of the result of Lê and Ramanujam [**L-R**], $H_{n+1}(F_0, F_{t_0})$ is free Abelian, since F_0 is obtained from F_{t_0} by attaching handles of index less than or equal to $n+1$. (One considers the function distance squared from the origin and lets the function grow from the small ball used to define F_{t_0} out to the ball used to define F_0. One hits no critical points of index greater than or equal to $n+2$.) Thus, $H_{n+1}(F_0, F_{t_0}) = 0$ and we have proved the first claim.

The second claim follows from the first, since $s \leqslant n-1$ guarantees that F_0 and F_{t_0} are simply-connected, and then we apply the Whitehead Theorem. $\square$

There are two more big results which we need to prove in this appendix – both deal with suspending singularities (see Chapter I.8). The first result is that there exists a universal system of privileged neighborhoods of a particularly nice form for the function $h+w^j$, where w is a variable disjoint from those of h. The second result says, with a few extra assumptions, that the constancy of the Milnor fibrations in the family f_t implies the constancy of the Milnor fibrations in the family $f_t + w^j$, where, again, w is disjoint from the variables of f_t. This second result seems reasonable since the result of Proposition II.8.1 is that the Milnor fibre of $f_t + w^j$ at the origin is homotopy-equivalent to one-point union of $j-1$ copies of the Milnor fibre of f_t at the origin. However, both of these results are technical nightmares.

Let $\mathcal{U}$ be an open neighborhood of the origin in $\mathbb{C}^{n+1}$ and let $h : (\mathcal{U}, \mathbf{0}) \to (\mathbb{C}, 0)$ be an analytic function. Let $j \geqslant 2$ and define $\tilde{h}(w, \mathbf{z}) := h(\mathbf{z}) + w^j$. We wish to show that the set $\{\mathbb{D}_\omega \times B_\epsilon^{2n+2} \mid 0 < \omega \ll \epsilon\}$ is a universal system of privileged neighborhoods for $\tilde{h}$ at the origin. Note that we may **not** use C.6 to conclude that $\{\mathbb{D}_\omega \times B_\epsilon^{2n+2} \mid 0 < \omega \ll \epsilon\}$ is even weakly privileged since the slice $V(w)$ contains the entire critical locus of $h+w^j$ and, hence, is certainly not prepolar for $h+w^j$. Of course, the actual argument is very similar to the proof of Proposition C.6.

Proposition C.14. *The set* $\{\mathbb{D}_\omega \times B_\epsilon^{2n+2} \mid 0 < \omega \ll \epsilon\}$ *is a universal system of privileged neighborhoods for* $h + w^j$ *at the origin.*

Proof. As $j \geqslant 2$, $\Sigma(h+w^j) = \{0\} \times \Sigma h$. Fix any good stratification, $\mathfrak{G}$, for $h + w^j$ at the origin in $\mathbb{C}^{n+2}$.

Let $\epsilon_0 > 0$ be so small that the critical locus of the map h inside $B_{\epsilon_0}^{2n+2}$ is contained in $V(h)$ and so small that, for all ϵ with $0 < \epsilon \leqslant \epsilon_0$,

$\{0\} \times \partial B_\epsilon^{2n+2}$ transversely intersects all strata of $\{0\} \times \Sigma V(h)$ inside $\{0\} \times \mathbb{C}^{n+1}$,

$\partial B_\epsilon^{2n+4}$ transversely intersects all strata of $\mathfrak{G}$ (we write $\partial B_\epsilon^{2n+4} \pitchfork \mathfrak{G}$), and

$\partial B_\epsilon^{2n+2}$ transversely intersects all strata of some good stratification for h at $\mathbf{0}$.

This last condition guarantees, for all small non-zero ζ, that

$$(*) \qquad \partial B_\epsilon^{2n+2} \pitchfork V(h - h(\zeta)).$$

Now fix an ϵ between 0 and ϵ_0. We wish to show that there exists $\omega_\epsilon > 0$ such that, for all ω with $0 < \omega \leqslant \omega_\epsilon$, we have:

a) $\overset{\circ}{\mathbb{D}}_\omega \times \partial B_\epsilon^{2n+2} \pitchfork \mathfrak{G}$;

b) $\partial \mathbb{D}_\omega \times \overset{\circ}{B}{}_\epsilon^{2n+2} \pitchfork \mathfrak{G}$;

c) $\partial \mathbb{D}_\omega \times \partial B_\epsilon^{2n+2} \pitchfork \mathfrak{G}$.

After we show this, it will still remain to show that, if $\mathbb{D}_{\omega_1} \times B_{\epsilon_1} \subseteq \mathbb{D}_{\omega_2} \times B_{\epsilon_2}$, then, for all small non-zero t, the inclusion

$$\mathbb{D}_{\omega_1} \times B_{\epsilon_1} \cap V(h + w^j - t) \hookrightarrow \mathbb{D}_{\omega_2} \times B_{\epsilon_2} \cap V(h + w^j - t)$$

is a homotopy-equivalence.

Proof of a): Clearly, as $\{0\} \times \partial B_\epsilon^{2n+2}$ transversely intersects all strata of $\{0\} \times \Sigma V(h)$ inside $\{0\} \times \mathbb{C}^{n+1}$, $\mathbb{C} \times \partial B_\epsilon^{2n+2}$ transversely intersects all singular strata of $V(h + w^j)$. Suppose, however, that no matter how small we pick $\omega > 0$, we still have a point in the smooth stratum, $S := V(h + w^j) - \{0\} \times \Sigma V(h)$, where S does not transversely intersect $\overset{\circ}{\mathbb{D}}_\omega \times \partial B_\epsilon^{2n+2}$.

Then, we would have a sequence $\mathbf{p}_i := (w_i, \mathbf{q}_i) \in \mathbb{C} \times \partial B_\epsilon$ contained in S such that $\mathbf{p}_i \to \mathbf{p} := (0, \mathbf{q}) \in \{0\} \times \partial B_\epsilon$, $T_{\mathbf{p}_i} S \subseteq T_{\mathbf{p}_i}(\mathbb{C} \times \partial B_\epsilon)$, $T_{\mathbf{p}_i} S$ converges to some $\mathcal{T}$, and $T_{\mathbf{p}_i}(\mathbb{C} \times \partial B_\epsilon) \to \mathbb{C} \times T_{\mathbf{q}}(\partial B_\epsilon)$. Let S' denote the stratum of $\mathfrak{G}$ containing $\mathbf{p}$.

By the Thom condition, $T_{\mathbf{p}} S' \subseteq \mathcal{T}$ (this is true because $\mathcal{T}$ comes from the smooth stratum – we are not assuming Whitney conditions hold between the strata). Hence,

$$T_{\mathbf{p}} S' \subseteq \mathcal{T} \subseteq \mathbb{C} \times T_{\mathbf{q}}(\partial B_\epsilon) = T_{\mathbf{p}}(\partial B_\epsilon^{2n+4}),$$

where this last equality is true because the w-coordinate of $\mathbf{p}$ is 0. But, this contradicts the fact that $\partial B_\epsilon^{2n+4} \pitchfork \mathfrak{G}$. This proves a).

Before we prove b) and c), note that if $w \in \partial \mathbb{D}_\omega$, then $w \neq 0$ and, hence, the only stratum of $\mathfrak{G}$ which $\partial \mathbb{D}_\omega \times \overset{\circ}{B}{}_\epsilon^{2n+2}$ and $\partial \mathbb{D}_\omega \times \partial B_\epsilon^{2n+2}$ intersect is the smooth stratum $S := V(h + w^j) - \{0\} \times \Sigma V(h)$.

Proof of b): Actually, we show, regardless of the size of $\omega > 0$, that $\partial \mathbb{D}_\omega \times \overset{\circ}{B}{}_\epsilon^{2n+2} \pitchfork S$.

For if not, we would have $\mathbf{p} := (w, \mathbf{q}) \in S$ such that $w \neq 0$ and

$$T_{\mathbf{p}} V(h + w^j) \subseteq T_{\mathbf{p}}(\partial \mathbb{D}_{|w|} \times \overset{\circ}{B}{}_\epsilon^{2n+2}).$$

This implies that

$$\frac{\partial h}{\partial z_0}_{|\mathbf{q}} = \cdots = \frac{\partial h}{\partial z_n}_{|\mathbf{q}} = 0,$$

i.e. that $\mathbf{q} \in \Sigma h$. Recalling that we chose ϵ such that $B_\epsilon \cap \Sigma h \subseteq V(h)$, we see that $h(\mathbf{q}) = 0$. However, this contradicts that $h(\mathbf{q}) = -w^j \neq 0$. This proves b).

Proof of c): Suppose not. Then, we would have a sequence $\mathbf{p}_i := (w_i, \mathbf{q}_i) \in S \cap (\mathbb{C} \times \partial B_\epsilon^{2n+2})$ with $w_i \neq 0$, $\mathbf{p}_i \to \mathbf{p} = (0, \mathbf{q}) \in \{0\} \times \partial B_\epsilon$, and such that

$$T_{\mathbf{p}_i} V(h + w^j) + T_{\mathbf{p}_i}(\partial \mathbb{D}_{|w_i|} \times \partial B_\epsilon) \neq \mathbb{C}^{n+2}.$$

This implies that $T_{\mathbf{q}_i} V(h - h(\mathbf{q}_i)) \subseteq T_{\mathbf{q}_i}(\partial B_\epsilon^{2n+2})$, while $h(\mathbf{q}_i) = -w_i^j$ approaches – but is unequal to – zero. This, however, is impossible by $(*)$. This proves c).

We must still prove the homotopy-equivalence statement. In a manner completely similar to the proofs of a), b), and c) above, one can easily show, using the Thom condition, that the following statements are true:

d) for all ϵ with $0 < \epsilon \leqslant \epsilon_0$, if ω_1 is between 0 and ω_ϵ, then for all ω_2 with $0 < \omega_2 \leqslant \omega_1$, there exists $\xi > 0$ such that

$$\begin{array}{c} (\mathbb{C} \times B_\epsilon^{2n+2}) \cap \Psi^{-1}((\mathbb{D}_\xi - 0) \times [\omega_2^2, \omega_1^2]) \\ \Big\downarrow \ \Psi := (h + w^j, |w|^2) \\ (\mathbb{D}_\xi - 0) \times [\omega_2^2, \omega_1^2] \end{array}$$

is a proper, stratified submersion and therefore

d′) $(\mathbb{D}_{\omega_2} \times B_\epsilon) \cap (h + w^j)^{-1}(\mathbb{D}_\xi - 0) \hookrightarrow (\mathbb{D}_{\omega_1} \times B_\epsilon) \cap (h + w^j)^{-1}(\mathbb{D}_\xi - 0)$

is a fibre-homotopy equivalence between total spaces (where the projection in each case is the obvious map $h + w^j$).

e) if $0 < \epsilon_2 < \epsilon_1 \leqslant \epsilon_0$, then for all small, non-zero ω, there exists $\xi > 0$ such that

$$\begin{array}{c} (\mathbb{D}_\omega \times \mathbb{C}^{n+1}) \cap \Phi^{-1}((\mathbb{D}_\xi - 0) \times [\epsilon_2^2, \epsilon_1^2]) \\ \Big\downarrow \ \Phi := (h + w^j, |\mathbf{z}|^2) \\ (\mathbb{D}_\xi - 0) \times [\epsilon_2^2, \epsilon_1^2] \end{array}$$

is a proper, stratified submersion and therefore

e′) $(\mathbb{D}_\omega \times B_{\epsilon_2}) \cap (h + w^j)^{-1}(\mathbb{D}_\xi - 0) \hookrightarrow (\mathbb{D}_\omega \times B_{\epsilon_1}) \cap (h + w^j)^{-1}(\mathbb{D}_\xi - 0)$

is a fibre-homotopy equivalence.

Suppose that we have $\mathbb{D}_{\omega_1} \times B_{\epsilon_1} \subseteq \mathbb{D}_{\omega_1} \times B_{\epsilon_1}$, where $0 < \epsilon_2 < \epsilon_1 \leqslant \epsilon_0$, $0 < \omega_2 < \omega_1 \leqslant \omega_{\epsilon_1}$, and $\omega_2 < \omega_{\epsilon_2}$. We shall show that, for all small $\xi > 0$,

$$(\mathbb{D}_{\omega_2} \times B_{\epsilon_2}) \cap (h + w^j)^{-1}(\mathbb{D}_\xi - 0) \hookrightarrow (\mathbb{D}_{\omega_1} \times B_{\epsilon_1}) \cap (h + w^j)^{-1}(\mathbb{D}_\xi - 0)$$

is a fibre-homotopy equivalence.

By applying e), we know that, for all small $\omega > 0$, there exists $\xi \neq 0$ such that e′) holds. On the other hand – by applying d) twice – for all small $\omega > 0$, there exists $\xi \neq 0$ such that

$$(\mathbb{D}_\omega \times B_{\epsilon_2}) \cap (h + w^j)^{-1}(\mathbb{D}_\xi - 0) \hookrightarrow (\mathbb{D}_{\omega_2} \times B_{\epsilon_2}) \cap (h + w^j)^{-1}(\mathbb{D}_\xi - 0)$$

and

$$(\mathbb{D}_{\omega_2} \times B_{\epsilon_1}) \cap (h + w^j)^{-1}(\mathbb{D}_\xi - 0) \hookrightarrow (\mathbb{D}_{\omega_1} \times B_{\epsilon_1}) \cap (h + w^j)^{-1}(\mathbb{D}_\xi - 0)$$

are fibre-homotopy equivalences.

The desired conclusion follows from the two homotopy-equivalences above together with e′). □

For the final results of this appendix, we return to the setting of families of analytic functions.

Again, $\mathcal{U}$ will denote an open neighborhood of the origin in $\mathbb{C}^{n+1}$, and $f_t : (\mathcal{U}, \mathbf{0}) \to (\mathbb{C}, 0)$ will be an analytic family. We continue with w being a variable disjoint from those of f_t and with $j \geqslant 2$. Recall from C.8 that $\mathfrak{T}_f$ denotes the Thom set of f at the origin.

We need the following easy lemma:

Lemma C.15. *If $V(t) \notin \mathfrak{T}_f$, then $V(t) \notin \mathfrak{T}_{f+w^j}$.*

Proof. This is completely trivial. We leave it as an exercise. □

Proposition C.16. *Suppose that $V(t) \notin \mathfrak{T}_f$ and that the family $f_t + w^j$ has the homology Milnor fibre lifting property. Then, $\Gamma^1_{f,t} = \emptyset$ near the origin and f_t has the homology Milnor fibre lifting property.*

Moreover, if $\dim_{\mathbf{0}} \Sigma f_0 \leqslant n - 2$, $V(t) \notin \mathfrak{T}_f$, and the family $f_t + w^j$ has the homotopy Milnor fibre lifting property, then f_t has the homotopy Milnor fibration lifting property.

Proof. The second claim follows immediately from the first claim, since the condition $\dim_{\mathbf{0}} \Sigma f_0 \leqslant n-2$ implies that the Milnor fibres are simply-connected. Also, since $V(t) \notin \mathfrak{T}_f$, we immediately have that $\Gamma^1_{f,t} = \emptyset$ near the origin. What we need to prove is that f_t has the homology Milnor fibre lifting property.

Fix a good stratification $\mathfrak{G}$ for f_0 at the origin. We must now make many choices.

1) Let $(B_{\epsilon_0}, \mathbb{D}_{\lambda_0})$ be a Milnor pair for f_0 such that
2) $B_{\epsilon_0} \cap \Sigma f_0 \subseteq V(f_0)$, and
3) ∂B_{ϵ_0} transversely intersects the strata of $\mathfrak{G}$.

From C.9, we know that the conormal condition holds, and so we may pick $\eta, \tau > 0$ such that

4) the map $G := (f, t)$ restricted to $\mathbb{C} \times \partial B_{\epsilon_0}$ has no critical values in $(\mathbb{D}_\eta - 0) \times \mathbb{D}_\tau$.

Using C.14, we may also choose $\omega_0, \xi_0 > 0$ such that

5) $(\mathbb{D}_{\omega_0} \times B_{\epsilon_0}, \mathbb{D}_{\xi_0})$ is a Milnor pair for $f_0 + w^j$, where
6) $\omega_0^j < \eta$, and
7) all of the obvious Whitney strata of $\mathbb{D}_{\omega_0} \times B_{\epsilon_0}$ transversely intersect all of the strata in the good stratification for $f_0 + w^j$ which is induced by $\mathfrak{G}$ (as given in Proposition 8.3).

Now, as $V(t) \not\in \mathfrak{T}_f$, Lemma C.15 tells us that $V(t) \not\in \mathfrak{T}_{f+w^j}$. Hence, $f_t + w^j$ satisfies the universal conormal condition and so, for all small $\nu \neq 0$ and all small t_1,

8) $\quad (\mathbb{D}_{\omega_0} \times B_{\epsilon_0}) \cap V(f_{t_1} + w^j - \nu)$ is diffeomorphic to $F_{f_0+w^j,\mathbf{0}}$.

We select t_1 so that

9) $\quad t_1$ is in $\mathbb{D}_\tau$,
10) $\quad \Gamma^1_{f,t} \cap (\mathbb{D}_{|t_1|} \times B_{\epsilon_0}) = \emptyset$, and
11) $\quad \Sigma f \cap (\mathbb{D}_{|t_1|} \times B_{\epsilon_0}) \subseteq V(f)$.

As t_1 is in $\mathbb{D}_\tau$, there exists λ'_0 such that

12) for all γ with $0 < \gamma < \lambda'_0$, $B_{\epsilon_0} \cap V(f_{t_1} - \gamma)$ is diffeomorphic to $F_{f_0,\mathbf{0}}$.
13) Now, let $(B_\epsilon, \mathbb{D}_\lambda)$ be a Milnor pair for f_{t_1} with
14) $\epsilon < \epsilon_0$ and $\lambda < \lambda'_0$.

Then, there exist $\omega, \xi > 0$ such that

15) $(\mathbb{D}_\omega \times B_\epsilon, \mathbb{D}_\xi)$ is a Milnor pair for $f_{t_1} + w^j$, where we assume that
16) $\omega^j < \min\{\lambda, \lambda'_0, \omega_0^j\}$ and
17) $\xi < \min\{\xi_0, \omega^j, \eta - \omega_0^j\}$, where $\eta - \omega_0^j > 0$ by 6).

Finally, we select ν in 8) so small that

18) $\quad 0 < |\nu| < \min\{\eta - \omega_0^j, \lambda - \omega^j, \xi\}$.

Now that we have made all of these choices, we are ready to begin the intuitive part of the proof.

We have the inclusions

$$F_{f_{t_1}+w^j,\mathbf{0}} \cong (\mathbb{D}_\omega \times B_\epsilon) \cap V(f_{t_1} + w^j - \nu) \xrightarrow{i} (\mathbb{D}_\omega \times B_{\epsilon_0}) \cap V(f_{t_1} + w^j - \nu)$$

$$\xrightarrow{l} (\mathbb{D}_{\omega_0} \times B_{\epsilon_0}) \cap V(f_{t_1} + w^j - \nu) \cong F_{f_0+w^j,\mathbf{0}},$$

where we are assuming that $l \circ i$ induces isomorphisms on homology. We will first show that l induces isomorphisms on homology and, hence, so does i. Actually, we will show that l is a homotopy-equivalence.

We accomplish this by showing that

$$(*) \qquad \begin{array}{c} ((\mathbb{D}_{\omega_0} - \overset{\circ}{\mathbb{D}}_\omega) \times B_{\epsilon_0}) \cap V(f_{t_1} + w^j - \nu) \\ \big\downarrow w \\ \mathbb{D}_{\omega_0} - \overset{\circ}{\mathbb{D}}_\omega \end{array}$$

is a proper, stratified submersion.

Critical points of the map in $(\mathbb{D}_{\omega_0} - \overset{\circ}{\mathbb{D}}_\omega) \times \overset{\circ}{B}_{\epsilon_0}$ occur where $\mathrm{grad}(f_{t_1}) = \mathbf{0}$; that is, at points $(w, t_1, \mathbf{z})$ such that $(t_1, \mathbf{z})$ is in $\Gamma^1_{f,t}$ or in Σf. By 10), $\Gamma^1_{f,t} \cap (\mathbb{D}_{t_1} \times B_{\epsilon_0})$ is empty and, by 11), $\Sigma f \cap (\mathbb{D}_{t_1} \times B_{\epsilon_0}) \subseteq V(f)$. But, if $f_{t_1} = 0$, then $w^j - \nu = 0$. However, this is impossible since $w \in \mathbb{D}_{\omega_0} - \overset{\circ}{\mathbb{D}}_\omega$ and thus we would have to have $|w^j| \geqslant \omega^j$ – but we know that $\omega^j > \xi > |\nu|$ by 17) and 18).

Now, we consider critical points of $(*)$ which occur on $(\mathbb{D}_{\omega_0} - \overset{\circ}{\mathbb{D}}_\omega) \times \partial B_{\epsilon_0}$. These occur at points $(w, \mathbf{p})$ where $T_\mathbf{p} V(f_{t_1} - f_{t_1}(\mathbf{p})) \subseteq T_\mathbf{p} \partial B_{\epsilon_0}$. However,

$$0 < |f_{t_1}(\mathbf{p})| = |w^j - \nu| \leqslant |w|^j + |\nu|,$$

where $0 < |w^j - \nu|$ by the argument of the preceding paragraph. But, $w \in \mathbb{D}_{\omega_0}$ and so $|w|^j + |\nu| \leqslant \omega_0^j + |\nu|$ which is $\leqslant \omega_0^j + \eta - \omega_0^j$ by 18). Hence, $0 < |f_{t_1}(\mathbf{p})| \leqslant \eta$, $t_1 \in \mathbb{D}_\tau$, and $T_\mathbf{p} V(f_{t_1} - f_{t_1}(\mathbf{p})) \subseteq T_\mathbf{p} \partial B_{\epsilon_0}$; this contradicts 4).

Therefore, the map $(*)$ is a proper, stratified submersion and, hence, is a locally trivial fibration. It follows at once that the inclusion, l, is a homotopy-equivalence and, thus, it follows that our earlier map

$$(\mathbb{D}_\omega \times B_\epsilon) \cap V(f_{t_1} + w^j - \nu) \xrightarrow{i} (\mathbb{D}_\omega \times B_{\epsilon_0}) \cap V(f_{t_1} + w^j - \nu)$$

induces isomorphisms on homology.

We wish now to show that i is obtained up to homotopy by wedging together $j-1$ copies of the suspension of the inclusion map $B_\epsilon \cap V(f_{t_1} - \nu) \hookrightarrow B_{\epsilon_0} \cap V(f_{t_1} - \nu)$ which, by 12), 13), and 18), is nothing more than the inclusion $F_{f_{t_1},\mathbf{0}} \hookrightarrow F_{f_0,\mathbf{0}}$. It would then follow that $F_{f_{t_1},\mathbf{0}} \hookrightarrow F_{f_0,\mathbf{0}}$ induces isomorphisms on homology since i does.

But, since $|w^j - \nu| \leqslant |w|^j + |\nu| \leqslant \omega^j + |\nu| \leqslant \min\{\xi, \lambda_0'\}$ by 18) and 14), we may proceed as in Proposition II.8.1 and find that projection by w realizes, up to homotopy:

$(\mathbb{D}_\omega \times B_\epsilon) \cap V(f_{t_1} + w^j - \nu)$ as the wedge of $j-1$ copies of the suspension of $F_{f_{t_1},\mathbf{0}}$,

$(\mathbb{D}_\omega \times B_{\epsilon_0}) \cap V(f_{t_1} + w^j - \nu)$ as the wedge of $j-1$ copies of the suspension of $F_{f_0,\mathbf{0}}$, and

the map i as the wedge of $j-1$ copies of the suspension of the map

$$F_{f_{t_1},\mathbf{0}} \cong B_\epsilon \cap V(f_{t_1} - \nu) \hookrightarrow B_{\epsilon_0} \cap V(f_{t_1} - \nu) \cong F_{f_0,\mathbf{0}}.$$

The conclusion follows. □

References

[A'C] N. A'Campo, *Le nombre de Lefschetz d'une monodromie*, Proc. Kon. Ned. Akad. Wet., Series A **76** (1973), 113–118.

[BBD] A. Beilinson, J. Berstein, and P. Deligne, *Faisceaux Pervers*, Astérisque **100**, Soc. Math. de France, 1983.

[BMM] J. Briançon, P. Maisonobe, and M. Merle, *Localisation de systèmes différentiels, stratifications de Whitney et condition de Thom*, Invent. Math. **117** (1994), 531–550.

[B-S] J. Briançon and J.P. Speder, *La trivialité topologique n'implique pas les conditions de Whitney*, C.R. Acad. Sci. Paris, Série A **280** (1975).

[Br] J. Brylinski, *Transformations canoniques, Dualité projective, Théorie de Lefschetz, Transformations de Fourier et sommes trigonométriques*, Soc. Math. de France, Astérisque **140** (1986).

[BDK] J. Brylinski, A. Dubson, and M. Kashiwara, *Formule de l'indice pour les modules holonomes et obstruction d'Euler locale*, C.R. Acad. Sci., Série A **293** (1981), 573–576.

[Ce] J. Cerf, *La stratification naturelle des espaces de fonctions différentiables réelles et le théorème de la pseudo-isotopie*, publ. I.H.E.S. **39** (1970), 187–353.

[Co1] D. Cohen, *unpublished note*, 1991.

[Co2] ———, *Cohomology and Intersection Cohomology of Complex Hyperplane Arrangements*, Dissertation, Northeastern University, 1992.

[Da] J. Damon, *Higher Multiplicities and Almost Free Divisors and Complete Intersections*, no. 589, Mem. Amer. Math. Soc. **123** (1996).

[De] P. Deligne, *Comparaison avec la théorie transcendante*, Séminaire de géométrie algébrique du Bois-Marie, SGA 7 II, Springer Lect. Notes **340** (1973).

[Di] A. Dimca, *On the Milnor fibration of weighted homogeneous polynomials*, Compositio Math. **76** (1990), 19–47.

[Fi] G. Fischer, *Complex Analytic Geometry*, Lecture Notes in Math., vol. 538, Springer-Verlag, 1976.

[Fu] W. Fulton, *Intersection Theory*, Ergebnisse der Math., Springer-Verlag, 1984.

[Gaf1] T. Gaffney, *personal communication*, 1991.

[Gaf2] ———, *Polar Multiplicities and Equisingularity of Map Germs*, Topology **32** (1993), 185–223..

[G-G] T. Gaffney and R. Gassler, *Segre Numbers and Hypersurface Singularities*, J. Algebraic Geom. **8** (1999), 695–736.

[G-K] T. Gaffney and S. Kleiman, *Specialization of Integral Dependence for Modules*, Invent. Math. **137** (1999), 541–574.

[Gas1] L. van Gastel, *Excess Intersections*, Thesis, University of Utrecht, 1989.

[Gas2] ———, *Excess Intersections and a Correspondence Principle*, Invent. Math. **103 (1)** (1991), 197–222.

[Gi] V. Ginsburg, *Characteristic Varieties and Vanishing Cycles*, Invent. Math. **84** (1986), 327–403.

[G-M1] M. Goresky and R. MacPherson, *Morse Theory and Intersection Homology*, Analyse et Topologie sur les Espaces Singuliers. Astérisque **101** (1983), Soc. Math. France, 135–192.

[G-M2] ———, *Stratified Morse Theory*, Ergebnisse der Math. 14, Springer-Verlag, Berlin, 1988.

[G-M3] ———, *Intersection homology II*, Inv. Math **71** (1983), 77–129.

[G-M4] ———, *Intersection homology theory*, Topology **19** (1980), 135–162.

[G-R1] H. Grauert and R. Remmert, *Coherent Analytic Sheaves*, Grund. math. Wiss. 265, Springer-Verlag, 1984.

[G-R2] ———, *Theory of Stein Spaces*, Grund. math. Wiss. 236, Springer-Verlag, 1979.

[Gr1] G. M. Greuel, *Constant Milnor number implies constant multiplicity for quasihomogeneous singularities*, Manuscr. Math. **56** (1986), 159–166.

[Gr2] ———, *Der Gauss Manin Zusammenhang isolierter Singularitäten von vollständigen Durchsnitten*, Math. Ann. **214** (1975), 235–266.

[H-L] H. Hamm and Lê D. T., *Un Théorème de Zariski du type de Lefschetz*, Ann. Sci. L'Ecole Norm. Sup. **6** (1973), 317–366.

[Ha] R. Hartshorne, *Residues and Duality*, Springer Lecture Notes 20, Springer-Verlag, 1966.

[H-M] J.-P. Henry and M. Merle, *Conditions de régularité et éclatements*, Ann. Inst. Fourier **37** (1987), 159–190.

[HMS] J.-P. Henry, M. Merle, and C. Sabbah, *Sur la condition de Thom stricte pour un morphisme analytique*, Ann. Sci. L'Ecole Norm. Sup. **17** (1984), 227–268.

[Hi] H. Hironaka, *Stratification and Flatness*, Real and Complex Singularities, Nordic Summer School (Oslo, 1976) (1977).

[Hu] D. Husemoller, *Fibre Bundles*, Grad. Text in Math. 20, Springer-Verlag, 1966.

[Io] I. N. Iomdin, *Variétés complexes avec singularités de dimension un*, Sibirsk. Mat. Z. **15** (1974), 1061–1082.

[Iv] B. Iverson, *Cohomology of Sheaves*, Universitext, Springer-Verlag, 1986.

[Ka] M. Kashiwara, *Systèmes d'équations micro-différentielles*, (Notes by T. M. Fernandes), Dépt. de Math., Univ. Paris-Nord **8** (1978).

[K-S1] M. Kashiwara and P. Schapira, *Microlocal Study of Sheaves*, Astérisque **128** (1985).

[K-S2] ———, *Sheaves on Manifolds*, Grund. math. Wiss. 292, Springer - Verlag, 1990.

[K-M] M. Kato and Y. Matsumoto, *On the connectivity of the Milnor fibre of a holomorphic function at a critical point*, Proc. of 1973 Tokyo manifolds conf. (1973), 131–136.

[Kl] S. Kleiman, *The transversality of a general translate*, Comp. Math. **28** (1974), 287–297.

[La] R. Lazarsfeld, *Branched Coverings of Projective Space*, Thesis, Brown University, 1980.

[Lê1] Lê D. T., *Calcul du Nombre de Cycles Évanouissants d'une Hypersurface Complexe*, Ann. Inst. Fourier, Grenoble **23** (1973), 261–270.

[Lê2] ———, *Complex Analytic Functions with Isolated Singularities*, J. Algebraic Geom. **1** (1992), 83–99.

[Lê3] ———, *Le concept de singularité isolée de fonction analytique*, Advanced Studies in Pure Math. **8** (1986), 215–227.

[Lê4] ______, *Ensembles analytiques complexes avec lieu singulier de dimension un (d'après I.N. Iomdin)*, Séminaire sur les Singularités (Paris, 1976–1977) Publ. Math. Univ. Paris VII (1980), 87–95.

[Lê5] ______, *The Geometry of the Monodromy Theorem*, in C. P. Ramanujam, a tribute, ed. K.G. Ramanathan, Tata Inst. Studies in Math. **8** (1978).

[Lê6] ______, *La Monodromie n'a pas de Points Fixes*, J. Fac. Sci. Univ. Tokyo, Sec. 1A **22** (1975), 409–427.

[Lê7] ______, *Morsification of* D*-Modules*, Bol. Soc. Mat. Mexicana (3) **4** (1998), 229–248.

[Lê8] ______, *Some Remarks on Relative Monodromy*, Real and Complex Singularities, Oslo 1976 (1977), 397–403.

[Lê9] ______, *Sur les cycles évanouissants des espaces analytiques*, C.R. Acad. Sci. Paris, Ser. A **288** (1979), 283–285.

[Lê10] ______, *Topological Use of Polar Curves*, Proc. Symp. Pure Math. **29** (1975), 507–512.

[Lê11] ______, *Topologie des Singularités des Hypersurfaces Complexes*, Astérisque **7 and 8** (1973), 171–192.

[Lê12] ______, *Une application d'un théorème d'A'Campo a l'equisingularité*, Indagat. Math. **35** (1973), 403–409.

[L-M] Lê D. T. and Z. Mebkhout, *Variétés caractéristiques et variétés polaires*, C.R. Acad. Sci. **296** (1983), 129–132.

[L-P] Lê D. T. and B. Perron, *Sur la Fibre de Milnor d'une Singularité Isolée en Dimension Complexe Trois*, C.R. Acad. Sci. **289** (1979), 115-118.

[L-R] Lê D. T. and C. P. Ramanujam, *The Invariance of Milnor's Number implies the Invariance of the Topological Type*, Amer. Journ. Math. **98** (1976), 67–78.

[Lê-Sa] Lê D. T. and K. Saito, *La constance du nombre de Milnor donne des bonnes stratifications*, C.R. Acad. Sci. **277** (1973), 793–795.

[L-T1] Lê D. T. and B. Teissier, *Cycles evanescents, sections planes et conditions de Whitney. II*, Proc. Symp. Pure Math. **40, Part 2** (1983), 65–103.

[L-T2] ______, *Variétés polaires locales et classes de Chern des variétiés singulières*, Annals of Math. **114** (1981), 457–491.

[Łoj] S. Łojasiewicz, *Ensembles semi-algébriques*, IHES notes (1965).

[Loo] E. Looijenga, *Isolated singular points on complete intersections*, London Math. Soc. Lect. Note Series, no. 77, 1984.

[Lo-St] E. Looijenga and J. Steenbrink, *Milnor number and Tjurina number of complete intersections*, Math. Ann. **271 no. 1**, 121–124.

[Mac1] R. MacPherson, *Chern classes for singular algebraic varieties*, Annals of Math. **100** (1974), 423–432.

[Mac2] ______, *Global Questions in the Topology of Singular Spaces*, Proc. Internat. Congress of Math., Warsaw (1983), 213–235.

[M-V] R. MacPherson and K. Vilonen, *Elementary construction of perverse sheaves*, Invent. Math. **84** (1986), 403–435.

[Mas1] D. Massey, *The Characteristic Polar Cycles of a Perverse Sheaf*, preprint, 1990.

[Mas2] ———, *Critical Points of Functions on Singular Spaces*, Top. and Appl. **103** (2000), 55–93.

[Mas3] ———, *Families Of Hypersurfaces with One-Dimensional Singular Sets*, Dissertation, Duke University (1986).

[Mas4] ———, *A General Calculation of the Number of Vanishing Cycles*, Top. and Appl. **62** (1995), 21–43.

[Mas5] ———, *Hypercohomology of Milnor Fibres*, Topology **35** (1996), 969–1003.

[Mas6] ———, *Lê Cycles and Hypersurface Singularities*, Lecture Notes in Mathematics, vol. 1615, Springer-Verlag, 1995.

[Mas7] ———, *The Lê-Ramanujam Problem for Hypersurfaces with One-Dimensional Singular Sets*, Math. Annalen **282** (1988), 33–49.

[Mas8] ———, *The Lê Varieties, I*, Invent. Math. **99** (1990), 357–376.

[Mas9] ———, *The Lê Varieties, II*, Invent. Math. **104** (1991), 113–148.

[Mas10] ———, *Local Morse Inequalities and Perverse Sheaves*, preprint, 1990.

[Mas11] ———, *Numerical Invariants of Perverse Sheaves*, Duke Math. J. **73** (1994), 307–369.

[Mas12] ———, *Prepolar Deformations and a new Lê-Iomdine Formula*, Pacific J. Math. **174** (1996), 459–469.

[Mas13] ———, *A Reduction Theorem for the Zariski Multiplicity Conjecture*, Proc. AMS **106** (1989), 379–383.

[Mas14] ———, *The Thom Condition along a Line*, Duke Math. J. **60** (1990), 631–642.

[M-S] D. Massey and D. Siersma, *Deformations of Polar Methods*, Ann. Inst. Fourier **42** (1992), 737–778.

[MSSVWZ] D. Massey, R. Simion, R. Stanley, D. Vertigan, D. Welsh, and G. Ziegler, *Lê Numbers, Matroid Identities, and the Tutte Polynomial*, J. Combin. Theory Ser. B **70** (1997), 118–133.

[Mat] J. Mather, *Notes on Topological Stability*, unpublished notes, Harvard Univ. 1970.

[Meb] Z. Mebkhout, *Local cohomology of analytic spaces*, Pub. Res. Inst. Math. Sc. **12** (1977), 247–256.

[Mer] M. Merle, *Variétés polaires, stratifications de Whitney et classes de Chern des espaces analytiques complexes*, Seminaire Bourbaki **600** (1982), 1–14.

[Mi1] J. Milnor, *Lectures on the h-cobordism Theorem*, Math. Notes 1, P.U.P., 1965.

[Mi2] ———, *Morse Theory*, Annals of Math. Studies, no. 51, P.U.P., 1963.

[Mi3] ———, *Singular Points of Complex Hypersurfaces*, Annals of Math. Studies, no. 77, P.U.P., 1968.

[M-O] J. Milnor and P. Orlik, *Isolated Singularities Defined by Weighted Homogeneous Polynomials*, Topology **9** (1969), 385–393.

[Ok] M. Oka, *On the homotopy type of hypersurfaces defined by weighted homogeneous polynomials*, Topology **12** (1973), 19–32.

[O-R] P. Orlik and R. Randell, *The Milnor fiber of a generic arrangement*, Arkiv für Mat. **31** (1993), 71–81.

[O-S] P. Orlik and L. Solomon, *Combinatorics and topology of complements of hyperplanes*, Invent. Math. **56** (1980), 167–189.

[O-T] P. Orlik and H. Terao, *Arrangements of Hyperplanes*, Grund. math. Wiss., vol. 300, Springer-Verlag, 1991.

[O'S] D. O'Shea, *Topologically Trivial Deformations of Isolated Quasihomogeneous Hypersurface Singularities are Equimultiple*, Proc. AMS **100** (1987), 260–262.

[P1] A. Parusiński, *A Generalization of the Milnor Number*, Math. Annalen (1988), 247–254.

[P2] ______, *Limits of Tangent Spaces to Fibres and the w_f Condition*, Duke Math. J. **72** (1993), 99–108.

[Ra] R. Randell, *On the Topology of Non-isolated Singularities*, Proc. Georgia Top. Conf., Athens, Ga., 1977 **99** (1979), 445–473.

[Sab1] C. Sabbah, *Proximité évanescente*, Compositio Math. **62** (1987), 283–328.

[Sab2] ______, *Quelques remarques sur la géométrie des espaces conormaux*, Astérisque **130** (1985), 161–192.

[Sai] M. Saito, *Mixed Hodge Modules*, Publ. RIMS, Kyoto Univ. **26** (1990), 221–333.

[Sak] K. Sakamoto, *The Seifert matrices of Milnor fiberings defined by holomorphic functions*, J. Math. Soc. Japan **26 (4)** (1974), 714–721.

[Sc-To] P. Schapira and N. Tose, *Morse Inequalities for R-Constructible Sheaves*, Adv. in Math. **93** (1992), 1–8.

[Se-Th] M. Sebastiani and R. Thom, *Un résultat sur la monodromie*, Invent. Math. **13** (1971), 90–96.

[Si1] D. Siersma, *A bouquet theorem for the Milnor Fibre*, preprint #24 of the European Singularity Project (1993).

[Si2] ______, *Isolated Line Singularities*, Proc. Symp. Pure Math. **40, Part 2** (1983), 485–496.

[Si3] ______, *The monodromy of a series of hypersurface singularities*, Comment. Math. Helvetici **65** (1990), 181–197.

[Si-Tr] Y. T. Siu and G. Trautmann, *Gap-Sheaves and Extension of Coherent Analytic Subsheaves*, Springer Lect. Notes **172**, Springer-Verlag, 1971.

[Sm] S. Smale, *Generalized Poincaré's Conjecture in Dimensions greater than 4*, Ann. Math. **64** (1956).

[Te1] B. Teissier, *A bouquet of bouquets for a birthday*, in Topological Methods in Modern Mathematics – A Symposium in Honor of John Milnor's Sixtieth Birthday, 1991, ed. L. Goldberg and A. Phillips. Holm (1993), 93–122.

[Te2] ______, *Cycles evanescents, sections planes et Conditions de Whitney*, in Singularités à Cargèse, Astérisque **7 et 8** (1973), 285–362.

[Te3] ______, *The Hunting of Invariants in the Geometry of Discriminants*, in Real and Complex Singularities, Oslo 1976, ed. P. Holm (1977), 565–677.

[Te4] ______, *Introduction to Equisingularity Problems*, Proc. Symp. Pure Math. **29** (1975), 593–632.

[Te5] ______, *Variétés polaires I: Invariants polaires des singularités d'hypersurfaces*, Invent. Math. **40 (3)** (1977), 267–292.

[Te6] ———, *Variétés polaires II: Multiplicités polaires, sections planes, et conditions de Whitney*, in Algebraic Geometry, Proc., La Rabida 1981, Springer Lect. Notes **961** (1982), 314–491.

[Te7] ———, *Variétés polaires locales et conditions de Whitney*, C. R. Acad. Sci. Paris **290** (1980), 799–802.

[Th] R. Thom, *Ensembles et Morphismes Stratifiés*, Bull. Amer. Math. Soc. **75** (1969), 240–284.

[Ti] M. Tibăr, *Bouquet Decomposition of the Milnor Fibre*, Topology **35** (1996), 227–241.

[Va1] J. P. Vannier, *Familles à paramètre de fonctions holomorphes à ensemble singulier de dimension zéro ou un*, Thèse, Dijon (1987).

[Va2] ———, *Sur les fibrations de Milnor de familles d'hypersurfaces à lieu singulier de dimension un*, Math. Ann. **287** (1990), 539–552.

[Ve] J. L. Verdier, *Catégories dérivées*, Etat 0, SGA $4\frac{1}{2}$, Lecture Notes in Math. **569** (1977), 262–311.

[Vo] W. Vogel, *Results on Bézout's Theorem*, Tata Lecture Notes 74, Springer-Verlag, 1984.

[W] H. Whitney, *Tangents to an Analytic Variety*, Ann. Math. **81** (1965), 496–549.

[Z] O. Zariski, *Open Questions in the Theory of Singularities*, Bull. AMS **77** (1971), 481–491.

Index

Editorial Information

To be published in the *Memoirs*, a paper must be correct, new, nontrivial, and significant. Further, it must be well written and of interest to a substantial number of mathematicians. Piecemeal results, such as an inconclusive step toward an unproved major theorem or a minor variation on a known result, are in general not acceptable for publication. Papers appearing in *Memoirs* are generally longer than those appearing in *Transactions*, which shares the same editorial committee.

As of February 1, 2003, the backlog for this journal was approximately 3 volumes. This estimate is the result of dividing the number of manuscripts for this journal in the Providence office that have not yet gone to the printer on the above date by the average number of monographs per volume over the previous twelve months, reduced by the number of volumes published in four months (the time necessary for preparing a volume for the printer). (There are 6 volumes per year, each containing at least 4 numbers.)

A Consent to Publish and Copyright Agreement is required before a paper will be published in the *Memoirs*. After a paper is accepted for publication, the Providence office will send a Consent to Publish and Copyright Agreement to all authors of the paper. By submitting a paper to the *Memoirs*, authors certify that the results have not been submitted to nor are they under consideration for publication by another journal, conference proceedings, or similar publication.

Information for Authors

Memoirs are printed from camera copy fully prepared by the author. This means that the finished book will look exactly like the copy submitted.

The paper must contain a *descriptive title* and an *abstract* that summarizes the article in language suitable for workers in the general field (algebra, analysis, etc.). The *descriptive title* should be short, but informative; useless or vague phrases such as "some remarks about" or "concerning" should be avoided. The *abstract* should be at least one complete sentence, and at most 300 words. Included with the footnotes to the paper should be the 2000 *Mathematics Subject Classification* representing the primary and secondary subjects of the article. The classifications are accessible from `www.ams.org/msc/`. The list of classifications is also available in print starting with the 1999 annual index of *Mathematical Reviews*. The Mathematics Subject Classification footnote may be followed by a list of *key words and phrases* describing the subject matter of the article and taken from it. Journal abbreviations used in bibliographies are listed in the latest *Mathematical Reviews* annual index. The series abbreviations are also accessible from `www.ams.org/publications/`. To help in preparing and verifying references, the AMS offers MR Lookup, a Reference Tool for Linking, at `www.ams.org/mrlookup/`. When the manuscript is submitted, authors should supply the editor with electronic addresses if available. These will be printed after the postal address at the end of the article.

Electronically prepared manuscripts. The AMS encourages electronically prepared manuscripts, with a strong preference for $\mathcal{A}_{\mathcal{M}}\mathcal{S}$-LaTeX. To this end, the Society has prepared $\mathcal{A}_{\mathcal{M}}\mathcal{S}$-LaTeX author packages for each AMS publication. Author packages include instructions for preparing electronic manuscripts, the *AMS Author Handbook*, samples, and a style file that generates the particular design specifications of that publication series. Though $\mathcal{A}_{\mathcal{M}}\mathcal{S}$-LaTeX is the highly preferred format of TeX, author packages are also available in $\mathcal{A}_{\mathcal{M}}\mathcal{S}$-TeX.

Authors may retrieve an author package from e-MATH starting from `www.ams.org/tex/` or via FTP to `ftp.ams.org` (login as `anonymous`, enter username as password, and type `cd pub/author-info`). The *AMS Author Handbook* and the *Instruction Manual* are available in PDF format following the author packages link from `www.ams.org/tex/`. The author package can be obtained free of charge by sending email to `pub@ams.org` (Internet) or from the Publication Division, American Mathematical Society, P.O. Box 6248, Providence, RI 02940-6248. When requesting an author package, please specify $\mathcal{A}\mathcal{M}\mathcal{S}$-LaTeX or $\mathcal{A}\mathcal{M}\mathcal{S}$-TeX, Macintosh or IBM (3.5) format, and the publication in which your paper will appear. Please be sure to include your complete mailing address.

Sending electronic files. After acceptance, the source file(s) should be sent to the Providence office (this includes any TeX source file, any graphics files, and the DVI or PostScript file).

Before sending the source file, be sure you have proofread your paper carefully. The files you send must be the EXACT files used to generate the proof copy that was accepted for publication. For all publications, authors are required to send a printed copy of their paper, which exactly matches the copy approved for publication, along with any graphics that will appear in the paper.

TeX files may be submitted by email, FTP, or on diskette. The DVI file(s) and PostScript files should be submitted only by FTP or on diskette unless they are encoded properly to submit through email. (DVI files are binary and PostScript files tend to be very large.)

Electronically prepared manuscripts can be sent via email to `pub-submit@ams.org` (Internet). The subject line of the message should include the publication code to identify it as a Memoir. TeX source files, DVI files, and PostScript files can be transferred over the Internet by FTP to the Internet node `e-math.ams.org` (130.44.1.100).

Electronic graphics. Comprehensive instructions on preparing graphics are available at `www.ams.org/jourhtml/graphics.html`. A few of the major requirements are given here.

Submit files for graphics as EPS (Encapsulated PostScript) files. This includes graphics originated via a graphics application as well as scanned photographs or other computer-generated images. If this is not possible, TIFF files are acceptable as long as they can be opened in Adobe Photoshop or Illustrator. No matter what method was used to produce the graphic, it is necessary to provide a paper copy to the AMS.

Authors using graphics packages for the creation of electronic art should also avoid the use of any lines thinner than 0.5 points in width. Many graphics packages allow the user to specify a "hairline" for a very thin line. Hairlines often look acceptable when proofed on a typical laser printer. However, when produced on a high-resolution laser imagesetter, hairlines become nearly invisible and will be lost entirely in the final printing process.

Screens should be set to values between 15% and 85%. Screens which fall outside of this range are too light or too dark to print correctly. Variations of screens within a graphic should be no less than 10%.

Inquiries. Any inquiries concerning a paper that has been accepted for publication should be sent directly to the Electronic Prepress Department, American Mathematical Society, P. O. Box 6248, Providence, RI 02940-6248.

Editors

This journal is designed particularly for long research papers, normally at least 80 pages in length, and groups of cognate papers in pure and applied mathematics. Papers intended for publication in the *Memoirs* should be addressed to one of the following editors. In principle the Memoirs welcomes electronic submissions, and some of the editors, those whose names appear below with an asterisk (*), have indicated that they prefer them. However, editors reserve the right to request hard copies after papers have been submitted electronically. Authors are advised to make preliminary email inquiries to editors about whether they are likely to be able to handle submissions in a particular electronic form.

Algebra to KAREN E. SMITH, Department of Mathematics, University of Michigan, 525 University, Suite 2832, Ann Arbor, MI 48109-1109; email: `kesmith@lsa.umich.edu`

Algebraic geometry to DAN ABRAMOVICH, Department of Mathematics, Boston University, 111 Cummington Street, Boston, MA 02215; e-mail: `abrmovic@bu.edu`

Algebraic topology and cohomology of groups to STEWART PRIDDY, Department of Mathematics, Northwestern University, 2033 Sheridan Road, Evanston, IL 60208-2730; email: `priddy@math.nwu.edu`

Combinatorics and Lie theory to SERGEY FOMIN, Department of Mathematics, University of Michigan, Ann Arbor, Michigan 48109-1109; email: `fomin@umich.edu`

Complex analysis and complex geometry to DUONG H. PHONG, Department of Mathematics, Columbia University, 2990 Broadway, New York, NY 10027-0029; email: `phong@math.columbia.edu`

***Differential geometry and global analysis** to LISA C. JEFFREY, Department of Mathematics, University of Toronto, 100 St. George St., Toronto, ON Canada M5S 3G3; email: `jeffrey@math.toronto.edu`

Dynamical systems and ergodic theory to ROBERT F. WILLIAMS, Department of Mathematics, University of Texas, Austin, Texas 78712-1082; email: `bob@math.utexas.edu`

***Geometric analysis** to TOBIAS COLDING, Courant Institute, New York University, 251 Mercer Street, New York, NY 10012; email: `colding@cims.nyu.edu`

Geometric topology, knot theory and hyperbolic geometry to ABIGAIL A. THOMPSON, Department of Mathematics, University of California, Davis, Davis, CA 95616-5224; email: `thompson@math.ucdavis.edu`

Harmonic analysis, representation theory, and Lie theory to ROBERT J. STANTON, Department of Mathematics, The Ohio State University, 231 West 18th Avenue, Columbus, OH 43210-1174; email: `stanton@math.ohio-state.edu`

***Logic** to THEODORE SLAMAN, Department of Mathematics, University of California, Berkeley, CA 94720-3840; email: `slaman@math.berkeley.edu`

Number theory to HAROLD G. DIAMOND, Department of Mathematics, University of Illinois, 1409 W. Green St., Urbana, IL 61801-2917; email: `diamond@math.uiuc.edu`

***Ordinary differential equations, and applied mathematics** to PETER W. BATES, Department of Mathematics, Michigan State University, East Lansing, MI 48824-1027; email: `peter@math.msu.edu`

***Partial differential equations** to PATRICIA E. BAUMAN, Department of Mathematics, Purdue University, West Lafayette, IN 47907-1395' email: `bauman@math.purdue.edu`

***Probability and statistics** to KRZYSZTOF BURDZY, Department of Mathematics, University of Washington, Box 354350, Seattle, Washington 98195-4350; email: `burdzy@math.washington.edu`

Real analysis and partial differential equations to DANIEL TATARU, Department of Mathematics, University of California, Berkeley, Berkeley, CA 94720; email: `tataru@math.berkeley.edu`

All other communications to the editors should be addressed to the Managing Editor, WILLIAM BECKNER, Department of Mathematics, University of Texas, Austin, TX 78712-1082; email: `beckner@math.utexas.edu`.

Titles in This Series

778 **David B. Massey,** Numerical control over complex analytic singularities, 2003
777 **Robert Lauter,** Pseudodifferential analysis on conformally compact spaces, 2003
776 **U. Haagerup, H. P. Rosenthal, and F. A. Sukochev,** Banach embedding properties of non-commutative L^p-spaces, 2003
775 **P. Lochak, J.-P. Marco, and D. Sauzin,** On the splitting of invariant manifolds in multidimensional near-integrable Hamiltonian systems, 2003
774 **Kai A. Behrend,** Derived ℓ-adic categories for algebraic stacks, 2003
773 **Robert M. Guralnick, Peter Müller, and Jan Saxl,** The rational function analogue of a question of Schur and exceptionality of permutation representations, 2003
772 **Katrina Barron,** The moduli space of $N = 1$ superspheres with tubes and the sewing operation, 2003
771 **Shigenori Matsumoto,** Affine flows on 3-manifolds, 2003
770 **W. N. Everitt and L. Markus,** Elliptic partial differential operators and symplectic algebra, 2003
769 **Jie Wu,** Homotopy theory of the suspensions of the projective plane, 2003
768 **R. Höpfner and E. Löcherbach,** Limit theorems for null recurrent Markov processes, 2003
767 **Po Hu,** S-modules in the category of schemes, 2003
766 **Su Gao and Alexander S. Kechris,** On the classification of Polish metric spaces up to isometry, 2003
765 **Robert Bieri and Ross Geoghegan,** Connectivity properties of group actions on non-positively curved spaces, 2003
764 **J. Spandaw,** Noether-Lefschetz problems for degeneracy loci, 2003
763 **Yasuyuki Kachi and Eiichi Sato,** Segre's reflexivity and an inductive characterization os hyperquadrics, 2002
762 **Leiba Rodman, Ilya M. Spitkovsky, and Hugo Woerdeman,** Abstract band method via factorization, positive and band extensions of multivariable almost periodic matrix functions, and spectral estimation, 2002
761 **Oliver Druet and Emmanuel Hebey,** The AB program in geometric analysis : Sharp Sobolev inequalities and related problems, 2002
760 **Markus Banagl,** Extending intersection homology type invarients to non-Witt spaces, 2002
759 **Donald M. Davis,** From representation theory to homotopy groups, 2002
758 **Alan Forrest, John Hunton, and Johannes Kellendonk,** Topological invariants for projection method patterns, 2002
757 **Douglas Bowman,** q-difference operators, orthogonal polynomials, and symmetric expansions, 2002
756 **José Ignacio Cogolludo-Agustín,** Topological invariants of the complement to arrangements of rational plane curves, 2002
755 **M. A. Mandell and J. P. May,** Equivariant orthogonal spectra and S-modules, 2002
754 **Edward L. Green, Idun Reiten, and Øyvind Solberg,** Dualities on generalized Koszul algebras, 2002
753 **Daniel Panazzolo,** Desingularization of nilpotent singularities in families of planar vector fields, 2002
752 **Linus Kramer,** Homogeneous spaces, Tits buildings, and isoparametric hypersurfaces, 2002
751 **Bruce Allison, Georgia Benkart, and Yun Gao,** Lie algebras graded by the root systems BC_r, $r \geq 2$, 2002